Wetting and Spreading Dynamics

Surfactant Science Series

Founding Editor: Martin J. Schick (1918–1998)

Series Editor: Arthur T. Hubbard

Books in the Surfactant Science series emphasize surfaces and interfaces, including basic principles, major developments, and important applications. A substantial proportion of all physical phenomena involve interfaces in one way or another, and the practical and commercial applications of interface science are numerous. The series covers experimental phenomena, behavior and properties, major advances, experimental approaches, essential instrumental methods, theoretical strategies, and important applications. The level of presentation is intended for readers having a basic scientific training, such as advanced science students encountering the book topic for the first time, as well as scientific professionals refreshing their knowledge of engineering aspects of the topic and interface science.

Colloids in Drug Delivery
Edited by Monzer Fanun

Applied Surface Thermodynamics: Second Edition
Edited by A. W. Neumann, Robert David, and Yi Y. Zuo

Colloids in Biotechnology
Edited by Monzer Fanun

Electrokinetic Particle Transport in Micro/Nano-fluidics: Direct Numerical Simulation Analysis
Shizhi Qian and Ye Ai

Nuclear Magnetic Resonance Studies of Interfacial Phenomena
Vladimir M. Gun'ko and Vladimir V. Turov

The Science of Defoaming: Theory, Experiment and Applications
Peter R. Garrett

Soil Colloids: Properties and Ion Binding
Fernando V. Molina

Surface Tension and Related Thermodynamic Quantities of Aqueous Electrolyte Solutions
Norihiro Matubayasi

Electromagnetic, Mechanical, and Transport Properties of Composite Materials
Rajinder Pal

Silicone Dispersions
Edited by Yihan Liu

Wetting and Spreading Dynamics, Second Edition
Victor M. Starov and Manuel G. Velarde

For more information about this series, please visit:
https://www.crcpress.com/Surfactant-Science/book-series/CRCSURFACSCI

Wetting and Spreading Dynamics

Second Edition

Victor M. Starov
Department of Chemical Engineering, Loughborough University
Loughborough, UK

Manuel G. Velarde
Instituto Pluridisciplinar, Universidad Complutense
Madrid, Spain

CRC Press
Taylor & Francis Group
Boca Raton London New York

CRC Press is an imprint of the
Taylor & Francis Group, an **informa** business

CRC Press
Taylor & Francis Group
6000 Broken Sound Parkway NW, Suite 300
Boca Raton, FL 33487-2742

First issued in paperback 2021

© 2020 by Taylor & Francis Group, LLC
CRC Press is an imprint of Taylor & Francis Group, an Informa business

No claim to original U.S. Government works

ISBN 13: 978-1-03-223602-5 (pbk)
ISBN 13: 978-1-138-58407-5 (hbk)

DOI: 10.1201/9780429506246

Library of Congress Cataloging-in-Publication Data

Names: Starov, V. M., author. | Velarde, Manuel G. (Manuel Garcâia), author.
Title: Wetting and spreading dynamics / Victor M. Starov, Manuel G. Velarde.
Description: Second edition. | Boca Raton : CRC Press, Taylor & Francis
Group, 2019. | Includes bibliographical references.
Identifiers: LCCN 2019007787 | ISBN 9781138584075 (hardback : alk. paper)
Subjects: LCSH: Wetting. | Surface chemistry. | Surface tension. | Interfaces
(Physical sciences) | Solid-liquid interfaces.
Classification: LCC QD506 .S7835 2019 | DDC 541/.33--dc23
LC record available at https://lccn.loc.gov/2019007787

Visit the Taylor & Francis Web site at
http://www.taylorandfrancis.com

and the CRC Press Web site at
http://www.crcpress.com

Contents

Preface to the First Edition

This book is for anyone who has recently started to be interested in, or is already involved in, research or applications of wetting and spreading, i.e., for newcomers and practitioners alike. Its contents are not a comprehensive and critical review of the existing research literature. Needless to say, it rather reflects the authors' recent scientific interests and understanding. The authors presume that the reader using this book has some knowledge in thermodynamics, fluid mechanics, and transport phenomena. Yet the book has been written in an almost self-contained manner, and it should be possible for a graduate student, scientist, or engineer with a reasonable background in differential equations to follow it. Although in various parts we have used the phrase "it can be shown ..." or the like, the authors have tried to go as deep into the details of derivation of results as required to make the book useful.

The term *wetting* commonly refers to the displacement of air from a solid surface. Throughout this book we shall be discussing wetting and spreading features of liquids, which partially (the most important example being water and aqueous solutions) or completely (oils) wet the solids or other liquids.

Wetting water films occur everywhere, even in the driest deserts or in the sauna and bathtub, although you might not see them with the naked eye because they are too thin or because they seem to disappear too quickly. Water is essential for life. It may very well be that without water, life would have not have started on Earth. In fact no life seems possible without fluids! Life, as we know it started in a little "pond," the "primordial soup" leading to the first replicating bio-related amino acids.

In the processes of wetting or spreading, three phases—air, liquid, and solids—meet along a line, which is referred to as a *three-phase contact line*. Recall the spreading drop and the drop edge, which is the three-phase contact line. In the vicinity of a three-phase contact line, the thickness of the droplet becomes very thin and, even more, virtually tends to zero. In a thin water layer, new very special surface forces come into play. These forces are well known in colloid science: forces in thin layers between interfaces of neighbor particles, droplets, and bubbles in suspensions and emulsions. Understanding of the importance of surface forces in colloid science has resulted in substantial progress in this area. In fact, it is the reason why colloid science is referred to nowadays as *colloid and interface science*.

Surface forces of the same nature act in thin liquid layers in the vicinity of the three-phase contact lines in the course of wetting and spreading. Surprisingly, the importance of surface forces has been much less recognized in wetting and spreading than it deserves. In Chapters 1 through 3 we will try to convince the reader that virtually all wetting and spreading phenomena are determined by the surface forces acting in a tiny vicinity of the three-phase contact line.

Water is, indeed, a strange liquid. For example, if you place a glass bottle full of pure water (H_2O) in the deep freezer, the bottle will break as water increases in volume while solidifying as ice, an anomalous property relative to other liquids. Life (fish) in frozen lakes would not be possible without the anomalous behavior of water around 4°C.

We shall see that a property of water relative to "surface" forces is key to understand its wetting and spreading features. We will also find that surface forces (frequently also referred to as *disjoining pressure*) have a very peculiar shape, in the case of water and aqueous solutions. This fact is critical for the existence of our life in a way which is yet to be understood.

Wetting and spreading are dramatically affected by SURFace ACTtive AgeNTS (in short, surfactants). Their molecules have a hydrophilic *head* (ionic or nonionic) with affinity for water and a hydrophobic *tail* (a hydrocarbon group), which is repelled by an aqueous phase. Fatty acids, alcohols, and some proteins (natural polymers), and washing liquids, powders, and detergents all act as surfactants. It is the reason why the kinetics of wetting and spreading of surfactant solutions is under investigation in this book.

On the other hand, a number of solid substrates—printing materials, textiles, hairs—when in contact with liquids are porous in different degrees. In spite of much experimental and practical experience in the area, only a limited number of publications are available in the literature that deals with fundamental aspects of the phenomenon. We show in this book that spreading kinetics over porous substrates differs substantially as compared with spreading over nonporous substrates.

Aiming at a logical progression in the problems treated with discussion at each level, building albeit not rigidly, upon the material that came earlier, the book can be divided into two parts: Chapters 1 through 3 form one part, and Chapters 4 and 5 constitute the other. Chapter 1 is key to the former in that its reading is a must for the understanding of Chapters 2 and 3. To a large extent Chapters 4 and 5 can be read independently from the preceding chapters, yet they are tied to each other and to the previous three.

Chapter 1 introduces surface forces and a detailed critical analysis of the current understanding of the Neumann–Young equation, the building block in most wetting and spreading research and in a number of publications. The surface forces are also frequently referred to in the literature as *colloidal* forces and *disjoining* pressure. All these terms are used as equivalents in this book, following appropriate clarification of concepts, terminology, and origins. Colloidal forces act in thin liquid films and layers when thickness goes down to about 10^{-5} cm $= 0.1$ μm $= 10^2$ nm. Below this thickness the surface forces or disjoining pressure become so increasingly powerful that they dominate all other forces (for example, capillary forces and gravity). Accordingly, surface forces determine the wetting properties of liquids in contact with solid substrates. One purpose of Chapters 1 through 3 is to show that progress in the area of equilibrium and dynamics of wetting demands due consideration of surface forces action in the vicinity of the three-phase contact line. Chapters 2 and 3 look sequentially at the equilibrium and kinetics or dynamics of wetting, showing that the action of surface forces determines all equilibrium and kinetics features of liquids in contact with solids. Note that Chapter 3 cannot be read and understood without reading the introduction to the chapter.

Colloidal forces or disjoining pressure are well known and widely used in colloid science to account for equilibrium and dynamics of colloidal suspensions and emulsions. The current theory behind colloidal forces between colloidal particles, drops, and bubbles is the DLVO theory, an acronym made after the names of Derjaguin (B.V.), Landau (L.D.), Verwey (E.J.W.) and Overbeek (J.Th.G.). The same forces act in the vicinity of the three-phase contact line, and their action is as important in this case as it is in the case of colloids. Unfortunately, most authors currently ignore the action of colloidal forces when discussing the equilibrium and dynamics of wetting. It is our belief that this has hampered progress in the area of wetting phenomena for decades.

Chapters 4 and 5 are devoted to a detailed discussion of some recent, albeit still fragmentary, developments regarding the kinetics of spreading over porous solid substrates, including the case of hydrophobic substrates in the presence of surfactants. Noteworthy are some new and universal spreading laws in the case of spreading over thin porous layers discussed in Chapter 4. Some arguments and theory in Chapter 5 are experiment-discussion oriented and heuristic or semiempirical in approach (Sections 5.4 and 5.5) and should be judged accordingly. To our understanding, little is well established about spreading over hydrophobic substrates in the presence of surfactants. Our treatment of the spontaneous adsorption of surfactant molecules on a bare hydrophobic substrate ahead of the moving liquid front, making an initially hydrophobic substrate partially hydrophilic, allows a good description of a number of phenomena. Yet the actual mechanism of transfer of surfactant molecules remains elusive. Further experimental and theoretical work is needed in view of wide technological interest and current use in the industry. We close the book with a few comments and warnings in a chapter of conclusions.

Victor M. Starov
Loughborough University
Leicestershire, United Kingdom

Manuel G. Velarde
Instituto Pluridisciplinar
Universidad Complutense
Madrid, Spain

Clayton J. Radke
University of California at Berkeley

Acknowledgments

In 1974 Victor M. Starov met Prof. Nikolay V. Churaev, the beginning of a collaboration that has continued for more than 30 years and for which author Starov would like to express very special thanks. Churaev involved Starov in the investigation of wetting and spreading phenomena in the former Surface Forces Department, Moscow Institute of Physical Chemistry (MIPCh), Russian Academy of Sciences, which was headed by Boris V. Derjaguin. This collaboration soon included a number of other colleagues from MIPCh; appreciation is extended to these, especially professors Georgy A. Martynov, Vladimir D. Sobolev, and Zinoviy M. Zorin.

In 1981, Starov took the position of head of the Department of Applied Mathematics, Moscow University of Food Industry. He organized a weekly seminar there, where virtually all problems presented in this book were either, solved, initiated, or at least discussed. These seminars were carried on until the Soviet Union collapsed. Author Starov would like to thank all members of the seminar but especially professors Anatoly N. Filippov and Vasily V. Kalinin, and Drs. Yury E. Solomentsev, Vladimir I. Ivanov, Sergey I. Vasin, and Vjacheslav G. Zhdanov.

In 1987, the University of Sofia celebrated its centennial. This book's first two authors, Victor M. Starov and Manuel G. Velarde, were honored by being chosen by Prof. Ivan B. Ivanov to be centennial lecturers at his university. Beyond being an honor, this was a lucky event in their lives. Both knew of Ivanov for quite some time but had not met him earlier nor had they worked together in the same field, although both had common interests in the interfacial phenomena. While in Sofia, hearing each other lecturing and discussing science "and beyond," they felt that it would be interesting to work together one day, particularly in exploring the consequences of surface tension and surface tension gradients, the latter of which, e.g., creates flow or alters an existing one (the Marangoni effect).

In 1991, Starov was able to visit with Manuel G. Velarde at the Instituto Pluridisciplinar of the Universidad Complutense, Madrid, Spain. Both were fortunate once more in being visited by Dr. Alain de Ryck, a young French scientist and brilliant experimentalist. He produced experiments where both Starov and Velarde were able to observe the striking role of the Marangoni effect in the spreading of a surfactant droplet over the thin aqueous layer. Later, the scientific relationship between the first two authors of this book was strengthened by the visit of Prof. Vladimir D. Sobolev, MIOCh, an outstanding scientist who went beyond being a highly skilled experimentalist. His work cemented the earlier mentioned scientific relationship and collaboration between Starov and Velarde. It was further enhanced when the former moved from Moscow to the Chemical Engineering Department, Loughborough University, United Kingdom, in 1999. There, Sobolev also worked with both Starov and Velarde, and this was the beginning of numerous Loughborough–Madrid exchanges involving also several younger colleagues: Drs. Serguei R. Kosvintsev, Pollina P. Prokopovich, Serguei A. Zhdanov, and Andre L. Zuev.

Then in 2001, the first two authors of this book jointly organized a summer school on wetting and spreading dynamics and related phenomena at El Escorial, Madrid, under the sponsorship of the Universidad Complutense Summer Programme. Economic support also came from the European Union (under the ICOPAC Network), the European Space Agency (ESA), Fuchs Iberica, L'Oreal, Inescop, and Unilever, Spain. Among the prestigious speakers from Bulgaria, France, Germany, Israel, the United States, and Spain was one of the invited lecturers, the third author of this book, Clayton J. Radke. We decided not to produce proceedings of that school, but soon after, the three future coauthors of this book started thinking of writing a joint monograph. Indeed, the present book is the result of our concern about the lack of systematized knowledge on wetting and spreading dynamics, i.e., the lack of a monograph for the use of basic and applied scientists, applied mathematicians, chemists, and engineers.

Two other schools are also worth mentioning. One on complex fluids, wetting, and spreading-related topics, coordinated by Velarde, took place in 1999 at La Rabida, Huelva, SW of Andalucia, Spain. The other course, much more focused on spreading problems, coordinated by Starov, was scheduled in

2003 at CISM (International Center for Mechanical Sciences) in Udine, Friuli-Venezia-Giulia, NE of Italy. There are proceedings of the latter ("Fluid mechanics of surfactant and polymer solutions," edited by Starov and Ivanov; Springer Verlag, 2004) but not of the former. In the past few years several other workshops, discussion meetings, and international conferences took place in Madrid and Loughborough on the subject.

The authors would like to express their gratitude to Nadezda V. Starova. Without her energy, endless patience, kindness, and expertise, this book most surely would have never been finished. We are also happy to thank Maria-Jesus Martin (Madrid) for her help in the preparation of the manuscript.

We wish to express our gratitude to the coauthors of our joint publications: Nikolay N. Churaev, Boris V. Derjaguin (deceased), Ivan B. Ivanov, Vladimir I. Ivanov, Vasiliy V. Kalinin, Olga A. Kiseleva, Serguei R. Kosvintsev, Georgy A. Martynov, David Quere, Alain de Ryck, Ramon G. Rubio, Victor M. Rudoy, Vladimir D. Sobolev, Serguei A. Zhdanov, Pavel P. Zolotarev, and Zinoviy M. Zorin,

We also would like to recognize the following colleagues, fruitful discussions with whom stimulated our research: Anne-Marie Cazabat, Pierre-Gilles de Gennes, Benoit Goyeau, George (Bud) Homsy, Dominique Langevin, Francisco Monroy, Alex T. Nikolov, Francisco Ortega, Len Pismen, Yves Pomeau, Uwe Thiele, Darsh T. Wasan.

Preparation of the manuscript was supported by a grant from the Royal Society, United Kingdom, which we would like to acknowledge. We wish to particularly acknowledge the support by Prof. John Enderby. The final revision of the manuscript was done while Manuel G. Velarde was Del Amo Foundation Visiting Professor with the Department of Mechanical Engineering and Environmental Sciences of the University of California at Santa Barbara. This was possible thanks to the hospitality of Prof. George M. Homsy. Last but not the least, we acknowledge the support for the research leading to this book which came from the Engineering and Physical Sciences Research Council, United Kingdom (Grants EP/D077869 and EP/D078814), and from the Ministerio de Educacion y Ciencia, Spain (Grants MAT2003-01517, BQU2003-01556, and VEVES).

Preface to the Second Edition

The second edition is significantly revised and updated relative to the first edition. First, two new chapters are added: Chapter 3, "Hysteresis of contact angles based on Derjaguin's pressure," and Chapter 7, "Kinetics of simultaneous spreading and evaporation." New sections are included in Chapters 2 and 5: Section 2.7 "Equilibrium of droplets on a deformable substrate: Influence of the Derjaguin's pressure" and Section 5.5 "Spreading of non-Newtonian liquids over dry porous layer: Complete and partial wetting cases" with Subsections "Complete wetting" and "Partial wetting."

The action of surface forces is one of the most important phenomena to properly account for all equilibrium, quasi-equilibrium, and kinetic properties of liquids in contact with solid or liquid substrates. In the first edition, we used the term "disjoining/conjoining pressure." However, in this second edition we decided to use instead "Derjaguin's pressure" to emphasize the significant contribution and leadership of Boris Derjaguin in this area of research.

The action of Derjaguin's pressure results in the formation of a transition region between the bulk liquid and solid or liquid substrate in front. As a result, there is no sharp three-phase contact line but instead an extended region, which is referred to in the book as the apparent three-phase contact line.

The main idea we would be happy to follow is to demonstrate the application of surface forces action through all eight chapters of the book. We followed this line in Chapters 1 through 4, where various aspects of equilibrium and kinetics are considered. At present, we do not need the consideration of Derjaguin's pressure in Chapter 5 where we discuss spreading over porous substrates; it looks like a macroscopic approach is enough for this purpose. Unfortunately, we were unable to include consideration of Derjaguin's pressure in Chapters 6 and 7 because the current state of the art does not permit accounting for it when dealing with problems of simultaneous spreading of surfactants and evaporation, leaving it as a matter for future research. In fact, we are even unable to describe the equilibrium of surfactant solution droplets on solid substrates! The action of Derjaguin's pressure in the vicinity of the apparent three-phase contact line is obviously required and it can only be guessed how it can be included.

As in the first edition, we have tried to make each section almost self-contained. However, it is impossible to understand the material presented in Chapters 1 through 4 without first reading Sections 1 and 2 in Chapter 1.

Acknowledgments

We take pleasure in expressing our gratitude once more to the colleagues already mentioned in the Acknowledgments of the first edition and now to the following colleagues with whom we had the pleasure of collaborating in the research that led to this new edition of our book: Sergey Semenov, Igor Kuchin, Omid Arjmandy-Tash, Anna Trybala, Nektaria Koursari, Gulraiz Ahmed, Diganta Das, Hezzie Agogo, Ramon G. Rubio, and Tzu Chieh Chao.

We would like to recognize support for our research by two Marie Curie ITN grants, MULTIFLOW and CoWet, from the EU; two European Space Agency grants, PASTA and MAP EVAPORATION; and grants from Proctor & Gamble and EPSRC, UK.

About the Authors

Victor M. Starov, PhD, DSc, Fellow of the Royal Society of Chemistry, Professor. Victor Starov has been a Professor at the Department of Chemical Engineering, Loughborough University, UK, since 1999. He received his PhD from the USSR Academy of Sciences in 1970 on "Capillary hysteresis in porous bodies." He received his DSc degree from St. Petersburg University in 1981 on "Equilibrium and kinetics of thin liquid layers in dispersed systems." Victor Starov is working on the influence of surface forces on kinetics of wetting and spreading over rigid and soft solids and the kinetics of spreading over porous and hydrophobic surfaces. He has published more than 300 scientific papers (http://scholar.google.co.uk/citations?hl=en&user=HxpVycEAAAAJ&view_op=list_works). Some of his research results are summarized in this book. He currently serves on the editorial board of 10 journals. He has participated in numerous organizing and scientific committees of international conferences and was a chairman of ECIC XVII in 2005. For more information please visit https://doi.org/10.1016/j.cis.2007.04.016 (Manuel G. Velarde. "Honorary Note," *Advances in Colloid and Interface Science*, 134–135, 1–2 (2007); and https://doi.org/10.1016/j.colsurfa.2016.11.028Get (N. M. Kovalchuk, R. Miller, Manuel G. Velarde. "Honorary Note," *Colloids and Surfaces A: Physicochemical and Engineering Aspects*, 521, 1–2 (2017).

Manuel G. Velarde, PhD Physics (UCM, Spain, 1968; ULB, Belgium, 1970), Honorary Doctor (Aix-Marseille, Saratov, Almeria), is an Emeritus Professor at the Instituto Pluridisciplinar, Universidad Complutense Madrid (Spain). He is a member of Academia Europaea and the European Academy of Sciences. He has published numerous papers and book chapters and six research frontier monographs and has edited several other books dealing with interfacial phenomena, wetting and spreading processes, waves and convective instabilities, and nonlinear dynamics as applied to various fields of science (patterns, waves, solitons, and chaos). He has over two decades of collaboration with V. M. Starov on wetting and spreading processes. At present he is mostly engaged in a theory of soliton-assisted electron transport (mechanical control of electrons at the nanolevel). He is the coinventor (with E. G. Wilson) of a novel field effect transistor, not using silicon, designed to operate with extremely low dissipation and huge mobility (UK Patent application published GB 2533105 A-15/6/2016). For further details see http://www.ucm.es/info/fluidos; see also R. G. Rubio et al., *Without Bounds: A Scientific Canvas of Nonlinearity and Complex Dynamics*, Springer, Berlin, Germany (2013) and R. Miller, R. G. Rubio and V. M. Starov, *Advances in Colloid and Interface Science*, 206 (2014).

1

Surface Forces and Equilibrium of Liquids on Solid Substrates

Introduction

In this chapter, we give a brief account of the theory and the experimental evidence of the action of *surface* or colloidal forces, which are the forces needed to account for phenomena occurring in a region where solid/liquid and/or liquid/air interfaces meet: very thin layers (referred to hereafter thin films). All these forces originate at the *microscopic* level, but we will look at the *macroscopic* manifestations of those forces. In particular, we emphasize the role of the so-called *Derjaguin's* pressure, which was referred to as "disjoining/conjoining" pressure in the first edition of this book.

Mostly, but not solely, we consider below two important liquid configurations: droplets on a solid substrate and liquid menisci in thin capillaries. In both cases, there is a region where solid/liquid/air meet. This region is referred to as the "apparent three-phase contact line." We will explain below why the term "apparent" is used. The surface forces or their manifestation, the Derjaguin's pressure, acts in the vicinity of the apparent three-phase contact line, and its action becomes dominant, for example, as a liquid profile approaches a solid substrate. If we consider the interaction between two liquid and/or solid particles or droplets, it is well recognized that their interaction is determined by surface or colloidal forces that are referred to as the DLVO, or modified DLVO, forces (DLVO stands for Derjaguin–Landau–Vervey–Overbeek—the scientists who introduced this theory) [1]. In the study of wetting and spreading processes, the surface forces of the same nature act in the vicinity of the apparent three-phase contact lines. Although both the nature of surface forces and the level of necessity are identical, the importance of surface forces in wetting/spreading phenomena seems to be less recognized in wetting and spreading than in colloid science.

Consideration of surface forces in the vicinity of the apparent three phase contact line is the most important direction in the area of wetting and spreading.

The relationship between the Derjaguin's pressure with the thickness of a liquid film is frequently referred to as the "Derjaguin's pressure isotherm" because it is, generally, measured at a given temperature. It is noteworthy that for water and aqueous solutions, the Derjaguin's pressure isotherm has a very special S-shaped form, which is very much different from all other liquids. Our life is highly tuned to the properties of water. However, to what extent does the S-shape of the Derjaguin's pressure isotherm of water affect life? This is an interesting question to be answered in the future.

We start with a discussion about the well-known and much-used Neumann–Young's equation in spreading and wetting dynamics (note that there are authors who refer to it as just Young's equation). We hope to convince the reader of the ill-founded thermodynamic support of the (historical) standard form of such a relationship. We argue and prove that the thermodynamically sound equation, the Derjaguin–Frumkin equation, is only possible if due account is taken of the Derjaguin's pressure. We see that the Derjaguin's pressure action either in the case of *complete* or *partial* wetting always leads to the formation of a thin liquid layer in the vicinity of the three-phase contact line. The latter results in a *microscopic* flow, which is determined by both the Derjaguin's pressure action and the topography of the surface (roughness, heterogeneity, chemical or otherwise). As a result, there is never a *real* three-phase contact line, but only an *apparent* macroscopic contact region.

In this chapter, we describe the three most used *components* of the Derjaguin's pressure. Finally, we consider at the heuristic level the *static* contact angle hysteresis when, say, a drop spreads on a *smooth* and *homogeneous* solid substrate and show that *microscopic* flow in the vicinity of the *apparent* three-phase contact line is unavoidable in the case of hysteresis. The complication introduced by such microscopic flow seems responsible for the present lack of a sound theory of the kinetics of spreading in the case of *partial* wetting in contrast to *complete* wetting, where the theory is well developed and has led to quite good agreement with experimental observations.

Note, Sections 1.1 and 1.2 provide information that is critical to understanding the material presented in Chapters 2–4. Please read these sections first.

1.1 Wetting and Neumann–Young's Equation

Why do droplets of different liquids deposited on identical solid substrates behave so differently? Why do identical droplets, for example, aqueous droplets, deposited on different substrates behave so differently?

Let us try to make a uniform layer of mercury on a glass surface. We cannot do this! Each time we try, the mercury layer will immediately form a droplet, which is a spherical cap with the contact angle bigger than $\pi/2$ (Figure 1.1). Note the contact angle is always measured inside the liquid phase (Figures 1.1 through 1.3). However, it is easy to make an oil layer (hexane or decane) on the same glass surface: An oil droplet can be deposited on the same glass substrate and it will spread out completely (Figure 1.3). In this case, the contact angle decreases with time down to the zero value over time.

Now let us try the same procedure with an ordinary droplet of tap water: An aqueous droplet deposited on the same glass substrate spreads out only partially down to some contact angle, θ, which is in between 0 and $\pi/2$ (Figure 1.2). That is, an aqueous droplet on a glass surface behaves in a way, which is intermediate between the behavior of mercury and oil.

These three cases (Figures 1.1 through 1.3) are referred to as non-wetting, partial wetting and complete wetting, respectively.

Now let us try to make a water layer on a Teflon surface. We will be unable to do this, exactly as in the case of mercury on a glass surface. That is, the same aqueous droplet can spread out partially on a glass substrate and does not spread at all on a Teflon substrate. The latter means that wetting or non-wetting is not a property of the liquid, but it is a property of the liquid–solid pair.

In a broader terms, complete wetting, partial wetting and non-wetting behavior are determined by the nature of both the liquid and the solid substrate.

FIGURE 1.1   Droplet on a solid substrate. Non-wetting case: contact angle is larger than $\pi/2$.

FIGURE 1.2   Droplet on a solid substrate. Partial wetting case: the contact angle is between 0 and $\pi/2$.

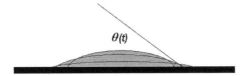

FIGURE 1.3 Droplet on a solid substrate. Complete wetting case: the droplet spreads out completely and only the dynamic contact angle can be measured, which tends to zero over time.

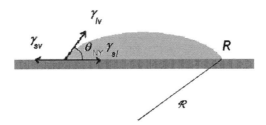

FIGURE 1.4 Droplet on a solid substrate. Interfacial tensions applied at the three-phase contact line. R is the radius of the droplet base, $\mathfrak{R}$ is the radius of the droplet curvature. The droplet is small enough that the gravity action can be neglected.

Let us consider the picture presented in Figure 1.4. Assume for the moment that the three-phase contact line is a real line, where three phases—liquid, solid and vapor—meet. We will show below that this assumption result in a contradiction with thermodynamics. Consideration of the forces in the tangential direction at the three-phase contact line results in the well-known Neumann–Young's rule, which connects three interfacial tensions—γ_{sl}, γ_{sv}, and γ_{lv}—with the value of the equilibrium contact angle, θ_{NY} (Figure 1.4), where γ_{sl}, γ_{sv}, and γ_{lv} are solid–liquid, solid–vapor, and liquid–vapor interfacial tensions, respectively:

$$\cos\theta_{NY} = \frac{\gamma_{sv} - \gamma_{sl}}{\gamma_{lv}}. \tag{1.1}$$

Note, the equilibrium contact angle in Eq. (1.1) is marked as θ_{NY}, and there is a good reason for that notation.

According to Figure 1.4, the complete wetting case corresponds to the case when all forces cannot be compensated in the tangential direction at any contact angle, that is, if $\gamma_{sv} > \gamma_{sl} + \gamma$. The partial wetting case, according to Eq. (1.1), corresponds to $0 < \cos\theta_{NY} < 1$, and, at last, the non-wetting case corresponds to $-1 < \cos\theta_{NY} < 0$. That is, Eq. (1.1) reduces complete wettability, partial wettability and non-wettability cases to the determination of three interfacial tensions, γ_{sl}, γ_{sv}, and γ_{lv}. It looks like everything is very easy and straightforward. However, unfortunately, the situation is far more complex than it looks.

Note, unfortunately two of the three interfacial tensions, γ_{sl} and γ_{sv}, cannot be measured experimentally.

Let us try to deduce Eq. (1.1) using a rigorous theoretical procedure based on the consideration of the excess free energy of the system presented in Figure 1.4. Let us adopt that the excess free energy of the small droplet (the gravity action is neglected) is as follows:

$$\Phi = \gamma_{lv}S + P_eV + \pi R^2(\gamma_{sl} - \gamma_{sv}), \tag{1.2}$$

where S is the area of the liquid–air interface; $P_e = P_a - P_l$ is the excess pressure inside the liquid; P_a is the pressure in the ambient air; P_l is the pressure inside the liquid; and R is the radius of the drop base. The last term in the right-hand side of Eq. (1.2) gives the difference between the energy of the part of the bare surface covered by the liquid drop as compared with the energy of the same solid surface without the droplet.

Note, that the excess pressure, P_e, is negative in the case of liquid droplets (concave liquid–air interface) and positive in the case of a meniscus in partially or completely wetted capillaries (convex liquid–air interface).

Under equilibrium conditions, the excess free energy (1.2) should reach the minimum value. It is shown below that the minimum of the excess free energy is reached then the contact angle reaches the value given by Neumann–Young's equation (1.1).

The highlighted text below can be omitted at the first reading.

Let $h(r)$ be the unknown profile of the liquid droplet, then the excess free energy (1.2) can be rewritten as

$$\Phi = 2\pi \int_0^R r\left(\gamma_{lv}\sqrt{1+h'^2} + P_e h + \gamma_{sl} - \gamma_{sv}\right)dr. \tag{1.3}$$

Now, we use one of the most fundamental principles: any profile, $h(r)$, in expression (1.3) should give the minimum of the excess free energy (1.2). Details of the procedure are given in Chapter 2 (Section 2.2).

Under equilibrium conditions, the excess free energy should reach its minimum value. The mathematical expressions for this requirement are the following conditions: (i) The first variation of the free energy, $\delta\Phi$, should be zero, the second variation; (ii) $\delta^2\Phi$, should be positive; and (iii) the transversality condition at the drop perimeter at the three-phase contact line, that is, at $r = R$, should be satisfied. In Section 2.2, these conditions are discussed in more detail and it is shown that one more condition should be fulfilled; however, now we ignore this extra condition because it is easy to check that this condition is always satisfied in the case of the excess free energy given by Eq. (1.3). Conditions (i) and (ii) are identical to those for a minimum of a regular functions. Condition (iii) is usually forgotten and is deduced using a different consideration.

Condition (i) results in the Euler equation, which gives the following equation for the drop profile:

$$\frac{\partial f}{\partial h} - \frac{d}{dr}\frac{\partial f}{\partial h'} = 0,$$

where

$$f = r\left[\gamma_{lv}\sqrt{1+h'^2} + P_e h + \gamma_{sl} - \gamma_{sv}\right],$$

or

$$\frac{\gamma}{r}\frac{d}{dr}\left(r\frac{h'}{\left[1+\left(h'\right)^2\right]^{\frac{1}{2}}}\right) = P_e. \tag{1.4}$$

The solution of the latter equation is a part of the sphere of the radius $\Re = -\dfrac{2\gamma_{lv}}{P_e}$ (Figure 1.4).

The second condition, (ii), gives

$$\frac{\partial^2 f}{\partial h'^2} > 0,$$

or

$$\frac{\gamma_{lv}}{\left(1+h'^2\right)^{3/2}} > 0,$$

which is always satisfied. The latter means that Eq. (1.4) actually gives a minimum value to the excess free energy (1.3).

Now, the third, the transversality condition, (iii), is as follows:

$$\left[f - h'\frac{\partial f}{\partial h'}\right]_{r=R} = 0$$

or

$$\left\{r\left[\gamma_{lv}\sqrt{1+h'^2} + P_e h + \gamma_{sl} - \gamma_{sv}\right] - h'\frac{r\gamma_{lv}h'}{\sqrt{1+h'^2}}\right\}_{r=R} = 0.$$

Taking into account that $h = 0$ at $r = R$, we conclude from the previous equation

$$\left\{\frac{\gamma_{lv}}{\sqrt{1+h'^2}} + \gamma_{sl} - \gamma_{sv}\right\}_{r=R} = 0. \tag{1.5}$$

Figure 1.4 shows that $h'\big|_{r=R} = -\tan\theta_{NY}$. Substitution of this expression into Eq. (1.5) results in Eq. (1.1).

In summary, application of the rigorous mathematical procedure to the excess free energy given by Eq. (1.2) results in

1. A spherical profile of the droplet with a radius of the curvature

$$\Re = -\frac{2\gamma_{lv}}{P_e} \tag{1.6}$$

2. The Neumann–Young's equation (Eq. 1.1) for the equilibrium contact angle θ_{NY}.

The foregoing consideration shows that the derivation of the Neumann–Young's equation (Eq. 1.1) is based on a firm theoretical basis if we adopt the expression for the free energy (1.3). This means that the Neumann–Young's equation (Eq. 1.1) is valid only in the case when the adopted expression for the excess free energy (1.3) is valid.

Let us ask ourselves a question: How many equilibrium states can a thermodynamic system have? The answer is well known: Either one or, in some special cases, two or even more states, which are separated from each other by potential barriers. According to (i) and (ii) we get an infinite continuous set of equilibrium states that are not separated from each other by potential barriers: The Neumann–Young's equation does not specify the equilibrium volume of the droplet, V, or the excess pressure inside the drop, P_e, which can be any negative value. Both the volume of the droplet and the excess pressure can be arbitrary. The latter means that the volume of the droplet is not specified: a droplet of any volume can be at the equilibrium.

That means the Neumann–Young's equation (Eq. 1.1) is in drastic contradiction with thermodynamics. Why is that? Where is the mistake? Definitely not in the derivation. That means we should go back to basics. Is the expression for the excess free energy (1.3) correct?

According to basic thermodynamics law the following three requirements should hold at equilibrium:

1. The liquid in the droplet must be at the equilibrium with its own vapor.
2. The vapor must be at equilibrium with the solid substrate.
3. The liquid in the droplet must be at the equilibrium with the solid covered by an adsorption layer.

Step by step, we will show that none of these three equilibria are taken into account by the expression for the excess free energy (1.3).

The first requirement (a) results in the equality of chemical potentials of the liquid molecules in the vapor and inside the droplet. This results in the following Kelvin's expression of the excess pressure, P_e:

$$P_e = \frac{R_g T}{v_m} \ln \frac{p_s}{p},\tag{1.7}$$

where v_m is the molar volume of the liquid; p_s is the pressure of the saturated vapor at the temperature T; R_g is the gas constant (do not confuse this with the radius of the drop base); and p is the vapor pressure, which is at the equilibrium with the liquid droplet. The latter equation determines the unique equilibrium excess pressure, P_e, and, hence, according to Eq. (1.6), the unique radius of the droplet, $\Re$.

Recall that the excess pressure inside the drop, P_e, should be negative (pressure inside the droplet is greater than the pressure in the ambient air). That means that the right-hand side of Eq. (1.7) should be negative also, but the latter is possible only if $p > p_s$, that is, the droplets can be at equilibrium only with oversaturated vapor! This is a big problem because the equilibration process takes hours, and it is necessary to keep the oversaturated vapor over the solid substrate under investigation for hours. To the best of our knowledge, nobody can do that experimentally. This means that it is difficult, if possible at all, to experimentally investigate equilibrium droplets on a solid substrate. A flood of investigations have been published on the equilibrium contact angles of droplets on solid substrates. The previous consideration shows that the contact angles measured are not at equilibrium: These contact angles are referred to as static advancing or static receding contact angles. Note, only in Chapter 3 will we be ready to clarify the subject completely and to express static advancing/receding contact angles via Derjaguin's pressure isotherm.

Unfortunately, this is not the end of problems with the Neumann–Young's equation (Eq. 1.1) because now we consider the requirements of equilibrium requirement (b). Let us assume that we can create, at least theoretically, an oversaturated vapor over the solid substrate and wait long enough until equilibrium is reached. Now the liquid molecules in the vapor are at equilibrium with the liquid molecules in the droplet. Note that the solid–liquid interfacial tension, γ_{sl}, differs from the solid–vapor interfacial tension, γ_{sv}. If they are not different, then, according to Eq. (1.1), the contact angle is equal to 9° (an intermediate case between partial wetting and non-wetting). In the case of partial wetting or complete wetting $\gamma_{sl} < \gamma_{sv}$. The latter means that the presence of liquid on the solid substrate results in lower surface tension as compared with the surface tension of the bare solid surface, γ_{sv}. Now, back to our theoretical case of the liquid droplet on the solid substrate at equilibrium with the oversaturated vapor. We should now take into account the equilibrium between the liquid vapor and the solid substrate: It is well known that liquid molecule adsorbs onto the solid substrate. Hence, the solid substrate *must be* covered by a layer of liquid molecules, and the presence of liquid molecules on the surface changes the initial surface tension, γ_{sv}. This means that the liquid molecules from the vapor must adsorb onto the solid substrate outside of the liquid droplet under consideration. The latter results in the formation of an adsorption liquid film on the surface and a new interfacial tension, γ_{hv}, where h is the thickness of the adsorbed layer and which is at equilibrium with the vapor pressure, p, in the ambient air.

One may say that a monolayer or, in the best case, several layers of the adsorbed liquid molecules are on the solid substrate and that the influence on the macroscopic droplet will be negligible. Let us show that even a small amount of liquid on the solid substrate is important. For this, let us consider the simple but important example of when the presence of only one monolayer changes drastically the wetting property. Let us take a microscope coverslip and place an aqueous droplet on this surface. The droplet will form a contact angle, which depends considerably on the type of the glass and, in some special case (which we consider now), it will be as small as 10°. Now, let us place a monolayer of an oil on the glass surface (recall that a monolayer means a layer with a thickness of 1 molecule, and it looks like nothing to worry about). Now again, let us place a water droplet on a new glass surface covered by a monolayer of oil. The droplet will form a contact angle that is higher than 90°. The presence of one tiny monolayer has changed partial wetting to non-wetting.

Now, back to our droplets on the solid surface at equilibrium with the oversaturated vapor. We understand now that the adsorption of vapor on the solid substrate is very important and that instead of the interfacial tension of the bare solid surface, γ_{sv}, we should use γ_{hv}. Interfacial tension will be discussed in detail in Chapter 2 (Section 2.1).

The previous consideration shows that to investigate equilibrium in liquid droplets, the following procedure should be followed:

- The solid substrate under investigation should be kept in the atmosphere of the oversaturated vapor until equilibrium adsorption of vapor on the solid substrate is reached, and a new interfacial tension, γ_{hv}, should be measured.
- After that, the droplet of the size that should be at equilibrium with the oversaturated vapor should be deposited and maintained until equilibrium is reached.
- The Neumann–Young's equation (Eq. 1.1) should now be rewritten as

$$\cos\theta_e = \frac{\gamma_{hv} - \gamma_{sl}}{\gamma_{lv}}. \tag{1.8}$$

The characteristic time scale of these processes depends on the liquid volatility and viscosity and is, in general, of hours of magnitude. To the best of the authors' knowledge, this kind of experiment has never before been attempted in the atmosphere of an oversaturated vapor. This means that equilibrium liquid droplets of volatile liquids probably have never been observed experimentally.

It is obvious for the same reasons as before that the thickness of the adsorbed layer, h, depends on the vapor pressure in the ambient air; that is, γ_{hv} is a function of the pressure in the ambient air, p, and, hence, according to Eq. (1.8), the contact angle changes with vapor pressure. Is this dependency strong or weak? The answer will be given in Chapter 2 (Sections 2.1 and 2.3).

Is this the end of the problems with the Neumann–Young's equation (Eq. 1.1)? Unfortunately, not, because we did not consider the last, but not the least, requirement of the equilibrium, requirement (c). In Figure 1.5, an equilibrium liquid droplet is presented in contact with an equilibrium adsorbed liquid film on a solid surface. What happens in the vicinity of the line where they meet?

Is the situation presented in Figure 1.5 possible? Such a sharp transition from the liquid droplet to the liquid film in front is impossible: The capillary pressure will be infinite on the line indicated by the arrow. Hence, there should be a smooth transition from the flat equilibrium liquid film on the solid surface to the spherical droplet, as shown in Figure 1.6, which presents such a smooth transition.

Let us call this region—where the transition from a flat film to a droplet takes place—a transition zone. The presence of the transition zone explains now why in the beginning we referred to the three-phase contact line as an "apparent three-phase contact line": There is no simple line; instead there is a whole transition zone.

The presence of this transition zone leads to much bigger problems because, as before, pure capillary forces cannot keep the liquid in this zone at equilibrium (Figure 1.6): The liquid profile is concave (hence, the capillary pressure under the liquid surface is greater than in the ambient air) to the right of the arrow in Figure 1.6, and the liquid profile is convex (hence, the capillary pressure under the liquid surface is less than in the ambient air) to the left of the arrow in Figure 1.6. However, the liquid profile depicted in Figure 1.6 is supposed to be at equilibrium, which is impossible in the presence of the

FIGURE 1.5 Cross section of an equilibrium liquid droplet (at oversaturation) in contact with an equilibrium-adsorbed liquid film on a solid substrate. What happens on the line (indicated by the arrow) where they meet?

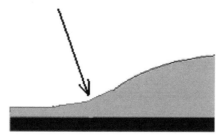

FIGURE 1.6 Transition zone from the flat equilibrium liquid film on a solid surface to the liquid droplet. The arrow indicates the point where to the right the liquid profile is concave and to the left the profile is convex.

pressure difference. This means that inside the transition zone some extra forces should come into play. We will come back to consider these extra forces below. The consideration of these forces was one of the motivations to replace the name "colloid science" with a new name: "colloid and interface science."

The above consideration shows that the Neumann–Young's equation can be probably used only in the case of nonvolatile liquids because we have too many problems with volatile liquids. Can a liquid really be nonvolatile? Usually low volatility means the liquid has large molecules, which means high viscosity and a correspondingly higher characteristic time scale for the equilibration process with an oversaturated vapor. Despite that, let us assume that the liquid is nonvolatile. In the case of partial wetting, as we already saw, at equilibrium, liquid droplets cannot be at equilibrium with a bare solid surface. At equilibrium, there should always be an adsorption layer of liquid molecules on the solid substrate in front of the droplet on the bare solid surface. If the liquid is volatile, then this layer is created by means of evaporation–adsorption. However, if the liquid is nonvolatile, the same layer should be created by means of flow from the droplet edge onto the solid substrate. As a result, the solid substrate is covered at equilibrium by an equilibrium liquid layer of thickness *h*. The thickness of the equilibrium liquid film, *h*, is determined (as we see below) by the potential of the surface forces action. The characteristic time scale for this process is hours because it is determined by the flow in the thinnest part in the vicinity of the apparent three-phase contact line where the viscous resistance is very high. During these hours, evaporation of the liquid from the droplet cannot be ignored, and we must return to the problem of volatility.

Let us assume, however, that after all the equilibrium film forms in front of the liquid droplet and that we waited long enough for the equilibrium. However, now again we have the three interfacial tensions—γ_{lv}, γ_{sl}, and γ_{vh}—which are liquid–vapor, solid–liquid, and solid substrate covered with the liquid film of thickness *h*–vapor interfacial tensions, respectively. We return to the same problem as in the case of a volatile liquid. We cannot measure the interfacial tension, γ_{vh}, and use it in Eq. (1.8). However, there is an answer to this problem, and the answer will be given in Section 2.1.

In view of the above discussion, hereafter we will use the term "apparent" three-phase contact line because there is no such line at the microscopic scale.

This means that even in the case of nonvolatile liquids, the applicability of the Neumann-Young's equation (Eq. 1.1) remains questionable. Unfortunately, despite this consideration, the Neumann–Young's equation is frequently used, and in some cases, it gives the correct quantitative description of the phenomenon. We would simply like to emphasize that it is necessary to bear in mind that this equation is erroneous from a thermodynamic point of view and that its application can result in erroneous conclusions.

1.2 Surface Forces and Derjaguin's Pressure

The presence of adsorbed liquid layers on a solid substrate is a result of the action of some special forces, which are referred to as surface forces.

Let us return to Figure 1.6 and consider the transition zone between the droplet and the flat liquid films in front of it. It looks like the profile presented in Figure 1.6 cannot be at equilibrium because the capillary pressure should change the sign inside the transition zone and it is in contradiction with the requirement of the constancy of pressure everywhere inside the droplet. Some additional forces are missing.

This problem has been under consideration by a number of scientists for over a century, and those efforts have resulted in a considerable reconsideration of the nature of the wetting phenomena.

A new class of phenomena has been introduced: surface phenomena, which are determined by the special forces acting in thin liquid films or in layers in the vicinity of the apparent three-phase contact line.

Surface forces are well known and widely used in colloid and interface science. They determine the stability and behavior of colloidal suspensions and emulsions. In the case of emulsions/suspensions, their properties and behavior (stability, instability, rheology, interactions, and so on) are completely determined by the surface forces acting between colloidal particles or droplets. This theory is widely referred to as the Derjaguin–Landau–Vervey–Overbeek, or DLVO, theory [1], named for the four scientists who developed the theory. There is no doubt that all colloidal particles have rough surfaces and, in a number of cases, even chemically inhomogeneous surfaces (e.g., living cells). Roughness and inhomogeneity of colloidal particles can substantially modify surface forces, that is, their nature, magnitude, range of action. However, the roughness and inhomogeneity of the surface of colloidal particles do not influence the main conclusion: All their interactions and properties are determined by the actions of the surface forces [2].

Surprisingly, it is quite a different story with wetting phenomena as compared with colloid and interface science: It is widely (and erroneously, as we shall see in this book) accepted that roughness and inhomogeneity of the solid substrate in contact with liquids can itself—without consideration of surface forces acting in a vicinity of the apparent three-phase contact line—explain wetting properties of liquids on solid substrates. As a result, the influence of surface forces on the kinetics of wetting and spreading is much less recognized than in the case of colloidal suspensions/emulsions despite their sharing the same nature of surface forces.

It has been established that the range of surface forces action is usually on the order of 0.1 μm [1]. Note that in the vicinity of the apparent three-phase contact line, $r = R$ (Figure 1.4), the liquid profile, $h(r)$, tends to zero thickness. The latter means that close to the apparent three-phase contact line, surface forces come into play and their influence cannot be ignored.

One manifestation of the action of surface forces is the Derjaguin's pressure. To explain the nature of the Derjaguin's pressure, let us consider the interaction of two thick, plain parallel surfaces divided by a thin liquid layer of thickness, h (e.g., an aqueous electrolyte solution). The surfaces are not necessarily of the same nature, as two important examples show: (i) surface 1 is air, 3 is a liquid film, and 2 is solid support; and (ii) surfaces 1 and 2 are air and 3 is a liquid film. The first case, (i), is referred to as a liquid film on a solid support, and it models the liquid layer in the vicinity of a three-phase contact line; the second case, (ii), is referred to as a free liquid film, and it models the film present in a foam and determines foam properties. A range of experimental methods can be used to measure the interaction forces between these two surfaces as a function of the thickness, h (Figure 1.7) [1,3,4].

If h is bigger than $\cong 10^{-5}$ cm $= 0.1°$μm, then the interaction force between phases 1 and 2 (Figure 1.7) can be neglected. However, if $h < 10^{-5}$ cm, then an interaction force appears. This force can depend on the thickness h in a very peculiar way. The interaction forces divided by the surface area of the plate has a dimension of pressure and is referred to as the Derjaguin's pressure [1]. Note that the term "disjoining pressure" is frequently used, but this is a bit misleading because the mentioned force can be either disjoining (repulsion between surfaces) or conjoining (attraction between surfaces). The term "Derjaguin's pressure" is used to recognize the contribution by Boris Derjaguin to the area of study.

Let us discuss the physical phenomena behind the existence of surface forces, taking a liquid–air interface as an example (Figure 1.8). It is obvious that the physical properties of the very first layer on the

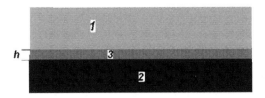

FIGURE 1.7 Interaction between two thick phases (1) and (2), which are possibly made of different materials, through a thin liquid layer, (3), in between. This interaction determines the Derjaguin's pressure in the thin liquid layer (3).

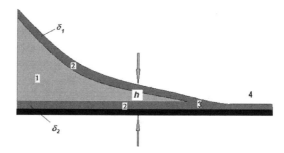

FIGURE 1.8 The liquid profile in the vicinity of the apparent three-phase contact line. (1) Bulk liquid, where boundary layers do not overlap; (2) boundary layer in the vicinity of liquid–air and liquid–solid interfaces of thickness δ_1 and δ_2, respectively; (3) a region, where boundary layers overlap; and (4) flat thin equilibrium film in front. The latter two regions are the ones where the Derjaguin's pressure acts.

interface differ substantially from the properties of the liquid in the bulk far from the interface. What can we say about the properties of the second, third and so on layers? It is understandable that the physical properties do not change jumpwise from the very first layer on the interface to the subsequent layers but that the change proceeds in a continuous way. The latter results in a formation of a special layer, which we refer to hereafter as a boundary layer, where all properties differ from the corresponding bulk properties. Do not confuse the introduced term "boundary layer" with a boundary layer as in hydrodynamics, they have nothing do to with each other!

Such boundary layers exist in proximity to any interface: solid–liquid, liquid–liquid, liquid–air. In the vicinity of the apparent three-phase contact line (Figure 1.8), those boundary layers overlap. The overlapping of boundary layers is the physical phenomenon which results in the existence of surface forces. The surface force per a unit area has a dimension of pressure and is referred to as the "Derjaguin's pressure" as mentioned above. Let the thickness of the two boundary layers be δ_1 and δ_2, respectively. In the vicinity of the three-phase contact line, the thickness of a droplet, h, is small enough, that is, $h \sim \delta_1 + \delta_2$, and, hence, boundary layers overlap (Figure 1.8), which results in the creation of the Derjaguin's pressure. The characteristic scale $\delta_i \sim 10^{-5}$ cm ($i = 1,2$) determines the characteristic thickness, where the Derjaguin's pressure acts. This thickness is referred to below as the range of the Derjaguin's (or surface forces) action, $t_s \sim \delta_1 + \delta_2$.

The main conclusion from the consideration above is that the pressure in thin layers close to the three-phase contact line is different from the pressure in the bulk liquid: It depends on the thickness of the layer, h, and the nature of the surface forces acting in the region 3 (Figure 1.8), and it varies with the thickness, h.

Below we review briefly the physical phenomena that result in the formation of the abovementioned surface forces and the Derjaguin's pressure (see [1] for details).

Components of the Derjaguin's Pressure

Several physical phenomena have been identified for the appearance of the Derjaguin's pressure. Below, we consider only three of them.

Molecular or Dispersion Component

Let us start with the most investigated molecular or dispersion component of surface forces. Note, in the case of complete wetting (only in this case) this component is the only component acting in thin liquid layers. Unfortunately, this component is used in the literature more frequently than others components (see below) even in the situations then this component is not the most important one.

It is well known that, at relatively large distances (but still in the range of angstroms, that is, 10^{-8} cm) all neutral molecules interact with each other, and the energy of this interaction is proportional to const/r^6, where r is the distance between molecules. Let us examine two surfaces made of different materials placed inside an aqueous electrolyte solution at a distance, h, from each other (Figure 1.7).

Calculation of the molecular contribution to the Derjaguin's pressure Π_m has been approached in two ways: from the approximation of interactions as pairwise additive forces and from a field theory of many-body interactions in condensed matter. The simpler and historically earlier approach is a theory based on summing individual London–van der Waals interactions between molecules pair-by-pair and was undertaken by Hamaker [1].

The more sophisticated, modern theory of Π_m was developed by I. E. Dzyaloshinsky, E. M. Lifshits, and L. P. Pitaevsky (see summary in [1]).

The following functional dependency was deduced based on a consideration of a fluctuating electro-magnetic field:

$$\Pi_m(h) = \begin{cases} \dfrac{A}{h^3}, & h < \lambda \\[2mm] \dfrac{B}{h^4}, & h > \lambda, \end{cases} \tag{1.9}$$

where λ is a characteristic wave length [1]. Hereafter, we use only expression (1.9) for the molecular component at $h < \lambda$ because the contribution of the Derjaguin's pressure at a "big" film or layer thickness at $h > \lambda$ is relatively small as compared with the first part at $h < \lambda$.

Parsegian and Ninham (see [1] for references) discovered a convenient technique for calculating the Hamaker constant. Precise measurements of Π_m in thick thin films are in good agreement with the theory predictions [1]. However, the latter theory does not apply to films so thin as to have dielectric properties that vary with thickness.

Using the approximation of direct summation of all molecular pairs in the system, the following expression was deduced for the molecular or dispersion components of the Derjaguin's pressure:

$$\Pi_m = -\frac{A_H}{6\pi h^3}, \quad A_H = A_{33} + A_{12} - A_{13} - A_{23}, \tag{1.10}$$

where A_H is referred to as the Hamaker constant, named after the scientist who carried out these calculations around a half-century ago [1]. The Hamaker constant, A_H, depends on the properties of the phases 1, 2 and 3. It is important that the Hamaker constant can be either positive (repulsion) or negative (attraction). The precise value of the Hamaker constant based on direct summation can be completely erroneous. This is the reason why a number of various approximations have been suggested for more accurate calculation of the Hamaker constants [1].

In the case of oil droplets on the glass surface, when the dispersion component is the only component of the Derjaguin's pressure acting in thin films, the dispersion interaction is repulsive, that is, the Hamaker constant is positive. Below we mostly consider the case of thin liquid films on solid substrates where the Hamaker constant is positive. Note that in this book the definition of the Hamaker constant is slightly different from Eq. (1.10)

$$\Pi_m = \frac{A}{h^3}, \quad A = -\frac{A_H}{6\pi} \tag{1.11}$$

and also note that the constant A alone is referred to below as the Hamaker constant. Note that a positive Hamaker constant A means a repulsion, and a negative one means an attraction.

The characteristic value of the Hamaker constant is $A \sim 10^{-14}$ erg (oil films on a glass, quartz or mica surfaces). The latter value of the Hamaker constant shows that at the thickness of the liquid layer $h \sim 10^{-7}$ cm, the dispersion component of the Derjaguin's pressure is $\Pi_m \sim 10^{-14}/10^{-21} = 10^7$ dyn/cm^2. Let us consider a small oil droplet of a radius $\Re \sim 0.1$ cm on a solid substrate (Figure 1.4); the surface tension of oils is about $\gamma \cong 30$ dyn/cm. The capillary pressure inside the spherical part of the droplet is $\frac{2\gamma}{\Re} \sim \frac{2\cdot30}{0.1} = 6\cdot10^2$ dyn/cm^2. The latter shows that in the vicinity of the three-phase contact line the capillary pressure is much smaller than the Derjaguin's pressure. Let us assume for a moment that the droplet shape remains spherical until contact with the solid substrate. However, as we have already shown, the capillary pressure is much smaller than the Derjaguin's pressure and cannot counterbalance the Derjaguin's pressure. The latter means that the Derjaguin's pressure action substantially distorts the spherical shape of droplets in the vicinity of the three-phase contact line. That means, droplets cannot remain in a spherical shape up to the contact line! See further consideration of the profile of liquid droplets in Section 2.3.

Before going further and discussing the next electrostatic component of Derjaguin's pressure, we should say a few words about the electrical double layer.

Electrical Double Layers

Neutral molecules of many salts, acids and alkalis dissociate into ions (cations and anions) in water, forming aqueous electrolyte solutions. For example, NaCl dissociates with the formation of a cation Na^+ and an anion Cl^-. Moreover, if we assume that water is completely pure without any salts, acids and so on, then the water molecules, H_2O, also dissociate according to the following dissociation reaction:

$$H_2O \leftrightarrow H^+ + OH^-.$$

That is, even in pure water both cations, H^+, and anions, OH^-, are always present. Note, the latter two ions H^+ and OH^- play the most important role in the kinetics of wetting and spreading of aqueous solutions because they determine the pH of the solution: $pH = -\log_{10} C_{H^+}$, where C_{H^+} is the concentration of H^+ in mol/L.

The total charge of cations is completely counterbalanced by the total charge of anions in the bulk of the liquid. These ions are referred to as free ions. All free ions can be transferred both convectively by the flow of water and by diffusion in the presence of a gradient of concentration of any ion either imposed or spontaneous. Ions can also be transferred under the action of an electric potential gradient (electromigration), either imposed or spontaneous.

In the aqueous electrolyte solution, the majority of solid surfaces acquire an electrical charge. Before speaking about the mechanism of the formation of this charge let us emphasize that these charges are mostly rigidly fixed on the solid surface and can usually be moved only with the solids. There are two main mechanisms for the formation of the charge of the solid surface in aqueous electrolyte solutions: the dissociation of surface groups (briefly discussed below) and the unequal adsorption of different types of ions.

A considerable number of solid surfaces have R–OH surface groups on the solid–liquid interface, where R– is the group rigidly connected to the solid. The –OH groups can dissociate in aqueous solutions, resulting in the formation of negatively charged groups, $R–O^-$, returning the H^+ ions into the solution. According to this or a similar mechanism, many solid surfaces (actually the majority) in aqueous solutions acquire a negative surface charge. This charge obviously strongly depends on the pH of the solution, that is, on the concentration of H^+ ions in the volume of solution. Note, pH = 7 corresponds to a neutral solution, pH < 7 an acidic solution, and pH > 7 an alkali solution.

In all processes, the free and bound ions behave in different ways: Free ions can freely move; the bound ions only move with the solid surface to which they are attached. Let us consider the distribution of ions in the close vicinity of a negatively charged surface in contact with an aqueous electrolyte solution, for example, NaCl. NaCl dissociates as NaCl→Na^+ + Cl^-. An electroneutral condition requires equal concentrations of cations and anions in the bulk solution far from the charged surface. However, close to the charged surface, according to the Coulomb's law, free Na^+ cations are attracted by the negatively charged solid surface, and the negatively charged Cl^- ions are repulsed by the same surface. As a result, the concentration of cations is higher near the surface, and the concentration of anions is lower than in the corresponding concentration in the bulk solution. Let us recall diffusion: The basic task of diffusion is to destroy all non-uniformity in the distribution of ions. In our case, diffusion will attempt to do exactly the opposite of the Coulomb's interaction: Diffusion acts to decrease the concentration of cations near the surface and to increase the concentration of anions. Because of these two opposing trends near the negatively charged surface, a layer of a finite thickness is created, inside which the concentration of cations reaches its maximum near the surface and monotonically decreases into the depths of the solution to its bulk value while the concentration of anions monotonically grows from its minimum value near the surface to its bulk value in the depth of solution. This layer, where the concentration of cations and anions differ from their bulk values, is referred to as the diffusive part of an electrical double layer. The characteristic thickness of the diffusive part of an electrical double layer is the Debye length, R_d. The characteristic value of the Debye length is $R_d = \frac{3 \cdot 10^{-8}}{\sqrt{C}}$ cm, where the electrolyte concentration, C, should be expressed in mol/L [1]. The latter expression shows that the higher electrolyte concentration, the thinner electrical double layer. For example, at $C = 10^{-4}$ mol/L: $R_d = 3 \cdot 10^{-6}$ cm (which is considered as a big thickness), while at $C = 10^{-2}$ mol/L: $R_d = 3 \cdot 10^{-7}$ cm (which is considered as a very small thickness). The electrical double layer is formed by two parts: The first part is the charged surface (usually negatively charged) with immobile ions, and the second part is the diffusive part. The electrical potential of the charged solid surface is referred to as the zeta-potential, ζ. A characteristic value of the ζ potential is equal to $R_g T/F = 25$ mV, where R_g is the universal gas constant; T is the absolute temperature in K, and F is the Faraday constant.

The difference in mobility of free mobile ions in the diffusive part of the double electrical layer and on the charged surface determines the nature of the electrokinetic phenomena, which are totally determined by the properties of electrical double layer.

Electrokinetic Phenomena

Currently, a number of electrokinetic phenomena have been discovered and investigated. Only one is briefly discussed here: the streaming potential. Let us consider the flow of an electrolyte solution in a capillary with negatively charged walls (for example, a glass or quartz capillary). In the initial state, the feed solution and the receiving solution have equal concentrations of electrolyte. The electrolyte solution starts to flow after a pressure difference is applied to both sides of the capillary. This flow involves mobile cations inside the electrical double layer near the solid, negatively charged walls of the capillary, into a convective motion, which is an electric current. As a result of the convective electric current, the concentration of cations increases in the receiving solution and an excess positive charge accumulates. This excess causes, in turn, the appearance of an electric potential difference between the beginning and the end of the capillary, generating an electric current in the direction opposite to the direction of the ion flow. This electric current destroys the emerging surplus of cations in the outflowing solution. The electric potential difference between the ends of the capillary in this case is referred to as a streaming potential. Let us note that the total electric current in the system is equal to zero, that is, there is no electric current in the system despite an electric potential difference between the ends of the capillary.

The Electrostatic Component of the Derjaguin's Pressure

Now we can continue examination of the next component of the Derjaguin's pressure: the electrostatic component.

Let us return to the examination of two charged surfaces (which are not necessarily of the same charge nature) in aqueous electrolyte solutions (Figure 1.9a and b). The surfaces are assumed to be charged equally or oppositely, that is, there is an electrical double layer near each of the surfaces. The sign of the charge of the diffusive part of each of the electrical double layers is opposite to the sign of the charge of its corresponding surface. If the width of clearance between the surfaces is $h \gg R_d$, the electrical double layers of surfaces do not overlap (Figure 1.9a) and there is no electrostatic interaction of the surfaces. However, if the thickness of the clearance, h, is comparable with the thickness of the electrical double layer, R_d, then the electrical double layers overlap, resulting in an interaction between the surfaces. If the

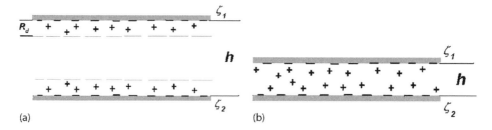

(a)   (b)

FIGURE 1.9 (a) ζ_1 and ζ_2 are negative. Distance between two negatively charged surfaces, h, is bigger than the thickness of the Debye layers, R_d. Electrical double layers do not overlap and there is no electrostatic interaction between these surfaces. ζ_1 and ζ_2 are electrical potentials of charged surfaces. (b) ζ_1 and ζ_2 are negative. Distance between two negatively charged surfaces, h, is smaller or comparable with the thickness of the electrical double layer, R_d. Electrical double layers of both surfaces overlap, which results in an interaction, which is repulsion in the case under consideration.

surfaces are equally charged, then their diffusive layers are equally charged as well, that is, repulsion appears as a result of their overlapping (the electrostatic component of the disjoining pressure is positive in this case).

If surfaces are oppositely charged, then because of the overlapping of the opposite charges, attraction appears, and the electrostatic component of the disjoining pressure is negative in this case (Figure 1.10a and b).

There are a number of approximate expressions for the electrostatic component of the Derjaguin's pressure [1]. For example, in the case of low ζ potentials of both surfaces, the following relation is valid [1]:

$$\Pi_e(h) = \frac{\varepsilon\kappa^2}{8\pi} \frac{2\zeta_1\zeta_2 \cos h\kappa h - \left(\zeta_1^2 + \zeta_2^2\right)}{\sin h^2\kappa h}, \tag{1.12}$$

where ε is the dielectric constant of water; and $1/\kappa = R_d$. The ζ potential is considered as low if the corresponding dimensionless potential $\frac{F\zeta}{R_g T} < 1$.

Note that in the case of oppositely charged surfaces, and at relatively small distances, the following expression for the electrostatic component of the Derjaguin's pressure is valid [1]:

$$\Pi_e(h) = -\frac{\varepsilon}{8\pi} \frac{\left(\zeta_1 - \zeta_2\right)^2}{h^2}, \tag{1.13}$$

which is always attraction. It is necessary to be very careful with the latter expression because in this case the attraction changes to repulsion at distances smaller than some critical distance [1].

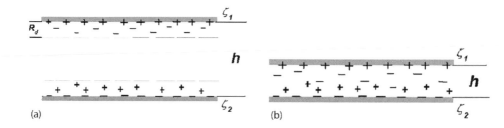

(a)   (b)

FIGURE 1.10 (a) $\zeta_1 > 0$ and $\zeta_2 < 0$. Distance between two charged surfaces bearing the opposite charges, h, is bigger than the thickness of the Debye layers, R_d. Electrical double layers do not overlap and there is no electrostatic interaction between these surfaces. ζ_1 and ζ_2 are electrical potentials of charged surfaces. (b) $\zeta_1 > 0$ and $\zeta_2 < 0$. Distance between two oppositely charged surfaces, h, is smaller or comparable with the thickness of the electrical double layer, R_d. Electrical double layers overlap, which results in an interaction, which is attraction in the case under consideration.

Equations (1.12) and (1.13) show that the Derjaguin's pressure does not vanish even in the case when only one of the two surfaces is charged (e.g., $\zeta_1 = 0$ and $\zeta_2 = 0$). The physical reason for this phenomenon is the deformation of the electrical double layer if the distance between the surfaces is smaller than the Debye radius.

The theory based on the calculation of the disjoining pressure based on those two components—that is, dispersion, $\Pi_m(h)$, and electrostatic, $\Pi_e(h)$—is referred to as the DLVO theory. According to the DLVO theory, the total Derjaguin's pressure is a sum of the two components, that is, $\Pi(h) = \Pi_m(h) + \Pi_e(h)$. The DLVO theory makes possible the explanation of a range of experimental data on the stability of colloidal suspensions/emulsions as well as the static and the kinetics of wetting. However, it was later understood that these two components alone are insufficient for explaining the phenomena in thin liquid films and layers and colloidal dispersions. There is a third important component of the Derjaguin's pressure, which becomes equally important in aqueous electrolyte solutions.

Structural Component of the Derjaguin's Pressure

This component of the Derjaguin's pressure is caused by the orientation of the water molecules in the vicinity of an aqueous solution–solid interface or an aqueous solution–air interface. Let us recall that water a molecule can be modeled as an electric dipole.

In the vicinity of a negatively charged interface, the positive parts of the water dipoles are attracted to the surface, forming a layer of water molecules at the surface with the negative parts of the dipoles directed outward. Then in the next layer, the water dipoles are facing the negatively charged parts of dipoles, which in its turn, results in the orientation of the next layer of dipoles and so on. However, thermal fluctuations tend to destroy this orientation (Figure 1.11).

Because of these two opposite trends, a finite layer forms where structure of water dipoles different from the completely random bulk structure of water. This layer is frequently referred to as "a hydration layer." If now we have two interfaces with hydration layers close to each of them (or even one of them), then at a close separation, comparable with the thickness of the hydration layer, these surfaces "feel each other." That is, the hydration layers overlap. This overlap results in either the attraction or the repulsion of the two surfaces.

Unfortunately, at present there is no firm theoretical background for the structural component of disjoining pressure, and we are unable to deduce theoretically in which case the structure formation results in an attraction and in which case in a repulsion. However, a qualitative consideration of the structural forces is presented in [5], see also a review on the subject [6]. The total structural force in this case can be presented in the following form [17,18]:

$$\Pi_S(h) = K_1 e^{-h/\lambda_1} + K_2 e^{-h/\lambda_2}, \tag{1.14}$$

where K_1, K_2 and λ_1, λ_2 are parameters related to the magnitude and the characteristic length of the structural forces. The subscripts 1 and 2 correspond to the short-range and long-range structural interactions, respectively. Currently the latter four constants can be extracted from experimental data only.

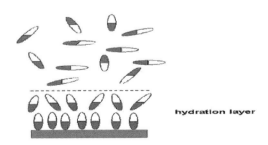

FIGURE 1.11 Formation of a hydration layer of water dipoles in a vicinity of a negatively charged interface. Darker part of water dipoles is positively charged, lighter part is negatively charged.

There is a clear physical meaning of only one parameter, $1/\lambda_1$, which is the correlation length of water molecules in aqueous solutions. The latter gives $1/\lambda_1 \sim 10\text{--}15$ Å, which is the characteristic thickness of the hydration layer.

Currently it is assumed [1,7], that the Derjaguin's pressure of thin aqueous films is equal to the sum of the three mentioned components:

$$\Pi(h) = \Pi_m(h) + \Pi_e(h) + \Pi_s(h). \qquad (1.15)$$

In Figure 1.12, the dependence of the Derjaguin's pressure on the thickness of a flat liquid film is presented for the cases of complete wetting [curve 1, which corresponds to a dispersion or molecular component of the Derjaguin's pressure, $\Pi_m(h)$, and partial wetting (curve 2, which corresponds to a sum of all three components of the Derjaguin's pressure, according to Eq. (1.15). The Derjaguin's pressure as presented by curve 1 in Figure 1.12 corresponds to the case of complete wetting, for example, oil droplets on a glass substrate, whereas curve 2 corresponds to the case of partial wetting, for example, aqueous electrolyte solutions on glass substrates.

In [1–4] a number of experimental data on the measurement of the Derjaguin's pressure are presented. The dependency (11) has been firmly confirmed in the case of oil thin films on glass, quartz, metal surfaces, corresponding to the case of complete wetting. Figure 1.13 presents the experimental data and calculations according to Eq. (1.15) for aqueous thin films.

In [1] all necessary details concerning the experimental data presented in Figure 1.13 are given. As shown in Chapter 2, the Derjaguin's pressure of flat thin films can be measured only at undersaturation, that is, at $\Pi(h) > 0$ if the stability condition $\Pi'(h) < 0$ is satisfied.

This shows that the Derjaguin's pressure can currently be measured only in a very limited interval if thicknesses are as shown in Figure 1.13. It is a major experimental and theoretical challenge to measure the Derjaguin's pressure in the whole range of film thickness.

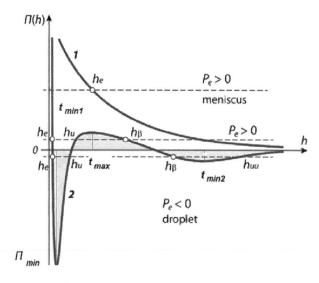

FIGURE 1.12 Types of isotherms of the Derjaguin's pressure, which are under consideration below: (1) Complete wetting, observed for oil films on quartz, glass, and metal surfaces [1,7]; and (2) partial wetting, observed for aqueous films on quartz, glass, and metal surfaces [1,7]. Excess pressure $P_e > 0$ corresponds to undersaturation (meniscus in thin capillaries); excess pressure $P_e < 0$ corresponds to oversaturation (droplets). h_e is a corresponding thermodynamically equilibrium flat thin film; h_u and h_{uu} are unstable films; h_β is a metastable equilibrium film. Note, relatively thick films in the region h_β are referred to as β-films, whereas much thinner films in the region of h_e are referred to as α-films. (From Derjaguin, B.V. et al., *Surface Forces*, Consultants Bureau, Plenum Press, New York, 1987.)

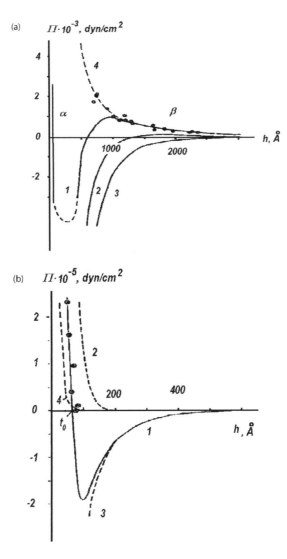

FIGURE 1.13 Calculated and experimentally measured isotherms of the Derjaguin's pressure, $\Pi(h)$, of water films on the quartz surface at $C = 10^{-5}$ mol/L concentration of KCl, pH = 7, and dimensionless ζ potential of the quartz surface equals to -6 [1]. (a) Within the region of large thicknesses: dimensionless ζ potential of the film–air interface equals -2.2 (curve 1), -1 (curve 2), and 0 (curve 3); and (b) within the region of small thicknesses: dimensionless ζ potential of the film–air interface equals to -2.2 (curve 1). The structural component, $\Pi_S(h)$, of the Derjaguin's pressure isotherm, and the electrostatic component, $\Pi_e(h)$, are shown by curves 2 and 3, respectively. Curves 4 in both part (a) and (b) were calculated according to Eq. (1.13).

1.3 Static Hysteresis of Contact Angle

The previous consideration shows that the situation with the Neumann–Young's equation (Eq. 1.1) is far more difficult than usually assumed. This equation is supposed to describe the equilibrium contact angle. We explained in Section 1.1 that the latter equation does not comply with any of three requirements of equilibrium: liquid–vapor equilibrium, vapor–solid equilibrium, and equilibrium of liquid with an adsorbed layer on the solid. This, along with the previous consideration, shows that it is difficult, if possible at all, to experimentally investigate equilibrium liquid droplets.

However, there is a phenomenon that from a practical point of view is far more important than the previous ones: the static hysteresis of contact angle.

The above derivation of Eq. (1.1) and further considerations show, that the latter equation (or its modifications) determines only one arguably unique equilibrium contact angle. Static hysteresis of contact angle results in an infinite number of "quasi-equilibrium contact angles" of the drop on the solid surface, not the unique contact angle, θ_e, but the whole range of contact angles in two intervals of contact angle from θ_r to θ_e and from θ_e to θ_a, where θ_r and θ_a are static receding and advancing contact angles, respectively.

Let us explain the meaning of static advancing and receding contact angles. For that purpose, let us consider a liquid droplet on a horizontal substrate, which is slowly pumped in or out through an orifice in the solid substrate (Figure 1.14). Let us assume that in some way an initial contact angle of the droplet was equal to the equilibrium one. Let us start carefully and slowly pump the liquid through an orifice in the center. The contact angle will grow; however, the radius of the drop base will not change until a critical value on the contact angle, θ_a, is reached. Further pumping will result in the drop spreading.

If we start from the same equilibrium contact angle and start to pump out the liquid through the same orifice, then again, the contact angle will decrease but the droplet will not shrink until the critical contact angle, θ_r, is reached. After that the droplet will start to recede.

For example, in the case of water droplets on a smooth homogeneous, specially treated-for-purity glass surface: $\theta_r \sim 0°–5°$, while θ_a is in the range of $40°–60°$.

It is usually believed that the static hysteresis of the contact angle is determined by the surface roughness and/or heterogeneity (Figure 1.15).

Figure 1.15b presents the magnified vicinity of the three-phase contact line of the same droplet as in Figure 1.15a. This picture gives a qualitative explanation of the phenomenon of the static hysteresis of contact angle widely adopted in the literature: the static hysteresis of contact angle is usually connected with multiple equilibrium positions on the drop edge on a rough surface. There is no doubt that a roughness and/or a chemical heterogeneity of the solid substrate contribute substantially to the contact angle hysteresis.

As mentioned above, the static hysteresis of the contact angle is usually related to the heterogeneity of the surface: either geometric (roughness) [8,9] or chemical [10,11]. In this case, it is assumed that at each point of the surface the equilibrium value of the contact angle of that point is established, depending only on the local properties of the substrate. As a result, a whole series of local thermodynamic equilibrium states can be realized, corresponding to a certain interval of values of the angle. The maximum value

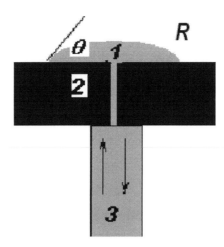

FIGURE 1.14 Schematic presentation of a liquid droplet on a horizontal solid substrate, the liquid is slowly pumped in or out through the liquid source in the drop center. R is the radius of the drop base; θ is the contact angle. (1) Liquid drop; (2) solid substrate with a small orifice in the center, and (3) liquid source (syringe).

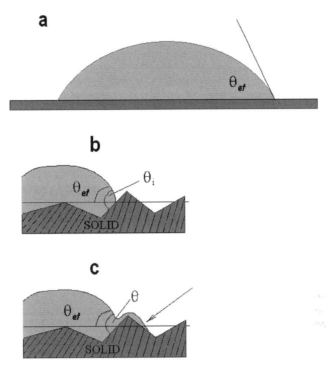

FIGURE 1.15 (a) Droplet on a solid substrate with a roughness that is invisible to the naked eye; (b) magnification of the apparent three-phase contact line; and (c) magnification on the apparent three-phase contact line with the rough surface covered by a liquid film that flows out from the droplet. An arrow indicates the zone where the microscopic flow occurs.

corresponds to the value of the advancing contact angle, θ_a, and the minimum value to the receding contact angle, θ_r.

According to this model, the dependency of the contact angle on the velocity of motion should be as presented in Figure 1.16.

There is no doubt that heterogeneity affects the wetting process. However, surface heterogeneity is apparently not the sole reason for hysteresis of the contact angle. This follows from the fact that not all the predictions made on the basis of this theory have turned out to be true [12,13]. Besides, hysteresis has been observed in cases involving quite smooth, uniform surfaces [14–19].

Even more than that, the static hysteresis of the contact angle is present even on surfaces that are definitely molecularly smooth: free liquid films [19,20].

Recall that in the vicinity of the apparent three-phase contact line, surface forces (the Derjaguin's pressure) substantially disturb the liquid profile and the picture presented in Figure 1.15b is impossible. Immediately after the droplet is deposited, the Derjaguin's pressure comes into play. This results in a coverage of the substrate in the vicinity of the apparent three-phase contact line by a thin liquid film. This means that the liquid edge is always in contact with the already wetted solid substrate and that the more realistic picture, Figure 1.15c, more adequately describes the situation in the vicinity of the apparent three-phase contact line. Equilibrium and hysteresis contact angles on rough surfaces have never before been considered from this point of view (i.e., taking into account the action of the Derjaguin's pressure in the vicinity of the apparent three-phase contact line) and are the subject of future investigations.

This means that the picture presented in Figure 1.16 cannot be realized on either a smooth or a rough substrate. This is why we consider static hysteresis on a completely smooth substrate.

A completely new concept of hysteresis of the contact angle on smooth homogeneous substrates was suggested in [7,21,22]. This new concept allows the calculation of both static advancing and static receding contact angles based on the Derjaguin's isotherm. This mechanism will be discussed in Chapter 3. Below we give a qualitative description of the phenomenon.

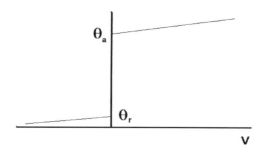

FIGURE 1.16   Idealized dependency of the contact angle on the velocity of advancing ($v > 0$) or receding ($v < 0$) three phase contact line, which cannot be experimentally observed because of a microscopic flow in a vicinity of the apparent three-phase contact line.

The picture presented in Figure 1.16 is in contradiction with the thermodynamics, which require a unique equilibrium contact angle, θ_e on smooth homogeneous substrates. This means that at any contact angle, θ, in the range $\theta_r < \theta < \theta_a$ and different from the equilibrium one, the liquid droplet cannot be at the equilibrium; instead it is in the state of very slow "microscopic" motion. More detailed observations and theoretical considerations show (see Chapter 4) that at any contact angle different from the equilibrium one, θ_e, the liquid droplet is in a state of slow microscopic motion in a tiny vicinity of the apparent three-phase contact line. The latter motion abruptly becomes "macroscopic" motion after a critical contact angle, θ_a or θ_r, is reached.

This means that the dependency presented in Figure 1.16 should be replaced by a more complicated, but realistic, dependency as shown in Figure 1.17.

The presence of the contact angle hysteresis shows that the actual equilibrium contact angle is very difficult to obtain experimentally even if we neglect the equilibrium with vapor and solid substrate.

Static Hysteresis of Contact Angles from the Microscopic Point of View: Surface Forces

On this stage we can explain qualitatively the nature of the hysteresis of contact angles via the S-shape of the isotherm of the Derjaguin's pressure (curve 2 in Figure 1.12) in the case of partial wetting. More details are given in Chapter 4.

Recall that equilibrium droplets on a solid substrate can exist only at equilibrium with oversaturated vapor; however, equilibrium liquid menisci in thin capillaries exist at undersaturation. That is, as explained above, the equilibrium of a liquid droplet is difficult, if possible at all, to investigate experimentally. However, the equilibrium of a liquid meniscus in a thin capillary can be relatively easy to create.

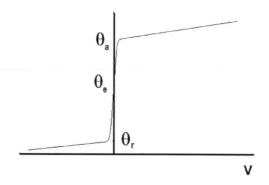

FIGURE 1.17   At any deviation from the equilibrium contact angle, θ_e, the liquid drop is in the state of a slow "microscopic" motion, which abruptly transforms to the state of a "macroscopic" motion after a critical contact angle, θ_a or θ_r, is reached.

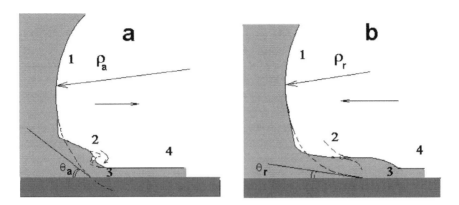

FIGURE 1.18 Hysteresis of contact angle in capillaries in the case of partial wetting (S-shaped isotherm of the Derjaguin's pressure). See Chapter 3 for details. (a) Advancing contact angle. (1) A spherical meniscus of radius, ρ_a; (2) transition zone with a "dangerous" point marked (see explanation in the text); (3) zone of flow; and (4) flat equilibrium films in front of the meniscus. Close to the marked point, a dashed line indicates the profile of the transition zone just after the contact angle reaches the critical value θ_a, which is the beginning of the "caterpillar" motion. (b) Receding contact angle. (1) A spherical meniscus of radius $\rho_r < \rho_a$; (2) transition zone with a "dangerous" point marked (see explanation in the text); (3) zone of flow; and (4) equilibrium films in front of the meniscus. Close to the marked point, dashed lines indicate the profile of the transition zone just after the contact angle reaches the critical value θ_r.

This is why the behavior of a liquid in thin capillaries was widely investigated by Russian scientists (mostly Derjaguin's group) [1]. Note, "thin capillary" means that the gravity action on the meniscus shape can be neglected. Below we consider the hysteresis of the contact angle in capillaries. Let us consider a meniscus in the case of partial wetting in a capillary (Figure 1.18a and b). Bear in mind that the capillary is in contact with a reservoir where the pressure, $P_a - P_e$, is maintained, that is, the pressure in the reservoir is lower than the atmospheric pressure, P_a.

If we increase the pressure in the reservoir and, hence, under the meniscus, then the meniscus does not move but instead changes its curvature to compensate for the excess pressure and, therefore, the contact angle increases accordingly. The meniscus does not move until some critical pressure and critical contact angle, θ_a, are reached. After a further increase in pressure, the meniscus starts to advance. A similar phenomenon takes place if we decrease the pressure under the meniscus: It does not recede until a critical pressure and corresponding critical contact angle, θ_r, are reached. This means that in the whole range of contact angles, $\theta_r < \theta < \theta_a$, the meniscus does not move macroscopically. It is obvious that on the smooth homogeneous solid substrate only one contact angle corresponds to the equilibrium position and all the rest do not. Based on that idea, Figure 1.17 presents a dependency of the contact angle on the velocity of motion, showing that all contact angles θ in the range $\theta_r < \theta < \theta_a$ correspond to a slow microscopic advancing or receding of the meniscus. This microscopic motion abruptly changes to macroscopic motion as soon as θ_r or θ_a are reached.

The explanation of the dependence presented in Figure 1.18 is based on the S-shaped isotherm of the Derjaguin's pressure in the case of partial wetting. This shape determines a very special shape of the transition zone in the case of an equilibrium meniscus (see Section 2.3). In the case of increasing pressure behind the meniscus (Figure 1.18a), a detailed consideration (Chapter 3) of the transition zone shows that, close to the "dangerous" point marked in Figure 1.18a, the slope of the profile becomes steeper with increasing pressure. In the range of very thin films (region 3 in Figure 1.18a), there is a zone of flow. Viscous resistance in this region is very high, which is why the advancing meniscus proceeds very slowly. After some critical pressure behind the meniscus is reached, the slope at the "dangerous" point reaches $\pi/2$, after which the flow occupies the region of the thick films in a stepwise fashion, and the fast "caterpillar" motion starts, as shown in Figure 1.18a. The meniscus instantaneously starts moving microscopically at the excess pressure $P \neq P_e$, that is, the excess pressure is different from its value at equilibrium. If the pressure under the meniscus is increased, then the meniscus starts moving microscopically in the beginning. All process can be subdivided into "fast macroscopic" and "slow microscopic" processes. The change of the meniscus curvature over the main part of the meniscus is a "fast macroscopic" process, and the flow in the relatively thick β-films is also a "macroscopic" process because the relative viscous resistance in these films is much smaller as compared with the flow in thin α-films. Hence, the idea is as follows: Until the flow takes place in relatively thin α-films, the meniscus is at a quasi-equilibrium state and in slow microscopic motion. As soon as the flow jumpwise occupies the zone of relatively thick β-films, the meniscus starts macroscopic motion.

Note that the thickness of β-films is around one order of magnitude greater than the thickness of α-films. Under the applied pressure difference, $\Delta p = P - P_e$, the flow rate, Q, in the film of thickness, h, is $Q = \frac{\Delta p}{6 \mu l} h^3$, where $l \sim \sqrt{H h_e}$ is the characteristic length of the transition zone between the bulk meniscus and the thin films in front (Chapter 2, Sections 2.3 and 2.4) is the dynamic viscosity. This means that the flow inside β-films is three orders of magnitude faster than in α-films. It is why the flow located inside α-films is referred to as "microscopic" flow, whereas the flow inside β-films is referred to as "macroscopic" flow.

In the case of decreasing pressure behind the meniscus, the event proceeds according to Figure 1.18b. In this case, again up to some critical pressure, the slope in the transition zone close to the "dangerous" point becomes more and more flat. In the range of very thin films (region 3 in Figure 1.18b), there is a zone of flow. Viscous resistance in this region again is very high, which is why the receding of the meniscus proceeds very slowly. After some critical pressure behind the meniscus is reached, the profile in the vicinity of the "dangerous" point shows discontinuous behavior, which is obviously impossible. This means that the meniscus will start to slide along the thick β-film. That is, the meniscus will move relatively quickly, leaving the thick β-film behind. The latter phenomenon (the presence of a thick β-film behind the receding meniscus of aqueous solutions in quartz capillaries) has been discovered experimentally (see discussion in Chapter 3) and supports our arguments explaining *static* contact angle hysteresis on smooth homogeneous substrates.

REFERENCES

1. Derjaguin, B. V., Churaev, N. V., and Muller, V. M. *Surface Forces*, Consultants Bureau, Plenum Press, New York, 1987.
2. Russel, W. B., Saville, D. A., and Schowalter, W. R. *Colloidal Dispersions*, Cambridge University Press, New York, 1999.
3. Exerowa, D., and Kruglyakov, P. *Foam and Foam Films: Theory, Experiment, Application*, Elsevier, New York, 1998.
4. Israelashvili, J. N. *Intermolecular and Surface Forces*, Academic Press, London, UK, 1991.
5. Donaldson, Jr., S. H., Røyne, A., Kristiansen, K., Rapp, M. V., Das, S., Matthew, A. G., Lee, D. W., Stock, P., Valtiner, M., and Israelachvili, J. Developing a general interaction potential for hydrophobic and hydrophilic interactions. *Langmuir*, 31 (7), 2051–2064 (2015).
6. Blake, T. D., and Haynes, J. M. Contact-angle hysteresis. *Prog. Surf. Membr. Sci.*, 6, 125 (1973).
7. Schwartz, A. M., Racier, C. A., and Huey, E. Contact angle, wettability, and adhesion. *Adv. Chem. Ser.*, 43, 250–267 (1964).
8. Holland, L. *The Properties of Glass Surfaces*, Chapman & Hall, London, UK, 1964, p. 364.
9. Zorin, Z. M., Sobolev, V. D., and Churaev, N. V. *Surface Forces in Thin Films and Disperse Systems* [in Russian], Nauka, Moscow, Russia, 1972, p. 214.
10. Romanov, E. A., Kokorev, D. T., and Churaev, N. V. Effect of wetting hysteresis on state of gas trapped by liquid in a capillary. *Int. J. Heat Mass Transfer*, 16, 549 (1973).
11. Neumann, A. W., and Good, R. J. Thermodynamics of contact angles. I. Heterogeneous solid surfaces. *J. Colloid Interface Sci.*, 38 (2), 341–358 (1972).
12. Chibowski, E. Surface free energy of a solid from contact angle hysteresis. *Adv. Colloid Interface Sci.*, 103 (2), (2003).
13. Platikanov, D., Nedyalkov, M., and Petkova, V. Phospholipid black foam films: Dynamic contact angles and gas permeability of DMPC bilayer films. *Adv. Colloid Interface Sci.*, 100–102, 185–203 (2003).
14. Petkova, V., Platikanov, D., and Nedyalkov, M. Phospholipid black foam films: Dynamic contact angles and gas permeability of DMPC+ DMPG black films. *Adv. Colloid Interface Sci.*, 104, 37 (2003).

15. Djikaev, Y. S., and Ruckenstein, E. The variation of the number of hydrogen bonds per water molecule in the vicinity of a hydrophobic surface and its effect on hydrophobic interactions. *Curr. Opin. Colloid Interface Sci.*, 16 (4), 272–284 (2011).
16. Kuchin, I., and Starov, V. Hysteresis of contact angle of sessile droplets on smooth homogeneous solid substrates via Derjaguin's pressure. *Langmuir*, 31, 5345–5352 (2015).
17. Arjmandi-Tash, O., Kovalchuk, N. M., Trybala, A., Kuchin, I. V., and Starov, V. Kinetics of wetting and spreading of droplets over various substrates. *Langmuir*, 33, 4367–4385 (2017).
18. Starov, V. M. Equilibrium and hysteresis contact angles. *Adv. Colloid Interface Sci.*, 39, 147–173 (1992).
19. Deryagin, B. V. Dependence of the contact angle on the microrelief or roughness of the wetted surface. *Dokl. Akad. Nauk SSSR.* [in Russian], 51, 357 (1946).
20. Johnson, R. E., and Dettre, R. H. *Surface and Colloid Science*, 2, Wiley, New York, 1969, p. 85.
21. Kwok, D. Y., and Neumann, A. W. Contact angle measurement and contact angle interpretation. *Adv. Colloid Interface Sci.*, 81 (3), 167–249 (1999).
22. Wenzel, R. Resistance of solid surfaces to wetting by water. *Ind. Eng. Chem.*, 28, 988–994 (1936).

2

Equilibrium Wetting Phenomena

Introduction

In this chapter we discuss equilibrium liquid shapes on solid substrates that demand thermodynamic equilibria: liquid–vapor, liquid–solid and vapor–solid: If any of these mentioned equilibrium conditions is not satisfied, then the liquid is not at the equilibrium. The vapor–solid equilibrium determines the presence of adsorbed liquid layers on solid surfaces for both *complete* and *partial* wetting. Now, as you are reading this text, everything around you is covered by a thin film of *water* and the thickness of this aqueous film depends on the humidity in the room. The adsorption is exactly at equilibrium with the surrounding humidity no matter how low or high it is.

The presence of liquid layers on solid substrates is determined by the action of surface forces (the Derjaguin's pressure) that were discussed in Chapter 1. One usually deals with the Derjaguin's pressure *isotherm* because it is usually measured at a constant temperature. In the case of water and aqueous solutions, the Derjaguin's pressure is S-shaped (Figure 1.12). Water and aqueous solutions are crucial for life. Is it that peculiar shape of the Derjaguin's pressure isotherm of water and aqueous solutions that in some way determines our existence?

That would mean that the peculiar shape of the Derjaguin's pressure isotherm of water and aqueous solutions in some unknown way determines the existence of our life. Now, we do not yet know in which way it does, but the peculiar shape of the Derjaguin's pressure isotherms of water and aqueous solutions tells us something about what we are unable to decode now. *This problem is to be answered in the future.*

In this chapter we investigate the influence of the combined action of the Derjaguin's pressure and capillary forces on the equilibrium shapes of liquids on solid substrates. Specifically, the equilibrium of liquid on plain solid surfaces (including deformable) and inside thin capillaries is investigated in this chapter. In all cases, the Derjaguin's pressure, which acts inside the apparent three-phase contact line, is considered. The whole liquid profile is subdivided into three interconnected regions: (i) the bulk region, where the thickness is big enough and where the action of the Derjaguin's pressure is neglected that capillary forces alone determine the liquid profiles; (ii) the transition region, where the Derjaguin's pressure and capillary forces become equally important; and (iii) the flat equilibrium film in front, which is determined completely by the Derjaguin's pressure action. All three regions are at equilibrium with each other. Important examples considered in this chapter include equilibrium liquid droplets and menisci on flat and cylindrical solid substrates and the shape of equilibrium droplets/bubbles in liquid. In all cases in this chapter, the gravity action is neglected.

2.1 Thin Liquid Films on Flat Solid Substrates

In this section, we consider the properties and stability of equilibrium thin liquid films on solid substrates under either *partial* or *complete* wetting conditions. We account for the Derjaguin's pressure action (i.e., surface forces action). As discussed in Chapter 1, the adsorption of liquid on solid substrates is a manifestation of surface forces action. However, before we start, let us recall that *partially* or *completely* wetted solid surfaces, at equilibrium, are always covered by a thin liquid film, which itself is at equilibrium with the vapor pressure, p, in the surrounding air. The free energy of such a solid substrate *covered by a thin liquid film* is lower than the free energy of the corresponding *bare* solid substrate. Hence, in all cases to be considered in Chapter 2, there is no real three-phase contact line at equilibrium because the whole solid surface is covered by a flat equilibrium thin liquid film. As mentioned in Chapter 1, this results in the formation of an *apparent* three-phase contact line, which is a transition zone between the bulk liquid and the equilibrium liquid films.

In this section, we consider the properties and stability of thin liquid films in the case of both partial and complete wetting. Note, the terms Derjaguin's pressure action and surface forces are used interchangeably in this section.

As we discussed in Chapter 1, the adsorption of liquid onto solid substrates is a manifestation of surface forces action. This means that the surface forces must be taken into consideration if we are to consider equilibrium states of liquid films on solid substrates.

The excess free energy, Φ, per unit area of a flat equilibrium liquid film of thickness, h_e, on a solid substrate at equilibrium with the vapor in the surrounding air is equal to

$$\Phi/S = \gamma + P_e h_e + f_D(h_e) + \gamma_{sl} - \gamma_{sv}, \tag{2.1}$$

where S is the surface covered by the liquid film; $f_D(h_e)$ is the potential of surface forces; $\gamma, \gamma_{sl}, \gamma_{sv}$ are liquid–air, solid–liquid, and liquid–vapor interfacial tensions, respectively; and P_e is the excess pressure, $P_e = P_a - P_l$, where P_l is the pressure inside the liquid film and P_a is the pressure in the ambient air. Note, according to the spontaneous adsorption of liquid molecules in the case of partial or complete wetting, this excess free energy should be negative; otherwise, the liquid molecules could adsorb, but the adsorption would be very small, which is seen in the case of non-wetting.

Because of the equilibrium of the liquid film with the vapor, the excess pressure, P_e, cannot be left as an arbitrary constant: It is determined by the equality of chemical potentials of liquid molecules in the film and in the vapor. This requirement results in the well-known Kelvin's equation:

$$P_e = \frac{R_g T}{v_m} \ln \frac{p_s}{p}, \tag{2.2}$$

where R_g, T, and v_m are the universal gas constant, the absolute temperature, and the liquid molar volume, respectively; and p_s and p are pressures of the saturated vapor and the vapor at which the liquid film is at equilibrium, respectively. Eq. (2.2) shows that the excess pressure, P_e, cannot be arbitrarily fixed: It is determined by the vapor pressure in the ambient air, p. Note again, Eq. (2.2) expresses the equality of the chemical potentials of water molecules in the vapor and liquid phases.

The excess free energy [Eq. (2.1)] is a function of one variable, h_e, which is the thickness of the equilibrium film. Hence, the usual conditions of the thermodynamic equilibrium should hold, giving a minimum to the excess free energy. Those conditions are as follows: $\frac{d\Phi}{dh_e} = 0, \frac{d^2\Phi}{dh_e^2} > 0$. The first requirement results in

$$P_e = \Pi(h_e), \tag{2.3}$$

and the second requirement yields

$$\frac{d\Pi(h_e)}{dh_e} < 0, \tag{2.4}$$

where $\Pi(h) = -\frac{df_D(h)}{dh}$ is referred to as the Derjaguin's pressure [1]. The Derjaguin's pressure, $\Pi(h)$, is the physical property that can be experimentally measured, and the discussion in this section is based on consideration of the Derjaguin's pressure. Using this definition, we can rewrite the excess energy equation, expressing $f(h)$ as $f_D(h) = \int\limits_{h}^{\infty} \Pi(h)dh$.

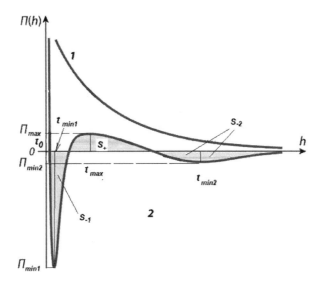

FIGURE 2.1 Two types of isotherms of the Derjaguin's pressure, which are under consideration below: (1) complete wetting; and (2) partial wetting. Isotherms for the partial wetting are observed for water films on most of surfaces, for example, on quartz, glass, metal surfaces. (From Deryaguin, B.V. et al., *Colloid J.*, 38, 789, 1976; Deryagin, B.V. and Churaev, N.V. et al., *Dokl. Akad. Nauk USSR.*, 207, 572, 1972.) Isotherms of type 1 are observed in cases of complete wetting only, for example, oils films on quartz, glass, metal surfaces. (From Deryagin, B.V. and Zorin, Z.M., *Sov. J. Phys. Chem.*, 29, 1755, 1955.)

Eq. (2.3) determines the thickness of the equilibrium liquid film, h_e, via the Derjaguin's pressure isotherm. Eq. (2.4) gives the well-known stability condition of flat equilibrium liquid films [1].

According to the stability condition (2.4), all flat equilibrium films are stable in the case of complete wetting (curve 1, Figure 2.1). However, in the case of partial wetting (curve 2, Figure 2.1), only films that range in thickness from 0 to t_{min1} (these films are referred to as α-films) and those that range in thickness between $t_{max} < h < t_{min2}$ (these films are referred to as β-films) are stable. It is in this section below that α-films are absolutely stable, whereas β-films are metastable and, hence, can exist only for a limited period of time. Thus, only α- and β-films can exist as flat films. Films in other regions cannot exist as flat films.

Note again, the S-shaped Derjaguin's pressure isotherms (curve 2, Figure 2.1) are characteristic shapes in the case of water and aqueous solutions. All properties of water and aqueous solutions are vitally important for life. This means that the special shape of Derjaguin's pressure of water and aqueous solutions, presented in Figure 2.1 in some way determines the existence of life. Now we do not know in which way it does, but the shape of curve 2 in Figure 2.1 tells us something about what we are unable to decode now.

It is possible to rewrite the expression for the excess free energy of the film [Eq. (2.1)] using the Derjaguin's pressure in the following way:

$$\Phi/S = \gamma + P_e h_e + \int\limits_{h_e}^{\infty} \Pi(h)dh + \gamma_{sl} - \gamma_{sv}. \tag{2.5}$$

This expression gives the excess free energy via a measurable physical dependency, $\Pi(h)$, which is the Derjaguin's pressure isotherm.

Now we can rewrite Eq. (1.2) from Chapter 1 Section 1.1 of the excess free energy of thin liquid films as

$$\Phi/S = \gamma_{svh_e} - \gamma_{sv}, \tag{2.6}$$

where

$$\gamma_{svh_e} = \gamma + P_e h_e + \int\limits_{h_e}^{\infty} \Pi(h)dh + \gamma_{sl} \tag{2.7}$$

is the "interfacial tension" (the excess free energy) of the solid substrate covered with the liquid film of thickness h_e.

Eq. (2.7) determines the unknown value of γ_{svh_e} in the Neumann-Young's Eq. (1.1) from Chapter 1 Section 1.1:

$$\cos\theta_e = \frac{\gamma_{svh} - \gamma_{sl}}{\gamma}. \tag{2.8}$$

The combination of Eqs. (2.7) and (2.8) results in

$$\cos\theta_e = \frac{\gamma + P_e h_e + \int\limits_{h_e}^{\infty} \Pi(h)dh}{\gamma} = 1 + \frac{P_e h_e + \int\limits_{h_e}^{\infty} \Pi(h)dh}{\gamma}. \tag{2.9}$$

Eq. (2.9) is the well-known Frumkin-Derjaguin equation for the equilibrium contact angle, which has been deduced using a different thermodynamic consideration [1]. This is a very important equation and is deduced in a different, more rigorous way in Section 2.3.

Because $-1 < \cos\theta_e < 1$, then using Eq. (2.9), we conclude that the integral in the right-hand side should be negative (assuming that $P_e h_e << \int\limits_{h_e}^{\infty} \Pi(h)dh$, which is usually satisfied). This requirement is satisfied in the case of partial wetting (see curve 2, Figure 2.1):

$$\int\limits_{h_e}^{\infty} \Pi(h)dh < 0 \tag{2.10}$$

This inequality is satisfied if

$$S_{-1} + S_{-2} > S_{+} \tag{2.11}$$

In the case of complete wetting, the right-hand side of Eq. (2.9) is always positive, that is, equilibrium droplets cannot exist on the solid substrate under either under- or oversaturation: They spread out completely and/or evaporate. However, equilibrium menisci (at undersaturation) can exist in capillaries. That is, the behavior of droplets and menisci is completely different in the case of complete wetting.

Using Eq. (2.9), we can rewrite the expression for the excess free energy of a flat liquid film [Eq. (2.1)] as

$$\Phi/S = \gamma\cos\theta_e + \gamma_{sl} - \gamma_{sv} = \gamma\left(\cos\theta_e - \cos\theta_{NY}\right). \tag{2.12}$$

Because of the spontaneous adsorption of liquid molecules onto solid substrates in the case of partial and complete wetting, the excess free energy [Eq. (2.12)] is negative, hence,

$$\cos\theta_e - \cos\theta_{NY} < 0, \quad \theta_e > \theta_{NY}. \tag{2.13}$$

That is, even if γ_{sl} and γ_{sv} can be measured, then the contact angle according to the original Neumann-Young's equation,

$$\cos\theta_{NY} = \frac{\gamma_{sv} - \gamma_{sl}}{\gamma}, \tag{2.14}$$

is smaller than the real equilibrium contact angle.

Frequently, a contact angle determined according to Eq. (2.14) is identified with a static advancing contact angle, θ_a. It is obvious that the static advancing contact angle is bigger than the equilibrium contact angle, $\theta_a > \theta_e$. If we compare now this inequality with Eq. (2.13), we can conclude that there is no justification for the identification of θ_{NY} and θ_a, because $\theta_{NY} < \theta_e < \theta_a$.

It is shown in Chapter 3 that the equilibrium contact angle in both cases (i.e., a droplet on a solid substrate and a meniscus in thin capillaries) is closer to the static receding contact angle, θ_r, than to the static advancing contact angle, θ_a.

Equilibrium Droplets on the Solid Substrate under Oversaturation ($P_e < 0$)

As already noted, the excess pressure, P_e, is negative at oversaturation according to Kelvin's law (Eq. 2.2). The thickness of equilibrium film/films is determined according to Eq. (2.3) at both under- and oversaturation.

Figure 2.2 shows that, in the case of complete wetting, there is no equilibrium flat film on a solid substrate under oversaturation: The line $P_e < 0$ does not intersect curve 1 in Figure 2.2. Hence, there are no equilibrium droplets on completely wettable solids at oversaturation: They are in the surrounding air.

However, in the case of partial wetting, Eq. (2.3) has three solutions (Figure 2.2). According to the stability condition of flat films [Eq. (2.4)], one of them corresponds to the stable equilibrium film of thickness, h_e, in the region of α-films; the second corresponds to the unstable film of thickness, h_u; and the third corresponds to the relatively thick β-film (Figure 2.2). This means that equilibrium droplets in the case of partial wetting are "sitting" on the stable equilibrium film of thickness h_e. (Recall that the third solution, h_β, is metastable).

However, even in the case of partial wetting, equilibrium droplets can exist on a solid substrate only during the limited interval of oversaturation, which is determined by $0 > P_e > \Pi_{min1}$ (Figure 2.2) or by using Eq. (2.2) in the following range of oversaturated pressures, p, over the solid substrate

$$1 < \frac{p}{p_s} < \exp\left(-\frac{v_m \Pi_{min1}}{R_g T}\right). \tag{2.15}$$

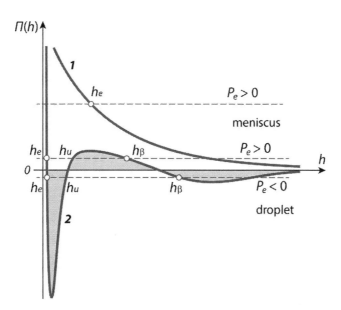

FIGURE 2.2 Two equilibrium flat films on solid substrate undersaturation: stable film of thickness h_e, metastable film of thickness h_β and unstable film of thickness h_u.

If Π_{min1} is in the range of (10^6 to 10^7) dyn/cm^2, then this inequality takes the following form:
$1 < \frac{p}{p_s} < 1 - \frac{v_m \Pi_{min1}}{RT} \approx 1.001 - 1.01$, that is, the equilibrium droplets in the case of partial wetting can exist only in the very limited interval of oversaturation on solid substrates. Beyond this interval, at higher oversaturation, as in the case of complete wetting, neither equilibrium liquid films nor droplets exist on the solid substrate. The critical oversaturation, $p_{cr} \frac{p_{cr}}{p_s} = \exp\left(-\frac{v_m \Pi_{min1}}{R_g T}\right)$, determined from Eq. (2.15) probably corresponds to the beginning of homogeneous nucleation, and at higher oversaturation, homogeneous nucleation is more favorable.

Let $\Re$ be the radius of the equilibrium droplet. According to the definition of the capillary pressure: $P_e = -\frac{2\gamma}{\Re}$. Hence, the radius of the equilibrium drops is $\Re = \frac{2\gamma}{-P_e}$. In the aforementioned narrow interval of oversaturation, the radius of the equilibrium drops changes from infinity at $p \rightarrow p_s$ to $\Re_{cr} = -\frac{2\gamma}{\Pi_{min1}}$ at $p = p_{cr}$. If $\Pi_{min1} \approx -10^6$ dyn/cm^2 and $\gamma \approx 72$ dyn, then $\Re_{cr} \approx \frac{144}{10^6} = 1.44 \ \mu m$; that is, the critical size is out of the range of the surface forces action and the droplet size is sufficiently big. However, if $\Pi_{min1} \approx -10^7$ dyn/cm^2, then $\Re_{cr} \approx \frac{144}{10^7} = 0.144 \ \mu m = 1440 \ \overset{o}{A}$, and the whole droplet is in the range of the surface forces action. In this case, the drop is so small that it does not have a spherical part anywhere (even on the very top) that is not disturbed by the surface forces.

Flat Films at the Equilibrium with Menisci ($P_e > 0$)

Eq. (2.3) and Figure 2.3 show that in the case of complete wetting, there is only one equilibrium flat film, h_e, which is stable according to the stability condition (2.4).

In the case of partial wetting (Figure 2.3), the solution of Eq. (2.3) is different when $P_e > \Pi_{max}$ and $P_e < \Pi_{max}$. If $P_e > \Pi_{max}$, Eq. (2.3) has only one solution, which is stable [according to the stability condition (2.4)] and is referred to as an α-film. In the second case, when $P_e < \Pi_{max}$ (Figure 2.3), Eq. (2.3) has three solutions: One of which corresponds to the stable equilibrium α-film with thickness h_e. The second solution of Eq. (2.3), h_u, is unstable according to the stability condition (2.4), and the third solution, h_β, is stable again according to the same stability condition (2.4). These films are referred to as β-films. Note, that the thickness of an equilibrium film in the case of complete wetting, h_c, is bigger than the thickness of a β-film, h_β, in the case of partial wetting.

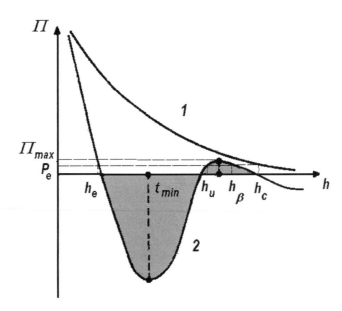

FIGURE 2.3 The Derjaguin's pressure isotherm in the case of (1) complete wetting; and (2) partial wetting. In thick capillaries $\left(H > \frac{\gamma}{\Pi_{max}}\right)$ there are three solutions of Eq. (2.3): h_e in the region of α-films, h_u, which cannot exist because it is unstable according to stability condition (2.4), and h_β in the region of thick β films.

Let us compare the excess free energy of flat α- and β-films, h_e and h_β. According to the definition (2.1), this difference is equal to

$$\Delta_{\alpha\beta} = \left[\Phi(h_\beta) - \Phi(h_\alpha) \right] / S = P_e \left(h_\beta - h_e \right) - \int_{h_e}^{h_\beta} \Pi(h)dh. \tag{2.16}$$

The difference $h_\beta - h_e$ is always positive (Figure 2.3). In the case of partial wetting, $S_{-1} > S_{+}$, according to Eq. (2.11). Hence, the integral in the right-hand side of Eq. (2.16) is negative. Thus, the excess free energy of β-films is higher than the excess free energy of α-films. This means that β-films are less stable than α-films, which is why β-films are referred to as metastable films and α-films are referred to as absolutely stable films. Even if a relatively thick β-film is created, then over time it will rupture and the result will be an α-film at the same P_e. The excess liquid volume will evaporate.

We must make some additional comments on α-films and β-films in the case of partial wetting. If we increase the vapor pressure over a partially wettable surface from $p = 0$ to the saturation pressure, p_s, then on the solid substrate one can observe the formation of only α-films, whose thickness changes correspondingly [according to Eq. (2.3) and Figure 2.3] from zero at $p = 0$ to $t_0 \approx 70\,\text{Å}$ [1]. However, β-films cannot be obtained during the adsorption process: They can be obtained only by decreasing the thickness of very thick films down to the equilibrium thickness of the β-film. This is why α-films are referred to as adsorption films (because they can be obtained during adsorption) and β-films are referred to as wetting films.

Let $\Re$ be a radius of the curvature of a meniscus in a flat capillary (a meniscus between two parallel plates). According to the definition of the capillary pressure, $P_e = \frac{\gamma}{\Re}$. Let us introduce $\Re_{max} = \frac{\gamma}{\Pi_{max}}$ (Figure 2.3), and consider $P_e > \Pi_{max}$ (Figure 2.3). We define a capillary as a "thin" capillary if $\Re < \Re_{max}$. In such a capillary, only thin α-films can exist at equilibrium with the meniscus, and equilibrium β-films do not exist at all in such thin capillaries. If, however, the capillary is "thick," that is, $\Re > \Re_{max}$, then in such a capillary, both α- and β-films can be at equilibrium with the meniscus. However, β-films are metastable, as shown above (see Eq. (2.16) and discussion below this equation).

If we adopt $\gamma \sim 70$ dyn/cm and $\Pi_{max} \sim 10^4$ dyn/cm^2 for estimations, then $\Re_{max} \sim 7 \times 10^{-3}$ cm.

The highlighted text below can be omitted at the first reading.

S-shaped isotherms of disjoining pressure in the special case $S_{-1} < S_{+}$.

Let us consider the case when the Derjaguin's pressure isotherm is S-shaped as in Figure 2.3, curve 2. However, let us assume that $\int_{h_e}^{\infty} \Pi(h)dh > 0$, that is, $S_{-} < S_{+}$ (Figure 2.3). In this case from Eq. (2.16) we conclude: $\Delta_{\alpha\beta}\big|_{P_e=0} = S_{-} - S_{+} < 0$. The latter means that at low P_e (or high humidity) β-films are more stable than α-films. It is easy to check using Eq. (2.16) that $\Delta_{\alpha\beta}$ is an increasing function of P_e, because $\frac{d\Delta_{\alpha\beta}(P_e)}{dP_e} > 0$. Hence, $\Delta_{\alpha\beta}$ can become positive at some value of P_e and after that thick β-films become less stable than thin α-films. The latter occurs if $\Delta_{\alpha\beta}(P_e)\big|_{P_e=\Pi_{max}} > 0$. In this case if P_e increases from zero (there thick β-films are more stable than thin α-films) it reaches a critical value, P_{cr}, such as $\Delta_{\alpha\beta} < 0$ at $0 < P_e < P_{cr}$, and $\Delta_{\alpha\beta} > 0$ at $P_{cr} < P_e < \Re_{max}$. The latter means that in the range $0 < P_e < \Pi_{cr}$ thick β-films are more stable than thin α-films, however, at $P_{cr} < P_e < \Pi_{max}$ α-films become more stable than β-films. This consideration shows that a cycle presented in Figure 2.4 with a spontaneous and reversible transitions from α-films to β-films (along DA) should take place at a decreasing of P_e along CD. Also, a *spontaneous reversible* transition from β-films to α-films (along BC) should take place at an increase of P_e along AB. Such spontaneous reversible transitions have been discovered by D Exerowa et al. [5–8]. In their experiments the Derjaguin's pressure isotherm was s-shaped, but the minimum value of the Derjaguin's pressure isotherm was positive, which means that condition $\int_{h_e}^{\infty} \Pi(h)dh > 0$, that is, $S_{-1} < S_{+}$, was satisfied.

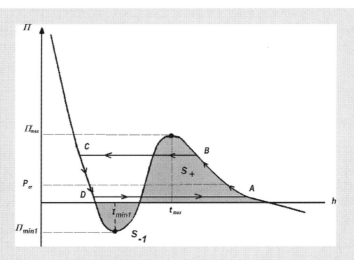

FIGURE 2.4   The S-shaped Derjaguin's pressure isotherm with $S_- < S_+$. Reversible transition from α- to β-films along DA at $P_e < P_{cr}$ and from β- to α-film along BC at $P_e > P_{cr}$.

2.2 Non-flat Equilibrium Liquid Shapes on Flat Solid Surfaces

In flat thin liquid films (oil and aqueous thin films and thin films of aqueous electrolyte and surfactant solutions, both free films and films on solid substrates), the Derjaguin's pressure acts alone and determines their thickness. However, if the film surface is curved or uneven, then both the Derjaguin's and capillary pressure act simultaneously. It is shown in this section that in the case of *partial* wetting their simultaneous action can result in existence of *non-flat* equilibrium shapes. For instance, due to the S-shaped Derjaguin's pressure isotherm, microdrops, micro-depressions and equilibrium periodic films can exist on flat solid substrates. We establish below the criteria for both the existence and the stability of such non-flat equilibrium liquid shapes. On the other hand, we see that a transition from thick β-films to thinner α-films can proceed transitorily via non-flat states with micro-depressions and wavy shapes, which both can be more stable than flat films in some range of hydrostatic pressures [9]. All mentioned non-flat equilibrium shapes on flat solid substrates are to be discovered experimentally.

The equilibrium contact angle of either an equilibrium drop or a meniscus in a capillary can be expressed via the Derjaguin's pressure isotherm as (see Section 2.3) (Figure 2.1):

$$\cos\theta_e = \frac{1 + \dfrac{1}{\gamma}\displaystyle\int_{h_e}^{\infty}\Pi(h)dh}{1 - \dfrac{h_e}{H}} \approx 1 + \frac{1}{\gamma}\int_{h_e}^{\infty}\Pi(h)dh \approx 1 - \frac{S_{-1} + S_{-2} - S_+}{\gamma}, \qquad (2.17)$$

where P_e is the excess pressure (negative in the case of drops and positive in the case of the meniscus); h_e is the equilibrium absolutely stable α-film (Figure 2.3); and H is the radius of the capillary in the case of meniscus and the maximum height in the case of drops.

Equation (2.17) shows that the partial wetting case corresponds to $S_{-1} + S_{-2} > S_+$, that is, the S-shaped isotherm 2 in Figure 2.1. Equation (2.17) shows that the equilibrium contact angle is completely determined by the shape of the Derjaguin's pressure in the case of molecular smooth substrates. There is no doubt that surface roughness influences the apparent value of the contact angle. However, it is obvious that the roughness cannot result in a transition from the non-wetting to the partial wetting case or from the partial wetting to the complete wetting case. That is why in this section only molecularly smooth solid substrates are under consideration. The influence of roughness and chemical heterogeneity is considered below in Sections 2.5 and 2.6.

The main idea of this section is to show that simultaneous action of the capillary pressure and the S-shaped Derjaguin's pressure isotherm results in the formation of non-flat equilibrium liquid shapes even on smooth homogeneous solid substrates. Again, we should emphasize that this is specific for water and aqueous solutions and, hence, is vitally important for life. However, we are still completely unaware of which way it is important for life.

Consideration of equilibrium non-flat liquid layers allows suggesting a new scenario of a rupture of thick meta-stable β-films and their transition to absolutely stable α-films.

General Consideration

The excess free energy, Φ, of a liquid layer, drop or meniscus on a solid substrate can be expressed in the following way:

$$\Phi = \gamma S + P_e V + \Phi_D - \Phi_{ref}, \tag{2.18}$$

where S, V, and Φ_D are excesses of the vapor–liquid interfacial area, the excess volume, and the excess energy associated with the surface forces action, respectively; γ is the liquid–vapor interfacial tension; P_e is the excess pressure (see Section 2.1); and Φ_{ref} is the excess free energy of a reference state (see below). The gravity action is neglected in Eq. (2.18). The excess pressure, P_e, is introduced as in Section 1.1. Equation (1.2) as

$$P_e = P_a - P_l, \tag{2.19}$$

where P_a is the pressure in the ambient air, and P_l is the pressure inside the liquid. This pressure is referred to below as the hydrostatic pressure inside the liquid. This means that $P_e > 0$ in the case of a meniscus, and $P_e < 0$ in the case of a drop. P_e is uniquely determined by the ambient vapor pressure, p, according to Kelvin's law, Eq. (2.2) (Section 2.1). Equation (2.2) shows that $P_e > 0$ corresponds to an undersaturation, while $P_e < 0$ corresponds to an oversaturation. This means that menisci can be at equilibrium at an undersaturation and drops can be at equilibrium at an oversaturation.

To simplify calculations below only two-dimensional equilibrium systems are under consideration. In this case, the excess free energy (2.17) can be rewritten as

$$\Phi = \int \left\{ \gamma \left(\sqrt{1 + h'^2} - 1 \right) + P_e(h - h_e) \int_h^\infty \Pi(h) dh - \int_{h_e}^\infty \Pi(h) \right\} dx, \tag{2.20}$$

where $h(x)$ is the liquid profile to be determined. Integration in this equation has taken over the whole space occupied by the system. Note, the excess free energy in Eq. (2.20) is selected as an excess over the energy of a reference state, which is the state of the same flat surface covered by a stable equilibrium α-film of thickness h_e. A selection of any reference state results in an additive constant in expression (2.20) and does not influence the result. However, the reference state is important from a mathematical point of view because (i) it makes the integral in (2.20) finite; otherwise it includes integration over an equilibrium α-film if there is infinite extension; (ii) it is important in the consideration of the liquid profiles in the vicinity of the apparent three-phase contact line (transversality condition).

Any liquid profile, $h(x)$, which gives the minimum value to the excess free energy, Φ, according to Eq. (2.20), describes an equilibrium liquid configuration. For the existence of the minimum of the excess free energy (2.20), the following four conditions should be satisfied:

A. $\delta\Phi = 0$.

B. $\delta^2\Phi > 0$, or $\frac{\partial^2 f}{\partial h'^2} > 0$.

where $f = \gamma \left(\sqrt{1 + h'^2} - 1 \right) + P_e(h - h_e) \int_h^\infty \Pi(h) dh - \int_{h_e}^\infty \Pi(h)$.

C. The solution of the Jacobi's equation, $u(x)$, should not vanish at any position x inside the region under consideration, except for boundaries of the region of integration in Eq. (2.20).

D. The transversality condition at the apparent three-phase contact line should be satisfied. The transversality condition provides the condition of a smooth transition from a non-flat liquid profile to a flat equilibrium film in front. The transversality condition reads $\left[f - h' \frac{\partial f}{\partial h'} \right]_{x=R} = 0$, where R is the position of the apparent three-phase contact line.

The transversality condition D can be rewritten using the above definition of f as

$$\left[\gamma \left(\sqrt{1+h'^2} - 1 \right) + P_e(h-h_e) + \int_h^\infty \Pi(h) dh - \int_{h_e}^\infty \Pi(h) - \frac{\gamma h'^2}{\sqrt{1+h'^2}} \right]_{x=R} = 0.$$

This condition shows that the three-phase contact line should be determined at the intersection of the liquid profile with the equilibrium liquid film of thickness, h_e, and not at the intersection with the solid substrate as usually assumed. This results in

$$\left[\left(\sqrt{1+h'^2} - 1 \right) - \frac{h'^2}{\sqrt{1+h'^2}} \right]_{x=R} = 0,$$

or

$$\left[\frac{1}{\sqrt{1+h'^2}} \right]_{x=R} = 1.$$

This is obviously satisfied only at

$$\left(h' \right)_{x=R} = 0 \text{ or } h'(h_e) = 0. \tag{2.21}$$

This transversality condition is discussed in Section 2.3. It is shown in Section 2.3 that this condition means

$$h' \to 0, \quad x \to \infty, \tag{2.22}$$

and the meaning of $x \to \infty$ is clarified, which is "tends to the end of the transition zone."

The first condition (A) results in the following equation:

$$\frac{\gamma h''}{(1+h'^2)^{3/2}} + \Pi(h) = P_e, \tag{2.23}$$

which should be referred to as the Derjaguin's equation because Derjaguin was the first to introduce it [1]. The first term in the left-hand side of Eq. (2.23) corresponds to the capillary pressure and the second term represents the Derjaguin's pressure action. If the thickness of the liquid is out of the range of the Derjaguin's pressure action, then Eq. (2.23) describes either a flat (at $P_e = 0$) liquid surface, a spherical drop profile (at $P_e < 0$), or a spherical meniscus profile (at $P_e > 0$).

The second condition (B) is always satisfied because $\frac{\partial^2 f}{\partial h'^2} = \frac{\gamma}{(1+h'^2)^{3/2}} > 0$. The third condition (C) results in the Jacobi's equation

$$\frac{d}{dx} \frac{\gamma u'}{(1+h'^2)^{3/2}} + \frac{d\Pi(h)}{dh} u = 0. \tag{2.24}$$

Direct differentiation of Eq. (2.23) results in

$$\frac{d}{dx} \frac{\gamma h''}{(1+h'^2)^{3/2}} + \frac{d\Pi(h)}{dh} \frac{dh}{dx} = 0.$$

The comparison of Eq. (2.23) with Eq. (2.24) shows that the solution of the Jacobi's equation is

$$u = const \cdot h'. \tag{2.25}$$

Hence, if $h'(x)$ does not vanish anywhere inside the system under consideration, then the system is stable; however, if $h'(x_0) = 0$ and x_0 is different from the ends of the system under consideration [i.e., within the range of integration in Eq. (2.20)], then the system is unstable.

The second-order differential Eq. (2.23) can be integrated once, which gives

$$\frac{1}{\sqrt{1+h'^2}} = \frac{C - P_e h - \int\limits_{h}^{\infty} \Pi(h)dh}{\gamma}, \tag{2.26}$$

where C is an integration constant to be determined. The important observation is that the right-hand side of this equation should be always positive.

In the case of a meniscus in a flat capillary, the integration constant, C, is determined from the following condition: at the capillary center: $h'(H) = -\infty$, which gives $C = P_e H$, where H is the half-width of the capillary (see Section 2.3). In the case of equilibrium droplet, the constant should be selected using a different condition at the drop apex, $h = H$: $h'(H) = 0$ (see Section 2.3), which results in $C = \gamma + P_e H$. An alternative way of selecting the integration constant C is using the transversality condition (2.21).

The integration constant, C, is selected in this section individually according to the boundary conditions in each case under consideration.

In the case of equilibrium liquid drops and menisci (see Section 2.3) they are supposed to be always at equilibrium with the flat films in front, with which they are in contact. Only the capillary pressure acts inside the spherical parts of drops or menisci and only the Derjaguin's pressure acts inside flat thin films. However, there is a transition zone between the bulk liquid (drops or menisci) and the flat thin film in front of them. In this transition zone, both the capillary pressure and the Derjaguin's pressure act simultaneously (see Section 2.3 for more detail). A profile of the transition zone between a meniscus in a flat capillary and a thin α-film in front of it in the case of partial wetting is presented in Figure 2.5. Figure 2.5 shows that the liquid profile is not always concave, but that its curvature changes inside the transition zone. This peculiar liquid shape in the transition zone alone determines the static hysteresis of the contact angle (see Chapter 3) and all other equilibrium and quasi-equilibrium macroscopic liquid properties on solid substrates (Chapters 2 and 3).

In the transition zone (Figure 2.5), all thicknesses are presented from very thick outside the range of the Derjaguin's pressure action to a thin α-film, h_e, in front. This means that the stability condition of flat films [Eq. (2.4), Section 2.1] can no longer be used because this condition is valid only in the

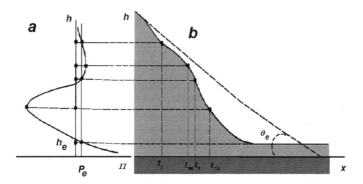

FIGURE 2.5 Partial wetting. Magnification of the liquid profile inside the transition zone in "thick capillaries." The S-shaped Derjaguin's pressure isotherm (left side, a) and the liquid profile in the transition zone (right side, b).

case of flat films. The more sophisticated Jacobi's condition (C) should be used instead; it shows that the transition zone is stable if $h'(x)$ does not vanish anywhere inside the transition zone.

The peculiar shape of the transition zone, where both the capillary pressure and the Derjaguin's pressure are equally important, prompts consideration of solutions of Eq. (2.26) in the case of S-shaped isotherm and to see if this equation has other stable solutions different from flat liquid films of a constant thickness. Each of such solutions (if any) corresponds to a non-flat liquid layer, whose stability should be checked using Jacobi's condition (C). Below we show that such non-flat equilibrium liquid shapes can exist.

This highlighted text below can be omitted at the first reading.

Microdrops: The Case Where $P_e > 0$ (The Case of Under-Saturation)

Below we consider the possibility of existence of microdrops, that is, drops with an apex in the range of influence of the Derjaguin's pressure. In this case, the drop does not have a spherical part even at the drop apex because its shape is distorted everywhere by the Derjaguin's pressure action.

The liquid profile, $h(x)$, is described by Eq. (2.26), which is obtained by the integration from Eq. (2.23) with the integration constant, C, to be determined.

According to the transversality condition (2.21), at $h = h_e$ gives $h'(h_e) = 0$, which means the drop edge approaches the equilibrium film of thickness h_e on the solid surface at a zero microscopic contact angle. This condition allows determining the integration constant in Eq. (2.26) as $C = \gamma + P_e h_e + \int_{h_e}^{\infty} \Pi(h)dh$. Hence, the drop profile is described by the following equation:

$$h' = -\sqrt{\frac{\gamma^2}{(\gamma - L(h))^2} - 1}, \tag{2.27}$$

where $L(h) = P_e(h - h_e) - \int_{h_e}^{h} \Pi(h)dh$. The expression under the square root in Eq. (2.27) should be positive, that is the following condition should be satisfied:

$$0 \leq L(h) \leq \gamma, \tag{2.28}$$

where the first equality corresponds to the zero derivative, and the second one corresponds to the infinite derivative.

Let h_+ be the apex of the microdrop. The top part of Figure 2.6 shows the S-shaped dependence of the Derjaguin's pressure isotherm, $\Pi(h)$, while the bottom part of Figure 2.6 shows the curve, $L(h)$, which has maximum or minimum at thickness, which are solutions of $P_e = \Pi(h)$. At the apex of the drop, when $h = h_+$, the first derivative should be zero, that is, $h'(h_+) = 0$ or, from Eq. (2.27),

$$L(h_+) = 0. \tag{2.29}$$

The origin is placed at the center of the drop. Below we consider only the situation that corresponds to the formation of microdrops at undersaturation, that is, at $P_e > 0$. Equilibrium macrodrops at oversaturation, that is, at $P_e < 0$, are considered in Section 2.3.

At $0 < P_e < \Pi_{max}$, equation $P_e = \Pi(h)$ has three roots (Figure 2.6), the smallest of which corresponds to the equilibrium flat α-film of thickness, h_e. The following conditions should be satisfied for the existence of microdrops: $h'' < 0$ at $h = h_+$ and $h'' > 0$ as $h \rightarrow h_e$ (Figure 2.7); hence, the following inequality should be satisfied: $h_u < h_+ < h_\beta$. At the drop apex $h'(h_+) = 0$, hence, according to Eq. (2.29)

$$P_e(h_+ - h_e) = \int_{h_e}^{h_+} \Pi(h)dh, \tag{2.30}$$

and the solution of this equation, h_+, should be in the following range: $h_u < h_+ < h_\beta$ (see the bottom part of Figure 2.6).

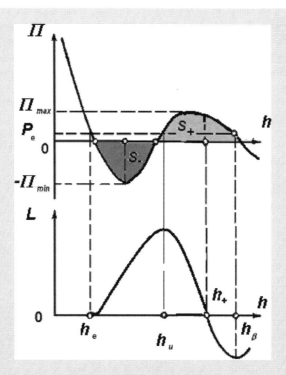

FIGURE 2.6 Determining of the microdrop apex. Upper part: S-shaped Derjaguin's pressure isotherm, bottom part: $L(h)$. $L(h_+) = 0$ determines the drop apex, h_+. Note, only part of the Derjaguin's isotherm is used: the part close to the second min is omitted because it is not important for this consideration.

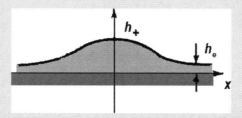

FIGURE 2.7 Profile of an equilibrium microdrop. Note, the apex of the microdrop is in the range of the Derjaguin's pressure action, that is, the drop does not have any spherical part even at the drop apex.

The left-hand side in Eq. (2.30) is positive, as also should be the right-hand side. Hence, a sufficient condition for the existence of equilibrium microdrops is as follows: $S_- < S_+$, that is, the Derjaguin's pressure isotherm should be S-shaped, but the equilibrium contact angle of macroscopic droplet should be equal to zero: That means such droplets spread out completely when deposited.

This means that equilibrium microdrops do not exist either in the case of partial wetting, when $S_- > S_+$ or in the case of a "regular" complete wetting, when the Derjaguin's pressure decreases in a monotonous way, as for example, $\Pi(h) = A/h^3$. However, conditions for the existence of equilibrium microdrops are satisfied in aforementioned experiments by Exerowa et al. [5–7].

Microscopic Quasi-equilibrium Periodic Films

In this section we consider the possibility of the existence of quasi-equilibrium periodic liquid films, situated completely in the range of the Derjaguin's pressure action (partial wetting, S-shaped

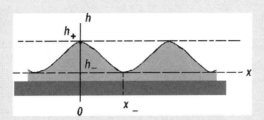

FIGURE 2.8 Equilibrium periodic film. h_+, maximum thickness; h_-, minimum thickness; and x_-, length of the half-period of the film. Is this film equilibrium?

Derjaguin's pressure isotherm). The undersaturation condition is again under consideration, that is, $P_e > 0$. It is explained below why these films are referred to as "quasi-equilibrium."

Let h_+ and h_- be the maximum and minimum heights of an equilibrium periodic film (Figure 2.8). Derivatives should be zero at $h = h_-$ and $h = h_+$, or $h'(h_-) = h'(h_+) = 0$. Using the second of these two conditions in Eq. (2.26), we can determine the integration constant, C, as $C = \gamma + P_e h_+ + \int\limits_{h_+}^{\infty} \Pi(h)dh$. Hence, the profile of the equilibrium periodic film is described by the following equation:

$$h' = -\sqrt{\frac{\gamma^2}{(\gamma - L_+(h))^2} - 1},$$ (2.31)

where

$$L_+(h) = -P_e(h_+ - h) + \int\limits_{h}^{h_+} \Pi(h)dh.$$ (2.32)

The following condition $0 \le L_+(h) \le \gamma$ should be satisfied for the expression under the square root in Eq. (2.31) to be positive.

The origin is placed at the position of the maximum (Figure 2.8) and Eq. (2.31) is written for a half-period of the periodic film from $x = 0$, which corresponds to the position of the maximum, to $x = x_-$, which corresponds to the position of the minimum (Figure 2.8). Notice that h_-, h_+ and x_- are to be determined.

Because we must have $h'(h_-) = 0$ at the position of the minimum of the film, it follows from Eq. (2.31) that

$$P_e(h_+ - h_-) = \int\limits_{h_-}^{h_+} \Pi(h)dh.$$ (2.33)

This condition relates the two unknown thicknesses h_- and h_+: $h_- = h_-(h_+)$.

Eq. (2.23) can be rewritten as $h'' = \frac{(1+h'^2)^{3/2}}{\gamma}(P_e - \Pi(h))$. Equation (2.23) and Figure 2.8 show that near the minimum, $h = h_-$, the liquid profile is convex, $h'' > 0$, that is, $P_e > \Pi(h_-)$; similarly, near the maximum, $h = h_+$, $h'' < 0$ if $P_e < \Pi(h_+)$.

At $P_e > 0$, for every pressure P_e, there exists either no solution at all of Eq. (2.33) or there exists an interval of values $h_{+\min} < h_+ < h_{+\max}$, where $h_{+\min}$ and $h_{+\max}$ are determined by the following conditions $h_-(h_{+\min}) = h_e$, $h_-(h_{+\max}) = h_\beta$.

Below, we give a method for determining the unique value of h_+, that is, the value which can be realized.

The excess free energy of a half-period, x_- (Figure 2.8), of the periodic film is given by the same relation (2.20), where we, however, omit the additive constant determined by the reference state.

This, is unimportant in this case. From Eq. (2.31) we can express $dx = -\dfrac{dh}{\sqrt{\frac{\gamma 2}{(\gamma - L_+(h))^2} - 1}}$. After substitution of this expression into Eq. (2.20), we arrive to

$$\Phi = \int_{h-}^{h+} \frac{\gamma\sqrt{1+h'^2} + \int_{h}^{\infty} \Pi(h)dh + P_e h}{\sqrt{\dfrac{\gamma^2}{\left[\gamma - L_+(h)\right]^2} - 1}} \, dh. \qquad (2.34)$$

This expression includes only one undetermined parameter, h_+, because h_- is expressed via h_+ according to Eq. (2.33). Only shapes with the minimum value of the excess free energy (2.34) can be realized, that is, the unknown h_+ should be determined using the following conditions:

$$\frac{\partial \Phi}{\partial h_+} = 0, \quad \frac{\partial^2 \Phi}{\partial h_+^2} > 0. \qquad (2.35)$$

Because the volume of the half-period per unit length of the periodic film is $V = \int_{x_c}^{x_+} h\,dx = V(h_+)$, conditions (2.35) are identical to the usual thermodynamic conditions of equilibrium: $\frac{\partial \Phi}{\partial V} = 0, \frac{\partial^2 \Phi}{\partial V^2} > 0$. The relations in Eq. (2.35) completely determine the shape of the periodic film.

The above suggested procedure (2.35) (minimization of the excess free energy) is consistent with the Euler's equation (Eq. 2.23), which minimizes the same excess free energy. Computer calculations below show that there is a unique h_+ value that satisfies the conditions for Eq. (2.35). This proves the thermodynamic stability of periodic films.

For calculations of the dependence of the excess free energy on h_+ according to Eq. (2.34), a Derjaguin's pressure isotherm should be selected. It is selected for the calculation as follows (Figure 2.9):

$$t_0 = 10^{-6} \text{ cm}, \quad t_{min} = 1.5 \cdot 10^{-6} \text{cm}, \quad t_2 = 2 \cdot 10^{-6} \text{cm}, \quad t_{max} = 3 \cdot 10^{-6} \text{ cm},$$

$$t_s = 5 \cdot 10^{-6} \text{ cm}, \quad \Pi_{max} = 10^6 \text{ dyn/cm}^2, \quad \Pi_{max} = 10^7 \text{ dyn/cm}^2, \quad \gamma = 72 \text{ dyn/cm}$$

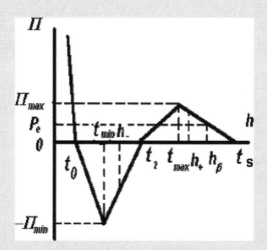

FIGURE 2.9 Isotherm of the Derjaguin's pressure used for calculations of the excess free energy of periodic films. A simplified Derjaguin's isotherm is used, there the part related to the second minimum is omitted to simplify the calculations.

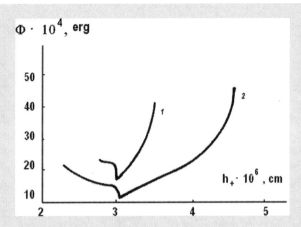

FIGURE 2.10   The excess free energy of periodic films, $\Phi(h_+)$, calculated according to Eq. (2.34) using the isotherm of Derjaguin's pressure presented in Figure 2.9. (1) $P_e = 0.7 \cdot 10^6$ dyn/cm^2, (2) $P_e = 0.2 \cdot 10^6$ dyn/cm^2.

This choice corresponds approximately to aqueous films on a quartz surface [1] and gives, according to Eq. (2.17), $\cos \theta_e \approx 1 - \frac{S_- - S_+}{\gamma} \approx 0.94$.

Using the adopted the Derjaguin's pressure isotherm (Figure 2.9), the dependence of the excess free energy on h_+ is calculated according to Eq. (2.34). The excess pressure, P_e, varies from 0 to $\Pi_{max} = 10^6$ dyn/cm^2. Two calculated dependences are shown in Figure 2.10. Each of these plots has a sharp minimum. The minimum value determines the unique h_+ value, which is realized at the possible equilibrium.

Equation (2.34), using the isotherm of the Derjaguin's pressure, is presented in Figure 2.9.

Let us compare the excess free energy of the corresponding β-film, F_β, of the same length as a half-period of the periodic film, $F_\beta = (\gamma + \int_{h_\beta}^{\infty} \Pi(h)dh + P_e h_\beta)x$, with the energy of a half-period of the periodic film. The comparison is presented in Table 2.1, which shows that at $P_e < 0.6 \cdot 10^{-6}$ dyn/cm^2, the excess free energy of β-films is lower than the corresponding energy of the periodic film (that is, β-films are more stable); however, at $P_e > 0.6 \cdot 10^{-6}$ dyn/cm^2, the free energy of the periodic film becomes lower.

This means that close to the maximum value of the Derjaguin's pressure isotherm, Π_{max}, periodic films have lower excess free energy than β-films, that is, the periodic films can be a transitional state before rupture of the β-films.

It was mentioned above that periodic films can exist only in the case of partial wetting, that is, if $S_- > S_+$.

TABLE 2.1

Comparison of the Excess Free Energy of Periodic Films
and β Films of Equal Length

$P_e \cdot 10^{-6}$ dyn/cm^2	$F \cdot 10^6$ dyn (the periodic film)	$F_\beta \cdot 10^6$ dyn
0.1	1130	1115
0.2	1212	1200
0.3	1303	1290
0.4	1299	1290
0.5	1446	1438
0.6	1633	1635
0.7	1727	1730
0.8	2227	2230
0.9	2250	2265

There is an interesting pure mathematical problem: the Jacobi condition, as mentioned above, is $h'(x)$ and should not change the sign in the whole range of definition, that is, for all for $_\infty < x < +\infty$. However, this condition is satisfied for only the half-period of periodic films. Can these periodic films be really be at equilibrium, and can they even exist? Periodic films are yet to be experimentally discovered.

The case, $p/p_s > 1$, when $P_e < 0$, can be treated similarly. It is possible to show that in this case the maximum thickness of periodic films can be outside the range of the Derjaguin's pressure action, that is, periodic films in this case are a periodic array of drops. In this case, there is no doubt that such a formation can experimentally exist. But the Jacobi condition is still satisfied for a half-period of such formation. Is it possible to resolve this mathematical problem?

Microscopic Equilibrium Depressions on β-Films

In this section an existence of equilibrium depressions on the surface of thick β-films is considered (Figure 2.11).

A minimum thickness of a depression is marked as h_- (Figure 2.11). The derivative should be zero at the top the depression, that is, $h'(h_\beta) = 0$. Using this condition, an integration constant in Eq. (2.26) can be determined as follows: $C = \gamma + P_e h_\beta + \int_{h_{\beta e}}^{\infty} \Pi(h)dh$. After that, Eq. (2.26) can be rewritten as

$$\frac{1}{\sqrt{1+h'^2}} = \frac{\gamma - L_-(h)}{\gamma},\qquad(2.36)$$

where $L_-(h) = -P_e(h_\beta - h) + \int_h^{h_\beta} \Pi(h)dh$. The right-hand side of Eq. (2.36) should be positive, and it gives the following restrictions: $0 \le L_-(h) \le \gamma$.

The derivative should be zero at the bottom of the depression, $h'(h_-) = 0$. This condition gives an equation to determine h_-:

$$L_-(h) = -P_e(h_\beta - h) + \int_h^{h_\beta} \Pi(h)dh = 0.\qquad(2.37)$$

The procedure for the selection of h_- is shown in Figure 2.12.

It is possible to show that (i) h_- always exists in the case of partial wetting, that is, if $S_- > S_+$, and (ii) equilibrium depressions have lower excess free energy than corresponding flat β-films above some critical value of P_e.

We can now suggest a new scenario for a transition from thick β-films to thin α-films in the case of partial wetting. At low values of P_e, β-films are more stable than equilibrium depressions or periodic films. However, above some critical value of P_e, β-films have higher excess free energy as compared with equilibrium depressions. That means isolated depressions develop on the β-film. At a further increase of P_e, their excess free energy exceeds the corresponding value of a periodic film and a transition from isolated depressions to a periodic film takes place. As mentioned above,

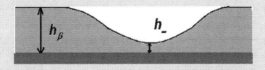

FIGURE 2.11 Schematic presentation of an equilibrium depression on the β-film with thickness h_β. h_-, minimum thickness of the depression.

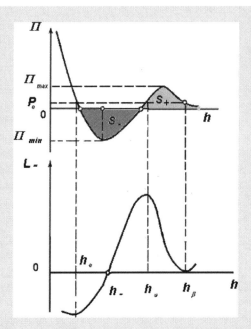

FIGURE 2.12 Determining of the minimum thickness of the depression, h_-. S-shaped Derjaguin's pressure isotherm (the upper part), function $L_-(h)$ (the bottom part). The part related to the presence of the second min on the
Derjaguin's isotherm is omitted, because it is not important in this part.

there is a critical value of P_e above which periodic films can no longer exist. This results in a transition from the periodic film to the α-film, with the equilibrium "sitting" on it. This transition can be
described as "a rupture." Residual microdrops cannot be at equilibrium with the α-film in the case
of partial wetting and they gradually disappear by evaporation and/or hydrodynamic flow.

Summarizing then, we have shown that in the case of the S-shaped Derjaguin's pressure isotherm, microdrops, micro-depressions and equilibrium periodic films are possible on flat solid
substrates. Criteria have been provided for both the existence and the stability of these *non-flat*
equilibrium liquid shapes. It has been suggested that transition from thick films to thinner films
goes via intermediate non-flat states like micro-depressions and/or periodic films, which are more
stable than flat films in some hydrostatic pressure range. Flat liquid films are unstable in the region
between α- and β-films and, hence, cannot be observed experimentally. However, the predicted
non-flat stable liquid layers (micro-depressions and periodic films) are in this unstable region.
Accordingly, experimental measurements of the profiles of these non-flat layers open the possibility of determining the Derjaguin's pressure isotherm in the *unstable* region.

2.3 Equilibrium Contact Angle of Menisci and Drops: Liquid Shape in the
Transition Zone from the Bulk Liquid to the Flat Films in Front

In this section we show that the Derjaguin's pressure action determines the peculiar shape of liquid inside
the *transition* zone from the bulk to the flat liquid films ahead for both menisci and drops. Thus, the equilibrium contact angle is determined using the Derjaguin's pressure isotherm.

In Figure 1.12, two types of the Derjaguin's pressure isotherm are presented, corresponding to the
two different wetting situations presented in Figures 1.2 (partial wetting) and 1.3 (complete wetting).
The Derjaguin's pressure isotherms can be directly measured, unfortunately, not in the whole range
of film thickness as we already discussed in Section 2.1, but only in the regions where flat films are

stable [see the stability condition of flat films (2.4). This means, that experimental measurements of the Derjaguin's pressure can be undertaken only in the case of flat and absolutely stable α-films and meta-stable β-films. In this section only the partial wetting case is under consideration. The complete wetting case is considered in Section 2.4.

There is no doubt that surface roughness influences the apparent value of the contact angle. However, it is obvious that the roughness cannot result in a transition from non-wetting to the partial wetting, or from partial wetting to complete wetting. That is why in the next section only molecularly smooth solid substrates are under consideration. Consideration of equilibrium liquid states on rough substrates when both capillary forces and surfaces forces are taken into account is a challenging subject to be developed in the future.

Equilibrium of Liquid in a Flat Capillary: Partial Wetting Case

To simplify the calculations, we will use a model of a two-dimensional capillary, that is, the capillary formed by two parallel plates. The excess free energy, Φ, of a liquid layer, a drop or a meniscus on a solid substrate can be expressed by Eq. (2.18) (see Section 2.2).

Equations (2.2) and (2.19) show that the case $P_e > 0$ corresponds to the case of meniscus or other non-flat liquid shapes (see Section 2.2) at equilibrium with undersaturated vapor and the case $P_e < 0$ corresponds to the case of drops or other non-flat liquid shapes (see Section 2.2) at equilibrium with oversaturated vapor.

The difference between volatile and nonvolatile liquids determines only the path and the rate of a transition to the equilibrium state but not the equilibrium state itself. Only equilibrium states are under consideration here, which is why it is not specified in this chapter whether a liquid is volatile or nonvolatile.

As mentioned in Chapter 1, all solid surfaces in contact with a volatile or a nonvolatile liquid at equilibrium are covered by a thin equilibrium liquid film. The thickness of this equilibrium film is determined by the surface forces action (the Derjaguin's pressure isotherm). That is, the choice of the reference state is uniquely determined if we want to consider a specific vicinity of the three-phase contact line at the state of equilibrium of bulk liquid in contact with solid substrate: The reference state is the state of solid substrate covered with the equilibrium liquid film. That is why, the reference state is selected as the plane-parallel film with the lowest possible equilibrium thickness (i.e., α-films, introduced in Section 2.1) and which is at equilibrium with the vapor pressure, p, in the ambient air. In this section two-dimensional equilibrium menisci in a flat chamber with a half-width, H, or two-dimensional equilibrium liquid drops are considered for simplicity. The extension of the derivation below to axial symmetry is briefly discussed in the end of this section.

According to this selection, Eq. (2.20) can be rewritten as

$$F = \int \left\{ \gamma \left[\sqrt{1 + h'^2} - 1 \right] + P_e \left(h - h_e \right) + f_D(h) - f_D(h_e) \right\} dx, \tag{2.38}$$

where h_e is the thickness of the equilibrium plane-parallel α-film, and $f_D(h)$ is the density of the energy of surface forces. Two substantial simplifications are adopted in the expressions for free energy (2.38): (i) the density of the energy of surface forces, $f_D(h)$, depends only on the films thickness, h, and is independent of the slope or curvature of the liquid profile, that is, it is independent of derivatives of the film thickness; and (ii) the interfacial tension remains its bulk value, γ. The first assumption, (i), means that only profiles with a relatively low slope can be described using such approximation although the expression "a relatively low slope" cannot be clarified at the current state of science in the area. It is an important problem to be resolved in the future.

The only attempt to consider a dependency of the surface forces, $f_D(h)$, on the first derivative of the liquid profile of dispersion forces was undertaken in [8]. However, calculations in [8] were based on a direct summation of molecular forces. This forces are well known as having a nonadditive nature [1]. This is probably why a controversial nonzero equilibrium contact angle has been predicted in the case of complete wetting [8] and which is why consideration of surface forces in the case of non-flat profiles remains a challenge, and we use assumption (i) even in cases when this assumption is not rigorously valid.

The two assumptions, (i) and (ii), are strongly interconnected: if the density of energy of surface forces, $f_D(h, h')$, depends on the derivative of the film profile, h', then tangential stress on the surface of the liquid is unbalanced. However, if we adopt both assumptions (i) and (ii), then at least from this point of view, we do not have any contradictions: Constant interfacial tension results in zero tangential stress under equilibrium conditions.

Let us briefly discuss what happens if the density of energy of surface forces, $f_D(h, h')$, depends on the derivative of the film profile, h'. In this case, Eq. (2.38) takes the following form:

$$\Phi = \int \left\{ \gamma \left[\sqrt{1+h'^2} - 1 \right] + P_e \left(h - h_e \right) + f_D(h, h') - f_D(h_e) \right\} dx.$$

At equilibrium, the first condition (A) (Section 2.2) must be satisfied and results in

$$\frac{\gamma h''}{(1+h'^2)^{3/2}} - \frac{\partial f_D}{\partial h} + \frac{d^2 f_D}{dh'^2} h'' + \frac{d^2 f_D}{dh dh'} h' = P_e.$$

Let us introduce the following functions:

$$\bar{a}(h, h') = \frac{d^2 f_D}{dh'^2} \left(1+h'^2\right)^{3/2}, \quad \bar{\Pi}(h, h') = -\frac{\partial f_D}{\partial h} + \frac{d^2 f_D}{dh dh'} h', \text{ then this equation takes the following form}$$

$$\frac{\left(\gamma + \bar{a}(h, h')\right) h''}{(1+h'^2)^{3/2}} + \bar{\Pi}(h, h') = P_e.$$

This means that (i) the "effective" surface tension, $\gamma + \bar{a}(h, h')$, depends on both thickness h and the slope h'; and (ii) the effective Derjaguin's pressure now depends on both mentioned values as the effective surface tension. The consequences of such dependences, as well as the physical meaning of these effective values, are yet to be understood. This is why we use the approximation adopted in Eq. (2.38), that is, the density of the energy of surface forces, $f_D(h)$, depends only on the films thickness, h, and is independent of the derivatives of the film thickness.

Integration in Eq. (2.38) has been taken over the whole space occupied by the flat meniscus.

Any liquid profile, $h(x)$, which gives the minimum value to the excess free energy, Φ, according to Eq. (2.23), describes an equilibrium liquid configuration on a planar surface. For the minimum to exist, the four conditions introduced in Section 2.2 should be satisfied [see conditions (A)–(D) and the discussion in Section 2.2].

The first requirement, (A), shows that the liquid profile gives minimum or maximum value to the excess free energy, Φ, while two other requirements, (B) and (C), prove that the profile provides minimum value to the excess free energy, Φ. It is necessary to note that both requirements (B) and (C) must be satisfied; only in this case does the excess free energy (2.4) have a minimum value.

The requirement (A) results in Euler's equation (Eq. 2.23) (which was first suggested by Derjaguin [1] and therefore should actually be referred to as the Derjaguin's equation), where the Derjaguin's pressure is introduced as $\Pi(h) = -\frac{df_D(h)}{dh}$.

If requirement (C) is not satisfied, then the solution of Eq. (2.23) does not give a stable solution.

Condition (C) shows that Eq. (2.23) can be integrated once, which gives Eq. (2.26). Note, the right-hand side of Eq. (2.26) should always be positive.

Meniscus in a Flat Capillary

In the case of a meniscus in a flat capillary, the integration constant, C (Eq. 2.26) is determined from the condition at the capillary center:

$$h'(H) = -\infty, \tag{2.39}$$

which gives $C = P_e H + \int_H^\infty \Pi(h)dh$, where H is the half-width of the capillary. Using this constant, Eq. (2.26) can then be rewritten as

$$\frac{1}{\sqrt{1+h'^2}} = \frac{P_e(H-h) - \int_h^H \Pi(h)dh}{\gamma}. \tag{2.40}$$

This equation describes an equilibrium profile of a meniscus in a flat capillary.

Let us consider the solution of Eq. (2.23) in more detail. This equation determines the liquid profile in three different regions:

a. A spherical meniscus, which is not disturbed by the surface forces action; that is, the disjoining pressure action can be neglected, and we arrive to a regular Laplace equation

$$\frac{\gamma h''}{(1+h'^2)^{3/2}} = P_e, \tag{2.41}$$

b. A flat equilibrium liquid film in front of the meniscus,

$$\Pi(h) = P_e \tag{2.42}$$

and

c. A transition zone in between, where both the capillary force and the Derjaguin's pressure act simultaneously.

Note, although Eq. (2.42) coincides with Eq. (2.3), we use Eq. (2.42) for convenience. Below we consider only "macroscopic capillaries." This means that the radius, H, is much bigger than the range of a surface forces action. Let the radius of the surface forces action be $t_s \approx 10^{-5}$ cm $= 1000$ Å $= 0.1$ μ $= 100$ nm, that is, at $h > t_s$: $\Pi(h) = 0$. In this case, Eq. (2.23) can be rewritten as Eq. (2.41) with boundary conditions

$$h(0) = H \tag{2.43}$$

and Eq. (2.39). The solution of Eq. (2.23) with boundary conditions Eqs. (2.39) and (2.43) gives a spherical profile

$$(H-h)^2 + \left(\frac{\gamma}{P_e} - x\right)^2 = \left(\frac{\gamma}{P_e}\right)^2, \tag{2.44}$$

that is, a spherical meniscus with the radius $\rho_e = \gamma/P_e$. This equation describes a profile of a spherical meniscus in the central part of the capillary, which is not disturbed by the Derjaguin's pressure action. Intersection of this profile with the thin equilibrium film of thickness h_e defines the apparent three-phase contact line and the macroscopic equilibrium contact angle, θ_e (Figure 2.13). A simple geometrical consideration shows that

$$P_e = \frac{\gamma \cos \theta_e}{H}. \tag{2.45}$$

At $h = h_e$: $h'^2 = 0$ and we conclude from Eq. (2.45)

$$1 = \frac{\frac{\gamma \cos \theta_e}{H}(H - h_e) - \int_{h_e}^H \Pi(h)dh}{\gamma}, \tag{2.46}$$

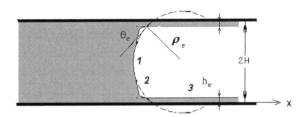

FIGURE 2.13 Profile of a meniscus in a flat capillary. (1) A spherical part of the meniscus of curvature ρ_e; (2) transition zone between the spherical meniscus and flat films in front; and (3) flat equilibrium liquid film of thickness h_e.

which allows determining the contact angle via the Derjaguin's pressure isotherm as

$$\cos\theta_e = \frac{1 + \dfrac{1}{\gamma}\displaystyle\int_{h_e}^{H}\Pi(h)dh}{1 - \dfrac{h_e}{H}} \approx 1 + \frac{1}{\gamma}\int_{h_e}^{\infty}\Pi(h)dh, \quad \text{at} \quad h_e < t_s \ll H. \tag{2.47}$$

This equation was first time deduced by Derjaguin and Frumkin [1] and is named after these two scientists as the Derjaguin–Frumkin's equation. Note, this equation was deduced using completely different arguments from those given in this section and for many years was considered as independent of Eq. (2.23). The derivation given in this section shows that the Derjaguin–Frumkin's equation (Eq. 2.47) is a direct consequence of Eq. (2.23). Note also, that the same equation for the contact angle was deduced in Section 2.1 in a different way.

Eq. (2.46) can be approximately rewritten (Figure 2.1)

$$\cos\theta_e \approx 1 - \frac{S_{-1} + S_{-2} - S_+}{\gamma}. \tag{2.48}$$

This equation shows that $\cos\theta_e < 1$ only if $S_{-1} + S_{-2} > S_+$ (Figure 2.1).

Now we can at last precisely define the term the "partial wetting": (i) S-shaped Derjaguin's pressure isotherm (curve 2 on Figure 2.1), and (ii) $S_{-1} + S_{-2} > S_+$.

Let us consider the case when $P_e < \Pi_{max}$, or $\frac{\gamma\cos\theta}{H} < \Pi_{max}$. Hence, $H > H_{cr}$, where $H_{cr} \sim \frac{\gamma}{\Pi_{max}}$. We refer to such capillaries as "thick capillaries." In the case of aqueous solutions, $\gamma \sim 72$ dyn/cm, $\Pi_{max} \sim 10^5$ erg, hence; $H_{cr} \sim 7\cdot10^{-4}$cm $\sim 10^{-3}$cm. Otherwise, a capillary is referred to as a "thin capillary"; that is, the capillary is "thin" if its thickness, H, is in the range $t_s \ll H < H_{cr}$, where t_s is the radius of the Derjaguin's pressure action. According to this definition "thin capillaries" are still big enough as compared with the radius of surface forces action, t_s. If the capillary radius is compared with the radius of surface forces action, t_s, then such capillary should be referred to as a microscopic capillary. Only thin and macroscopic capillaries are under consideration in this book.

In the case of thick macroscopic capillaries, that is, $H > H_{cr}$, where $H_{cr} \sim \frac{\gamma}{\Pi_{max}}$, Eq. (2.42) has three solutions, one of which corresponds to the stable equilibrium α-film with thickness h_e. The excess free energy of α-films is equal to zero according to our choice in Eq. (2.38). The second solution of Eq. (2.42), in this case, h_u, is unstable according to the stability condition (2.4) (Section 2.1), and the third solution, h_β, is a β-film, which is also stable according to the same stability condition (2.4) (Section 2.1). It was shown in Section 2.1 that β-films have higher excess free energy as compared with α-films; that is, β-films are less stable and eventually rupture to form thinner and absolutely stable α-films.

However, in "thin capillaries" Eq. (2.42) has only one solution (not shown in Figure 2.3), which is an absolutely stable α-film.

Meniscus in a Flat Capillary: Profile of the Transition Zone

Let us estimate the length of the transition zone, L. Inside the transition zone, the capillary pressure and the Derjaguin's pressure are of the same order of magnitude of $P_e \sim \frac{\gamma}{H}$. According to Eq. (2.23), the capillary pressure inside the transition zone can be estimated as $\frac{\gamma h''}{\sqrt{1+h'^2}} \sim \frac{\gamma h_e}{L^2}$ or $\frac{\gamma h_e}{L^2} \sim \frac{\gamma}{H}$. From this estimation we conclude

$$L \sim \sqrt{h_e H}. \tag{2.49}$$

In the case of $h_e \sim 10^{-6}$ cm and $H \sim 0.01$ cm, this estimation gives $L \sim 1$ μm. Note, that the same estimation of the length of the transition zone is also valid in the case of droplets.

Now let us rewrite Eq. (2.23) in the following way:

$$h'' = \frac{1}{\gamma}(1+h'^2)^{3/2}[P_e - \Pi(h)]. \tag{2.50}$$

Equation (2.50) shows that the sign of the second derivative is determined by the difference $P_e - \Pi(h)$. In the case of "thick capillaries," that is, $H > H_{cr}$, Figure 2.5 shows that

$h'' > 0$, in the following range of thickness: $h_\beta < h < H$: the profile is concave;

$h'' <$, at $h_u < h < h_\beta$: the profile is convex;

$h'' > 0$, at $h_e < h < h_u$: the profile is concave again.

This means that the profile of the liquid inside the transition zone does not remain concave all the way through the transition zone; instead, the curvature changes at two inflection points: $h(x_\beta) = h_\beta$ and $h(x_u) = h_u$ (Figure 2.5). The magnification of the liquid profile inside the transition zone is schematically shown in Figure 2.5.

Now an important question arises: flat thin films in the range of thickness from t_{min1} to t_{max} and from t_{min2} to t_s (Figure 1.12) are unstable according to the stability condition (2.4) (Section 2.1). We would like to emphasize that the mentioned condition is the stability condition of *flat films*. As discussed in Section 2.2 and in this section, the stability condition (C) (Section 2.2) of non-flat liquid layers is completely different and according to Eq. (2.25) is satisfied inside the transition zone (h' is negative everywhere in Figure 2.5). Nobody should expect any "convergence" of the two stability conditions, Eq. (2.4) (Section 2.1) of flat films and (C) (Section 2.2) of non-flat films: They are completely different. A qualitative physical explanation of the stability of the transition zone inside the "dangerous" range of thickness from t_{min1} to t_{max} and from t_{min2} to t_s is as follows. The extent of the "dangerous region," for example, from x_{max} to x_{min1} (Figure 2.5) is small enough that any fluctuation inside this "dangerous region" is damped by neighboring stable regions on both sides (Figure 2.5).

The liquid profile inside the transition zone in the case of thin capillaries, that is, $H < H_{cr}$, is much simpler (Figure 2.14) and does not have any inflection points, which is different from the case of "thick capillaries." In this case according to Eq. (2.50), the liquid profile is always concave.

Note that in all cases under consideration there is no real three-phase contact line at the equilibrium because the whole solid surface is covered by a flat equilibrium liquid film. This is why we refer to it as an "apparent contact line." The transversality condition (D) (Section 2.2) at the apparent three-phase contact line results in Eq. (2.21) (Section 2.2)

$$h'(h_e) = 0. \tag{2.51}$$

Let us consider this condition in more detail in Appendix 1. This consideration shows that, in general cases (except for a very special model isotherm of the Derjaguin's pressure), the transition from a non-flat transition zone to a flat equilibrium film goes very smoothly and asymptotically as $x \to \infty$ (clarified in Appendix 1). That is, in a general case, there is no final point there the transition zone ends; instead, it asymptotically approaches the flat film.

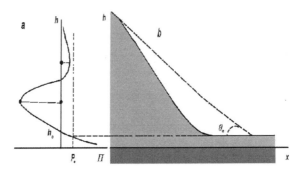

FIGURE 2.14 Magnification of the transition zone in the case of partial wetting in "thin capillaries." (a) Derjaguin's pressure isotherm; and (b) transition region.

A more detailed description of the transition zone between the meniscus in a flat capillary and the equilibrium thin film in front can be found in [10,11].

Partial Wetting: Macroscopic Liquid Drops

Recall that $P_e < 0$ in this part because the liquid drops can be at the equilibrium only with oversaturated vapor. A "macroscopic drop" means that the drop apex, H, is outside the range of the surface forces (or the Derjaguin's pressure) action. Microscopic drops, that is, the drops with the apex is in the range of the Derjaguin's pressure action, are considered in Section 2.2.

The equilibrium film/films are determined according to Eq. (2.42). Note, that in the case of complete wetting, there are no equilibrium flat films on solid substrates because the line $P_e < 0$ does not intersect curve 1 in Figure 2.1. Hence, there are no equilibrium droplets on completely wettable solids under oversaturation.

However, in the case of partial wetting, Eq. (2.42) has four solutions (Figures 1.12 and 2.17, on the left side). According to the stability condition of flat films (2.4) Section 2.2, the first solution corresponds to the stable equilibrium film of thickness, h_e. This means that equilibrium droplets in the case of partial wetting are "sitting" on a stable equilibrium film of thickness, h_e.

However, even in the case of partial wetting, equilibrium droplets can exist on the solid substrate only in a limited interval of oversaturation, which is determined by the following inequality: $0 > P_e > \Pi_{\min 1}$ (Figure 1.12) or by using Eq. (2.2) in the following range of oversaturated pressure over the solid substrate

$$1 < \frac{p}{p_s} < \exp\left(-\frac{v_m \Pi_{\min}}{RT}\right).$$

(2.52)

If $-\Pi_{\min}$ is in the range 10^6–10^7 dyn/cm^2, then this inequality takes the following form

$$1 < \frac{p}{p_s} < 1 - \frac{v_m \Pi_{\min}}{RT} \approx 1.001 - 1.01,$$

that is, the equilibrium droplets exist only in a very limited interval of oversaturation. Beyond this interval at higher oversaturation, neither equilibrium liquid films nor droplets exist on the solid substrate, as in the case of complete wetting. The critical oversaturation, p_{cr}: $\frac{p_{cr}}{p_s} = \exp\left(-\frac{v_m \Pi_{\min}}{RT}\right)$, determined using Eq. (2.52) probably corresponds to the beginning of homogeneous nucleation and heterogeneous nucleation is more favorable below that level.

The radius of the curvature of an equilibrium drop is $\Re_e = \frac{2\gamma}{-P_e}$. In the abovementioned narrow interval of oversaturation, see Eq. (2.52) the radius of equilibrium drops changes from infinity at $P \to p_s$ to $\Re_{cr} = \frac{2\gamma}{-\Pi_{\min}}$. If $\Pi_{\min} \approx 10^6$ dyn/cm^2 and $\gamma \approx 72$ dyn then $\Re_{cr} \approx \frac{144}{10^6} = 1.44$ µm, that is, the critical size is out of the range of the surface forces action. However, if $\Pi_{min} \approx 10^7$ dyn/cm^2, then $\Re_{cr} \approx \frac{144}{10^7} = 0.144\ \mu m = 1440\ \overset{\circ}{A}$, and the whole droplet is in the range of the surface forces action. That is, in this case the drop is so small that it does not have anywhere (even on the very apex) that is non-disturbed by the action of the Derjaguin's pressure spherical part.

Below we consider now only two-dimensional drops for simplicity. Three-dimensional axisymmetric drops and menisci in cylindrical capillaries are considered briefly at the end of this section.

On the drop apex, H, the derivative vanishes, $h'(H) = 0$. Using this condition and Eq. (2.26) we arrive to the following integration constant $C = \gamma + P_e H + \int_H^\infty \Pi(h)dh$. In this case, Eq. (2.26) transforms as follows:

$$\frac{1}{\sqrt{1+h'^2}} = \frac{\gamma + P_e(H-h) - \int_h^H \Pi(h)dh}{\gamma}. \tag{2.53}$$

This equation describes the profile of an equilibrium liquid drop on a flat solid substrate at $P_e < 0$.

As in the case of a meniscus, the whole profile of a droplet can be subdivided into three parts:

A spherical part of the drop

A transition zone, where both capillary pressure and the Derjaguin's pressure act simultaneously

A region of flat equilibrium liquid films in front of the drop

Outside the range of the Derjaguin's pressure action, we can neglect the surface forces action in Eq. (2.53), and this equation then describes the profile of a spherical drop:

$$\frac{1}{\sqrt{1+h'^2}} = \frac{\gamma + P_e(H-h)}{\gamma}. \tag{2.54}$$

This equation describes the profile of a spherical droplet that is not disturbed by surface forces action. The intersection of this profile with the thin equilibrium film of thickness, h_e, defines the apparent three-phase contact line and the macroscopic equilibrium contact angle, $h'(h_e) = -tg\theta_e$. Substitution of this expression into Eq. (2.54) results in $P_e = -\frac{\gamma(1-\cos\theta_e)}{H}$. Casting this expression into Eq. (2.53) at $h = h_e$ results in the following definition of the contact angle in the case of drops on a flat substrate:

$$\cos\theta_e = 1 + \frac{\frac{1}{\gamma}\int_{h_e}^H \Pi(h)dh}{1 - \frac{h_e}{H}} \approx 1 + \frac{1}{\gamma}\int_{h_e}^\infty \Pi(h)dh, \quad \text{at} \quad t_s \ll H, \tag{2.55}$$

which is similar to Eq. (2.47) in the case of the meniscus. At $h_e \ll H$, expressions for the equilibrium contact angles of a meniscus (2.47) and droplet (2.55) coincide. However, it is necessary to note that expressions for the equilibrium contact angle in the case of menisci (Eq. 2.47) and drops (Eq. 2.55) are still different: Integration in these expressions, even in the case of "thick capillaries" and "big drops" ($t_s \ll H$), starts from different values of h_e. This value is always bigger in the case of drops than in the case of menisci (Figure 1.12). This results in different values of equilibrium contact angles for these two cases.

Let us rewrite Eq. (2.53) for the drop profile in the identical form using the transversality condition at $h = h_e(h'(h_e) = 0)$. This results in

$$h' = -\sqrt{\frac{\gamma^2}{(\gamma - L(h))^2} - 1}, \tag{2.56}$$

where

$$L(h) = P_e(h - h_e) - \int_{h_e}^{h} \Pi(h)dh. \tag{2.57}$$

The expression under the square root in Eq. (2.56) should be positive, that is, the following condition should be satisfied:

$$0 \le L(h) \le \gamma, \tag{2.58}$$

where the first equality corresponds to the zero derivative and the second one corresponds to the infinite derivative. Note, beyond the radius of surface forces action, t_s, that is, at $h > t_s$: $L(h) = P_e(h - h_e)$ becomes a straight line (bottom part in Figure 2.15).

It is important to mention that the droplet is under equilibrium with the oversaturated vapor in the surrounding air. This means the droplet volume cannot be fixed and, thus, neither can the droplet height, H. That is, the drop height, H, is to be determined as below.

The top part of Figure 2.15 shows the S-shaped dependence of the Derjaguin's pressure isotherm, $\Pi(h)$, whereas the bottom part shows the dependency of $L(h)$, which has a maximum or minimum at thickness and which are solutions of $P_e = \Pi(h)$. In Figure 2.15, we consider only the case $P_e < \Pi_{min2}$. Other cases can be considered in the same way.

At the apex of the drop, when $h = H$, the first derivative should be zero, that is, $h'(H) = 0$ or, from Eq. (2.58):

$$L(H) = 0. \tag{2.59}$$

In this section macrodrops, that is, drops with their apex outside the range of the Derjaguin's pressure action (partial wetting), are under consideration. In this case, max $L(h) = L(h_u) > 0$ and $L(h) \to -\infty$ as $h \to +\infty$ (Figure 2.15, bottom part). Therefore, Eq. (2.59) always has a solution, H, which is in the following range: $h_u < H < \infty$ (Figure 2.15, bottom part).

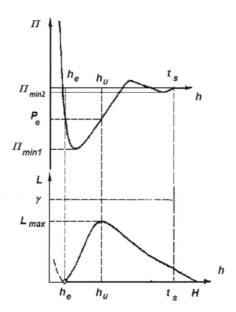

FIGURE 2.15 Determining the droplet height H. Top part: Derjaguin's pressure isotherm; bottom part: shape of function $L(h)$ [see definition (2.58)].

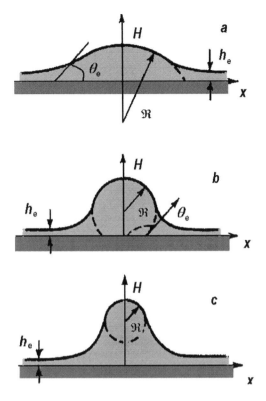

FIGURE 2.16 Determining the equilibrium contact angle of droplets. (a) partial wetting; (b) non-wetting; and (c) complete non-wetting.

If the value of H lies outside the range of the influence of the Derjaguin's pressure, then it becomes possible to determine the equilibrium macroscopic contact angle of the drop. Depending on the value of the radius of the curvature in the central part of the drop, $\Re$, three different possibilities can occur (Figure 2.16a–c): (a) $\Re > H - h_e$, which corresponds to the contact angle $0 < \theta_e < \pi/2$, the partial wetting case; (b) $\Re < H - h_e$ but $2\Re > H - h_e$, which corresponds to the macroscopic contact angle $\pi/2 < \theta_e < \pi$, the non-wetting case; (c) if $2\Re < H - h_e$, then despite the apex of the drop being outside the range of influence of the Derjaguin's pressure, it is impossible to determine the macroscopic contact angle because there is no intersection of the circle of radius R with the solid surface (Figure 2.16c). The last case can be referred to as "the complete non-wetting" and can be referred to as $\theta_e > \pi$, in a similar way as in the case of the complete wetting, when "$\cos \theta_e > 1$" (see Section 2.4).

It is interesting to note that probably the cases (b) and (c) (Figure 2.16) have never been observed experimentally. It means either that such kinds of the Derjaguin's pressure isotherm do not exist in the nature or that such cases are yet to be discovered.

It is possible to check [using Eq. (2.55)] that in the case of partial wetting, $\theta_e(P_e)$, dependence increases with decreasing P_e, that is, the drop "elevates" itself above the solid surface as P_e decreases and at $P_e = \Pi_{min1}$, the drop separates itself from the solid surface and goes into the surrounding air. This corresponds to a transition from a heterogeneous to a homogeneous nucleation.

Profile of the Transition Zone in the Case of Droplets

Equation (2.50) and Figure 2.17a (left-hand side) shows that in the case of small droplets ($P_e < \Pi_{min2}$) droplets have only one inflection point on the drop profile, at $h(x_u) = h_u$. Hence, the drop profile inside the transition zone is shown in Figure 2.17a (right-hand side). However, in the case of large droplets (that is,

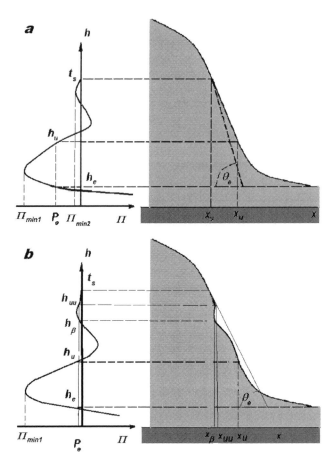

FIGURE 2.17   The drop profile inside the transition zone. Left: Derjaguin's pressure isotherm; right: the drop profile. (a) Small droplets; and (b) big droplets.

$0 > P_e > \Pi_{min2}$), the droplet profile is more sophisticated, as shown in Figure 2.17b: The droplet profile has two inflection points, at $h(x_u) = h_u$ and $h(x_{uu}) = h_u$. This case is considered in more detail in Section 3.2.

Deviations from the spherical profile start immediately as surface forces step into action at $h < t_s$.

Axisymmetric Drops

In this case, the expression for excess free energy takes the following form:

$$\Phi = 2\pi \int r \left\{ \gamma \left[\sqrt{1+h'^2} - 1 \right] + P_e(h-h_e) + f_D(h) - f_D(h_e) \right\} dr,$$

which gives the following equation for the liquid profile of an axisymmetric drop:

$$\frac{\gamma}{r} \frac{d}{dr} \left(\frac{rh'}{\sqrt{1+h'^2}} \right) + \Pi(h) = P_e, \tag{2.60}$$

where $P_e = -\frac{2\gamma}{\Re}$, which results in $P_e = -\frac{2\gamma(1-\cos\theta_e)}{H}$ for drops. Note the multiplier, 2, in these expressions.

Unfortunately, Eq. (2.60) cannot be integrated once as in the case of two dimensional menisci and drops. However, this equation can be rewritten as

$$\frac{\gamma h''}{\left(1+h'^2\right)^{3/2}}+\frac{\gamma}{r}\,\frac{h'}{\sqrt{1+h'^2}}+\Pi(h)=P_e. \qquad (2.61)$$

The first term on the left-hand side of Eq. (2.61) is due to the first curvature [the same term as in the case of two-dimensional drops or menisci, Eq. (2.23)], and the second term is due to the second curvature, which is shown below to be small as compared with the first term inside the transition zone. The characteristic length of the transition region, L, is given by Eq. (2.49): $L \sim \sqrt{h_e H}$. This expression shows that $L \ll H$. Let us estimate the ratio of the second term to the first term on the left-hand side of Eq. (2.61) inside the transition zone:

$$\frac{\gamma}{r}\,\frac{h'}{\sqrt{1+h'^2}}\bigg/\frac{\gamma h''}{\left(1+h'^2\right)^{3/2}} \sim \frac{h'}{rh''} \sim \frac{h/L}{Hh/L^2}=\frac{L}{H}\sim\sqrt{\frac{h_e}{H}}\ll 1.$$

This estimation shows that the second term on the left-hand side of Eq. (2.61) is small as compared with the first term and can therefore be neglected inside the transition region. After that, Eq. (2.61) in the same way as Eq. (2.23), can be integrated once and Eq. (2.26) is recovered. Outside the region of the surface forces action, Eq. (2.60) can be easily solved. This solution is the "outer solution," whereas the previous solutions are "an inner solutions." Matching these two asymptotic solutions gives the real profile in the case of axial symmetry (see this procedure in the case of complete wetting in Section 2.4).

Meniscus in a Cylindrical Capillary

In this case, the expression for excess free energy is as follows:

$$\Phi=\int\left\{2\pi\gamma\left[(H-h)\sqrt{1+h'^2}-(H-h_e)\right]+\pi P_e\left[(H-h_e)^2-(H-h)^2\right]+2\pi H\left[f_D(h)-f_D(h_e)\right]\right\}dx,$$

where H is the radius of the cylindrical capillary. The same minimization procedure results

$$\frac{\gamma h''}{\left(1+h'^2\right)^{3/2}}+\frac{\gamma}{H-h}\frac{1}{\sqrt{1+h'^2}}+\frac{H}{H-h}\Pi(h)=P_e.$$

Note that in this case the Derjaguin's pressure is $\frac{H}{H-h}\Pi(h)$, which is different from the Derjaguin's pressure of flat films, $\Pi(h)$. We discuss this in Section 2.9.

APPENDIX 1

Let us assume that the transition zone profile does not tend asymptotically to the equilibrium thickness, h_e, but meets the film at the final point $x = x_0$. In this case in a vicinity of this point, we approximate the Derjaguin's pressure isotherm by a linear dependency $\Pi(h)\approx\Pi(h_e)-a(h-h_e)$, where $a=-\Pi'(h_e)$ is a positive value: h_e is a stable flat liquid film and the derivative of the Derjaguin's pressure should be negative and $\Pi(h_e) = P_e$. The liquid profile in this region has a low slope, which means Eq. (2.23) can be rewritten as

$$\gamma h''+a(h-h_e)=0.$$

The solution of this equation is

$$h(x) = h_e + C_1 \exp(\alpha x) + C_2 \exp(-\alpha x), \tag{A1.1}$$

where $\alpha = \sqrt{\dfrac{a}{\gamma}}$, C_1 and C_2 are two integration constants.

At $x = x_0$, according to Eq. (2.21), the following two boundary conditions should be satisfied:

$$h(x_0) = h_e$$

$$h'(x_0) = 0.$$

Equation (1.1) and these boundary conditions result in the following system of two algebraic equations for the determination of the two integration constants C_1 and C_2:

$$C_1 \exp(\alpha x_0) + C_2 \exp(-\alpha x_0) = 0,$$

$$C_1 \exp(\alpha x_0) - C_2 \exp(-\alpha x_0) = 0$$

The only solution of this system is $C_1 = C_2 = 0$, which is obviously a contradiction. Hence, the only possibility is that $C_1 = 0$, and the liquid profile has the following form if $h \to h_e$, $h(x) = h_e + C_2 \exp(-\alpha x)$.

That is, the liquid profile in the transition zone tends asymptotically to the equilibrium thickness, h_e, and does not meet the equilibrium flat film in any final point, x_0.

Note, that in the case when $\Pi(h) = \infty$, at $h < t_0$, our assumption on linearization of the Derjaguin's pressure isotherm is no longer valid and this is the only one very special case when the transition zone profile meets the equilibrium flat liquid film at the final point, x_0. The upper limit in integration in Eq. (2.1) should be replaced by x_0.

Note that $\alpha = \sqrt{\dfrac{a}{\gamma}}$ gives a new scale of the transition zone, which is $1/\alpha$. It is possible to check that this new scale and the previous one given by Eq. (2.49) are of the same order of magnitude. Indeed

$$\frac{1}{\alpha} = \sqrt{\frac{\gamma}{\Pi'(h_e)}} \sim \sqrt{\frac{\gamma}{\Pi(h_e)/h_e}} = \sqrt{\frac{\gamma h_e}{P_e}} \sim \sqrt{\frac{\gamma h_e}{\gamma/H}} = \sqrt{h_e H} = L.$$

2.4 Profile of the Transition Zone between a Wetting Film and the Meniscus of the Bulk Liquid in the Case of Complete Wetting

The profile of a liquid in the transition zone between a capillary meniscus and a wetting film is calculated below for two types of the Derjaguin's pressure isotherms (both in the case of *complete* wetting). As discussed in Section 2.3, wetting films ahead of the meniscus are separated from the capillary meniscus by a transition zone where the surface forces and the capillary forces act simultaneously. Because measurements of equilibrium contact angles and surface curvature of bulk liquids should be carried out outside such transition zone, its size and profile are of interest. Moreover, the shape itself of the liquid profile in the transition zone supplies information on the Derjaguin's pressure isotherm of liquid films on a given solid substrate.

Next we consider, for simplicity, a transition zone under equilibrium conditions between a capillary meniscus placed between two parallel plates and wetting films in front (Figure 2.18). The width of the capillary, $2H$, is assumed to be much wider than the thickness of the equilibrium flat film, h_e. In the case under consideration, the thickness of the liquid layer, $h(x)$, is a function of a single coordinate, x, directed along the capillary surface. It has been already shown in Section 2.2 that the

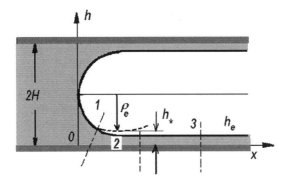

FIGURE 2.18 Complete wetting case. Schematic representation of a circular capillary meniscus (1), transition zone; (2), wetting films; and (3) in a flat capillary. Continuation of a spherical meniscus (broken line) does not intersect neither solid walls of the capillary or thin liquid film of thickness h_e in front of the meniscus. The radius of the curvature of the meniscus, ρ_e, is smaller than the half-width H.

meniscus profile comes to the flat liquid film at a zero contact angle according to the transversality condition (2.21) and the condition for that follows from Eq. (2.26) as

$$\gamma = P_e(H - h_e) - \int_{h_e}^{\infty} \Pi(h)dh,$$

where the equilibrium thickness, h_e, is determined as

$$\Pi(h_e) = P_e.$$

In the case of complete wetting and the Derjaguin's pressure isotherms of the type $\Pi(h) = A/h^n$, these two equations result in

$$\gamma = P_e(H - h_e) - \frac{A}{(n-1)h_e^{n-1}},$$

and

$$\frac{A}{h_e^n} = P_e,$$

where n is the exponent in the expression for the isotherm; γ is the surface tension of the bulk liquid; and H is the half-width of the capillary. That is, we have two equations with two unknown values, P_e and h_e. These two equations allow determining the equilibrium pressure, P_e, via the thickness of the equilibrium flat film, h_e. This results in

$$P_e = \frac{\gamma}{H - \frac{n}{n-1}h_e} = \frac{\gamma}{H - h_*}, \quad h_* = \frac{n}{n-1}h_e > h_e. \tag{2.62}$$

Recall that the equilibrium pressure is equal to $P_e = \frac{\gamma}{\rho_e} = \frac{\gamma}{H - h_*}$, according to Eq. (2.62). This means that the radius of the curvature of the meniscus in the case of complete wetting $\rho_e = H - h_* < H$ and the continuation of the spherical meniscus intersects neither the capillary walls nor the flat liquid films in front of the meniscus (Figure 2.18).

According to Sections 2.2 and 2.3, the profile of the meniscus, transition zone and flat wetting films in front is described by Eq. (2.23), which in the case under consideration becomes

$$\frac{\gamma h''}{\left(1+h'^2\right)^{3/2}} + \frac{A}{h^n} = P_e, \tag{2.63}$$

where $h(x)$ is the local liquid profile; and $\frac{A}{h^n}$ is the local Derjaguin's pressure isotherm. Inside the transition zone (Figure 2.18), the liquid profile has a very low slope, that is, $(dh/dx)^2 << 1$ is satisfied and the Derjaguin's pressure isotherm, $\frac{A}{h^n}$, of flat films can be safely used at each point of the liquid layer of varying thickness. As above, in this section we assume that the surface tension of the film is the same as that of the bulk liquid.

The consideration below is carried out for the case of complete wetting.

Because the liquid profile inside the transition zone satisfies the condition $(dh/dx)^2 << 1$, the low slope approximation can be used. Introducing the dimensionless variables: $\xi = h/h_e$ and $y = [x - (H - h_*)]/l$, where l is a length scale along x to be determined, which is the length scale of the transition zone. It is shown below that y is a local variable inside the transition zone. Using this notation in Eq. (2.63), we arrive to

$$\xi'' + \frac{1}{\xi^n} = 1, \tag{2.64}$$

where $\xi = \xi(y)$, and the length scale is selected as

$$l = \sqrt{h_e\left(H - h_*\right)x_+}. \tag{2.65}$$

Note that this selection is in excellent agreement with our previous estimation of the length of the transition zone in Section 2.3 (Eq. 2.49). We will use the notation L to mark the value of the precisely calculated extension of the transition zone according to Eq. (2.65).

The thickness of the equilibrium flat film, h_e, is determined as before from the condition $A/h_e^n = P_e$.

According to the Jacobi's condition the dependency $\xi(y)$ is a monotonic one. Therefore, because the independent variable, y, does not appear explicitly in Eq. (2.64), we can introduce a new unknown function $\xi' = $ function (ξ). Taking into account that $\xi'(1) = 0$ Eq. (2.64) can be rewritten as

$$\xi' = -\sqrt{2(\xi-1) + \frac{2}{(n-1)}\left(\frac{1}{\xi^{n-1}} - 1\right)}. \tag{2.66}$$

Equation (2.66) is solved below for the most important cases: $n = 3$ and $n = 2$. Films that obey the $\Pi = A/h^3$ law (that is, $n = 3$) correspond to the case where the Derjaguin's pressure of the film is determined by dispersion forces [1]. For nonpolar liquids on solid dielectrics, we can adopt $A = 10^{-14}$ erg [1] and $\gamma = 30$ dyn/cm.

In the case $n = 3$, it follows from Eq. (2.66) that

$$\xi' = -\frac{\xi-1}{\xi}\sqrt{2\xi+1}.$$

Upon integration, the solution of this equation is

$$\sqrt{2\xi+1} + \frac{1}{\sqrt{3}}\ln\frac{\sqrt{2\xi+1}-\sqrt{3}}{\sqrt{2\xi+1}+\sqrt{3}} = -\left(y+C\right), \tag{2.67}$$

where C is the integration constant to be determined.

If $\xi(y) \gg 1$ in Eq. (2.67), that is, in the region where the transition zone and the meniscus meet each other, we conclude from this equation that

$$\xi(y) \approx \frac{(y+C)^2}{2}. \tag{2.68}$$

To determine the unknown constant C, we consider the meniscus shape corresponding to large h values for which it can be assumed in Eq. (2.63) that the Derjaguin's pressure can be neglected, which results in

$$\frac{\gamma h''}{\left[1+(h')^2\right]^{3/2}} = P_e = \frac{\gamma}{H - h_*}.$$

Integration of this equation with the boundary condition, $h(0) = H$, results in the following solution for the spherical meniscus:

$$h(x) = H - \sqrt{\left(H - h_*\right)^2 - \left(x - \left(H - h_*\right)\right)^2}.$$

Using the same local variables for the transition zone, $\xi(y)$ and $y \ll 1$, in this equation we conclude

$$\xi(y) \approx \frac{n}{n-1} + \frac{y^2}{2}. \tag{2.69}$$

Comparison of Eqs. (2.68) and (2.69) results in $C = 0$. That is, Eq. (2.67) can be rewritten now as

$$\sqrt{2\xi+1} + \frac{1}{\sqrt{3}}\ln\frac{\sqrt{2\xi+1}-\sqrt{3}}{\sqrt{2\xi+1}+\sqrt{3}} = -y. \tag{2.70}$$

Let us call an "ideal profile," $\xi_i(y)$, the following function:

$$\xi_i(y) = \begin{cases} \dfrac{n}{n-1} + \dfrac{y^2}{2}, & y < 0 \\[2mm] 1, & y > 0 \end{cases}. \tag{2.71}$$

The real profile, calculated according to Eq. (2.70) and the "ideal profile," according to Eq. (2.71) at $n = 3$, are presented in Figure 2.19. Figure 2.19 shows that the extent of the transition zone can be roughly estimated as from $y = -1.2$ to $y = 1.7$. That is, the total length of the transition zone, L, in dimensional units is $L = 2.9l = 2.9\sqrt{h_e(H - h_*)}$. Figure 2.19 shows that the maximum deviation of the real profile from the ideal profile is located at $y = 0$, that is, at the position minimum of the continuation of the spherical meniscus, and the maximum deviation is around 2.5 h_e. Figure 2.19 also shows that the deviation of the real from ideal profile is roughly symmetrical from both sides from the position of the maximum deviation at $y = 0$.

An isotherm $\Pi = B/h^2$ corresponds to thick β-films of water, which are stabilized by the ionic-electrostatic component of the Derjaguin's pressure [1]. In this case, integration of Eq. (2.66) at $n = 2$ results in

$$\sqrt{\xi} + \frac{1}{2}\ln\frac{\sqrt{\xi}-1}{\sqrt{\xi}+1} = -\frac{y}{\sqrt{2}}, \tag{2.72}$$

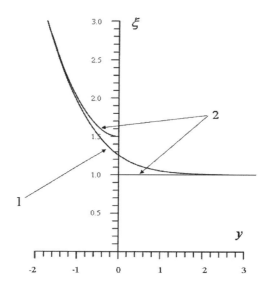

FIGURE 2.19 Real profile (1) inside the transition zone calculated according to Eq. (2.70) and the ideal profile (2) according to Eq. (2.71) at $n = 3$.

where the integration constant is equal to zero, which is concluded in precisely the same way as in the case $n = 3$.

Table 2.2 shows that although the absolute thickness of the transition zone, L, decreases with a decrease of the capillary radius, H, the relative transition zone thickness, L/H, increases with decreases of H. The L values vary within a range from 37 μm for thick films ($h_e \sim 500$ Å) to 0.2 μm for thin films ($h_e \sim 30$ Å).

In the case $n = 2$, calculations were made using Eq. (2.72) and adopting $B = 2 \cdot 10^{-7}$ dyn and $\gamma = 72$ dyn/cm (Table 2.3) and the "ideal profile," according to (2.71) at $n = 2$, are presented in Figure 2.20. Figure 2.20 shows that the extent of the transition zone can be roughly estimated as from $y = -2.4$ to $y = 1.9$. That is, the total length of the transition zone, L, in dimensional units is $L = 4.3\, l = 4.3\sqrt{h_e(H - h_*)}$. Figure 2.20 shows that the maximum deviation of the real profile from the ideal one is located at $y = 0$, that is, at the position minimum of the continuation of the spherical meniscus, and the maximum deviation is around $2.5h_e$; however, now the deviation is greater from the meniscus than from the flat film. Figure 2.20 also shows that the deviation of the real profile from the ideal one is not symmetrical from both sides from the position of the maximum deviation at $y = 0$ but decreases more rapidly from the flat liquid film side and extends farther into the depth from the meniscus side. This behavior is different from the case $n = 3$.

TABLE 2.2

Characteristics of the Transition Zone for the Case of the $\Pi = A/h^3$ Isotherm

H, cm	0.3	0.2	0.1	$5 \cdot 10^{-2}$	10^{-2}	10^{-3}	10^{-4}
h_e, Å	445	405	322	256	150	70	32
L, cm	$3.7 \cdot 10^{-3}$	$2.88 \cdot 10^{-3}$	$1.81 \cdot 10^{-3}$	$1.14 \cdot 10^{-3}$	$3.92 \cdot 10^{-4}$	$0.85 \cdot 10^{-4}$	$1.81 \cdot 10^{-5}$

TABLE 2.3

Characteristics of the Transition Zone for the Case of the $\Pi = A/h^2$ Isotherm

H, cm	1	0.5	0.2	0.1	10^{-2}	10^{-3}	10^{-4}
h_e, Å	5250	3730	2640	1660	525	166	53
L, cm	$3.62 \cdot 10^{-2}$	$2.1 \cdot 10^{-2}$	$1.15 \cdot 10^{-2}$	$6.44 \cdot 10^{-3}$	$1.15 \cdot 10^{-3}$	$2.04 \cdot 10^{-4}$	$3.64 \cdot 10^{-5}$

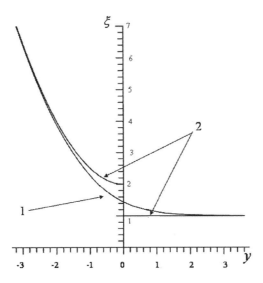

FIGURE 2.20 Real profile (1) inside the transition zone calculated according to Eq. (2.72) and the ideal profile (2) according to Eq. (2.71) at $n = 2$.

The extent of the transition zone, L, regions is larger at $n = 2$ than at $n = 3$ at equal H values: the maximum L values are about 362 μm for thick film in wide slots ($H = 1$ cm, $h_e \sim 0.525$ μm, $n = 2$) and 0.36 μm for thin films ($H = 1$ μm, $h_e = 53$ A, $n = 3$).

Thus, the transition zone is very much extended for low capillary meniscus pressures and thick liquid films. The radius of the meniscus curvature must be studied outside the transition zone, that is, at a distance larger than L from the apparent three-phase contact line.

2.5   Thickness of Equilibrium Wetting Films on Rough Solid Substrates

Let us consider now the case of equilibrium thin liquid films on rough solid substrates in the case of *complete* wetting. In such a case, one can only measure the mean thickness of the film, $\bar{h}$. It appears that the mean thickness, $\bar{h}$, of equilibrium wetting films on *rough* solid surfaces is larger than the corresponding thickness of a flat equilibrium film, h_e, on a *smooth* substrate. It also appears that the mean thickness, $\bar{h}$, approaches the value h_e at high and low film thicknesses when these are smaller or bigger relative to the characteristic scale of the roughness of the solid substrate, α. For $\bar{h} \gg α$, the effect of the roughness is made negligible and the surface of the film at the interface with the air becomes practically smooth. When $\bar{h} \ll α$, the film copies the surface of the substrate, maintaining a constant value of the film per unit area of the rough surface. Hence, the maximum deviation of the average thickness of liquid films on a rough substrate from the corresponding thickness on a flat substrate should be expected when $\bar{h} \approx α$.

When a wetting film of a uniform thickness covers a curved surface, its equilibrium with the vapor of the same liquid is determined by Eq. (2.23) in Section 2.2, which we rewrite as

$$\gamma K + \Pi(h) = \frac{R_g T}{v_m} \ln \frac{p_s}{p} = P_e, \tag{2.73}$$

where γK is the capillary pressure, due to the local curvature, K, of the surface of the film; γ is the surface tension; $\Pi(h)$ is the Derjaguin's pressure, which is a function of the local thickness of the film h; R_g is the gas constant; T is the temperature in K; v_m is the molar volume of the liquid; p is the equilibrium pressure of the vapor above the film; and p_s is the pressure of the saturated vapor of the same liquid.

Let us make a further examination for a one-component liquid in the case of compete wetting, where the Derjaguin's pressure isotherm, $\Pi(h)$, is determined only by the dispersion forces. The isotherm of the Derjaguin's pressure of a flat film in this case has the same form as in the Section 2.4.

$$\Pi(h) = A / h^n, \tag{2.74}$$

where $n = 3$ for small and $n = 4$ for large thicknesses of the film. The Hamaker constant A is characterized based on the spectral characteristics of the film and the substrate [12].

In the general case, the Derjaguin's pressure in thin films on a curved substrate, $\Pi_r(h)$, is different from the corresponding Derjaguin's pressure in films on a flat substrate, $\Pi(h)$. Thus, for example, for films at the internal surface of a capillary of radius, r, the Derjaguin's pressure, Π_r, (in the approximation $h << r$) is given by [13]

$$\Pi_r = \Pi(h) \left[1 + \frac{h}{2r} \frac{\int_0^\infty \Delta_{32} \cdot \Delta_{21}\left(\Delta_{32} + \Delta_{21}\right)d\xi}{\int_0^\infty \Delta_{32} \cdot \Delta_{21} d\xi} \right], \tag{2.75}$$

where $\Pi(h)$ is the Derjaguin's pressure of a flat film of the same thickness on a flat substrate; $\Delta_{ik} = \frac{\varepsilon_i - \varepsilon_k}{\varepsilon_i + \varepsilon_k}$, $\varepsilon(i\xi)$ is the dielectric permeability, which is a function of the angular frequency ξ, taken at the imaginary axis [14]. The subscripts 1, 2, and 3 relate, respectively, to the gas, the film, and the solid substrate. At $h/r \to 0$, $\Pi_r \to \Pi$. Quantitative evaluations for films of decane on the surface of quartz show that the contribution of the second term on the right-hand side of Eq. (2.75) is relatively small. Thus, with $h/r \le 0.2$, the difference between Π_r and Π does not exceed 2.5%.

Thus, under the condition of the smallness of the curvature, h/r, the isotherm (2.74) of the Derjaguin's pressure of flat films can be used with sufficient precision.

Real surfaces, as a rule, have a roughness. In this case, the local thickness of the film is a function of the coordinate, and a mean value of the thickness of the film, $\bar{h}$, should be used. The problem is how significantly the mean thickness, $\bar{h}$, differs from the thickness of a flat film on an ideally smooth surface of the same nature, and how these differences affect the roughness (or topology) of the surface.

Let us consider a simplified model of a one-dimensional roughness (Figure 2.21), where the profile of the surface is a function of one coordinate, x. Let $H_s(x)$ be the equation of the surface of the substrate, and

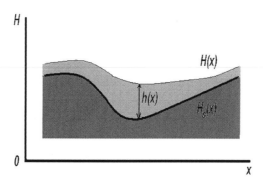

FIGURE 2.21   Calculations of the thickness of wetting films on a rough cylindrical surface. $H_s(x)$, profile of the solid substrate; $H(x)$, the liquid profile; and $h(x) = H(x) - H_s(x)$, the film thickness.

$H(x)$ be the equation of the surface of the film, forming a boundary with the air (Figure 2.21). The local thickness of the film is determined as

$$h(x) = H(x) - H_s(x).$$ (2.76)

Using these notations, Eq. (2.73) takes the following form:

$$\frac{\gamma H''}{\left[1+(H')^2\right]^{3/2}} + \frac{A}{h^n} = P_e,$$ (2.77)

where H' and H'' are the first and the second derivatives of $H(x)$, and P_e is determined by the vapor pressure in the ambient air according to Eq. (2.73).

On a plane substrate, the liquid curvature is zero and Eq. (2.77) gives

$$\frac{A}{h_e^n} = P_e.$$ (2.78)

Substitution of Eq. (2.78) into Eq. (2.77) results in

$$\frac{d}{dx} \frac{\gamma H''}{\left[1+(H')^2\right]^{1/2}} = -\left(\frac{A}{h^n} - \frac{A}{h_e^n}\right).$$ (2.79)

Let us introduce the average values of any function, φ, of a random variable as follows:

$$\bar{\varphi} = \lim_{X \to \infty} \frac{1}{X} \int_0^X \varphi(x) dx \approx \frac{1}{X} \int_0^X \varphi(x) dx, \quad X \gg \lambda,$$

where λ is a characteristic scale of surface roughness in x direction; and the overbar means averaged over a random substrate.

Let us integrate both sides of Eq. (2.79) from 0 to X, where X is much bigger than the characteristic scale of the roughness in the x direction. We assume that the surface is statistically homogeneous, that is, there is no preferable positive or negative curvature of the liquid film. We can subdivide the whole interval of integration from 0 to X into a big number, N, of subintervals of a small length, $\lambda = X/N$. After that

$$\frac{1}{X} \int_0^X K dx = \frac{1}{X} \sum_{i=0}^N \int_{x_i}^{x_{i+1}} K dx = \frac{1}{X} \sum_{i=0}^N \left(\sin\theta(x_{i+1}) - \sin\theta(x_i)\right) =$$

$$= \frac{1}{\lambda}\left(\frac{1}{N}\sum_{i=0}^N \sin\theta(x_{i+1}) - \frac{1}{N}\sum_{i=0}^N \sin\theta(x_i)\right) = \frac{1}{\lambda}\left(\overline{\sin\theta} - \overline{\sin\theta}\right) = 0,$$

$$x_i = \Delta \cdot i, \quad i = 0,1,2,...N, \quad x_N = X$$

where θ is the local slope of the liquid profile. Hence, in the case of random and statistically homogeneous roughness, the averaged value of the left-hand side of Eq. (2.79) vanishes, and the average of the right-hand side results in

$$\frac{1}{X}\int_0^X\left(\frac{A}{h^n}-\frac{A}{h_e^n}\right)dx=0. \tag{2.80}$$

Equation (2.80) can be rewritten in the following form:

$$\frac{1}{X}\int_0^X\frac{A}{h^n}dx=P_e,$$

that is, the measured averaged Derjaguin's pressure on a rough substrate coincides with the Derjaguin's pressure on a corresponding flat substrate. Hence, this expression can be rewritten as

$$\frac{\overline{A}}{h^n}=P_e. \tag{2.81}$$

Recall a well-known theorem from the probability theory. Let us consider a concave function, φ, of random variable h, then

$$\overline{\varphi(h)}>\varphi(\overline{h}),$$

that is, the average of the concave function is larger than the function of the average. The Derjaguin's pressure isotherm of the type under consideration is a concave function of h because the second derivative is positive: $\left(\dfrac{A}{h^n}\right)''=\dfrac{n(n-1)A}{h^{n-2}}>0$. Application of this theorem results in

$$\frac{\overline{A}}{h^n}>\frac{A}{\overline{h}^n}.$$

Comparison of this inequality and (2.81) results in

$$\frac{A}{h_e^n}>\frac{A}{\overline{h}^n},$$

and, hence,

$$\overline{h}>h_e. \tag{2.82}$$

This inequality shows that on a rough solid substrate, the measured average thickness of the liquid film is bigger than the equilibrium thickness of the equilibrium film on a corresponding flat substrate.

Equation (2.82) is only a qualitative result. Below we try to investigate quantitatively how the thickness of the film on a rough substrate influences the Derjaguin's pressure measurements using a model roughness. For this, we consider below a model rough surface of the following kind:

$$H_s(x)=\alpha\cos kx, \tag{2.83}$$

that is, a periodic roughness with an amplitude, α, and the wave length, $1/k$. This allows us to restrict our consideration to x in the following range: $-\dfrac{\pi}{k}<x<\dfrac{\pi}{k}$.

In the case of the model roughness (2.83), we consider only half of the period. Hence, the following boundary conditions are satisfied:

$$H'(0)=0, \tag{2.84}$$

and

$$H'(X) = 0, \tag{2.85}$$

where $X = \dfrac{\pi}{k}$ is the half-period.

To simplify the calculation, we assume that $\delta = k\alpha \ll 1$. Then, the model roughness has a low slope, and the solution of Eq. (2.77) can be expanded in a series in terms of the small parameter δ:

$$H = h_e + H_1 + H_2 + \dots , \tag{2.86}$$

where $H_i(x) \sim \delta^i$, $i = 1, 2 \dots$. We limit ourselves here to the first two terms of the expansion because the nonlinearity of the curvature is of the third order of smallness. Using these notations, we can write

$$\frac{A}{h^n} = \frac{A}{h_e^n} - \frac{nA}{h_e^{n+1}}(H_1 + H_2 + \dots - H_s) + \frac{n(n+1)A}{2h_e^{n+2}}(H_1 + H_2 + \dots - H_s)^2 + \dots$$

Substitution of this expression into integral (2.80) and collecting only terms of the first and second order in the small parameter δ, we conclude

The first order

$$\int_0^X (H_1 - H_s)\, dx = 0, \tag{2.87}$$

and collecting two terms of the second order

$$-\frac{nA}{h_e^{n+1}} \int_0^X H_2\, dx + \frac{n(n+1)A}{h_e^{n+2}} \int_0^X (H_1 - H_s)^2\, dx = 0.$$

This equation can be rewritten as

$$\int_0^X H_2\, dx = \frac{(n+1)}{h_e} \int_0^X (H_1 - H_s)^2\, dx. \tag{2.88}$$

Let us now determine the averaged thickness as

$$\bar{h} = h_e + \frac{1}{X} \int_o^X (H_1 + H_2 + \dots - H_s)\, dx \approx h_e + \frac{1}{X} \int_o^X (H_1 - H_s)\, dx + \frac{1}{X} \int_o^X H_2\, dx.$$

According to Eq. (2.87), the first term on the right-hand side of this equation is equal to zero and the second term is given by Eq. (2.88). Hence, this equation can be rewritten as

$$\bar{h} = h_e + \frac{(n+1)}{Xh_e} \int_0^X (H_1 - H_s)^2\, dx. \tag{2.89}$$

The integral in the right-hand side of this equation is always positive. Therefore, we come to the same conclusion as in the general case [see inequality (2.82) above], that is, the averaged film thickness on a rough substrate is always bigger than the thickness on the corresponding film on a flat substrate.

Equation (2.89) shows that we must determine only the first function, $H_1(x)$, to calculate the second correction to the average thickness, $\bar{h}$.

Substituting the expansion Eq. (2.86) into Eq. (2.77) and collecting terms of an identical order, we obtain the following system

$$\frac{A}{H_0^n} = P_e, \tag{2.90}$$

Equation (2.90) shows that $H_0 = \left(\frac{A}{P_e}\right)^{1/n}$ is a constant that is equal to the thickness of the equilibrium film on a flat substrate, h_e. The next equation is

$$\gamma H_1'' - \frac{nA}{h_e^{n+1}} \cdot \left(H_1 - H_s\right) = 0, \tag{2.91}$$

with the periodic boundary conditions

$$H_1'(0) = H_1'\left(\frac{\pi}{k}\right) = 0, \tag{2.92}$$

Introducing the notation $a^2 = \dfrac{nA}{\gamma h_e^{n+1}}$, instead of (2.91), we arrive to

$$H_1'' - a^2 H_1 + \alpha a^2 \cdot \cos \cdot kx = 0. \tag{2.93}$$

Solution of Eq. (2.93) with the boundary conditions (2.92) is

$$H_1 = \frac{\alpha a^2 \cdot \cos \cdot kx}{a^2 + k^2}. \tag{2.94}$$

For a crest of the sinusoid, we get from Eq. (2.94) $h\left(\frac{\pi}{k}\right) = h_e - \frac{\alpha k^2}{a^2 + k^2}$, and for a trough, $h(0) = h_e + \frac{\alpha k^2}{a^2 + k^2}$, that is, the film is thicker in a depression than on a convexity.

Using Eq. (2.94), we conclude $H_1 - H_s = -\frac{\alpha k^2 \cdot \cos \cdot kx}{a^2 + k^2}$. Substitution of this expression into Eq. (2.89) results in the following second approximation of the mean thickness of the film:

$$\bar{h} = h_e + \frac{\alpha^2 k^4 \gamma P_e^{1/n} (n+1)}{2 A^{1/n} \left(\dfrac{nP_e^{1+1/n}}{\gamma A^{1/n}} + k^2\right)^2}. \tag{2.95}$$

This equation shows that $\bar{\Delta} = \bar{h} - h_e$ tends to zero, at both $P_e \to \infty$ $\left(\text{or } \frac{p}{p_s} \to 0\right)$ and $P_e \to 0$ $\left(\text{or } \frac{p}{p_s} \to 1\right)$. The difference in the thickness of the films $\bar{\Delta} = \bar{h} - h_e$ on a rough and smooth substrate goes via a maximum at some value of $P_{e\max}$, which can be determined by direct differentiation of the second term on the right-hand side of Eq. (2.95), which results in

$$P_{e\max} = \left(\frac{\gamma k^2}{n(2n+1)}\right)^{n/(n+1)} \cdot A^{1/(n+1)}. \tag{2.96}$$

At $\bar{h} \gg \alpha$, the effect of the roughness is damped and the surface of the film at the interface with the gas becomes practically smooth. At $\bar{h} \ll \alpha$, the film copies the surface of the substrate. Hence, the maximum deviation of the thickness of liquid films on a rough substrate from the corresponding thickness on a flat substrate should be expected at $\bar{h} \approx \alpha$.

A qualitative form of the function $\bar{h}\,(P_e)$ is shown in Figure 2.22 (curve 2). The straight line 1 illustrates the Derjaguin's pressure isotherm of flat films according to Eq. (2.95). The maximal deviations of isotherm (curve 2) from isotherm (line 1) correspond to the values of $P_{e\,max}$.

Because polished surfaces, as a rule, have grooves left by the solid grains of the polishing pastes, we must expect a qualitative picture for these surfaces, as presented in Figure 2.22. This was experimentally observed in [15] for wetting films of tetradecane on polished steel surfaces, where the qualitative picture presented in Figure 2.22 was experimentally observed in the case of Derjaguin's pressure isotherm for the complete wetting, $n = 3$ (Figure 2.23).

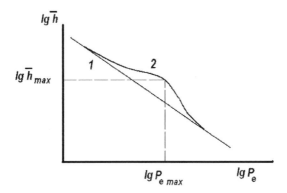

FIGURE 2.22 The Derjaguin's pressure isotherm $P_e\,(h_e) = A/h_e^n$. Sketch of deviations of the measured average liquid film thickness, $\bar{h}$, from the predicted film thickness on a flat surface, h_e. (1) Thickness of the film on a flat substrate; and (2) average film thickness on a rough substrate.

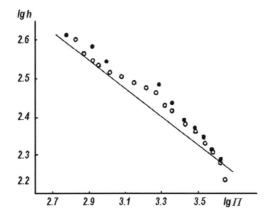

FIGURE 2.23 Experimental data on average film thickness of tetradecane on steel, obtained by continuous thinning of the film (open symbols) and subsequent thickening (closed symbols). Solid line according to the Derjaguin's pressure isotherm $\Pi = \frac{A}{h^3}$. (From Shishin, V.A. et al., *Colloid J.*, 39, 520, 1977.)

2.6   Equilibrium Films on Locally Heterogeneous Surfaces: Hydrophilic Surface with Hydrophobic Inclusions

Equilibrium liquid films on heterogeneous solids, that is, solids where solid *hydrophilic* surfaces include *hydrophobic* spots, are under consideration in this section [16,17]. Changes in the profile of an equilibrium film over the spots are expected, and its eventual breakdown is predicted. The combined action of the Derjaguin's pressure and capillary forces should allow predicting the critical width of the hydrophobic spot on such heterogeneous substrates before the film's breakdown. It is shown that this critical size depends on the parameters of the Derjaguin's pressure isotherms of the hydrophilic and hydrophobic parts and the relative vapor pressure in the surrounding medium. Two different model Derjaguin's pressure isotherms for the *hydrophilic* and the *hydrophobic* parts of the substrate are considered below. Complete wetting is assumed for the hydrophilic part and partial wetting for the hydrophobic one or otherwise "less hydrophilic" spot. The equilibrium profiles should be calculated according to the spot size and the value θ_e of the equilibrium contact angle of the hydrophobic spot. It is shown below that the critical width of a hydrophobic spot decreases by an order of magnitude with an increase of the contact angle of the more hydrophobic spot from 10° to 90°.

Let us consider a flat hydrophilic surface covered with a sufficiently thick equilibrium film in the presence of a hydrophobic strip of width $2L$ (Figure 2.24a). The origin of the x-axis corresponds to the middle of the strip.

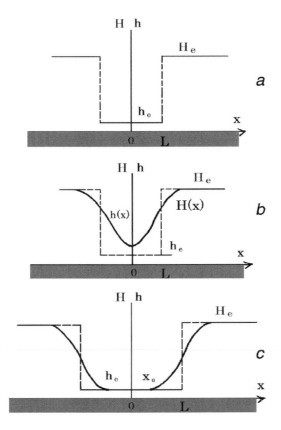

FIGURE 2.24   Schematic diagram of an equilibrium film on a solid surface containing a hydrophobic spot at $-L < x < L$. (a) Reference system: stepwise film profile, H_e and h_e equilibrium film thickness on the corresponding hydrophilic and hydrophobic parts if each part is unbounded; (b) depression formation over a hydrophobic spot when the width L of a hydrophobic part is smaller than a critical value L_c; and (c) rupture of a film if the width of the hydrophobic part $L > L_c$ and formation of a thin film on the hydrophobic part.

As a reference system, the state was chosen when each part of the surface is covered with a corresponding equilibrium film and an interaction between the films is absent. The equilibrium thickness of each of the films, H_e and h_e, is determined by the corresponding Derjaguin's pressure isotherms, $\Pi(H)$ and $\pi(h)$ respectively:

$$\Pi(H_e) = \pi(h_e) = P_e = \frac{R_g T}{v_m} \cdot \ln \frac{p_s}{p}, \tag{2.97}$$

where P_e is the excess pressure of the film as compared with a bulk liquid at the same temperature, T; R_g is the gas constant; v_m is the molar volume of the liquid; and p and p_s are the equilibrium and saturated vapor pressures, respectively.

Between the two idealized states of the films shown in Figure 2.24a a transition zone is formed, the possible shape of which is shown in Figure 2.24b and c. Consideration of the variations of the free energy of the system (similar to Section 2.1) results in the following set of differential equations enabling calculation of the profile of the transition zone for hydrophobic and hydrophilic regions of the surface, respectively:

$$\gamma h'' + \pi(h) = P_e, \quad \text{at} \quad 0 < x < L,$$

$$\gamma H'' + \Pi(H) = P_e, \quad \text{at} \quad x > L, \tag{2.98}$$

where γ is the surface tension of the liquid; h', H' and h'', H'' are the first and the second derivatives of the film thickness over x within the zones $0 < x < L$ and, $x > L$, respectively.

Equation (2.98) are applicable in the case of low-slope profiles, when $(h')^2 \ll 1$ and $(H') \ll 1$. The first terms in the left-hand side of Eq. (2.98) determine the local capillary pressure, and the second terms are the local Derjaguin's pressure. The pressure, P_e, sets the chemical potential of the system and relative vapor pressure according to Eq. (2.97). In the case of flat films, the capillary terms are equal to zero.

Let us now formulate the boundary conditions that are used for solving Eq. (2.98). The first condition characterizes the equilibrium state of the thick hydrophilic film far from the hydrophobic spot:

$$H(x) \to H_e \text{ at } x \to \infty. \tag{2.99}$$

The second condition reflects the symmetry of the system

$$h'(0) = 0. \tag{2.100}$$

To match the profiles at $x = L$, the following two conditions of continuity of the liquid profile should be used:

$$h(L) = H(L); \tag{2.101}$$

$$h'(L) = H'(L). \tag{2.102}$$

As discussed in Section 2.1, not all solutions of Eq. (2.98) with boundary conditions (Eqs. 2.99 through 2.102) describe stable equilibrium profiles; only those which satisfy the Jacobi's condition (C) Section 2.2 do. This condition in application to the problem under discussion has the following form:

$$\gamma(U')' + \pi'(h)U = 0, \text{ at } 0 < x < L;$$

$$\gamma(V')' + \Pi'(H)V = 0, \text{ at } x > L, \tag{2.103}$$

with the following boundary condition:

$$U(0) = 0,$$

and continuity at $x = L$, where $U(x)$ and $V(x)$ are the Jacobi functions. The requirement of the stability of the solution is as follows: The solution of Eq. (2.103) does not vanish anywhere except at the points $x = 0$ and $x = \infty$. The solution of Eq. (2.103) is found below.

After differentiating Eqs. (2.98) over x, we obtain:

$$\left(\gamma h''\right)' + \pi'(h)h' = 0, \text{ at } 0 < x < L;$$

$$\left(\gamma H''\right)' + \Pi'(H)H' = 0, \text{ at } x > L. \tag{2.104}$$

Comparing these equations with Eq. (2.103), we conclude that $U(x) = \text{const} \cdot h'(x)$ and $V(x) = \text{const} \cdot H'(x)$. This means that profiles $h(x)$ and $H(x)$ must behave in a monotonous way inside the corresponding zones as shown in Figure 2.24b and c. Non-monotonous behavior results in the loss of stability.

Further calculations are made using simplified expressions for isotherms of the Derjaguin's pressure (Figure 2.25) consisting of linear parts:

$$\pi(h) = \begin{cases} \infty, & h < h_e, \\ a(h - t_s), & h_e < h < t_s, \\ 0, & h > t_s \end{cases} \tag{2.105}$$

$$\Pi(h) = \begin{cases} A(t_s - H), & 0 < h < t_s, \\ 0, & h > t_s. \end{cases} \tag{2.106}$$

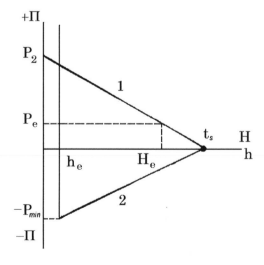

FIGURE 2.25 Simplified forms of the Derjaguin's pressure isotherms of the films formed on the hydrophilic surface, $\Pi(h)$ (curve 1) and on its hydrophobic part, $\pi(h)$ (curve 2). It is adopted (for simplification) that radius of surface forces action of both the Derjaguin's isotherms is equal.

The thickness, t_s, characterizes the range of surface forces action and is selected to be identical for both Derjaguin's pressure isotherms. The equilibrium thickness, h_e, is independent of the vapor pressure in the surrounding air, p, according to adopted isotherm Eq. (2.105). The film thickness H_e depends, according to Eq. (2.106), on the vapor pressure in the surrounding medium. The parameters $A = \dfrac{P_2}{t_s}$ and $a = \dfrac{P_{min}}{t_s - h_e}$ determine the slope of the corresponding isotherm. For the films on hydrophilic surface (curve 1, Figure 2.25), the isotherm ranges from $\Pi = 0$ (when $\dfrac{p}{p_s} = 1$ and $H_e = t_s$) to $\Pi = P_2$, when $H_e = 0$. At pressure P_e (between 0 and P_2), the film thickness equals H_e is determined as $A(t_s - H_e) = P_e$ (Figure 2.25). Such a form of the isotherm corresponds to the case when only repulsion forces $(\Pi > 0)$ act in the film and complete wetting takes place. Equilibrium films on a hydrophobic surface (curve 2 in Figure 2.25) have a smaller thickness h_e, which is adopted to be independent of relative vapor pressure in the range of P_e higher than $-P_{min}$. At $h > h_e$, attractive forces act in the films $(\Pi < 0)$, making the films unstable in this region of thickness. Stable films in the system are considered only undersaturation, that is, at $\dfrac{p}{p_s} < 1$ and $P_e > 0$.

The less hydrophilic (or "hydrophobic") spot may be characterized by the value of the contact angle, θ_e, that a droplet of the liquid forms on this substrate. The contact angle, θ_e, is calculated based on the equation deduced in Section 2.1 (Eq. 2.9) using the Derjaguin's pressure isotherm $\pi(h)$ of the films on the hydrophobic substrate:

$$\cos \theta_e = 1 + \frac{1}{\gamma} P_e h_e + \frac{1}{\gamma} \cdot \int_{h_e}^{t_s} \pi(h)\,dh \approx 1 + \frac{1}{\gamma} \cdot \int_{h_e}^{t_s} \pi(h)\,dh. \tag{2.107}$$

Substituting the model Derjaguin's pressure isotherm, Eq. (2.105), into Eq. (2.107), we obtain:

$$\cos \theta_e = 1 - \frac{a(t_s - h_e)^2}{2\gamma}. \tag{2.108}$$

In the framework of the model adopted, we can characterize the state of the "hydrophobic" surface by the contact angle, θ_e, which is calculated using Eq. (2.108). This equation includes the parameters a, t_s, and h_e of the isotherm as well as the surface tension, γ, of the liquid.

We would like to analyze two possible situations that are schematically shown in Figure 2.24b and c. In the first case, the transition zones between hydrophilic and "hydrophobic" parts overlap, and the film thickness in the middle (at $x = 0$) is higher than the equilibrium film thickness of the "hydrophobic" spot, h_e. In the second case, in the middle of a wider "hydrophobic" spot, the film thickness is equal to the equilibrium value, h_e, and deviation from this thickness starts only at $x > x_0$. In this case according to the transversality condition discussed in Section 2.2 (condition (D), the condition $h'(x_0) = 0$ holds at $x = x_0$.

Let us consider the first case, Figure 2.24b. Equation (2.98), which determines the film profile, takes the following form using the Derjaguin's pressure isotherms given by Eqs. (2.105) and (2.106):

$$\gamma h'' + a(h - t_s) = P_e, \tag{2.109}$$

$$\gamma H'' + A(t_s - H) = P_e. \tag{2.110}$$

The solution of Eq. (2.110) that satisfies the boundary condition (Eq. 2.99) is:

$$H = H_e + C_2 \exp\left[-(x - L)\left(\frac{A}{\gamma}\right)^{1/2}\right], \tag{2.111}$$

where C_2 is an integration constant. Solution of Eq. (2.109) is different for the first and the second cases (Figure 2.24b and c, respectively). In the first case (Figure 2.24b), the solution that satisfies the symmetry condition (Eq. 2.100) has the form:

$$h = t + \frac{P_e}{a} + C_1 \cdot \cos \cdot \left[x \sqrt{\frac{a}{\gamma}} \right]. \tag{2.112}$$

The integration constants, C_1 and C_2, should be determined using boundary conditions Eqs. (2.101) and (2.102) at $x = L$ at the border between the hydrophilic and "hydrophobic" zones. It follows from the Jacobi's conditions that the profile within the "hydrophobic" zone (between $x = -L$ and $x = L$) is stable when the following restriction is satisfied: $L\left(a/\gamma\right)^{1/2} < \pi$. In this case the solutions of Eq. (2.110) vanish only at the point $x = 0$. This restriction gives an estimation of the critical width of the hydrophobic spot when the stability condition is violated. Therefore, this condition gives the critical size of the hydrophobic spot:

$$L < 3.14 \cdot \left(\gamma/a\right)^{1/2}. \tag{2.113}$$

From two boundary conditions (Eqs. 2.101 and 2.102) and Eqs. (2.111) and (2.112), two algebraic equations are obtained, which are used for determining the unknown constants C_1 and C_2:

$$C_2 - \frac{P_e}{A} = \frac{P_e}{a} + C_1 \cos \cdot \beta, \tag{2.114}$$

$$C_2 \left(A/a\right)^{1/2} = C_1 \sin \beta, \tag{2.115}$$

where $\beta = L\left(a/\gamma\right)^{1/2}$.

Considering Eq. (2.113), the constant β can range between *0* and π, and, hence, sin $\beta > 0$. Therefore, parameters C_1 and C_2 have the identical sign. Substitution of the expression for C_2 from Eq. (2.114) into Eq. (2.115) results in

$$C_1 = -P_e \cdot \left[(1/A) + (1/a) \right] / \left[\cos \beta - \left(a/A\right)^{1/2} \sin \beta \right]. \tag{2.116}$$

The profile of the liquid within the hydrophobic zone must have monotonously increasing thickness at x between 0 and L (Figure 2.24b), that means the value of the parameter C_1 must be negative. However, this is possible only when $\cos \beta > \frac{a}{A} \sin \beta$. This leads to a stronger restriction of the critical width of the hydrophobic zone as compared with (2.113): $L < \left(\gamma/a\right)^{1/2} \arctan \left(A/a\right)^{1/2}$. Because of $\arctan \left(A/a\right)^{1/2} < \frac{\pi}{2}$, a more accurate definition of the critical width of the hydrophobic zone when the wetting film ruptures is

$$L = \left(\gamma/a\right)^{1/2} \frac{\pi}{2}. \tag{2.117}$$

Further analysis shows that an even stronger limitation of the critical L values exists, which follows from the condition that film thickness at $x = 0$ cannot be lower than h_e, where h_e is the equilibrium film thickness in the center of the hydrophobic spot at any given pressure, P_e. From Eq. (2.112) for the film profile and from Eq. (2.116) for the parameter C_1 at $x = 0$, it follows the expression for the thickness of the film in the center, h_0

$$h_0 = t - \frac{P_e}{a} - P_e \left[(1/A) + (1/a) \right] / \left[\cos \beta - \left(a/A\right)^{1/2} \sin \beta \right]. \tag{2.118}$$

Hence, this equation and the condition $h_0 = h_e$ finally determine the critical length of the hydrophobic spot, L_c:

$$\cos\beta_c - (a/A)^{1/2}\sin\beta_c = \left[t - \frac{P_e}{a} - h_e \right] / \left(\frac{P_e}{A} + \frac{P_e}{a} \right), \qquad (2.119)$$

where $\beta_c = L_c(\gamma/a)^{1/2}$.

In distinction from the previous approximations (Eqs. 2.113 and 2.117) the critical width of the hydrophobic zone, L_c, according to Eq. (2.119) depends on parameters of both isotherms, a and A, as well as on the relative vapor pressure in the surrounding media that is characterized by the pressure P_e.

The profiles of a transition zone beyond the hydrophobic spot, at $x > L$, are calculated using Eqs. (2.111) and (2.115). At $x = L$, when the two profiles are matched according to boundary conditions (2.101) and (2.102).

Let us find the critical width of the hydrophobic zone using a simplified definition given by Eq. (2.117). In this case, the value of L_c may be calculated dependent on the degree of surface "hydrophobicity" that is characterized by the contact angle. Substituting the expression for $(1-\cos\theta_e)$ from Eq. (2.108) into Eq. (2.117), we conclude

$$L_c = \pi(t_s - h_e)/2\left[2(1-\cos\theta_e) \right]^{1/2} \approx 1.1 \cdot t_s / (1-\cos\theta_e)^{1/2}. \qquad (2.120)$$

Results of calculations of the dependence of $\frac{L_c}{t_s}$ on the contact angle θ_e are shown in Figure 2.26. The results show that the values of the critical width, L_c, decrease with an increasing contact angle and fall sharply at $\theta_e > 30°$. Supposing the range of surface forces action, t_s, is of the order of 10^{-6} cm, we may conclude that the critical width of a "hydrophobic" spot, when film ruptures in the center, decreases from $L_c \approx 10^{-5}$ cm at $\theta_e = 10°$ to $L_c \approx 10^{-6}$ cm at $\theta_e = 180°$. Note, Eq. (2.119) shows that at $\frac{p}{p_s} \to 1$, the wetting film thickness, H_e, approaches its highest value, t_s, and the critical width tends to decrease.

The prediction of the theory is in line with experimental investigations of wetting film stability on heterogeneous methylated glass surfaces [18]. Film rupturing is sensitive to the contact angle values at $\theta_e < 45°$. At larger values of contact angle, the effect is practically independent on the degree of "hydrophobicity" of the "hydrophobic" spot.

In a similar way, the length x_0 (the extension of the part of a hydrophobic spot covered with an equilibrium film with the thickness h_e) may be calculated (Figure 2.24c):

$$x_0 = \left[\beta - \varepsilon(P_e) \right](\gamma/a)^{1/2}, \qquad (2.121)$$

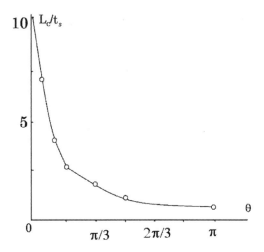

FIGURE 2.26 Calculated according to Eq. (2.24) dependence of the critical width of a hydrophobic spot, L_c, at which the film in the center ruptures, on the value of contact angle θ_e that characterizes the "hydrophobic" spot on the surface.

where the function $\varepsilon\left(P_e\right)$ is determined from the following equation:

$$\cos\varepsilon - \left(a/A\right)^{1/2}\sin\varepsilon = \left[\left(P_e/A\right)+\left(P_e/a\right)\right]/\left[t+\left(P_e/a\right)-h_e\right]. \qquad (2.122)$$

Therefore, the solution obtained opens the possibility of studying the critical size of a hydrophobic spot, L_c, dependent on the parameters of the Derjaguin's pressure isotherms. When the width L of a "hydrophobic" spot is smaller than L_c, a depression cavity is formed over the spot (Figure 2.24b) and the equilibrium liquid profile, $h(x)$, which describes a profile of a relatively thick film even over the "hydrophobic" spot. However, at $L \geq L_c$ the thick wetting film ruptures and part of the "hydrophobic" surface becomes almost "dry." A further increase in the dimension of the hydrophobic spot leads to expansion of the "dry" part of the surface (Figure 2.24c).

In view of the foregoing discussion, we can say that the presence of more hydrophobic spots on smooth hydrophilic substrates results in the formation of depressions, where the film thickness is lower than the film thickness on the rest of the substrate. Hence, the presence of more "hydrophobic" spots results in a lower mean thickness of the film relative to the thickness on a uniform hydrophilic substrate. Thus, the surface roughness and the presence of "hydrophobic" spots on the surface influence the mean thickness of the equilibrium film in opposite ways: The presence of roughness results in an increase of the mean film thickness (Section 2.5), whereas the presence of hydrophobic inclusions leads to the contrary.

2.7 Equilibrium of Droplets on a Deformable Substrate: Influence of the Derjaguin's Pressure

In this section the equilibrium of liquid droplets on soft deformable substrates is investigated. The Derjaguin's pressure action in the vicinity of the apparent three-phase contact line is taken into account. The main requirement of the equilibrium is that, at equilibrium, the excess free energy of the system should be at minimum. Based on consideration of the excess free energy, a system of two interconnected differential equations is deduced in this section. This system of interconnected differential equations describes both the droplet profile (everywhere including the transition zone) and the profile of the deformed substrate. It is shown that the combined Derjaguin's, capillary pressure action and elasticity of the deformable substrate determine both the droplet profile and the substrate deformation. A simplified linear Derjaguin's pressure isotherm and simple Winkler's model to account for the substrate deformation are used, allowing for the deduction of analytical solutions for both the liquid profile and the substrate deformation. The apparent equilibrium contact angle of the liquid droplet with the deformable substrate is calculated and its dependency on the system parameters is investigated as in [19].

The important problem, which until now has been completely ignored in the literature, is as follows: Does the solution of the deduced system of differential equations really correspond to the minimum of the excess free energy? Now, step by step, the first variation of the excess free energy results in the interconnected system of two differential equations (2.137) and (2.138). The positive second variation of the excess free energy gives the necessary condition of the deduced profiles; the first and second variations simply confirm that if there are equilibrium profiles of both the droplet shape and the deformable substrates, then these profiles should be solutions of the deduced system of differential equations. However, a solution of the deduced system of differential equations does not necessarily correspond to the equilibrium. In mathematical terms, zero first variation and the positive second variation are necessary, but not sufficient, conditions of equilibrium. Hence, if we obtained the solution of the mentioned system of interconnected differential equation, then the mentioned two conditions, with first and second variations, do not guarantee that the deduced solution corresponds to equilibrium. For this purpose, it is necessary to check one extra condition, which is the Jacobi's sufficient condition, to determine if it is really satisfied for the deduced solution. In mathematical terms, only Jacobi's condition gives a sufficient condition of equilibrium. It is shown below that the solution obtained in this section does satisfy the Jacobi's condition; that is, it provides the minimum of excess free energy [20]. In the future, any solution of the problem of equilibrium of droplets on deformable substrates should be checked to see that the obtained solution complies with Jacobi's condition.

Introduction

The equilibrium of a liquid droplet on a solid substrate is frequently described based on Young's equation [21]. It was explained in Chapter 1 that this equation is in contradiction with thermodynamics. However, there is also another problem with Young's equation: This simplified equation involves the balance of the horizontal forces, but it leaves the vertical force unbalanced. This is possible in the case of a rigid substrate, but it should be reconsidered in the case of deformable substrates. It was shown in [19] that the Derjaguin's pressure action in the vicinity of the apparent three-phase contact line results in a deformation of a deformable solid substrate and accounts for the total balance of all forces.

Note that the direct application of Young's equation leads to a deformation singularity at the three-phase contact line, that is, the substrate deformation goes to infinity [22–26]: These investigations revealed that all the equilibrium properties (e.g., contact angle, droplet radius) of the system under consideration rely upon the selected artificial length parameter that determines a width of zone near the contact line where surface tension is applied. It is important to note that neither of these solutions was checked to see if they satisfied the Jacobi's sufficient condition. Until this condition is checked, it is impossible to claim that any of the deduced solutions really provide the minimum of excess free energy, that is, that they correspond to the equilibrium.

It is shown below that the Derjaguin's pressure, which is a real physical phenomenon and which is frequently ignored in the literature dealing with droplets on deformable substrates, determines the deformation of deformable surfaces and the shape of the deposited droplets [19]. It is shown below that the combined action of the Derjaguin's pressure, capillarity and elasticity of the deformable substrate determines an apparent contact angle on deformable substrates, and it is verified that the obtained solution really corresponds to the equilibrium using the Jacobi's sufficient condition.

The problem of equilibrium of a droplet on a deformable substrate has gained a lot of interest over last decades. A number of experimental studies [27–36] were conducted to investigate deformation of soft substrates near the apparent contact line by deposited liquid droplets. The first attempt to consider the Derjaguin's pressure in the vicinity of the three-phase contact line was undertaken in [37] and which was followed by publications [38,39]. However, further theoretical understanding of the problem is required because the real physical phenomenon, the Derjaguin's pressure action in the vicinity of the three-phase contact line, has been frequently ignored: this resulted in an artificial singularity at the three-phase contact line. Below a mathematical model is presented that incorporates the effect of both capillary and Derjaguin's pressure on substrate deformation.

Derjaguin's Pressure and Deformation of Soft Substrates

In the case of partial wetting, that is, if the equilibrium contact angle, $0 < \theta_e < /2$, the shape of the transition zone between the bulk of the liquid droplet and the thin film on a rigid substrate can be expressed via the Derjaguin's pressure isotherm, (h) (see Chapters 1 and 2). The equilibrium excess pressure, P_e, inside the droplet can be expressed according to Kelvin's equation and the equilibrium contact angle of a two-dimensional droplet can be expressed using the Derjaguin's pressure isotherm as follows (see Chapter 1):

$$\cos\theta_{e,2D} = 1 + \frac{\Pi(h_e)h_e}{\gamma} + \frac{1}{\gamma}\int\limits_{h_e}^{\infty}\Pi(h)dh. \tag{2.123}$$

The S-shaped Derjaguin's pressure isotherm, $\Pi(h)$, that is, the case of partial wetting, is described by Eq. (2.123). As a first step, the transition zone between a two-dimensional droplet on a deformable substrate and a flat film having, $0 < \theta_e < /2$, is examined qualitatively following [37]. The liquid droplet forms a wedge changing to flat equilibrium film of thickness, h_e, away from the droplet (Figure 2.27).

The origin is taken at $x = 0$, which is a point on the profile lying beyond the influence of surface forces (Figure 2.27). The profile of the transition zone, $h(x)$, can be calculated (if we neglect the soft substrate

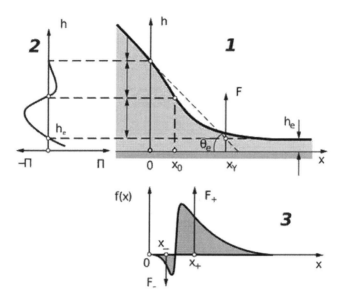

FIGURE 2.27 (1) Profile of transition zone, $h(x)$ between bulk liquid and flat wetting film; (2) the S-shaped Derjaguin's pressure isotherm, $\Pi(h)$; and (3) profile of normal pressures acting on substrate; $x_{\pm}$ are the positions where the vertical forces are exerted. The second min on the Derjaguin's pressure isotherm is omitted to simplify the picture.

deformation in the zeroth approximation) according to the following equation, which includes the influence of both capillary and Derjaguin's pressure (Section 2.3):

$$\frac{\gamma h''}{\left[1+\left(h'\right)^2\right]^{3/2}} + \Pi\left(h\right) = P_e, \tag{2.124}$$

where $h' = dh/dx$; $h'' = d^2h/dx^2$; and P_e is the excess pressure inside the droplet; γ is the liquid–air interfacial tension. In the region of the flat equilibrium film, $h'' = 0$ and $\Pi\left(h_e\right) = P_e$. In the bulk of the liquid, beyond the influence of the surface forces, $\Pi = 0$ and $P_e = -\gamma/\Re$, where $\Re$ is the radius of the curvature of the droplet. In the case of a planar wedge (Figure 2.27), $\Re = \infty$ and $P_e = 0$. Therefore, it can be concluded from Eq. (2.124):

$$\Pi\left(h\right) = -\frac{\gamma h''}{\left[1+\left(h'\right)^2\right]^{3/2}}. \tag{2.125}$$

The resultant force on the substrate is given by the following equation:

$$F = \int_0^\infty \Pi(h)dx. \tag{2.126}$$

Substituting the Derjaguin's pressure isotherm $\Pi(h)$ according to Eq. (2.125) into Eq. (2.126) results in:

$$F = -\gamma \int_0^\infty \left\{ \frac{h''}{\left[1+\left(h'\right)^2\right]^{\frac{3}{2}}} \right\} dx = \gamma \frac{h'(0)}{\sqrt{\left[1+\left(h'(0)\right)^2\right]}} = \frac{\gamma \tan\theta_e}{\sqrt{\left[1+\tan^2\theta_e\right]}}, \tag{2.127}$$

$$= \gamma \sin\theta_e.$$

Boundary conditions used in the expression Eq. (2.127) are, $h'(\infty) = 0$ and $h'(0) = \tan\theta_e$. This shows that the integration performed over the local values leads to the same expression as the vertical component of the surface tension from the Young's equation. In contrast to Young's derivation, the force is not exerted at a specific point; instead, it is distributed over the region where the Derjaguin's pressure acts, that is, the transition zone. It is why we refer to this zone as the "apparent contact line." Based on this preliminary consideration, a new mathematical model is derived below in this section.

Mathematical Model and Derivation

In this section a mathematical model is deduced for a liquid droplet on a deformable substrate. A simple Winkler's model for the deformable solid deformation is used below. According to the Winkler's model, there is a linear relationship between the local deformation and the applied local stress [40,41].

Deformation in the deformable substrate is local and is directly proportional to the applied pressure, P. According to the Winkler's model,

$$h_s = -KP, \tag{2.128}$$

where K is the elasticity coefficient and h_s is the local deformation of the substrate due to the presence of the applied pressure, P, from the droplet above (Figure 2.28).

Let P_{air} be the pressure in the ambient air. Under the action of the pressure from the ambient air, the soft solid deformation is:

$$h_{se} = -KP_{air}. \tag{2.129}$$

The deformed substrate is covered by an equilibrium liquid thin film, which is calculated according to combination of the well-known Kelvin's equation and Derjaguin's pressure isotherm (see Chapter 1):

$$\Pi(h_e) = P_e = \frac{R_g T}{v_m} \ln \frac{p_s}{p}, \tag{2.130}$$

where, v_m is the molar volume of the liquid; T is the temperature in K; R_g is the gas constant; and p is the vapor pressure, p which is higher than the saturated pressure, p_s. Recall that a droplet can be at the equilibrium with oversaturated vapor only according to Kelvin's equation.

The excess free energy of the equilibrium thin film on the deformed solid per unit area is given by

$$\frac{\Phi_{e,film}}{S_{film}} = \gamma + \gamma_s + P_e h_e + \frac{h_{se}^2}{2K} + \int_{h_e}^{\infty} \Pi(h)\,dh, \tag{2.131}$$

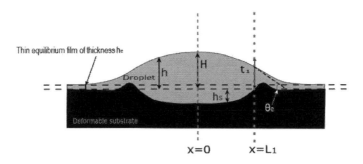

FIGURE 2.28 Schematic diagram of the liquid droplet on a deformable substrate.

where, $P_e = P_{air} - P_{liquid}$, γ and γ_s are liquid–vapor and solid–liquid interfacial tensions. The free energy (2.131) should be subtracted from the free energy of the droplet on the deformable substrate; otherwise the excess free energy of the system is infinite. Hence, the excess free energy of the droplet on a deformable solid substrate is as follows (Figure 2.28):

$$\Phi - \Phi_{e,\,film} = \gamma \Delta S + \gamma_s \Delta S_s + \Delta V + F_{surface\ forces} + F_{deformation},\tag{2.132}$$

where Δ means "as compared to a flat equilibrium film."

Eq. (2.132) can be rewritten as

$$\Phi - \Phi_{e,\,film} = \int_0^\infty f\left(h,h',h_s,h_s'\right)dx,\tag{2.133}$$

where

$$f\left(h,h',h_s,h_s'\right) = \begin{bmatrix} \gamma\sqrt{1+h'^2(x)} - \gamma + \gamma_s\sqrt{1+h_s'^2(x)} - \gamma_s + \\ P_e\left(h-h_s\right) - P_e h_e + \dfrac{h_s^2}{2K} - \dfrac{h_{se}^2}{2K} \\ + \displaystyle\int_{h-h_s}^\infty \Pi(h)dh - \int_{h_e}^\infty \Pi(h)dh \end{bmatrix}.\tag{2.134}$$

In Eqs. (2.133) and (2.134), x is the tangential coordinate; and γ_s the solid–liquid interfacial tension. The expression under the integral in Eq. (2.133) tends to zero as the distance, x, tends to infinity.

Under equilibrium conditions, the excess free energy (2.133) should reach the minimum value. To satisfy this condition, the first variation of excess free energy (2.133) should be zero, which results in two Euler equations for the droplet and deformable substrate profiles:

$$\frac{d}{dx}\left(\frac{\partial f}{\partial h'}\right) - \frac{\partial f}{\partial h} = 0,\tag{2.135}$$

$$\frac{d}{dx}\left(\frac{\partial f}{\partial h_s'}\right) - \frac{\partial f}{\partial h_s} = 0.\tag{2.136}$$

Substitution of the expression for f from Eq. (2.134) into Eqs. (2.135) and (2.136) results in the following system of second-order differential equations:

$$\frac{\gamma h''}{\left(1+h'^2\right)^{3/2}} + \Pi\left(h-h_s\right) = P_e,\tag{2.137}$$

$$\frac{\gamma_s h_s''}{\left(1+h_s'^2\right)^{3/2}} - \Pi\left(h-h_s\right) - \frac{h_s}{K} = -P_e.\tag{2.138}$$

Equations (2.137) and (2.138) form a system of two differential equations for two unknown profiles: the liquid droplet, $h(x)$, and deformed solid substrate, $h_s(x)$. It is easy to check that the second variation of the excess free energy (2.133) is positive [42], that is, the necessary condition of minimum is satisfied.

The low-slope approximation case, $h'^2 \ll 1$, $h_s'^2 \ll 1$, which is valid at the small contact angles, is considered below.

In the case of a low-slope approximation, $h'^2 \ll 1, h_s'^2 \ll 1$, the profile of the droplet, $h(x)$, and profile of the substrate, $h_s(x)$, satisfy the following set of second-order ordinary differential equations:

$$\gamma h'' + \Pi(h - h_s) = P_e, \tag{2.139}$$

$$\gamma_s h_s'' - \Pi(h - h_s) - \frac{h_s}{K} = -P_e. \tag{2.140}$$

Equation (2.139) is different from the usual capillary equation for the droplet profile on a rigid substrate, because now the Derjaguin's pressure term depends on the profile of the deformable substrate, $h_s(x)$, which is determined according to Eq. (2.140). Hence, Eqs. (2.139) and (2.140) are coupled and in general can be solved only numerically; however, the problem can be simplified even further to obtain an analytical solution. For this purpose, a simplified Derjaguin's pressure isotherm [linear function of h (Figure 2.29)] is adopted:

$$\Pi(h) = \begin{cases} P_1 - ah & \text{at } h \leq t_1 \\ 0 & \text{at } h > t_1 \end{cases}. \tag{2.141}$$

P_1 and t_0 are defined in Figure 2.29; and t_1 is the range of surface forces action. The corresponding length from the origin to the point of height t_1 is L_1, see Figure 2.28. The slope a of the $\Pi(h)$ dependency is given by

$$a = \frac{P_1 - P_e}{h_e}. \tag{2.142}$$

The selected linear dependency of the Derjaguin's pressure isotherm $\Pi(h)$ on h according to Eq. (2.141) was successfully used earlier [43] and still captures the essential properties of the Despite considerable simplification of the Derjaguin's pressure isotherm, (i) it satisfies the stability condition, $\Pi'(h) < 0$ when $h < t_1$; (ii) the influence of surface forces is short ranged and radius of its action is defined by t_1; and (iii) it corresponds to the partial wetting case at the proper choice of the Derjaguin's pressure parameters. Although a similar type of isotherm has been used in past to determine the shape of the transition zone [43], it has

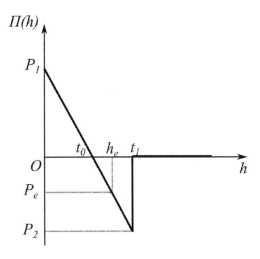

FIGURE 2.29 Simplified Derjaguin's pressure isotherm adopted below for calculations.

not been used for droplets on deformable substrates. The equilibrium of droplets on deformable substrates, including the Derjaguin's pressure action, which corresponds to real systems, is to be considered in the future.

In the case of an adopted simplified isotherm of the Derjaguin's pressure (2.141), the governing Eqs. (2.139) and (2.140) can be rewritten as:

$$\gamma h'' + a\left(t_0 - h + h_s\right) = P_e,$$
(2.143)

and

$$\gamma_s h_s'' - a\left(t_0 - h + h_s\right) - \frac{h_s}{K} = -P_e,$$
(2.144)

with the boundary conditions:

$$h'(0) = h_s'(0) = 0,$$
(2.145)

$$h(x) \to h_e, h_s(x) \to 0, at\ x \to \infty.$$
(2.146)

Spherical Region: $h - h_s > t_1$

For the bulk of the liquid droplet, that is, the spherical region, $h - h_s > t_1$,

$$\gamma h_{sp}'' = P_e,$$
(2.147)

$$\gamma_s h_{s,sp}'' - \frac{h_{s,sp}}{K} = -P_e,$$
(2.148)

where the extra subscript sp denotes the spherical region of the droplet. Using the boundary conditions (2.145) the solution of Eqs. (2.147) and (2.148) is

$$h_{sp}(x) = \frac{P_e x^2}{2\gamma} + H_e,$$
(2.149)

and

$$h_{s,sp}(x) = KP_e + C_1\left(e^{\lambda_s x} + e^{-\lambda_s x}\right),$$
(2.150)

where $\lambda_s = \sqrt{\dfrac{1}{K \cdot \gamma_s}}$, H_e is the apex of the droplet and C_1 is an integration constant.

Using the boundary condition, $h_{sp}(L_1) - h_{s,sp}(L_1) = t_1$, results in

$$\frac{P_e L_1^2}{2\gamma} + H_e - KP_e - C_1\left(e^{\lambda_s L_1} + e^{-\lambda_s L_1}\right) = t_1.$$
(2.151)

Transitional Region: $h - h_s \leq t_1$

The equation for the droplet's profile from Eq. (2.143) is

$$h'' - \frac{a\left(h - h_s\right)}{\gamma} = \frac{P_e - at_0}{\gamma}.$$
(2.152)

The substrate deformation profile can be simplified from Eq. (2.144)

$$h_s'' - h_s\left(\frac{1 + aK}{\gamma_s K}\right) + \frac{ah}{\gamma_s} = -\left(\frac{P_e - at_0}{\gamma_s}\right).$$
(2.153)

The unknown solution is presented in the following form:

$$y(x) = h(x) + \alpha h_s(x), \qquad (2.154)$$

where $y(x)$ is a new unknown function that is a linear combination of both profiles, h_s, the deformed substrate, and h, the droplet; and α is a constant to be determined.

Let us multiply Eq. (2.153) by α and add the resulting equation to Eq. (2.152), using (2.154) the resulting equation is

$$y'' - a\left(\frac{1}{\gamma} - \frac{\alpha}{\gamma_s}\right)y + h_s\left[a\alpha\left(\frac{1}{\gamma} - \frac{\alpha}{\gamma_s}\right) + \frac{a}{\gamma} - \frac{\alpha}{\gamma_s}\left(\frac{1+aK}{K}\right)\right] = (P_e - at_0)\left(\frac{1}{\gamma} - \frac{\alpha}{\gamma_s}\right). \qquad (2.155)$$

The aim is to get an equation that includes only one unknown function, $y(x)$. Hence, the constants in square brackets must be set zero, that is,

$$\alpha^2 + \left(-\frac{\gamma_s}{\gamma} + \frac{(1+aK)}{aK}\right)\alpha - \frac{\gamma_s}{\gamma} = 0, \qquad (2.156)$$

which is a quadratic equation in α, which can be determined from Eq. (2.145) and has two roots: where $\alpha_1 > 0$ and $\alpha_2 < 0$. Substitution of these values into Eq. (2.155) results in two new equations for two unknown functions, y_i, where $i = 1,2$:

$$y_i'' - a\left(\frac{1}{\gamma} - \frac{\alpha_i}{\gamma_s}\right)y_i = (P_e - at_0)\left(\frac{1}{\gamma} - \frac{\alpha_i}{\gamma_s}\right), i = 1,2. \qquad (2.157)$$

Let us introduce for convenience a new tangential coordinate $z = x - L_1$ in the transition region. The solution of Eq. (2.157) results in

$$y_1 = h_e + C_2 e^{z\lambda_1} + C_3 e^{-z\lambda_1}$$

and

$$y_2 = h_e + C_4 e^{z\lambda_2} + C_5 e^{-z\lambda_2}.$$

Because of the boundary conditions (2.146), two integration constants, $C_2 = C_4 = 0$, and these equations become

$$y_1 = h_e + C_3 e^{-z\lambda_1} \qquad (2.158)$$

$$y_2 = h_e + C_5 e^{-z\lambda_2} \qquad (2.159)$$

where $h_e = \frac{-(P_e - at_0)}{a}$, $\lambda_1 = \sqrt{a\left(\frac{1}{\gamma} - \frac{\alpha_1}{\gamma_s}\right)}$ and $\lambda_2 = \sqrt{a\left(\frac{1}{\gamma} - \frac{\alpha_2}{\gamma_s}\right)}$, $y_1 = h + \alpha_1 h_s$ and $y_2 = h + \alpha_2 h_s$. At the position, $x = L_1$, profiles of both the droplet and the deformable substrate and their derivatives should be continuous. From Eq. (2.158) at $z = 0$ and $x = L_1$ for the spherical region according to Eqs. (2.149) and (2.150)

$$y_1(0) = h_e + C_3$$

$$= h_{sp}(L_1) + \alpha_1 h_{s,sp}(L_1) \tag{2.160}$$

$$= \frac{P_e L_1^2}{2\gamma} + H_e + \alpha_1 \left[KP_e + C_1 \left(e^{\lambda_s L_1} + e^{-\lambda_s L_1} \right) \right]$$

The similar procedure for derivatives results in

$$y_1'(0) = -\lambda_1 C_3$$

$$= h_{sp}'(L_1) + \alpha_1 h_{s,sp}'(L_1) \tag{2.161}$$

$$= \frac{P_e L_1}{\gamma} + \alpha_1 \left[\lambda_s C_1 \left(e^{\lambda_s L_1} - e^{-\lambda_s L_1} \right) \right]$$

From Eqs. (2.160) and (2.161),

$$C_3 = \frac{P_e L_1^2}{2\gamma} + H_e + \alpha_1 \left[KP_e + C_1 \left(e^{\lambda_s L_1} + e^{-\lambda_s L_1} \right) \right] - h_e \tag{2.162}$$

$$-C_3 = \frac{1}{\lambda_1} \left\{ \frac{P_e L_1}{\gamma} + \alpha_1 \left[\lambda_s C_1 \left(e^{\lambda_s L_1} - e^{-\lambda_s L_1} \right) \right] \right\} \tag{2.163}$$

The same procedure for y_2 and its derivative results in

$$y_2(0) = h_e + C_5$$

$$= h_{sp}(L_1) + \alpha_2 h_{s,sp}(L_1)$$

$$= \frac{P_e L_1^2}{2\gamma} + H_e + \alpha_2 \left[KP_e + C_1 \left(e^{\lambda_s L_1} + e^{-\lambda_s L_1} \right) \right],$$

or

$$C_5 = \frac{P_e L_1^2}{2\gamma} + H_e + \alpha_2 \left[KP_e + C_1 \left(e^{\lambda_s L_1} + e^{-\lambda_s L_1} \right) \right] - h_e. \tag{2.164}$$

The same procedure for derivatives results in

$$y_{12}'(0) = -\lambda_2 C_5 = h_{sp}'(L_1) + \alpha_2 h_{s,sp}'(L_1) = \frac{P_e L_1}{\gamma} + \alpha_2 \left[\lambda_s C_1 \left(e^{\lambda_s L_1} - e^{-\lambda_s L_1} \right) \right],$$

which can be rewritten as

$$-C_5 = \frac{1}{\lambda_2} \left\{ \frac{P_e L_1}{\gamma} + \alpha_2 \left[\lambda_s C_1 \left(e^{\lambda_s L_1} - e^{-\lambda_s L_1} \right) \right] \right\}. \tag{2.165}$$

Five equations (Eqs. 2.151, 2.162 through 2.165) include five unknown constants (C_1, C_3, C_5, H_e and L_1) to be determined. All these equations are linear in respect to four unknowns, C_1, C_3, C_5, and H_e. The solution of these linear equations results in the following equation of only one unknown, L_1:

$$
\left[-\frac{P_e L_1^2}{2\gamma} - \alpha_1 \cdot \frac{P_e L_1^2}{2\gamma} + \alpha_1 \cdot t_1 + h_e - \frac{P_e L_1}{\gamma \cdot \lambda_1} - \frac{\alpha_1 \cdot \lambda_s \cdot M}{\lambda_1} \cdot \frac{P_e L_1^2}{2\gamma} + \frac{\alpha_1 \cdot \lambda_s \cdot M \cdot K \cdot P_e}{\lambda_1} \right.
$$
$$
\left. + \frac{\alpha_1 \cdot \lambda_s \cdot M \cdot t_1}{\lambda_1} \right] \cdot \frac{1}{\left(1 + \alpha_1 + \alpha_1 \cdot \lambda_s \cdot \dfrac{M}{\lambda_1} \right)} - \left[-\frac{P_e L_1^2}{2\gamma} - \alpha_2 \cdot \frac{P_e L_1^2}{2\gamma} + \alpha_2 \cdot t_1 + h_e - \frac{P_e L_1}{\gamma \cdot \lambda_2} \right.
$$
$$
\left. - \frac{\alpha_2 \cdot \lambda_s \cdot M}{\lambda_2} \cdot \frac{P_e L_1^2}{2\gamma} + \frac{\alpha_2 \cdot \lambda_s \cdot M \cdot K \cdot P_e}{\lambda_2} + \frac{\alpha_2 \cdot \lambda_s \cdot M \cdot t_1}{\lambda_2} \right] \cdot \frac{1}{\left(1 + \alpha_2 + \alpha_2 \cdot \lambda_s \cdot \dfrac{M}{\lambda_2} \right)} = 0, \tag{2.166}
$$

where

$$
M = \frac{\left(e^{\lambda_s L_1} - e^{-\lambda_s L_1} \right)}{\left(e^{\lambda_s L_1} + e^{-\lambda_s L_1} \right)}. \tag{2.167}
$$

L_1(cm) was calculated from Eq. (2.166) in [19,20]. It was shown in [19] that the transition region and deformation of the soft substrate are determined by three different length scales (Figure 2.30) [19], and all of these length scales are determined by the elasticity of the substrates and the parameters of the Derjaguin's pressure isotherm [19].

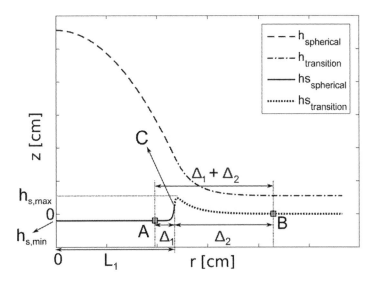

FIGURE 2.30 Schematic diagram of droplet on a deformable substrate. The macroscopic contact angle is determined as an intersection of the continuation of the spherical part with the plane of initial substrate.

Equilibrium Contact Angle

The intersection of the spherical profile with the initial non-deformable substrate defines the apparent three-phase contact line with a macroscopic equilibrium contact angle (Figure 2.28). If $x = L$ is the length of the spherical region, then

$$h_{sp}(L) = 0$$

and

$$h'_{sp}(L) = -\tan\theta_e \cong -\theta_e$$

From the spherical profile (2.149)

$$L = \sqrt{\frac{-2\cdot\gamma H_e}{P_e}}$$

Combining these equations results in the following expression for equilibrium contact angles θ_e:

$$\theta_e = \sqrt{\frac{2\cdot H_e(-P_e)}{\gamma}}$$

$$(2.168)$$

It is important to note that θ_e is determined by both P_e, which is an independent variable, and H_e, which depends on the elasticity coefficient $K(cm^3/dyn)$.

The dependency between equilibrium contact angle, θ_e [°], and the elasticity coefficient, $K(cm^3/dyn)$ is presented in Figure 2.31.

As shown in Figure 2.31, equilibrium contact angle decreases with the elasticity coefficient $K(cm^3/dyn)$ as expected, which is in agreement with [19].

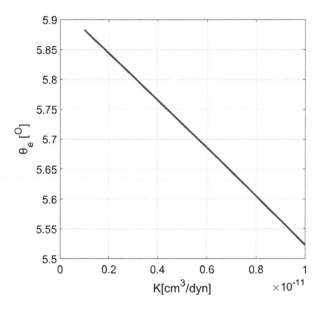

FIGURE 2.31 Apparent contact angle as a function of elasticity of the substrate.

Jacobi's Condition

According to [42], the four Jacobi equations for four unknown functions, $u_{ji}(x)$, where $i, j = 1,2$ are as follows. At $0 < x < L_1$, which is in the area where the Derjaguin's pressure does not act (Figure 2.28),

$$\gamma u_{1i}'' = 0 \tag{2.169}$$

$$\gamma u_{2i}'' - \frac{u_{2i}}{2K} = 0 \tag{2.170}$$

where $i = 1,2$. At $x > L_1$, the same functions satisfy the following equations [where the Derjaguin's pressure is considered according to Eq. (2.141)]:

$$\gamma \cdot u_{11}' - (a \cdot u_{11} - a \cdot u_{22}) = 0, \tag{2.171}$$

$$\gamma_s \cdot u_{22}'' - \left(-a \cdot u_{11} + a \cdot + \frac{u_{22}}{2K} \right) = 0, \tag{2.172}$$

Linear differential equations for unknown functions u_{12} and u_{21} are omitted. Note that Eqs. (2.171) and (2.172) are linear differential equations with respect to all unknown Jacobi's functions, $u_{ji}(x)$, where $i, j = 1,2$ and they should be subjected to the following initial conditions [42]:

$$\left\{ \begin{array}{c} u_{11}(0) = u_{12}(0) = u_{21}(0) = u_{22}(0) = 0 \\ u_{11}'(0) = 1 \,, \; u_{22}'(0) = 1 \\ u_{12}'(0) = u_{21}'(0) = 0 \end{array} \right\} \tag{2.173}$$

Because of the linearity of Eqs. (2.169) and (2.172) and zero initial conditions for functions $u_{12}(x)$ and $u_{21}(x)$, these two functions are equal to zero identically. Hence, the only two unknown nonzero functions are $u_{11}(x)$ and $u_{22}(x)$. The procedure for the solution of Eqs. (2.171) and (2.172) is similar to that for Eqs. (2.152) and (2.153).

The Jacobi's condition is as follows [42]: $D = \begin{vmatrix} u_{11} & u_{12} \\ u_{21} & u_{22} \end{vmatrix}$ should not change sign for all $x > 0$. Bearing in mind that $u_{12}(x) = u_{21}(x) = 0$ and considering that D should be positive, the Jacobi's condition can be rewritten as

$$D(x) = \begin{vmatrix} u_{11} & 0 \\ 0 & u_{22} \end{vmatrix} = u_{11}u_{22} > 0, \text{ at } x > 0. \tag{2.174}$$

The solution of Eqs. (2.169) and (2.172) for $u_{11}(x)$, $u_{22}(x)$ in two regions $0 < x < L_1$ [the spherical part of the droplet (Figure 2.28)] and $x > L_1$ (the transition region, presented in [20]). The solution is as follows:
Spherical droplet: $0 < x < L_1$, $h - h_s > t_1$ (Figure 2.28):

$$u_{11}(x) = x; \; u_{22}(x) = \frac{1}{\lambda_s} \cdot \sinh(\lambda_s x), \tag{2.175}$$

where $\lambda_s = \sqrt{\frac{1}{K \cdot \gamma_s}}$, which means that $D = u_{11} \cdot u_{22} = x \frac{1}{\lambda_s} \cdot \sinh(\lambda_s \cdot x) > 0$ at $0 < x < L_1$.

Inside the transition zone, $x > L_1$, $h < t_1$ (Figure 2.28),

$$
\begin{aligned}
u_{11}(z) = \frac{1}{(\beta_2 - \beta_1)} \cdot \Bigg\{ & \left(\frac{\beta_2 \cdot \beta_1 \cdot \sinh(\lambda_s \cdot L_1) \cdot \cosh(\lambda_1 \cdot z)}{\lambda_s} \right) \\
& + \left(\beta_2 \cdot L_1 \cdot \cosh(\lambda_1 \cdot z) \right) \\
& + \left(\frac{\beta_2 \cdot \beta_1 \cdot \cosh(\lambda_s \cdot L_1) \cdot \sinh(\lambda_1 \cdot z)}{\lambda_s} \right) \\
& + \left(\frac{\beta_2 \cdot \sinh(\lambda_1 \cdot z)}{\lambda_1} \right) - \left(\beta_1 \cdot L_1 \cdot \cosh(\lambda_2 \cdot z) \right) \\
& - \left(\frac{\beta_1 \cdot \sinh(\lambda_2 \cdot z)}{\lambda_2} \right) \\
& - \left(\frac{\beta_1 \cdot \beta_2 \cdot \cosh(\lambda_s \cdot L_1) \cdot \sinh(\lambda_2 \cdot z)}{\lambda_2} \right) \\
& - \left(\frac{\beta_1 \cdot \beta_2 \cdot \sinh(\lambda_s \cdot L_1) \cdot \cosh(\lambda_2 \cdot z)}{\lambda_s} \right) \Bigg\}
\end{aligned}
\tag{2.176}
$$

$$
\begin{aligned}
u_{22}(z) = \frac{1}{(\beta_1 - \beta_2)} \cdot \Bigg\{ & \left(\frac{\beta_1 \cdot \sinh(\lambda_s \cdot L_1) \cdot \cosh(\lambda_1 \cdot z)}{\lambda_s} \right) \\
& + \left(L_1 \cdot \cos h(\lambda_1 \cdot z) \right) \\
& + \left(\frac{\beta_1 \cdot \cos h(\lambda_s \cdot L_1) \cdot \sinh(\lambda_1 \cdot z)}{\lambda_1} \right) + \left(\frac{\sin h(\lambda_1 \cdot z)}{\lambda_1} \right) \\
& - \left(\frac{\beta_2 \cdot \sin h(\lambda_s \cdot L_1) \cdot \cos h(\lambda_2 \cdot z)}{\lambda_s} \right) \\
& - \left(L_1 \cdot \cosh(\lambda_2 \cdot z) \right) \\
& - \left(\frac{\beta_2 \cdot \cosh(\lambda_s \cdot L_1) \cdot \sinh(\lambda_2 \cdot z)}{\lambda_2} \right) - \left(\frac{\sinh(\lambda_2 \cdot z)}{\lambda_2} \right) \Bigg\}
\end{aligned}
\tag{2.177}
$$

where $z = x - L_1$, $\varepsilon = \dfrac{\gamma_s}{\gamma}$,

$$
\beta_{1,2} = \frac{-\left(-\varepsilon + 1 + \dfrac{1}{a \cdot K}\right) \mp \sqrt{\left(-\varepsilon + 1 + \dfrac{1}{a \cdot K}\right)^2 + 4 \cdot \varepsilon}}{2}
$$

$$\beta_1 > 0 \text{ and } \beta_2 < 0; \lambda_1 = \sqrt{\frac{a}{\gamma_s}(\varepsilon - \beta_1)}, \lambda_2 = \sqrt{\frac{a}{\gamma_s}(\varepsilon - \beta_2)}.$$

It was shown in [20] that both $u_{11}(z)$ and $u_{22}(z)$ are positive for the region $x > L_1$. Hence, $D = u_{11}(z) \cdot u_{22}(z) > 0$ at $x > L_1$. This means that D is positive at all $x > 0$. That is, the Jacobi's condition is satisfied, and the solutions obtained for both the liquid profile and the deformable substrate profile are really equilibrium ones. These results prove that for the obtained solution for a droplet deposited on a deformable substrate, the excess free energy is at its minimum value and deduced solutions for the droplet profile and profile of the deformed substrate correspond to the equilibrium.

2.8 Deformation of Fluid Particles in the Contact Zone

The hydrostatic pressure in thin liquid films intervening between two drops/bubbles differs from the pressure inside the drops/bubbles. This difference is caused by the action of both capillary and surface forces. The manifestation of the surface forces action is the Derjaguin's pressure, which has a special S-shaped form in the case of partial wetting (aqueous thin films and thin films of aqueous electrolyte and surfactant solutions). The Derjaguin's pressure acts solely in flat thin liquid films and determines their thickness. If the film surface is curved, then both the Derjaguin's and the capillary pressure act simultaneously. A theory is developed below enabling one to calculate the shape of a liquid interlayer between emulsion droplets or between gas bubbles of different radii under equilibrium conditions considering both the local Derjaguin's pressure of the interlayer and the local curvature of its surfaces [44–47].

The model of solid non-deforming particles is frequently used, when carrying out an analysis of the forces acting between colloidal particles. However, real droplets/bubbles and even soft solid particles within the contact zone can deform. This changes the conditions of their equilibrium. In this case, interaction is not limited only to the zone of a flat contact, but it is extended onto the surrounding parts within the range of action of the surface forces. For elastic solid particles, such a problem was treated in [48,49]. Below, the case of droplets/bubbles is considered (e.g., emulsions, gas bubbles in a liquid), whose shape changes especially easily under the influence of surface forces (Figure 2.32).

Below, we take into account both the finite thickness of a liquid interlayer between droplets/bubbles and variation in its thickness in the transition zone between the interlayer and the equilibrium bulk liquid phase. We have already mentioned that there is a problem in using an approach of thickness-dependent interfacial tension: If we try to use the Navier-Stokes equation for the description of the flow/equilibrium in thin liquid films, a thickness-dependent surface tension results in an unbalanced tangential stress on the surface of thin films. This is why this approach is used in this book.

Below we use the same approach as in previous sections of Chapter 2, which considers the interlayer thickness and the effect of the transition zone between the thin interlayer and the bulk liquid. This effect is equivalent to the line tension, which is considered in Section 2.9. A low-slope approximation and constant surface tension approximations are used below. As was shown in Sections 2.1 through 2.3, it is possible to use the equation considering both the Derjaguin's pressure and the capillary pressure in the interlayer.

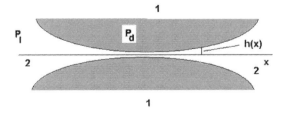

FIGURE 2.32 Two identical drops/bubbles: (1) bubbles at equilibrium in a surrounding liquid; and (2) $h(x)$ is a half-thickness of the liquid film between two drops/bubbles.

Two Identical Cylindrical Drops/Bubbles

First, let us consider a case of two identical cylindrical drops/bubbles (Figure 2.33). It is assumed for a simplicity that the radius of action of surface forces is limited to a certain distance, t_s. Beyond this distance, the surface of droplets retains a constant curvature radius, R, and is not disturbed by surface forces. The interacting droplets are considered as being surrounded by a liquid with constant pressure, P_l.

The thickness of the liquid interlayer, $2h(x)$, varies from $2h_0$ on the axis of symmetry at $x = 0$–$2h = t_s$ at $x = x_0$ (Figure 2.33).

For the drop profiles not disturbed by the surface forces, we use the following notations: h at $y < B$, and H for $y > B$ (Figure 2.33). The liquid in droplets is assumed incompressible, which leads to the condition of the constancy of the volume of droplets per unit length: $V = 4\int_0^{R_0}\left(H_S^0 - h_S^0\right).dx = 2\pi\,(R^0)^2 = \text{const}$, where superscript 0 marks an undisturbed isolated droplet prior to contact.

Note, that in general the system of two droplets is thermodynamically unstable with regard to coalescence and Ostwald ripening. Therefore, the following calculations give the conditions of the meta-stable equilibrium of droplets separated by a thin interlayer of the surrounding liquid.

The excess free energy of the two cylindrical drops (per unit length of cylindrical droplets), Φ, is equal to $\Phi = \gamma S + \Phi_D + P_e V$, where S is the total interfacial area per unit length, γ, is the interfacial tension, Φ_D is the excess free energy per unit length determined by the surface forces action, P_e is the excess pressure, and V is the volume per unit length. In the case of cylindrical drops/bubbles S, Φ_D and V are as follows:

$$S = 4\int_0^R \sqrt{1 + H_S'^2}\,dx + 4\int_{x_0}^R \sqrt{1 + h_S'^2}\,dx + 4\int_0^{x_0} \sqrt{1 + h'^2}\,dx$$

$$\Phi_D = 2\int_0^{x_0}\left[\int_{2h}^\infty \Pi(h)dh\right]dx$$

$$V = 4\int_{x_0}^R \left(H_S - h_S\right)\cdot dx + 4\int_0^{x_0}\left(H_S - h\right)dx = 2\pi\,R_0^2 = \text{const} \;, \qquad (2.178)$$

where $h(x)$, $h_s(x)$, $H_s(x)$, x_0 and radius, R, are as determined in Figure 2.33; and t_s is the radius of surface forces action.

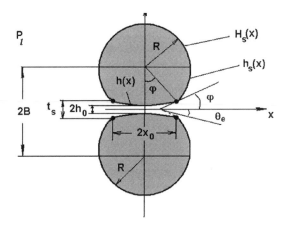

FIGURE 2.33 The equilibrium profile of interlayer, $h(x)$, between two cylindrical droplets of the same radius, R.

Substitution of Eq. (2.178) into the excess free energy Φ results in

$$\Phi = 4\gamma \left(\int_0^R \sqrt{1 + H_s'^2}\, dx + \int_{x_0}^R \sqrt{1 + h_s'^2}\, dx + \int_0^{x_0} \sqrt{1 + h'^2}\, dx \right) + 2 \int_0^{x_0} \left[\int_{2h}^\infty \Pi(h)\, dh \right] dx$$

$$+ P_e \left[4 \int_{x_0}^R \left(H_s - h_s \right) \cdot dx + 4 \int_0^{x_0} \left(H_s - h \right) dx \right].$$

A variation of the excess free energy with respect to $h(x)$, $h_s(x)$, $H_s(x)$ and the two values, x_0 and R, results in the following equations:

$$\gamma \cdot h'' \left[1 + h'^2 \right]^{-3/2} + \Pi(2h) = -P_e$$

$$\gamma \cdot h_s'' \left[1 + h_s'^2 \right]^{-3/2} = -P_e \tag{2.179}$$

$$\gamma \cdot H_s'' \left[1 + H_s'^2 \right]^{-3/2} = P_e,$$

where the last two equations give the equation of a circle of radius R. This immediately determines the unknown excess pressure as $P_e = -\dfrac{\gamma}{R}$. After that, Eq. (2.179) takes the following form:

$$\gamma \cdot h'' \left[1 + h'^2 \right]^{-3/2} + \Pi(2h) = \frac{\gamma}{R}, \tag{2.180}$$

where the first term in the left-hand side is due to the capillary pressure; and the second term is determined by the local value of the Derjaguin's pressure, $\Pi(2h)$.

The boundary conditions for Eq. (2.180) are as follows:

$$h(x_0) = h_s(x_0), \quad h'(x_0) = h_s'(x_0), \quad h(x_0) = t_s/2, \tag{2.181}$$

$$h_s(R) = H_s(R); \quad h_s'(R) = -H_s'(R) = \infty, \tag{2.182}$$

$$h'(0) = H_s'(0) = 0. \tag{2.183}$$

Recall that h_s describes the profile of the droplet at $2h_s > t_s$.

If the droplets are located at distance, $2h_0 > t_s$, then $\Pi(2h) = 0$; and in this case, the solution of Eq. (2.180) gives the profile $h_s(x)$, which corresponds to the circular form of the cross section of non-deformed droplets that do not interact with one another. However, at $2h_0 < t_s$, the interlayer is under the effect of both surface forces, whose contribution is determined by the term $\Pi(2h)$ and the capillary forces, depending on the local curvature of the interlayer surfaces.

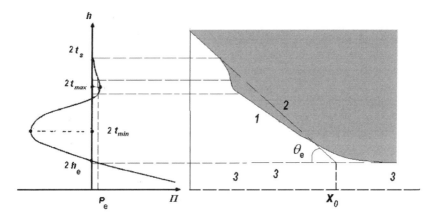

FIGURE 2.34 Partial wetting. The S-shaped Derjaguin's pressure isotherm (left side) and the liquid profile in the transition zone (right side). Magnification of the upper part of the transition zone between the drop/bubble and thin liquid interlayer. (1) Real deformed profile of the drop/bubble; (2) ideal spherical profile, when the influence of the Derjaguin's pressure has been ignored; and (3) thin liquid interlayer of thickness $2h_e$.

Equation (2.180), with boundary conditions (2.181) and (2.182) and with the condition of constancy of volume (2.178), provides a solution to the problem: It enables one to determine the profile of an interlayer, $h(x)$, between the droplets at the known isotherm of the Derjaguin's pressure, $\Pi(2h)$.

An example of the profile of the droplets in the transition zone in the case of S-shaped Derjaguin's pressure isotherm is shown in Figure 2.34. Note, in general the thickness in the central part between drops is different from the equilibrium thickness, h_e, and it is why it is referred to in Figure 2.33 as h_0. These two thicknesses coincide if the central part between the two drops is flat.

The solution thus obtained may be verified in the following way: At the equilibrium state, the total force of interaction of droplets per unit of their length, F, should be equal to zero. According to Section 2.8, this force is equal to

$$F = 2\int_0^{x_0} \Pi(2h)\cdot dx. \tag{2.184}$$

Substituting the expression for $\Pi(2h)$ from Eqs. (2.179) and (2.180) into Eq. (2.184) and then carrying out integration, we obtain

$$F = 2\gamma\left\{x_0 - \frac{h'(x_0)}{\sqrt{1+\left[h'(x_0)\right]^2}}\right\}. \tag{2.185}$$

In view of the boundary condition (2.181), at the point $x = x_0$, both values of $h(x) = h_s(x)$ and their derivatives, $h' = h'_s$ are equal. This enables one to express $h'(x_0)$ through the central angle φ (Figure 2.33) and values x_0 and R:

$$h'(x_0) = \tan\phi = x_0\left(R^2 - x_0^2\right)^{-1/2}. \tag{2.186}$$

Substituting this expression for $h'(x_0)$ into Eq. (2.185), we obtain $F = 0$, as it should be at the equilibrium. Thus, as one should expect that the conditions of equilibrium given by Eq. (2.180) and the boundary conditions (2.181) and (2.183) correspond to zero interaction forces between the droplets.

Note, we do not check the Jacobi's condition in the case under consideration because it is $h' > 0$, which obviously satisfies the Jacobi's condition.

It should be noted that angle φ has a value that is very close to that of the contact angle θ_e, which is to be determined at the point of intersection of the continuation of the undisturbed profile of a droplet with axis x (Figure 2.33). The values of θ_e and φ practically coincide when the interlayer thickness is small as compared with R and when x_0 is not too small. This enables one to use Eq. (2.186) for calculating the contact angles.

Thus, derived values of x_0, B, θ_e, and the droplet profile, $h(x)$, give the full solution of the problem, where the distance between the center of droplets, B, (Figure 2.33) is

$$B = t_s + 2\left(R^2 - x_0^2\right)^{1/2} = t_s + \left(2x_0/\tan\varphi\right). \tag{2.187}$$

Interaction of Cylindrical Droplets of Different Radii

Let us now consider a more complicated case of interaction of cylindrical droplets of different radii, $R_2 > R_1$ (Figure 2.35a).

Applying the same method of minimization of excess free energy, we obtain the following equations:

$$\gamma \cdot h_1'' \left[1 + \left(h_1'\right)^2\right]^{-3/2} + \Pi(t) = \gamma/R_1, \tag{2.188}$$

$$\gamma \cdot h_2'' \left[1 + \left(h_2'\right)^2\right]^{-3/2} - \Pi(t) = -\gamma/R_2, \tag{2.189}$$

where $h_1(x)$ and $h_2(x)$ are measured from an arbitrary plane perpendicular to the axis of symmetry, $t(x) = h_1(x) - h_2(x)$ is the thickness of the interlayer. Equations (2.188) and (2.189) enable one to determine the two profiles, $h_1(x)$ and $h_2(x)$. Equations (2.188) and (2.189), with the boundary conditions

$$h_1'(0) = h_2'(0) = 0; \; t\left(x_0\right) = t_s, \tag{2.190}$$

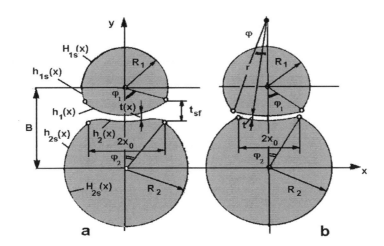

FIGURE 2.35 The equilibrium profile of interlayer $t(x) = h_1.(x) - h_2(x)$ between two droplets of different radii R_1, and R_2 (a) in the general case; and (b) in the simplified case, when the transition region is neglected.

$$h_1'(x_0) = \tan \varphi_1 = x_0 \left(R_1^2 - x_0^2 \right)^{-1/2},$$

$$h_2'(x_0) = \tan \varphi_2 = x_0 \left(R_2^2 - x_0^2 \right)^{-1/2} \tag{2.191}$$

and with the conditions of constancy of volumes, give a solution to the problem.

After the addition of Eqs. (2.188) and (2.189), we obtain

$$\left(\gamma / R_1 \right) - \left(\gamma / R_2 \right) = \Delta P_k, \tag{2.192}$$

which means that the curvature of the whole interlayer equalizes the capillary pressure drop between the droplets.

In a similar way the same problem may be solved for two droplets of different composition with different interfacial tensions, γ_1 and γ_2, of the first and the second droplet, respectively. In this case, even for the droplets of the same radius, the interlayer overall proves to be curved owing to the appearance of a capillary pressure drop, $\Delta P_k = (\gamma_1 - \gamma_2)/R$.

In the case where the interlayer between the droplets is not strongly curved, the term $\left(h_i^2 \right)^2 \ll 1, i = 1,2$, may be neglected. Then, we obtain from Eqs. (2.188), (2.189), and (2.192) and after introducing $t(x) = h_1(x) - h_2(x)$:

$$\gamma \cdot t'' + 2 \cdot \Pi(t) = \gamma \left(\frac{1}{R_1} + \frac{1}{R_2} \right). \tag{2.193}$$

This equation should be subjected the following boundary conditions:

$$t'(0) = 0;\ t(x_0) = t_s;\ t'(x_0) = x_0 \left[\left(R_1^2 - x_0^2 \right)^{-1/2} + \left(R_2^2 - x_0^2 \right)^{-1/2} \right], \tag{2.194}$$

and coupled with the conditions of constancy of volumes, which determine the unknown radii R_1 and R_2. These conditions and Eq. (2.193) enable one to calculate $t(x)$, determining the variable thickness of the interlayer.

Thereafter, on substituting the known dependence, $\Pi\left[t(x) \right] = \Pi(x)$, into Eq. (2.189), it is possible to obtain from

$$\gamma \cdot h_2'' - \Pi(x) = -\gamma / R_2 \tag{2.195}$$

the profile $h_2(x)$ of the lower surface of the interlayer. The boundary conditions for Eq. (2.195) are as follows:

$$h_2'(0) = 0;\ h_2(x_0) = R \cos \varphi_2;\ h_2'(x_0) = \tan \varphi_2. \tag{2.196}$$

It this expression, it has been considered that the angle φ_2 is determined from Eq. (2.191). The sum $h_2(x) + t(x)$ gives the profile of the upper surface of the interlayer.

Calculation of the interaction of cylindrical droplets of different radii ($R_2 > R_1$) can be simplified if we assume that the interlayer is of a constant thickness. This means that the effect of a transition zone is neglected, which is justified only at $x_0 \gg t_s$.

In this case, the curvature of each surface of the interlayer is constant (Figure 2.35b), and Eqs. (2.188) and (2.189) may be rewritten in the following way:

$$(\gamma/r) + \Pi(t) = \gamma/R_1, \tag{2.197}$$

$$[\gamma/(r+t)] - \Pi(t) = -\gamma/R_2. \tag{2.198}$$

This system of equations enables one to determine two unknown values: t, the interlayer thickness, and r, the radius of curvature of its surface on the side of the smaller droplet. Summing and subtracting by terms Eqs. (2.197) and (2.198), we obtain (at $r \gg t$):

$$r \sim 2R_1 R_2/(R_2 - R_1), \tag{2.199}$$

$$\Pi(t) \sim \gamma/R_1 - \gamma/r \sim \gamma \cdot (R_2 + R_1)/2R_1 R_2. \tag{2.200}$$

If the shape of the Derjaguin's pressure isotherm, $\Pi(t)$, is known, then Eq. (2.201) determines the equilibrium thickness, $t = $ const, of the curved interlayer.

At $R_2 \gg R_1$ Eqs. (2.199) and (2.200) results in:

$$r \sim 2R_1 \tag{2.199'}$$

$$\Pi(t) \sim \gamma/2R_1. \tag{2.200'}$$

Let us note that Eqs. (2.199) and (2.200) may be derived by another method using the concept of the Derjaguin's pressure:

$$\Pi = P_d - P_l, \tag{2.201}$$

where P_d is the pressure under the interlayer surface, and P_l is the pressure in the bulk phase with which the interlayer is at equilibrium.

On the side of a smaller droplet, we have $P_d = P_l + (\gamma/R_1) - (\gamma/r)$; on the side of a larger droplet, $P_d = P_l + (\gamma/R_2) + \gamma(r+h)$. As the Derjaguin's pressure does not depend on which side of the interlayer it is determined, then from Eq. (2.201) it follows:

$$\Pi = (\gamma/R_1) - (\gamma/r) = (\gamma/R_2) + [\gamma/(r+t)], \tag{2.202}$$

which coincides with Eqs. (2.199) and (2.200).

However, determination of r and t does not completely solve the problem because the position of the center of the interlayer curvature remains unknown. Its position can be determined by minimizing the value of the free energy of the system

$$\Phi = 2\gamma \cdot R_2 \left(\pi - \varphi_2\right) + 2\gamma \cdot R_1 \left(\pi - \varphi_1\right) + 2r\varphi \left[2\gamma + \int_t^\infty \Pi(\xi) d\xi\right], \tag{2.203}$$

and by considering the condition of the constancy of the volume of droplets

$$R_1^2 \left(\pi - \varphi_1 + \frac{1}{2}\sin 2\varphi_1\right) + r^2 \left(\varphi - \frac{1}{2}\sin 2\varphi\right) = \pi \cdot R_{10}^2 = \text{const}, $$

$$R_2^2 \left(\pi - \varphi_2 + \frac{1}{2}\sin 2\varphi_2\right) - r^2 \left(\varphi - \frac{1}{2}\sin 2\varphi\right) = \pi \cdot R_{20}^2 = \text{const}, \tag{2.204}$$

where R_{10} and R_{20} are the radii of the undisturbed droplets (at $h_0 > t_s$).

It is possible to express all the values via $x_0 : x_0 = R_1 \sin\varphi_1$; $x_0 = R_2 \sin\varphi_2$; $x_0 = r \sin\varphi$, then using condition $\partial\Phi / \partial x_0 = 0$ to determine the value of x_0 corresponding to the equilibrium position of droplets. In carrying out this procedure, the values of t and r can be expressed through R_1, R_2, and γ, in accordance with Eqs. (2.199) and (2.200).

In conclusion, let us consider the equilibrium conditions for the most general case of two spherical droplets of different radii and compositions. Using the method of minimization of the excess free energy described in this section, we obtain

$$\gamma_1 \left\{ \frac{h_1''}{\left[1 + \left(h_1' \right)^2 \right]^{3/2}} + \frac{h_1'}{r \left[1 + \left(h_1' \right)^2 \right]^{1/2}} \right\} + \Pi(t) = \frac{2\gamma_1}{R_1}, \qquad (2.205)$$

$$\gamma_2 \left\{ \frac{h_2''}{\left[1 + \left(h_2' \right)^2 \right]^{3/2}} + \frac{h_2'}{r \left[1 + \left(h_2' \right)^2 \right]^{1/2}} \right\} - \Pi(t) = -\frac{2\gamma_2}{R_2}, \qquad (2.206)$$

where r is now the radial coordinate.

The boundary conditions for Eqs. (2.205) and (2.206) are given by Eqs. (2.190) and (2.191), where x_0 should be replaced by r_0. In this case, conditions of the constancy of the volume of droplets can be written as

$$V_1 = 2\pi \int_{r_0}^{R_1} \left(H_{1s} - h_{1s} \right) \cdot r\, dr + 2\pi \int_0^{r_0} \left(H_{1s} - h \right) r\, dr = \frac{4}{3} \pi R_{10}^3 = \text{const.}$$

$$V_2 = 2\pi \int_{r_0}^{R_2} \left(H_{2s} - h_{2s} \right) \cdot r\, dr + 2\pi \int_0^{r_0} \left(H_{2s} - h \right) r\, dr = \frac{4}{3} \pi R_{20}^3 = \text{const}$$

At $(h')^2 \ll 1$, Eqs. (2.205) and (2.206) can be simplified. On summing these equations, we obtain

$$t'' + \frac{t'}{r} + \left(\frac{1}{\gamma_1} + \frac{1}{\gamma_2} \right) \cdot \Pi(t) = 2 \left(\frac{1}{R_1} + \frac{1}{R_2} \right), \qquad (2.207)$$

where $t(r) = h_1(r) - h_2(r)$.

Solution of Eq. (2.207) describes the variations in the thickness, $t(r)$, of the interlayer between the droplets. The distance between the centers of the droplets can be obtained using the same methods as used in the case of cylindrical droplets.

Thus, the use of the isotherms of the Derjaguin's pressure of thin interlayers enables one to solve the problem of the equilibrium of droplets when they are in close contact with one another and to calculate the shape of the deformed droplets and of the interlayer.

Shape of a Liquid Interlayer between Interacting Droplets: Critical Radius

As shown in this section, the shape of a liquid interlayer between interacting droplets depends on their size, interfacial tensions, and the shape of the Derjaguin's pressure. The calculations presented below describe the shape of an interlayer between droplets under equilibrium conditions.

In carrying out further calculations, we assume that the interlayer thickness, $h(r)$, varies within the contact region but not very sharply, that is, the approximation $(h')^2 \ll 1$ can be used. This enables one to use the isotherm of the Derjaguin's pressure of flat interlayers, $\Pi(h)$, in equations of equilibrium, as well as to simplify the expression for the local curvature of the interlayer surfaces.

Let us consider the interactions of two identical spherical droplets (Figure 2.33). In this case, the equation of the interlayer profile, $h(r)$, can be derived by solving Eq. (2.207)

$$\frac{\gamma}{2}\left(h'' + \frac{h'}{r} \right) + \Pi(h) = 2\gamma/R = P_e, \tag{2.208}$$

where $h' = dh/dr$; $h'' = d^2h/dr^2$; γ is the interfacial tension; and R is the radius of droplets.

For carrying out quantitative calculations, we use the model isotherm of the Derjaguin's pressure, $\Pi(h)$, in the following form as it was used in Section 2.7 (Figure 2.36):

$$\Pi(h) = \begin{cases} 0, h \geq t_s \\ a(t_0 - h), \quad 0 \leq h \leq t_s. \end{cases} \tag{2.209}$$

Such an isotherm (Figure 2.36) provides a possibility for obtaining an analytical solution of the problem, and at the same time it possesses the main properties of real isotherms, corresponding to the attraction of droplets at large separations, that is, $t_0 < h < t_s$, and to their repulsion at small distances, at $h < t_0$. Here, the thickness t_s corresponds to the radius of action of surface forces. Beyond its limits, at $h > t_s$, the interaction forces vanish: $\Pi = 0$. Parameter $a = (\Pi_1 + \Pi \min)/t_s$ determines the slope of the isotherm.

The solution of Eq. (2.208) together with isotherm (2.209) has the following form:

$$h(r) = t_0 - (2\gamma/aR) + A \cdot I_0(z), \tag{2.210}$$

where $z = r(2a/\gamma)^{1/2}$, and I_0 is the Bessel's function of an imaginary variable. The constant A is determined from the boundary condition, $h(r_0) = t_s$, where r_0 is the radius of the zone of deformation (Figure 2.34), and $z_0 = r_0(2a/\gamma)^{1/2}$, which results in

$$A = \left[t_s - t_0 + (2\gamma/aR) \right] / I_0(z_0). \tag{2.211}$$

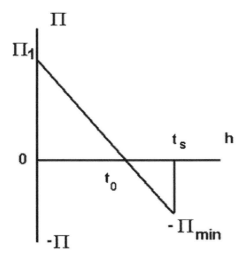

FIGURE 2.36 The Derjaguin's pressure isotherm, $\Pi(h)$, as used in calculations.

Substituting Eq. (2.211) into Eq. (2.210), we obtain an equation determining the profile of the interlayer between droplets:

$$h(r) = t_0 - \frac{2\gamma}{aR} + \left(t_s - t_0 + \frac{2\gamma}{aR}\right) \cdot \frac{I_0(z)}{I_0(z_0)}. \tag{2.212}$$

Accordingly, the minimum distance between the surfaces of droplets is equal to

$$h_0 = h(0) = t_0 - \frac{2\gamma}{aR} + \left(t_s - t_0 + \frac{2\gamma}{aR}\right) \cdot \frac{1}{I_0(z_0)}. \tag{2.213}$$

In Eqs. (2.212) and (2.213), the value of r_0 remains still to be determined. For this, let us use the condition (2.184):

$$F = \int_0^{r_0} \Pi(h) \cdot 2\pi r \cdot dr = 0. \tag{2.214}$$

Substituting into Eq. (2.214) the equation of isotherm (2.209) and replacing $h(r)$ by its expression from Eq. (2.212), we obtain the following expression:

$$r_0^2 = \frac{4aR}{\gamma} \left(t_{sf} - t_0 + \frac{2\gamma}{aR}\right) \int_0^{r_0} r \frac{I_0(z)}{I_0(z_0)} \cdot dr. \tag{2.215}$$

In the case when $z_0 \geq 5$ (sufficiently large drops), the integral in Eq. (2.215) is equal to $r_0 \left(\gamma/2a\right)^{1/2}$. In this case, Eq. (2.215) results in the following expression for the radius of the contact zone:

$$r_0 = (2a\gamma)^{1/2} \left[\frac{R(t_s - t_0)}{2} + \frac{1}{a}\right]. \tag{2.216}$$

Using Eq. (2.216), the relative extension of the contact zone, where the effect of the surface forces is pronounced, can be expressed as

$$\frac{r_0}{R} = (2a \cdot \gamma)^{1/2} \left(\frac{t_s - t_0}{2\gamma} + \frac{1}{aR}\right). \tag{2.217}$$

At $1/aR \ll (t_s - t_0)/2\gamma$, or $R \gg 2\gamma/a(t_s - t_0)$, the ratio r_0/R tends to the value $(t_s - t_0)(a/2\gamma)^{1/2}$, which is independent of the radius of the drop, R. This means there is geometric similarity for all the large fluid droplets that are deformed by the contact interaction.

For the droplets of the smaller radius, R, one may use another approximation for $I_0(z)$, which is valid at $z < 1$: $I_0(z) = 1 + (z^2/4)$. In this case, integration in Eq. (2.215) yields

$$r_0^2 = \frac{2R(t_s - t_0)}{1 - (aR/2\gamma) \cdot (t_s - t_0)}. \tag{2.218}$$

Let us compare this expression with the solution for solid spheres. Their equilibrium state (Figure 2.37b) can also be characterized by the interaction region of radius r_0, within which the surface forces exert

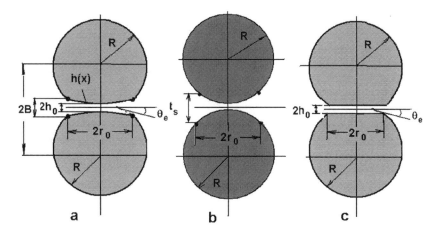

FIGURE 2.37 The schematic representation of the contact interaction of (a, c) the fluid and (b) the solid particles. In the case (c) the thickness in the center coincides with the equilibrium thickness h_e.

their effect. The profile of the solid sphere can be represented as $h = h_0 + 2R \cdot \left[1 - \sqrt{1 - \left(r/R \right)^2} \right]$. At $r_0 \ll R$, its approximate form may be used:

$$h \cong h_0 + \left(r^2/R \right).$$ (2.218)

The minimum distance, h_0, between the solid particles can be determined from the same condition (2.214) and the known the Derjaguin's approximation [1]:

$$F\left(h_0 \right) = \pi \cdot R \int_{\infty}^{h_0} \Pi \left(h \right) \cdot dh.$$ (2.219)

Replacing the lower integration limit by t_s and using the model Derjaguin's pressure isotherm Eq. (2.209), after integration in Eq. (2.219), we obtain the following expression: $h_0 = 2t_0 - t_1$. On substituting this expression into Eq. (2.218) and taking into account that $h(r_0) = t_s$, we get

$$r_0^2 = 2R\left(t_s - t_0 \right).$$ (2.220)

The same value of r_0 is also obtained by solving Eq. (2.218) for "small" fluid droplets. This means that very small emulsion droplets and gas bubbles practically behave as solid particles: they practically do not deform in the contact zone.

Let us call the critical size of fluid droplets, R^*, such that at $R \ll R^*$, and their interaction does not differ from that of solid spheres. Comparing Eqs. (2.218) and (2.220), note that these coincide under the condition that the second term in the denominator of Eq. (2.218) is much smaller than unity. This condition allows one to determine R^* as

$$R^* = 2\gamma / \left(t_s - t_0 \right) \cdot a.$$ (2.221)

The distance between the centers of droplets can be determined using Eq. (2.187) and simple geometrical considerations:

$$B = t_s + 2\left(R^2 - r_0^2 \right)^{1/2}.$$ (2.222)

Accordingly, the contact angle, θ_e, can also be determined at the point of intersection between the non-deformed surface of a droplet and the r axis (Figure 2.37a):

$$\cos\theta_e = B/2R = (t_s/2R) + \left[1 - (r_0/R)^2\right]^{1/2}. \tag{2.223}$$

This expression holds only for relatively large droplets. As has been pointed out above, Eq. (2.221) small droplets $(R < R^*)$ are practically non-deformable, and their undisturbed, circular profile does not intersect the r axis. In the case of small droplets, like that of complete wetting (see Section 2.4), the contact angle is absent.

Let us now numerically calculate the profiles of fluid droplets in the contact zone using Eq. (2.212) and compare them with two other known models: the model of solid non-deformable particles (Figure 2.37b) and the model of a flat interlayer between similar droplets (Figure 2.37c) (in this model, the effect of a transition zone between the surrounding bulk medium and the flat portion of the interlayer is not taken into account). In the case of flat interlayers let us use the equations of equilibrium of flat thin films, see Eq. (2.47) in Section 2.3, which was modified for the case of two droplets/bubbles:

$$2 \cdot \gamma (\cos\theta_e - 1) = \int_{2h_e}^{\infty} \Pi(h) \cdot dh + 2P_e \cdot h_e, \tag{2.224}$$

where θ_e is the equilibrium contact angle determined at the point of intersection between the continuation of the non-deformed part of a sphere and the surface of the equilibrium flat film. In that case, the thickness of the flat interlayer, $2h_e$, is determined using the Derjaguin's pressure isotherm, $\Pi(h)$, as $\Pi = P_e$, where $P_e = 2\gamma/R$ is the capillary pressure drop at the spherical interface. The contact area of droplets at $h_e \ll R$ is equal to $\pi r_0^2 = \pi R^2 \cdot \sin^2\theta_e$ where r_0 is the radius of the contact zone.

Substituting Eq. (2.209) of the Derjaguin's pressure isotherm into Eq. (2.224), we obtain the following expression:

$$\cos\theta_e = 1 - (a/4\gamma) \cdot \left[(t_s - t_0)^2 - (2\gamma/aR)^2\right] + (1/R) \cdot \left[t_0 - (2\gamma/aR)\right]. \tag{2.225}$$

The calculations were carried out for the model Derjaguin's pressure isotherm, $\Pi(h)$, (Eq. 2.209) and using the following parameters: $\gamma = 30$ dyn/cm, $t_s = 3 \times 10^{-6}$ cm, $t_0 = 2 \times 10^{-6}$ cm, $\Pi_1 = \Pi(0) = 3 \times 10^6$ dyn/cm^2, and $\Pi_{min} = \Pi(t_s) = -1.5 \times 10^6$ dyn/cm^2, which gives $a = 1.5 \times 10^{12}$ dyn/cm^3. According to Eq. (2.224), the adopted values correspond to the contact angle $\theta_e = 9°$.

The left-hand part of the plots in Figure 2.38 relates to the spherical droplets having radius $R = 10^{-3}$ cm, the right-hand part to $R = 10^{-4}$ cm. As appears from the left-hand part the large-sized droplets deform considerably, thus forming the practically flat contact zone. Profiles 1 and 2 (Figure 2.38) are close to each other, and the transition zone occupies a rather small region immediately near the contact perimeter. Now, deviations from the profile of solid particles are very large (curve 3, Figure 2.38). Thus, in the case of $R \gg R^*$, the conditions of the droplets equilibrium may be described with the framework of the theory of flat interlayers, that is, based on Eq. (2.224).

A decrease in the size of droplets (Figure 2.38, right-hand side of the graph) makes the profiles of solid particles (curve 3) and of droplets (curve 1) approach one another. In the central part of the contact region, the flat interlayer region reduces, while the differences from the profile of the flat interlayers (curve 2) increase. A further decrease in the droplet radius causes a still greater approaching of the profiles of the fluid and the solid particles to one another and the disappearance of the flat portion of the interlayer. As has been shown above Eq. (2.221) at $R \ll R^*$, one may use the known solutions for the interaction of solid spheres.

In this connection, it is important to evaluate values of R^*. This allows one to determine the regions of applicability of different solutions; namely, flat interlayers for $R \gg R^*$, solid particles for $R \ll R^*$, and, finally, the approach developed in this section for the intermediate values of R.

As appears from Eq. (2.221), the critical radius R^* depends on the parameters of the Derjaguin's pressure isotherm (a, t_s, and t_0) and the interfacial tension. Under otherwise equal conditions, a decrease in the interface tension reduces the values of R^*, thus limiting the region of the applicability of the theory of solid spheres.

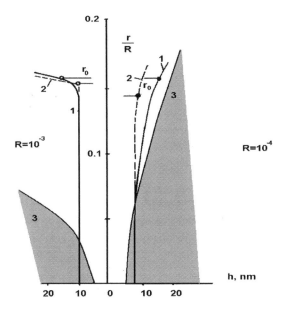

FIGURE 2.38 The profiles. $h(r/R)$, of the contact zone of identical spherical particles, $R = 10^{-3}$ cm (to the left), and $R = 10^{-4}$ cm (to the right), calculated while (1) taking into account the transition zone, and in the approximation of (2) a flat interlayer and (3) solid particles.

Let us evaluate R^*, assuming $\gamma = 50$ dyn/cm, and $(t_s - t_0) = 10^{-6}$ cm. The values of parameter a will vary, which, in accordance with Eq. (2.225), is equivalent to a change in the contact angle, θ_e:

$$\cos\theta_e = 1 - (a / 4\gamma) \cdot (t_s - t_2)^2. \tag{2.226}$$

The calculations show that the values of R^* decrease as θ_e increases. Thus, at small values of θ_e, $R^* = 10^{-3}$ cm; at $\theta_e = 5°$–$6°$, $R^* = 10^{-4}$ cm; at $\theta_e = 20°$–$30°$, $R^* = 5 \times 10^{-6}$ cm; and at $\theta_e = 90°$, $R^* = 5 \times 10^{-7}$ cm. In this way, an increase in the contact angle, corresponding to the enhancing of the droplets interaction, causes a decrease in the critical radius, R^*. Now, this means that in the case of a strong interparticle interaction, the approach of solid spheres becomes ever less applicable; yet in the case of weakly interacting droplets, that approach may be used even for relatively large-size droplets. The solutions obtained allow one to evaluate R^* and choose corresponding equations for calculating the equilibrium shape of the droplets within the contact zone.

2.9 Liquid Profiles on Curved Interfaces, Effective Derjaguin's Pressure. Equilibrium Contact Angles of Droplets on Outer/Inner Cylindrical Surfaces and Menisci inside Cylindrical Capillaries

In this section we obtain *effective* Derjaguin's pressure isotherms for liquid films of uniform thickness on inner and outer cylindrical surfaces and on the surface of spherical particles. This effective Derjaguin's pressure is expected to depend on the surface curvature. From its expression, we can calculate the equilibrium contact angles of drops on the outer surface of a cylinder and the menisci inside cylindrical capillaries. We see that the mathematical expression for contact angle is almost independent of liquid geometry. However, there is a substantial difference: In the expressions for equilibrium contact angles (according to geometry) the value of the film of uniform thickness with which the bulk liquid (drops or menisci) is at equilibrium are substantially different. This thickness determines the lower limit of the integral in the expression for the equilibrium contact angle.

Liquid Profiles on Curved Surface: Derivation of Governing Equations

Excess free energy, Φ, of the liquid droplet on an outer surface of a cylindrical capillary of a radius a is as follows:

$$\Phi = \int_{0}^{\infty} \left\{ 2\pi\gamma \left[(a+h)\sqrt{1+h'^2} - (a+h_e) \right] + \pi P_e \left[\left((a+h)^2 - a^2 \right) - \left((a+h_e)^2 - a^2 \right) \right] \right.$$
$$\left. + 2\pi a \left[\int_{h}^{\infty} \Pi(h)dh - \int_{h_e}^{\infty} \Pi(h)dh \right] \right\} dx \tag{2.227}$$

where, we selected a reference state as the outer surface of the cylindrical capillary covered by the equilibrium liquid film of the thickness h_e; and x is in the direction parallel to the cylinder axis (Figure 2.39).

The condition of the equilibrium according to Section 2.2 condition (A), which is the first variation of the excess free energy (2.227), results in the following equation describing the liquid profile on the surface of the cylindrical capillary:

$$\frac{\gamma h''}{\left(1+h'^2\right)^{3/2}} - \frac{\gamma}{(a+h)\sqrt{1+h'^2}} + \frac{a}{a+h}\Pi(h) = P_e. \tag{2.228}$$

Note, that this equation is different from Eq. (2.228) in Section 2.2 not only because of the presence of the second curvature, $-\frac{\gamma}{(a+h)\sqrt{1+h'^2}}$, but also because of a difference in the definition of the Derjaguin's pressure, which is now $\frac{a}{a+h}\Pi(h)$ instead of $\Pi(h)$ as in Section 2.2. This difference results in a substantial consequence as shown below.

Note that the excess pressure, P_e, is determined by the vapor pressure in the surrounding air and given by Eq. (2.2) (Section 2.1). In the case of droplets on the cylindrical capillary, as in the case of droplets on flat solid substrates, the equilibrium is possible only at oversaturation, that is, at $P_e < 0$.

For the equilibrium film of the uniform thickness, h_e, we conclude from Eq. (2.228):

$$-\frac{\gamma}{(a+h_e)} + \frac{a}{a+h_e}\Pi(h_e) = P_e. \tag{2.229}$$

Let us introduce the effective Derjaguin's pressure as

$$\Pi_{eff}(h) = -\frac{\gamma}{(a+h)} + \frac{a}{a+h}\Pi(h). \tag{2.230}$$

Below we show that the introduced effective Derjaguin's pressure provides the correct stability condition. For that purpose, we consider the excess free energy per unit length of the capillary, Φ_e, of the equilibrium film of a uniform thickness, h_e, which is

$$\Phi_e = 2\pi\gamma(a+h_e) + \pi P_e \left[(a+h_e)^2 - a^2 \right] + 2\pi a \int_{h_e}^{\infty} \Pi(h)dh + 2\pi a(\gamma_{sl} - \gamma_{sv}). \tag{2.231}$$

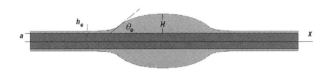

FIGURE 2.39 Cross section of an axisymmetric liquid droplet on outer surface of a cylinder of radius a. H, the maximum height of the droplet; and h_e, the thickness of an equilibrium film of the uniform thickness.

According to requirements of the equilibrium, the following conditions should be satisfied:

$$\frac{d\Phi_e}{dh_e} = 0,$$ (2.232)

$$\frac{d^2\Phi_e}{dh_e^2} > 0.$$ (2.233)

The first condition of the equilibrium Eq. (2.232) results in

$$\gamma + P_e(a + h_e) - a\Pi(h_e) = 0$$

This equation can be rewritten using the definition of the effective Derjaguin's pressure Eq. (2.230) as

$$\Pi_{eff}(h_e) = P_e$$ (2.234)

The second condition (2.233) gives

$$P_e - a\Pi'(h_e) > 0.$$ (2.235)

Let us check that the introduction of the effective Derjaguin's pressure isotherm according to Eq. (2.230) satisfies the stability condition Eq. (2.235). Indeed

$$\frac{d\Pi_{eff}}{dh_e} = -\frac{a}{(a+h_e)^2}\left[\Pi(h_e) - \frac{\gamma}{a}\right] + \frac{a}{a+h_e}\Pi'(h_e).$$

Substituting the expression for $\Pi(h_e)$ from Eq. (2.234), we conclude

$$\frac{d\Pi_{eff}}{dh_e} = -\frac{a}{(a+h_e)^2}\left[\left(\frac{\gamma}{a} + P_e\frac{a+h_e}{a}\right) - \frac{\gamma}{a}\right] + \frac{a}{a+h_e}\Pi'(h_e) = \frac{1}{a+h_e}\left[a\Pi'(h_e) - P_e\right] < 0,$$ according to condition (2.235).

Hence, we conclude

$$\frac{d\Pi_{eff}}{dh_e} < 0.$$ (2.236)

This means that the introduced effective Derjaguin's pressure according to Eq. (2.230) possesses all necessary properties according to Eqs. (2.234) and (2.236).

In the case of a uniform film on a spherical particle of radius, a, we get an expression for the excess free energy that is like Eq. (2.229):

$$\Phi_e = 4\pi\gamma(a + h_e)^2 + \frac{4\pi}{3}P_e\left[(a+h_e)^3 - a^3\right] + 4\pi a^2\int_{h_e}^{\infty}\Pi(h)dh + 4\pi a^2(\gamma_{sl} - \gamma_{sv}).$$ (2.237)

Let us introduce the effective Derjaguin's pressure isotherm in this case as

$$\Pi_{eff}(h) = \frac{a^2}{(a+h)^2}\Pi(h) - \frac{2\gamma}{a+h}.$$ (2.238)

The same procedure results in the following two equilibrium conditions (2.234) and (2.236). This means that the effective Derjaguin's pressure isotherm defined according to Eq. (2.238) really describes the stability of uniform films on spherical particles.

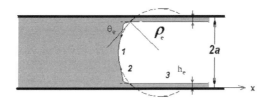

FIGURE 2.40 Profile of a meniscus in a cylindrical capillary of radius a. (1) A spherical part of the meniscus of curvature r_e; (2) transition zone between the spherical meniscus and flat films in front; and (3) flat equilibrium liquid film of thickness h_e.

Now, in the case of liquid layers inside the inner part of the capillary of radius a, we have the following expression for the excess free energy:

$$\Phi = \int_0^\infty \left\{ 2\pi\gamma\left[(a-h)\sqrt{1+h'^2} - (a-h_e) \right] + \pi P_e\left[\left(a^2-(a-h)^2\right) - \left(a^2-(a-h_e)^2\right) \right] \right. $$
$$\left. + 2\pi a\left[\int_h^\infty \Pi(h)dh - \int_{h_e}^\infty \Pi(h)dh \right] \right\} dx \tag{2.239}$$

which is similar to the expression for excess free energy on the outer cylindrical surface (Figure 2.40).

Exactly in the same way as in the case of the outer cylindrical surface, we deduce the following equation for the liquid profile:

$$\frac{\gamma h''}{\left(1+h'^2\right)^{3/2}} + \frac{\gamma}{(a-h)\sqrt{1+h'^2}} + \frac{a}{a-h}\Pi(h) = P_e. \tag{2.240}$$

Note again that the resulting Eq. (2.240) is different from the both corresponding Eqs. (2.228) (liquid on the outer cylindrical surface) and Eq. (2.23) in Section 2.2 (for a flat surface).

Let us introduce the effective Derjaguin's isotherm in this case as

$$\Pi_{eff}(h) = \frac{a}{a-h}\Pi(h) + \frac{\gamma}{(a-h)}. \tag{2.241}$$

The corresponding expression for the excess free energy per unit length of a uniform film on the inner cylindrical surface is

$$\Phi_e = 2\pi\gamma(a-h_e) + \pi P_e\left[a^2 - (a-h_e)^2 \right] + 2\pi a\int_{h_e}^\infty \Pi(h)dh + 2\pi a(\gamma_{sl} - \gamma_{sv}). \tag{2.242}$$

This expression and the definition (2.241) result in the conditions (2.234) and (2.236), which describe the stability of the film of a uniform thickness on the inner surface of the cylindrical capillary.

Equilibrium Contact Angle of a Droplet on an Outer Surface of Cylindrical Capillaries

The droplet profile is described by Eq. (2.228). Let H be the maximum height of the droplet in the center, that us, $h(0) = H$. Let us introduce a new unknown function $u = \frac{1}{\sqrt{1+h'^2}}$ in this equation and integrate Eq. (2.228) once, which results in

$$\frac{1}{\sqrt{1+h'^2}}=1+\frac{P_e\left(\dfrac{H^2}{2}+aH-ah-\dfrac{h^2}{2}\right)-a\displaystyle\int_h^\infty\Pi(h)dh}{\gamma(a+h)}, \tag{2.243}$$

where condition $h'(H)=0$ is considered. If we neglect the Derjaguin's pressure in the right-hand side of this equation, we get the "outer solution," which describes the drop profile not distorted by the Derjaguin's pressure action:

$$\frac{1}{\sqrt{1+h'^2}}=1+\frac{P_e\left(\dfrac{H^2}{2}+aH-ah-\dfrac{h^2}{2}\right)}{\gamma(a+h)}. \tag{2.244}$$

If we continue the "outer solution" (2.244) to the intersection with the surface of the cylinder, we get $h'(0)=-\tan\theta_e$, where θ_e is the equilibrium contact angle to be determined. Using this condition in the "outer solution" (2.244), we conclude

$$\cos\theta_e=1+\frac{P_e\left(\dfrac{H^2}{2}+aH\right)}{\gamma a},$$

or

$$P_e=\frac{(\cos\theta_e-1)2\gamma a}{H^2+2aH}<0. \tag{2.245}$$

This expression shows that the equilibrium droplets in the outer surface of a cylinder can be at equilibrium only at oversaturation as droplets on a flat substrate.

Now from the whole Eq. (2.243), we conclude that the local profile tends asymptotically to the film of the uniform thickness, h_e. That is locally, the profile satisfies the condition $h'(h_e)=0$. Using this condition, we conclude from Eq. (2.243)

$$0=\frac{P_e\left(\dfrac{H^2}{2}+aH-ah_e-\dfrac{h_e^2}{2}\right)-a\displaystyle\int_{h_e}^\infty\Pi(h)dh}{\gamma(a+h_e)},$$

or

$$P_e\left(\frac{H^2}{2}+aH-ah_e-\frac{h_e^2}{2}\right)-a\int_{h_e}^\infty\Pi(h)dh=0, \tag{2.246}$$

where the equilibrium thickness of the uniform film is determined from the following equation

$$\Pi_{eff}(h_e)=-\frac{\gamma}{(a+h_e)}+\frac{a}{a+h_e}\Pi(h_e)=P_e. \tag{2.247}$$

Substitution of the expression (2.245) into Eq. (2.246) results in the following equation for the determination of the equilibrium contact angle

$$\cos\theta_e = 1 + \cfrac{1}{1 - \cfrac{2ah_e + h_e^2}{H^2 + 2aH}} \, \frac{1}{\gamma} \int\limits_{h_e}^{\infty} \Pi(h)dh. \tag{2.248}$$

If we omit the small term, $\dfrac{2ah_e + h_e^2}{H^2 + 2aH}$, in this equation, we arrive to

$$\cos\theta_e \approx 1 + \frac{1}{\gamma} \int\limits_{h_e}^{\infty} \Pi(h)dh. \tag{2.249}$$

The form of this equation is identical to Eqs. (2.47) (meniscus in a flat capillary) and (2.55) (droplet on a flat substrate) deduced in Section 2.3. However, there is a substantial difference between Eqs. (2.47) and (2.55) in Section 2.3 and Eq. (2.249): the lower limit of integration in those equations, which is the thickness of the uniform film, is substantially different in each of those equations.

Equilibrium Contact Angle of a Meniscus inside Cylindrical Capillaries

In this case, the meniscus profile is described by Eq. (2.240). We introduce a new unknown function $u = \frac{1}{\sqrt{1+h'^2}}$, as in the case of a droplet on a cylindrical surface. After one integration, Eq. (2.240) takes the form

$$\frac{1}{\sqrt{1+h'^2}} = \frac{\dfrac{P_e}{2}(a-h)^2 - a\int\limits_{h}^{\infty} \Pi(h)dh}{\gamma(a-h)}, \tag{2.250}$$

where we already take into account the condition in the center of the capillary $h'(a) = -\infty$. If we neglect the Derjaguin's pressure in this equation, we arrive to

$$\frac{1}{\sqrt{1+h'^2}} = \frac{P_e(a-h)}{2\gamma}, \tag{2.251}$$

which describes the spherical meniscus profile. Continuation of this profile to the intersection with the capillary surface results in a similar way to the previous case of droplets:

$$\cos\theta_e = \frac{P_e a}{2\gamma},$$

or

$$P_e = \frac{2\gamma \cos\theta_e}{a} > 0, \tag{2.252}$$

as expected.

Now from the whole Eq. (2.250), we conclude that the local profile tends asymptotically to the film of the uniform thickness, h_e. That is, locally, the profile satisfies the condition $h'(h_e) = 0$. Using this condition and Eq. (2.252) we conclude from Eq. (2.250)

$$\cos\theta_e = \frac{1}{1-\frac{h_e}{a}} + \frac{1}{\left(1-\frac{h_e}{a}\right)^2} \frac{1}{\gamma} \int\limits_{h_e}^{\infty} \Pi(h)dh, \tag{2.253}$$

where the thickness of the uniform film is determined from

$$\Pi_{eff}(h_e) = \frac{a}{a-h_e}\Pi(h_e) + \frac{\gamma}{(a-h_e)} = P_e. \tag{2.254}$$

If we omit the small terms in which $\frac{h_e}{a} \ll 1$, in Eq. (2.253), we arrive to the same functional dependency of the $\cos\theta_e$ as earlier [Eqs. (2.249), or (2.47) and (2.55) in Section 2.3].

Let us emphasize the significant difference, also discussed in Section 2.3, between the equilibrium of drops and menisci: In the case of drops (no difference between drops on a flat surface or an outer cylindrical surface), the droplets can be at equilibrium with oversaturated vapor. The external super-saturated vapor pressure in the ambient air may be arbitrary inside the narrow limits determined in Section 2.3, Eq. (2.52). The drop size adjusts to the imposed pressure at equilibrium. The situation is very different in the case of an equilibrium meniscus in a capillary (no difference in a flat chamber or inside a cylindrical capillary): They can be at equilibrium with undersaturated vapor. There is only a single vapor pressure allowing the equilibrium of the meniscus, this equilibrium at undersaturated pressure is determined by the capillary radius. At all other vapor pressure, the meniscus cannot be at equilibrium.

Derjaguin's Pressure of Uniform Films in Cylindrical Capillaries (2.254)

It is possible to show that with a decreasing capillary radius, a, the region of stable states of α-films is narrowed, but their thickness is greater than the corresponding thickness of films on a planar surface. Also narrowed is the region of existence of β-films, with the appearance of both a lower limit, t_1, but also an upper limit, t_2, of film stability. In narrower cylindrical pores, only thin α-films are stable, and β-films disappear completely.

These conclusions are supported by published experimental data. The existence of an upper limit of stability for β-films in cylindrical capillaries was experimentally discovered in [50]. It has been also experimentally observed in [51] that in glass cylindrical capillaries with radius $a > 0.4$ μm, thin α-films are formed with thickness $h \approx 50–60$ Å. However, in thinner capillaries with $a = 0.2–0.3$ μm (with a corresponding reduction of p/p_s), thicker β-films appear with a thickness of $h \approx 300–400$ Å. Both of these two experimental observations correspond to the isotherm $\Pi_{eff}(h)$ (2.254). In conclusion of this section, we examine the case of complete wetting, when the isotherm $\Pi(h)$ of a planar film can be represented by $\frac{A}{h^3}$.

According to the stability condition, the critical thickness, h_*, can be calculated when the film of a uniform thickness loses its stability. Based on Eq. (2.254), we can conclude that $h_* = (3Aa^2/\gamma)^{1/4}$. At $h \geq h_*$, the films on the inner capillary surface loses stability, and the liquid changes over into a more stable state, forming a capillary condensate. For $a \sim 10^{-4}$ cm, $A \sim 10^{-14}$ erg, and $\gamma \sim 30$ dyn/cm, we obtain a thickness h_* of the order of 10^{-6} cm. Thus, the condition $h \ll a$ is fulfilled over the entire interval of film thicknesses that is physically realizable in a capillary. Even with $a \sim 10^{-6}$ cm, the values of h_* are no greater than 10^{-7} cm.

As an example, Figure 2.41 shows curves plotted based on Eq. (2.254) for the thickness of a decane film, h, on the inner surface of quartz capillaries as a function of the relative vapor pressure, p/p_s. For this purpose, in Eq. (2.254) we used a disjoining pressure isotherm as given by $\frac{A}{h^3}$, and P_e is replaced by its values from Eq. (2.2) from Section 2.1, which results in

$$\frac{a}{a-h}\left(\frac{A}{h^3} + \frac{\gamma}{a}\right) = \frac{RT}{v_m}\ln\frac{p_s}{p}. \tag{2.255}$$

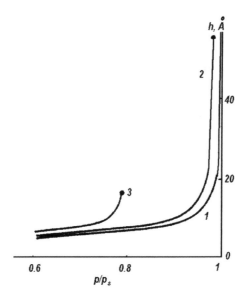

FIGURE 2.41 Adsorption isotherms, $h\ (p/p_s)$, for films of decane in quartz capillaries with radius $r = (1)\ \infty$; $(2)\ 10^{-5}$ cm; and $(3)\ 10^{-6}$ cm.

In the calculations below it is assumed, the same as previously, that $A = 8.5 \cdot 10^{-15}$ erg, $\gamma = 23$ dyn/cm, $v_m = 195$ cm³/mole, and $T = 293$ K.

The curve 1 in Figure 2.41 is the isotherm $h(p/p_s)$ at $a = \infty$, that is, for a planar surface. Curves 2 and 3 were plotted for capillaries with radii $a = 10^{-5}$ and 10^{-6} cm, respectively. Whereas on the planar surface $h \to \infty$ when $p/p_s \to 1$ (curve 1), we find that in the capillaries the isotherm breaks off at $p/p_s = 0.98$ (curve 2) and $p/p_s = 0.79$ (curve 3). Break-off of the isotherms corresponds to loss of film stability in accordance with Eq. (2.255). Note, in the entire region of the physically realizable film thicknesses, $h \leq h_*$, the condition $h << a$ is satisfied. That is, the capillary condensation starts much earlier than the capillary is filled with the liquid.

Even though the thickness of adsorbed films with equal values of p/p_s is bigger in a capillary with a smaller capillary radius, the region of their existence is diminished quite substantially. When $h > h_*$, the films lose stability, and the capillary is filled with condensate.

2.10 Line Tension

The presence of the transition zone between a drop or a bubble and thin liquid interlayers can be described in terms of *line* tension, τ, a concept first introduced by Gibbs [52]. This concept was introduced by analogy with surface tension: According to Gibbs [52], there is no sharp transition between the liquid and vapor (transition zone); this transition zone is replaced by a plane of tension with excess surface energy, γ [52]. By analogy, the transition zone between a drop/bubble/meniscus and the thin liquid film in front may be replaced by a *three-phase contact line* with an excess *linear* energy τ [52]. In contrast to surface tension defined as always positive, the value of the line tension may be positive or negative. When positive, it causes the wetting perimeter to contract, whereas it expands if the line tension is negative [44–47].

A number of attempts have been undertaken to improve Neumann–Young's equation and to make it more theoretically justified. The most important of them is the introduction of line tension, τ. In Section 2.3 was shown that the drop profile cannot keep its spherical shape up to the three-phase contact line. It was shown in the Section 2.3 that the surface forces action results in a substantial deviation of the drop shape in the vicinity of the three-phase contact line from a spherical shape. It results in a formation of a transition zone where the influence of the Derjaguin's pressure is important and cannot be ignored. However, if we still want to consider a spherical droplet, then the existence of the transition region can be effectively considered by replacing the whole transition region with an additional free energy located on the three-phase contact line. In a way, this consideration is like the introduction of an interfacial tension. If the line tension is considered, then the excess free energy of the droplet on the solid substrate, Φ, should be written as

$$\Phi = \gamma S + P_e V + \pi r_0^2 (\gamma_{sl} - \gamma_{svh}) + 2\pi r_0 \tau, \tag{2.256}$$

where r_0 is the radius of the droplet base. Note, that according to Section 2.1, we used the interfacial tension of the solid substrate covered by an equilibrium liquid film, γ_{svh}, but not the corresponding interfacial tension of a bare solid substrate, γ_{sv}. In Section 2.1 we explained that this surface tension cannot be used at equilibrium and that these two interfacial tensions can differ considerably. Using the expression for the droplet profile, $h_{id}(r)$, Eq. (2.256) can be rewritten as

$$F = 2\pi \int_0^{r_0} r \left[\gamma \sqrt{1 + \left(h'_{id} \right)^2} + P \cdot h + \gamma_{sl} - \gamma_{svh} \right] dr + 2\pi r_0 \tau, \tag{2.257}$$

where τ is line tension; and $h_{id}(r)$ is the idealized droplet profile, with an extension of the spherical part up to the intersection with the surface of the equilibrium flat film. The first two conditions of the excess free energy minimum result in the identical equation for the spherical drop profile [i.e., Eq. (2.41) in Section 2.3]. However, the transversality condition (D), in Section 2.2, takes the following form

$$\left\{ \frac{\gamma}{\sqrt{1 + h'^2}} + \gamma_{sl} - \gamma_{svh} \right\}_{r=r_0} + \frac{d\tau}{dr_0} r_0 + \tau = 0. \tag{2.258}$$

If we now introduce a new real equilibrium contact angle, which takes into account line tension as θ_e and uses the previous definition of the contact angle, which is referred now as $\theta_{e\infty}$ according to Eq. (2.47), then Eq. (2.257) can be rewritten as

$$\gamma \left(\cos\theta_e - \cos\theta_{e\infty} \right) + \frac{d\tau}{dr_0} + \frac{\tau}{r_0} = 0. \tag{2.259}$$

Note, that the derivative in this equation is usually neglected. Neglecting the derivative of the line tension in Eq. (2.259) results in

$$\cos\theta_e = \cos\theta_{e\infty} - \frac{\tau}{\gamma r_0}. \tag{2.260}$$

If the line tension is negative, then the influence of line tension results in a bigger contact angle as compared with predictions according to Eq. (2.260).

In general, case Eq. (2.259) can be rewritten as

$$m\gamma \cdot \left(\cos\theta_e - \cos\theta_{e\infty} \right) = \pm \left(\frac{\tau}{r_0} + \frac{\partial \tau}{\partial r_0} \right), \tag{2.261}$$

where m and plus or minus depends on the system geometry: $m = 1$ and "$-$" correspond to the drop on the solid substrate, $m = 2$ and "$+$" correspond to the two identical drops/bubbles in contact (Figure 2.33); and contact angle $\theta_{e\infty}$ in the case of big cylindrical drops. In the case of liquid drops on the flat solid substrate, positive values of τ cause an increase in the values of contact angles θ_e, whereas negative values result in their decrease. In the case of flat films in contact with a concave meniscus, the influence of the line tension τ is inversed, because a minus sign should be used now in Eq. (2.261).

For water and aqueous electrolyte solutions, the line tension values are of about 10^{-6} to 10^{-5} dyn [53] and below. Thus, the terms in the right-hand side of Eq. (2.261) become noticeable at $r_0 \leq 10^{-4}$ cm.

Important, the above cited value of the line tension can be Eq. (2.261) used only at equilibrium. It is impossible to experimentally measure equilibrium liquid droplets on solid substrates because they should be at the equilibrium with oversaturated vapor, as it was explained in Section 2.3. This means that everything in Eq. (2.261) is usually either far from equilibrium or under a quasi-equilibrium condition (as caused by the hysteresis of contact angle, see Chapter 3). In this case the value of line tension can be many orders of magnitude higher than 10^{-6} to 10^{-5} dyn. However, this line tension should be referred to as "dynamic line tension." To the best of our knowledge, there has not yet been any attempt to introduce or investigated the "dynamic line tension."

The values of the line tension, τ, for drops on solid substrates has been calculated as a difference between the values of $\gamma \cdot \cos\theta_e$ and $\gamma \cdot \cos\theta_{e\infty}$, calculated in two different ways: (i) neglecting the transition zone and (ii) taking it into account. Because the line tension arises due to the existence of the transition zone, this difference is associated only with the additional terms in the right-hand side of Eq. (2.260).

An expression for τ was obtained in the case of a model isotherm of the Derjaguin's pressure [53], and the line was estimated as being $10^{-5} \div 10^{-6}$ dyn and negative. de Feijter and Free [54] considered the transition zone between Newton black film (a different name for α-films) and bulk liquid. According to their estimations, the line tension value is also negative and has the same order of magnitude. T. Kolarov and Z.M. Zorin [55] measured the line tension value using Sheludko's cell for measurements of the properties of free liquid films. An aqueous solution of 0.1% NaCl with SDS at 0.05% concentration (CMC = 0.2%) was used. They calculated the line tension for this system using the Neumann-Young' equation (Eq. 2.261) and found the value of line tension to be $-1.7 \cdot 10^{-6}$ dyn. That is, negative, and in good agreement with theoretical predictions [53].

However, D. Platikanov et al. [56] carried out experimental measurements of line tension dependency on salt concentration and presented experimental evidence of line tension sign change. Below we present a theory, which is modified as compared with that presented in [53], and it explains the experimentally discovered change in the line tension sign.

The highlighted text below can be omitted at the first reading.

Let us consider two identical drops/bubbles in contact (Figure 2.33) under equilibrium conditions. It was explained in Section 2.3 that it is important to properly select the reference state. This reference state is introduced as "a flat equilibrium liquid film of the thickness $2h_e$." The reason for that was presented in Sections 2.1 and 2.3. From a mathematical point of view, it means the addition/subtraction of a constant to/from the excess free energy. This constant does not influence the final equation, which describes the profile in the transition zone. However, as we saw in Section 2.3, this choice is essential for transversality conditions at the apparent three-phase contact line, and this choice is also important for the definition of the line tension. This means that the choice of the reference state is very important, and below we use the same choice of the reference state as used in Section 2.1, which is a uniform flat equilibrium film of thickness $2h_e$, where h_e is the half-thickness of the equilibrium film. Using that choice, the excess free energy, Φ, of the system (curve 1 in Figure 2.34) has the following form:

$$\Phi = 2\pi \int_0^R r \left[2\gamma \left(\sqrt{1 + {h'}^2} - 1 \right) + \int_{2h}^{\infty} \Pi(h)\,dh - \int_{2h_e}^{\infty} \Pi(h)\,dh + 2P_e \cdot (h - h_e) \right] dr, \qquad (2.262)$$

where $h(x)$ is the half-thickness of the liquid layer; h_e is a half-thickness of the flat equilibrium thin liquid film; $P_e = 2\gamma/R$ is the excess pressure; $\Pi(h)$ is the Derjaguin's pressure; and $h(R) = H$ is the position of the drop's end. The lower limit of integration corresponds to the end of the transition zone, $h = h_e$. We can use infinity as the upper limit of integration instead of R because at this stage we are not interested in the upper part of the drop/bubble. Note, the sign of the excess pressure now is opposite to that used in Section 2.8 because now the excess pressure, P_e, as compared with the pressure inside the droplets is positive.

Under the equilibrium condition the system is at the minimum free energy state, that is, conditions (A)–(D) should be satisfied (Section 2.2). This results in the equation for determination of the liquid profile in the transition zone:

$$\frac{\gamma}{r}\frac{d}{dr}\left(r\frac{\cdot h''}{\left[1+\left(h'\right)^2\right]^{\frac{1}{2}}}\right)+\Pi\left(2h\right)=P_e. \tag{2.263}$$

The transversality condition (D) results in (see Appendix 1 in Section 2.3)

$$h' \to 0, \tag{2.264}$$

at the end of the transition zone, which means a smooth transition from the transition zone to the flat thin film.

Let us introduce an ideal profile of the liquid interlayer in the transition zone, $2h_{id}$, which is a spherical part up to the intersection with the equilibrium liquid interlayer of thickness, $2h_e$. The excess free energy of such an ideal profile differs from the exact excess free energy given by Eq. (2.262) because the presence of the transition zone is ignored in the case of the ideal profile. This means that the line tension, τ, should be introduced to compensate for the difference:

$$\Phi = 2\pi\int_{r_0}^{\infty} r\left[2\gamma\left(\sqrt{1+\left(h_{id}'\right)^2}-1\right)-\int_{2h_e}^{\infty}\Pi(h)dh+2P_e\cdot\left(h_{id}-h_e\right)\right]dr+2\pi r_0\tau, \tag{2.265}$$

Under equilibrium conditions, the same equilibrium conditions (A)–(D) should be satisfied, which gives an equation for the ideal liquid profile in the transition zone

$$\frac{\gamma}{r}\frac{d}{dr}\left(r\frac{h_{id}'}{\left[1+\left(h_{id}'\right)^2\right]^{\frac{1}{2}}}\right)=P, \tag{2.266}$$

and the transversality condition, which in the case of ideal liquid profile with excess free energy given by Eq. (2.265), is as follows:

$$-\left[f_{id}-h_{id}'\frac{\partial f_{id}}{\partial h_{id}'}\right]_{r=r_0}+\frac{d(r_0\tau)}{dr_0}=0, \tag{2.267}$$

where $f_{id}=r\left[2\gamma\left(\sqrt{1+h_{id}'^2}-1\right)-\int_{2h_e}^{\infty}\Pi(h)dh+2P\cdot(h_{id}-h_e)\right]$. Substitution of this expression into condition (2.267) results in the following equation at $r=r_0$ and $h=h_e$:

$$\left[2\gamma\left(\sqrt{1+h_{id}'^2}-1\right)-\int_{2h_e}^{\infty}\Pi(h)dh-\frac{2\gamma h_{id}'^2}{\sqrt{1+h_{id}'^2}}\right]_{r=r_0}-\left(\frac{d\tau}{dr_0}+\frac{\tau}{r_0}\right)=0.$$

After a rearrangement, this condition becomes

$$2\gamma \cdot \left(\cos\theta_e - \cos\theta_{e\infty} \right) = \left(\frac{\tau}{r_0} + \frac{\partial \tau}{\partial r_0} \right),$$

(2.268)

where

$$2\gamma \cos\theta_{e\infty} = 2\gamma + \int\limits_{2h_e}^{\infty} \Pi(h)\,dh.$$

(2.269)

Equation (2.268) coincides with Eq. (2.261) if select $m = 2$ and "+". Equation (2.269) gives the expression for the contact angle of a big cylindrical drop.

Excess free energy given by Eqs. (2.262) and (2.265) should be equal. This gives the following definition of the line tension, τ:

$$\int\limits_{r_0}^{\infty} r\left[2\gamma \left(\sqrt{1+\left(h'_{id}\right)^2} - 1 \right) - \int\limits_{2h_e}^{\infty} \Pi(h)\,dh + 2P \cdot \left(h_{id} - h_e \right) \right] dr + r_0\tau$$

$$= \int\limits_{0}^{\infty} r\left[2\gamma \left(\sqrt{1+h'^2} - 1 \right) + \int\limits_{2h}^{\infty} \Pi(h)\,dh - \int\limits_{2h_e}^{\infty} \Pi(h)\,dh + 2P \cdot (h - h_e) \right] dr.$$

(2.270)

This equation presents an exact definition of the line tension, τ, in contrast to Eq. (2.268), where the value of line tension is unknown. In Eq. (2.270), the real liquid profile, $h(r)$, is the solution of Eq. (2.263) and the ideal liquid profile, $h_{id}(r)$, is the solution of Eq. (2.266).

Dependency of the line tension on the radius, r_0, was investigated in [53] in the case of a model Derjaguin's pressure isotherm. Below we focus on the absolute value of the line tension and a possible comparison with experimental data.

For this, let us consider the line tension in the simplest possible case: contact of two identical cylindrical drops/bubbles. In this case, the corresponding excess free energies (2.262) and (2.265) take the following form:

$$\Phi = \int\limits_{0}^{\infty} \left[2\gamma \left(\sqrt{1+h'^2} - 1 \right) + \int\limits_{2h}^{\infty} \Pi(h)\,dh - \int\limits_{2h_e}^{\infty} \Pi(h)\,dh + 2P_e \cdot (h - h_e) \right] dx,$$

(2.271)

and

$$\Phi = \int\limits_{x_0}^{\infty} \left[2\gamma \left(\sqrt{1+\left(h'_{id}\right)^2} - 1 \right) - \int\limits_{2h_e}^{\infty} \Pi(h)\,dh + 2P_e \cdot \left(h_{id} - h_e \right) \right] dx + \tau,$$

(2.272)

where now Φ is an excess free energy per unit length.

From Eq. (2.271), we conclude

$$\frac{\gamma \cdot h''}{\left[1 + \left(h'\right)^2 \right]^{\frac{3}{2}}} + \Pi(2h) = P.$$

(2.273)

Equation (2.273) describes the whole range of the liquid profile, including the lower bulk part of the drop/bubble, the flat thin liquid interlayer in front of it, and the transition zone in between. The boundary conditions for Eq. (2.273) are

$$h(R) = H, \quad h'(R) = -\infty, \tag{2.274}$$

$$h \to h_e, \quad x \to \infty$$
$$h'\big|_{h=h_e} = 0, \quad x \to \infty. \tag{2.275}$$

We can integrate Eq. (2.273) once using boundary condition Eq. (2.274), which yields

$$\frac{\gamma}{\left[1 + \left(h'\right)^2\right]^{\frac{1}{2}}} = L_e(h), \tag{2.276}$$

where $L_e(h) = P_e \cdot (H - h) - \int_{2h}^{\infty} \Pi(h)dh$. Equation (2.276) can be rewritten as

$$h' = -\sqrt{\frac{\gamma^2}{L_e^2(h)} - 1}. \tag{2.277}$$

The left-hand side of Eq. (2.276) is always positive and less than γ. That means the same should be true for the right-hand side of Eq. (2.276):

$$0 \le L(h) \le \gamma, \tag{2.278}$$

where $L(h) = \gamma$, if $h = h_e$ and $L(h) = 0$, if $h = H$.

Condition Eq. (2.275) results in

$$P_e \cdot (H - h_e) = \gamma + \int_{2h_e}^{\infty} \Pi(h)dh. \tag{2.279}$$

The capillary pressure can be expressed as before:

$$P_e = \frac{\gamma}{R}, \tag{2.280}$$

where R is the radius of the curvature of the cylindrical drop. Simple geometrical considerations show

$$R = \frac{H}{\cos \theta_e}. \tag{2.281}$$

With the help of this condition, we can conclude that

$$P_e = \frac{\gamma \cdot \cos \theta_e}{H}. \tag{2.282}$$

Using Eqs. (2.279) and (2.282) we conclude

$$\cos\theta_e = \frac{1+\dfrac{1}{2\gamma}\cdot\displaystyle\int_{2h_e}^{\infty}\Pi(h)dh}{1-\dfrac{h_e}{H}},\tag{2.283}$$

$$\frac{\gamma\cdot\cos\theta_e}{H}=\Pi\left(2h_e\right),\tag{2.284}$$

these two equations are modifications of our previous consideration in the case of drops/menisci (see Section 2.3). If $h_e/H \ll 1$, the contact angle in Eq. (2.283) is referred to as $\theta_{e\infty}$ and coincides with that given by Eq. (2.269).

In the case of partial wetting, the contact angle is in the following range $0\le\theta_e\le\dfrac{\pi}{2}$, or $0<\cos\theta_e<1$. Using condition (2.278), this inequality can be rewritten as

$$-\gamma<\int_{2h_e}^{\infty}\Pi\left(h\right)dh<-\frac{\gamma\cdot h_e}{H}.$$

Hence, the integral, $\int_{2h_e}^{\infty}\Pi\left(h\right)dh$, should be negative in the case of partial wetting.

In the case of the ideal profile, we should use Eq. (2.272), which results in

$$\frac{\gamma\cdot h_{id}''}{\left[1+\left(h_{id}'\right)^2\right]^{\frac{3}{2}}}=P_e.\tag{2.285}$$

The boundary conditions for Eq. (2.285) are

$$h_{id}\left(R\right)=H,\qquad h_{id}'\left(R\right)=-\infty.\tag{2.286}$$

Equation (2.285) can be integrated once using boundary conditions (2.286), which yields:

$$h_{id}'=-\sqrt{\frac{\gamma^2}{L_{id}^2}-1},\tag{2.287}$$

where $L_{id}\left(h_{id}\right)=P(H-h_{id})$.

Line tension τ can be expressed using Eqs. (2.271) and (2.272) as

$$\tau=\int_0^{\infty}\left[2\gamma\left(\sqrt{1+h'^2}-1\right)+\int_{2h}^{\infty}\Pi\left(h\right)dh-\int_{2h_e}^{\infty}\Pi\left(h\right)dh+2P\cdot(h-h_e)\right]dx$$

$$-\int_{x_0}^{\infty}\left[2\gamma\left(\sqrt{1+\left(h_{id}'\right)^2}-1\right)-\int_{2h_e}^{\infty}\Pi\left(h\right)dh+2P\cdot\left(h_{id}-h_e\right)\right]dx.$$

Using Eqs. (2.277) and (2.287), we can rewrite this equation as

$$\tau=2\int_0^{\infty}\left[\gamma\sqrt{1+h'^2}-L(h)\right]dx-2\int_{x_0}^{\infty}\left[\gamma\sqrt{1+\left(h_{id}'\right)^2}-L_{id}(h_{id})\right]dx.\tag{2.288}$$

Using Eqs. (2.277) and (2.287), we can switch from integration over x to integration over thickness, h. This transformation of Eq. (2.288) gives

$$\tau = 2\int_{h_e}^{\infty} (\sqrt{\gamma^2 - L^2(h)} - \sqrt{\gamma^2 - L_{id}^2(h)})\, dh.$$ (2.289)

Note that integration in the right-hand side is over h from h_e to infinity, which does not depend on the profile (real or ideal profile) but only on the integration limits.

Equation (2.289) can be rewritten as

$$\tau = \frac{2}{\gamma} \int_{h_e}^{\infty} \frac{2P(H-h)\displaystyle\int_{2h}^{\infty} \Pi(h)\, dh - \left(\displaystyle\int_{2h}^{\infty} \Pi(h)\, dh\right)^2}{\sqrt{1 - L^2(h)/\gamma^2} + \sqrt{1 - L_{id}^2(h)/\gamma^2}}\, dh.$$ (2.290)

In the case of $h \ll H$ inside the whole transition zone, the expressions for $L(h)$ and $L_{id}(h)$ can be rewritten as

$$L(h) = \gamma\left(\cos\theta_{e\infty} - \varepsilon(h)\right), \quad \varepsilon(h) = \frac{1}{\gamma}\int_{2h}^{\infty} \Pi(h)\, dh$$

$$L_{id}(h) = \gamma\cos\theta_{e\infty}.$$

Using this expression, Eq. (2.290) can be rewritten as

$$\tau = \frac{2\gamma}{\sin\theta_{e\infty}} \int_{h_e}^{\infty} \frac{2\cos\theta_{e\infty}\varepsilon(h) - \varepsilon^2(h)}{1 + \sqrt{1 + \dfrac{2\cos\theta_{e\infty}\varepsilon(h) - \varepsilon^2(h)}{\sin^2\theta_{e\infty}}}}\, dh.$$ (2.291)

In the case of small contact angles, $\varepsilon(h) \ll 1$, this equation takes the following form:

$$\tau = \frac{2\gamma}{\sin\theta_{e\infty}} \int_{h_e}^{\infty} \frac{2\cos\theta_{e\infty}\varepsilon(h) - \varepsilon^2(h)}{1 + \sqrt{1 + \dfrac{2\cos\theta_{e\infty}\varepsilon(h) - \varepsilon^2(h)}{\sin^2\theta_{e\infty}}}}\, dh.$$ (2.292)

It is possible to show that $0.5 < \dfrac{1}{1 + \sqrt{1 + \frac{2\cos\theta_{e\infty}\varepsilon(h)}{\sin^2\theta_{e\infty}}}} < 1$ in the case under consideration. Let the mean

value of this expression be ω, where $0.5 < \omega < 1$, then Eq. (2.292) takes the fowwllowing form

$$\tau = \frac{4\omega}{\tan\theta_{e\infty}} \int_{2h_e}^{\infty} \left(\int_{2h}^{\infty} \Pi(h)\, dh\right) dh \approx \frac{4\omega}{\theta_{e\infty}} \int_{2h_e}^{\infty} \left(\int_{2h}^{\infty} \Pi(h)\, dh\right) dh.$$ (2.293)

Equation (2.269) can be rewritten as $\cos\theta_{e\infty} = 1 + \varepsilon(h_e)$, or

$$\sin\theta_{e\infty} \approx \theta_{e\infty} = 2\sqrt{-\varepsilon(h_e)} = 2\sqrt{-\frac{1}{\gamma}\int_{2h_e}^{\infty} \Pi(h)\, dh}.$$ (2.294)

The combination of Eqs. (2.293) and (2.294) results in

$$\tau = \frac{2\omega}{\sqrt{-\dfrac{1}{\gamma}\displaystyle\int_{2h_e}^{\infty}\Pi(h)dh}} \int_{2h_e}^{\infty}\left(\int_{2h}^{\infty}\Pi(h)dh\right)dh = \frac{2\omega\sqrt{\gamma}}{\sqrt{-\displaystyle\int_{2h_e}^{\infty}\Pi(h)dh}} \int_{2h_e}^{\infty}\left(\int_{2h}^{\infty}\Pi(h)dh\right)dh. \qquad (2.295)$$

Below we compare the experimental data by D. Platikanov, M. Nedyalkov's [56] to the theory predictions according to Eq. (2.295) (at $\omega = 1$).

Below, we use only dispersion and electrostatic components of the Derjaguin's pressure. The following expressions were used for the different components of the Derjaguin's pressure:

For the dispersion component,

$$\Pi_d = -\frac{A}{h^3}, \qquad (2.296)$$

where $A \approx 10^{-14}$ dyn $\cdot$ cm is the Hamaker constant, and for the electrostatic component,

$$\Pi_{el} = 64 \cdot c \cdot R_g \cdot T \cdot \tan h^2(\bar{\psi}/4) \cdot \exp(-\kappa \cdot h), \qquad (2.297)$$

where $\cdot c$, R_g, T, F, $\bar{\psi} = \dfrac{F\psi}{RT}$, $\kappa = \sqrt{\dfrac{8\pi F^2 c}{\varepsilon_w RT}}$ are concentration of the electrolyte, the universal gas constant, temperature in K, Faraday number, dimensionless zeta potential of the film surfaces, and the inverse Debye length, respectively; and ε_w is the dielectrics constant of water. Hence, the total Derjaguin's pressure is

$$\Pi(h) = -\frac{A}{h^3} + 64 \cdot c \cdot R_g \cdot T \cdot \tan h^2(\bar{\psi}/4) \cdot \exp(-\kappa \cdot h). \qquad (2.298)$$

Unfortunately, the Derjaguin's pressure in the form given by Eq. (2.298) does not allow any equilibrium liquid films at low thickness (α-films). To overcome this problem, we introduce a cutoff thickness, t_* (Figure 2.42).

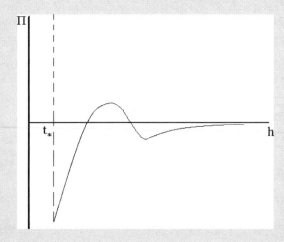

FIGURE 2.42   Model Derjaguin's pressure isotherm according to Eq. (2.298) with cutoff thickness, t_*.

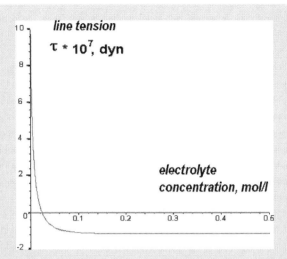

FIGURE 2.43 Calculated dependency of line tension on electrolyte concentration.

According to this choice, the equilibrium thickness, $2h_e$, does not depend on the pressure inside the drops/bubbles and is always equal to t_*.

D. Platikanov et al. [56] determined both the contact angle and the line tension on the NaCl concentration in the range of 0.2–0.45 mol/L. These two dependencies are discussed below.

The Comparison with the Experimental Data and Discussion

Details of experimental measurement and the system under consideration are given in [56]. The line tension of free liquid film between two bubbles was investigated in the range of NaCl concentrations of 0.2–0.45 mol/L. The film and bubble surfaces were stabilized by surfactants [56]. The zeta-potential of the film surface was $\psi = 17$ mV or $\bar{\psi} = 0.68$ according to [56].

The cutoff thickness, t_*, was used as a fitting parameter. We used experimental values of the contact angle from [56] on a salt concentration to determine the cutoff thickness, t_*, according to Eq. (2.294). A reasonable agreement between experimental dependency of the contact angle on the salt concentration and the calculated according to Eq. (2.294) was attained. The fitted dependency of the contact angle on the salt concentration was much weaker than the original experimental data [56]. However, we tried to compare our calculation of line tension according to Eq. (2.295) and the corresponding experimental data of line tension from [56] using the already calculated cutoff thickness, t_*. The calculated dependency of line tension on the electrolyte concentration is shown in Figure 2.43.

The following conclusions can be made based on the consideration of the dependency presented in Figure 2.43:

- Line tension dependency on the salt concentration is in qualitative agreement with experimental dependency in [56], that is, line tension goes from positive to negative values as the electrolyte concentration increases.
- The absolute values of the calculated line tension were found to be considerably different from the corresponding experimental values; that is, the calculated electrolyte concentration (0.022 mol/L) at which the line tension switches from positive to negative values does not match the experimental value (0.36 mol/L).

- The calculated line tension remains almost constant in the range of electrolyte concentrations used in [56].
- Line tension decreases much faster with an increase of electrolyte concentration (Figure 2.43) than experimental values [56].

The discrepancy between the measured and calculated line tension dependencies can be caused by one or both of the following reasons:

- Only dispersion and electrostatic components of Derjaguin's pressure were used in comparison with the experimental data. It looks like these two components are not enough to adequately describe the behavior of thin liquid films and the transition region in the system under consideration. The influence of both the structural component (caused by the water dipoles orientation in a vicinity of free film surfaces) and the steric component (caused by the direct interaction of head of the surfactant molecules on the film surfaces) cannot be ignored. The theory of these components of the Derjaguin's pressure is yet to be developed.
- According to the definition of line tension (2.290), it is determined by the equilibrium liquid profile in the transition region from the flat thin liquid interlayer to the bulk surface of bubbles. Note that in the case of partial wetting, which was under consideration, the static hysteresis of the contact angle (see Chapter 3) is unavoidable. This phenomenon can substantially influence the comparison.

In [56] considerable efforts were undertaken to reach the equilibrium state. We would like to emphasize again that if the liquid profile in the transition region from thin liquid interlayer to the bulk drop/bubble interface is not at equilibrium, then values of line tensions can differ considerably from theoretically predicted equilibrium values.

In the case of quasi-equilibrium (no macroscopic motion but possible microscopic motion inside the transition zone; see Chapter 3), it is necessary to introduce a new "dynamic line tension." We believe that this is a challenging area of the research in the future.

2.11 Capillary Interaction between Solid Bodies

In this section we consider the capillary interaction between two solid plates partly immersed in liquid, which can be, one or both, *completely* wetted, *partially* wetted or *non-wetted* at all by the liquid. We derive an expression for the force of interaction between the solid plates at *large* separation distances. It is shown that if one of the plates is wetted and the other is not, then a critical separation exists such that below the separation the plates attract each other, whereas at larger separations, there is repulsion. It is shown also that there is a critical angle of relative inclination which separates the regions of attraction and repulsion [57]. We not specify the kind of contact angle because it can be either a static *advancing* or a static *receding* contact angle, depending on the way the system is formed. For example, if both plates are immersed into the liquid from above, then both contact angles are static advancing contact angles.

The force of interaction between the plates is calculated below at large separations $L >> a$, where $a = \sqrt{\gamma/\rho g}$ is the capillary length, γ is the surface tension, ρ is the density of the liquid, and g the gravity acceleration. We consider only the case of two partially (or complete) wettable plates, that is, $\theta_1 < \pi/2$, $\theta_2 < \pi/2$, because the case of two non-wettable plates, $\theta_1 > \pi/2$, $\theta_2 > \pi/2$, can be treated in a similar way and also results in an attraction between plates as in the case of partially wettable plates.

Note that in the case of partial or complete wetting, the height of the liquid between plates is higher than outside plates, that is, $h_1 > H_1$ and $h_2 > H_2$ (Figure 2.44).

The principle of the frozen state [58] is used to calculate the force of interaction between the plates. Let us imagine that the liquid between plates 1 and 2 (Figure 2.44) has solidified above the level of the free liquid surface. The resulting solid body is divided by a plane parallel to plates 1 and 2, passing

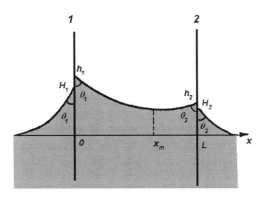

FIGURE 2.44 Liquid profile between two partially wettable plates, partially immersed into the liquid. Contact angles are $\theta_1 \neq \theta_2$, respectively.

through $x = x_m$, where $\left.\dfrac{dh}{dx}\right|_{x=x_m} = 0$. The displacement of plate 2 through dL while plate 1 is fixed changes the excess free energy of the system, Φ, by $d\Phi = \dfrac{\rho g l}{2} h_m^2 dL$, where l is the width of the plates. Because $d\Phi = FdL$, where F is the force of between the plates, we get

$$F = \frac{\rho g}{2} l h_m^2. \tag{2.299}$$

According to the previous consideration, the shape of a liquid surface between plates in the gravity field can be described by

$$\frac{\gamma h''}{\left(1+h'^2\right)^{3/2}} = \rho g h,$$

or

$$\frac{h''}{\left(1+h'^2\right)^{3/2}} = \frac{h}{a^2}.$$

Close to the position x_m, the liquid profile has a low slope, that is, $h'^2 \ll 1$, and, hence, in the vicinity of this position, the shape of the liquid surface can be described by the linearized equation $h'' = h/a^2$, hence close to $x = x_m$,

$$h(x) = c_1 \exp(-x/a) + c_2 xp\left(-(L-x)/a\right). \tag{2.300}$$

The condition $\left.\dfrac{dh}{dx}\right|_{x=x_m} = 0$ results in

$$h_m^2 = 4c_1 c_2 \exp(-L/a), \tag{2.301}$$

where $h_m = h(x_m)$.

The exact solution for the shape of a liquid surface close to an isolated plate is deduced in Appendix 2:

$$x = \frac{a}{2}\ln\frac{\left(\sqrt{2}-\sqrt{1+\sin\theta}\right)\left(1+\sqrt{1-\frac{h^2}{4a^2}}\right)}{\left(\sqrt{2}+\sqrt{1+\sin\theta}\right)\left(1-\sqrt{1-\frac{h^2}{4a^2}}\right)} + a\sqrt{2}\left(\sqrt{1+\sin\theta}-\sqrt{2-\frac{h^2}{2a^2}}\right). \tag{A2.8}$$

Using this expression at $x \gg a$, we get $h(x) \approx c_\infty(\theta)\exp(-x/a)$, where $c_\infty(\theta) = 4a\sqrt{\frac{\sqrt{2}-\sqrt{1+\sin\theta}}{\sqrt{2}+\sqrt{1+\sin\theta}}}$.

Substituting $c_i = c_\infty.(\theta_i)$, $i = 1,2$ from this equation into Eqs. (2.300) and (2.301) gives an expression for the force of interaction between the two plates at $L \gg a$:

$$F = 32\rho ga^2 l\sqrt{\frac{\sqrt{2}-\sqrt{1+\sin\theta_1}}{\sqrt{2}+\sqrt{1+\sin\theta_1}} \cdot \frac{\sqrt{2}-\sqrt{1+\sin\theta_2}}{\sqrt{2}+\sqrt{1+\sin\theta_2}}}\, e^{-L/a}, \tag{2.302}$$

In the case of two identical pates ($\theta_1 = \theta_2 = \theta$), this expression results in

$$F = 32\rho ga^2 l\frac{\sqrt{2}-\sqrt{1+\sin\theta}}{\sqrt{2}+\sqrt{1+\sin\theta}}\, e^{-L/a}, \tag{2.302'}$$

and in the case of complete wetting ($\theta = 0$)

$$F = 5.5\rho ga^2 l e^{-L/a} \tag{2.302''}$$

This equations show, that two partially wettable plates attract each other and that the force of attraction between such plates decays exponentially at $L \gg a$.

We can similarly calculate the force of interaction between any two bodies for which the surface between the bodies, along which the height of capillary rise of the liquid has a minimum, is a plane. Examples are two identical spherical or cylindrical particles. In [59], the profile was calculated for a liquid close to a cylinder partially immersed in the liquid. Using this expression for the shape of the surface of the meniscus at a cylinder of radius R at large separations, we get the following expression:

$$F = 4\rho ga^2 R\cos^2\theta\int_{L/2a}^{\infty} K_0^2(x)\,d\sqrt{x^2-L^2/4a^2} \approx \frac{\pi^{3/2}\rho ga^2 R\cos^2\theta}{\sqrt{L/a}}\cdot\exp\{-L/a\}, \tag{2.303}$$

where K_0 is a cylindrical function. This expression shows that the force between two partially wetted cylinders is also attraction and that this force decays faster as compared with the case of two plates.

Below an expression deduced in Appendix 2 for the height of capillary rise of the meniscus at the plates on the free liquid side is used:

$$H_i = a\sqrt{2(1-\sin\theta_i)}. \tag{A2.5}$$

We now consider the interaction between wettable and non-wettable plates. Let there be an isolated plate 1 with contact angle $\theta_1 < \pi/2$. At some distances, L_k, we draw an intersecting plate 2 (Figure 2.45).

The angle between the surface of the unperturbed meniscus and the intersecting plate $\theta_2 > \pi/2$. If we replace the intersecting plane by a real plate with contact angle θ_2, the surface of the meniscus between the plates will obviously be unchanged; the meniscus at the second plate on the side of the free liquid surface will sink to a height, H_2, according to Eq. (A2.5). Thus, there are no forces acting along the x axis at plates 1 and 2. We denote the distance at which the angle between the intersecting plane and the meniscus is equal to the wetting angle θ_2 as $L_k(\theta_1, \theta_2)$. It is easy to check that $L_k(\theta_1, \theta_2)$ exists only if

$$\theta_1 + \theta_2 < \pi, \theta_1 < \frac{\pi}{2}, \theta_2 > \frac{\pi}{2}. \tag{2.304}$$

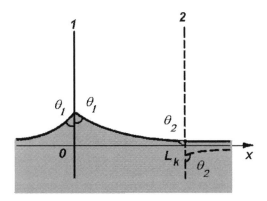

FIGURE 2.45 Capillary interaction between wettable and non-wettable plates.

If $\theta_1 + \theta_2 \to \pi$, $L_k \to 0$, and if $\theta_2 \to \frac{\pi}{2}$, $L_{k-} \to \infty$. If the condition $\theta_1 + \theta_2 < \pi$ holds, we can easily determine $L_k(\theta_1, \theta_2)$ by substituting $H_2 = a\sqrt{2(1-\sin\theta_2)}$ into the equation for the unperturbed meniscus at plate 1 (A2.8), which results in

$$L_k(\theta_1,\theta_2) = \frac{a}{2}\ln\frac{\left(\sqrt{2}-\sqrt{1+\sin\theta_1}\right)\left(\sqrt{2}+\sqrt{1+\sin\theta_2}\right)}{\left(\sqrt{2}+\sqrt{1+\sin\theta_1}\right)\left(\sqrt{2}-\sqrt{1+\sin\theta_2}\right)} + a\sqrt{2}\left(\sqrt{1+\sin\theta_1}-\sqrt{1+\sin\theta_2}\right). \quad (2.305)$$

The angle between the meniscus and the intersecting plate when $L < L_k$ is greater than θ_2, and consequently displacement of plate 2 from the position $L = L_k$ toward $L < L_k$ causes the liquid to rise between the plates and thus sets up attraction between the two plates. Similarly, displacement of plate 2 from the position $L = L_k$ toward $L > L_k$ causes the liquid to fall between the plates and consequently sets up repulsion. Thus, $L = L_k(\theta_1, \theta_2)$ separates two regions of interaction between the plates: attraction when $L < L_k$ and repulsion when $L > L_k$, that is, $L = L_k$ is a state of unstable equilibrium.

> *Important conclusion: According to condition Eq. (2.304) any completely wettable plate on a liquid surface is attracted to a non-wettable plate at sufficiently small distances.*

At separations $L \gg a$ (when $L > L_k$) between plates, one of which is partially wetted while the other is not, the liquid surface between them must intersect the level $h = 0$.

Using, the principle of the frozen state, the force of interaction between the plates in this case is

$$F = -\gamma l(1-\cos\alpha),$$

where $\alpha = arctan\, h'(x_0)$; and $h(x_0) = 0$. Close to $x = x_0$, the liquid profile, $h(x)$, as in the preceding case becomes $h = c_1\exp(-x/a) - c_2\exp(-(L-x)/a)$, whence

$$\cos\alpha = \frac{1}{\sqrt{1+h'^2(x_0)}} \approx 1 - \frac{1}{2}h'^2(x_0) = 1 - \frac{2c_1c_2^2}{a^2}\exp(L/a),$$

and for the interaction force, F, we get the following expression: $F = -\frac{2\gamma l c_1 c_2}{a^2}e^{-L/a}$.

As before, by expressing c_1, c_2 from the equations for unperturbed surfaces, we finally get

$$F = -32\gamma l\sqrt{\frac{\sqrt{2}-\sqrt{1+\sin\theta_1}}{\sqrt{2}+\sqrt{1+\sin\theta_1}} \cdot \frac{\sqrt{2}-\sqrt{1+\sin\theta_2}}{\sqrt{2}+\sqrt{1+\sin\theta_2}}}\, e^{-L/a}. \quad (2.306)$$

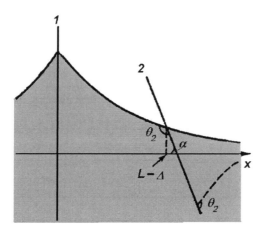

FIGURE 2.46 Critical inclination of plates.

Note, that this force is a repulsion force, which is completely different from the case of two partially wettable plates according to Eq. (2.302).

Now let us place a non-wettable plate 2 at an arbitrary position $L > L_k$ at an angle $\alpha(L) > 0$ (Figure 2.46), choosing $\alpha(L)$ so that the meniscus at the point $x = L - \Delta$ would not be disturbed by the presence of plate 2.

Figure 2.46 shows that the angle $\alpha(L)$ must be selected as

$$\alpha(L) = \pi - \theta_2 - \arctan h'(L - \Delta),\tag{2.307}$$

where Δ according to Figure 2.46 is given by

$$\Delta = \cot an\alpha(L)\, a\sqrt{2\left[1 + \sin\left(\frac{\pi}{2} - \arctan h'(L - \Delta)\right)\right]} = \cot an\alpha(L)\, a\sqrt{2\left[1 + \frac{1}{\sqrt{1 + h'^2(L - \Delta)}}\right]}.\tag{2.308}$$

At $L > L_k$ and $L \gg a$, $h' \to 0$ and, hence, according to Eq. (2.307),

$$\alpha(\infty) = \pi - \theta_2.\tag{2.309}$$

Thus, there are no forces acting on plate 1 when plate 2 is positioned at an angle α_k; this applies equally to plate 2. Inclination of plate 2 at an angle smaller than $\alpha(L)$ causes the liquid to rise between the plates (i.e., it sets up attraction between them), but an inclination at an angle bigger than $\alpha(L)$ causes the liquid to fall between the plates (i.e., it sets up repulsion between two plates). Thus, the angle $\alpha(L)$ separates the regions of attraction and repulsion between the plates, defining a state of unstable equilibrium.

Equations (2.307) and (2.308) can be used for numerical calculation of the unknown wetting angle of plate 2 at all separations, whereas Eq. (2.306) can be used at large separations only.

In conclusion, we note that positioning two non-wettable plates at an angle $\beta = \theta_2 - \pi/2$ from the vertical (Figure 2.47a) will constrain the liquid to a horizontal position between them, that is, the liquid between the plates forms a convex meniscus when the inclination angle is smaller than β and a concave meniscus when the inclination angle is bigger than β_k. That is, in this case, the liquid will rise between the plates, although it does not wet either plate.

Similarly, positioning wettable plates at an angle $\omega = \pi/2 - \theta_1$ (Figure 2.47b) will constrain the liquid to a horizontal surface between them, that is, when the inclination angle is bigger than

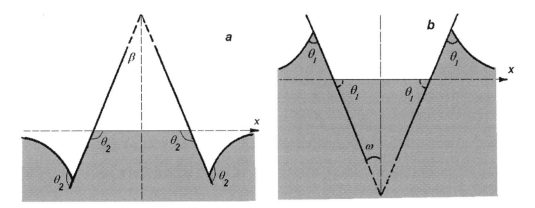

FIGURE 2.47 (a) Capillary rise between non-wettable plates. (b) Capillary interaction between wettable plates at inclination.

ω, the liquid will sink between wettable plates, forming a convex meniscus although it wets both plates.

This consideration and Figure 2.47a and b exemplify the fact that capillary imbibition of a liquid into a porous body depends significantly on the angle of opening of the pores and its change along the pore axis, that is, the pore distribution function with respect to a derivative of the radius [60]. The liquid cannot move beyond those places where the meniscus in the pore becomes flat, that is, where the effective pore radius is equal to infinity. This phenomenon is one of the reasons for capillary hysteresis in porous bodies and in the presence of places with trapped air.

In summary, we have shown that the capillary attraction or repulsion between solid bodies depends not only on their wetting features but also on their separation and on the angle of mutual relative inclination.

APPENDIX 2

Equilibrium liquid shape close to a vertical plate

Let us consider a vertical plate partially immersed into a liquid of density, ρ, and surface tension γ.

In this case, only capillary and gravity forces act, and the equation, which describes the liquid profile, is as follows:

$$\frac{\gamma h''}{\left(1+h'^2\right)^{3/2}} = \rho g h, \tag{A2.1}$$

where g is the gravity acceleration. Equation (A2.1) is the differential equation of the second order; hence, two boundary conditions should be specified. These conditions are

$$h'(0) = -\cot an\,\theta, \tag{A2.2}$$

$$h(x) \to 0, \quad x \to \infty. \tag{A2.3}$$

Condition (A2.3) is used below as

$$h'\big|_{h=0} = 0. \tag{A2.3'}$$

Using the capillary length, $a = \sqrt{\dfrac{\gamma}{\rho g}}$, Eq. (A2.1) can be rewritten as

$$\frac{h''}{\left(1+h'^2\right)^{3/2}} = \frac{h}{a^2}.$$

(A2.1′)

Multiplication of this equation by h' and integration with x results in

$$\frac{1}{\left(1+h'^2\right)^{1/2}} = C - \frac{h^2}{2a^2},$$

where C is an integration constant. Using the boundary condition (A2.3′) we conclude that $C = 1$,

$$\frac{1}{\left(1+h'^2\right)^{1/2}} = 1 - \frac{h^2}{2a^2}.$$

(A2.4)

The left-hand side of this equation is positive and so also should be right-hand side, which results in the restriction $h \le a\sqrt{2}$. We show below (see Eq. (A2.6)) that $H = a\sqrt{2}$ corresponds to the maximum possible elevation (Figure 2.48) in the case of complete wetting, that is, at $\theta = 0$.

Using boundary condition (A2.2), we conclude from Eq. (A2.4) that

$$H = a\sqrt{2(1-\sin\theta)},$$

(A2.5)

which gives $H = a\sqrt{2}$ in the case of complete wetting.

From Eq. (A2.4), we conclude

$$h' = -\frac{h\sqrt{2 - \dfrac{h^2}{2a^2}}}{a\sqrt{2}\left(1 - \dfrac{h^2}{2a^2}\right)},$$

(A2.6)

with boundary condition

$$h(0) = H = a\sqrt{2(1-\sin\theta)}.$$

(A2.7)

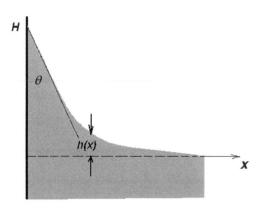

FIGURE 2.48 Liquid profile close to the vertical wall. θ, contact angle; and H, the maximum elevation.

Integration of Eq. (A2.6) with boundary condition (A2.7) results in

$$-\frac{x}{a} = \int_H^h \frac{\left(1 - \dfrac{h^2}{2a^2}\right)}{h\sqrt{1 - \dfrac{h^2}{4a^2}}} dh,$$

or

$$\frac{x}{a} = \frac{1}{2}\ln \frac{\left(\sqrt{2} - \sqrt{1+\sin\theta}\right)\left(1 + \sqrt{1 - \dfrac{h^2}{4a^2}}\right)}{\left(\sqrt{2} + \sqrt{1+\sin\theta}\right)\left(1 - \sqrt{1 - \dfrac{h^2}{4a^2}}\right)} + \sqrt{2}\left(\sqrt{1+\sin\theta} - \sqrt{2 - \dfrac{h^2}{2a^2}}\right). \qquad (A2.8)$$

Equation (A2.8) gives an implicit dependence of the liquid profile h on x.

REFERENCES

1. Derjaguin, B. V., Churaev, N. V., and Muller, V. M. *Surface Forces*, Plenum Press, New York, 1987.
2. Deryaguin, B. V., Starov, V. M., and Churaev, N. V. Profile of transition zone between a wetting film and meniscus of a bulk liquid. *Colloid J.*, USSR Academy of Sciences, 38, 789 (1976).
3. Deryagin, B. V., and Churaev, N. V. Isotherm of the disjoining pressure of water films on quartz surface. *Dokl. Akad. Nauk USSR.*, 207, 572 (1972).
4. Deryagin, B. V., and Zorin, Z. M. Investigation of the surface condensation and vapour adsorption close to saturation conditions with micropolarization method. *Sov. J. Phys. Chem.*, 29, 1755 (1955).
5. Exerowa, D. Chain-melting phase transition and short-range molecular interactions in phospholipid foam bilayers. *Adv. Colloid Interface Sci.*, 96, 75 (2002).
6. Exerowa, D., and Kruglyakov, P. M. *Foam and Foam Films*, Elsevier, New York, 1998.
7. Cohen, R., Exerowa, D., Kolarov, T., Yamanaka, T., and Muller, V. Thickness transitions in lysolecithin foam films. *Colloids Surf.*, 65, 201 (1992).
8. Miller, C. A., and Neogi, P. *Interfacial Phenomena*, New York, Marcel Dekker, 1985; Miller C. A., and Ruckenstein, E. The origin of flow during wetting of solids. *J. Colloid Interface. Sci.*, 48, 368 (1974).
9. Eremeev, G. G., and Starov, V. M. Capillary hysteresis and characteristics of structure of isotropic porous media. *J. Phys. Chem. USSR* (in Russian), 47 (11), 2921 (1973).
10. Derjaguin, B. V., and Starov, V. M. Capillary interaction between solid bodies. *Colloid J.*, USSR Academy of Sciences, 39 (3), 383 (1977).
11. Kuchin, I. V., Matar, O. K., Craster, R. V., and Starov, V. M. Influence of the disjoining pressure on the equilibrium interfacial profile in transition zone between a thin film and a capillary meniscus. *J. Colloid Interface Sci. Comm.*, 1, 18–22 (2014).
12. Churaev, N. V. Calculation of Hamaker constants for bodies interacting through a liquid interlayer. *Colloid J.*, USSR Academy of Sciences, 34, 988 (1972).
13. Parsegian, V. A. Formulae for the electrodynamic interaction of point particles with a substrate. *Mol. Phys.*, 37, 1503 (1974).
14. Dzyaloshinskii, I. E., Lifshitz, E. M., and Pitaevskii, L. P. *Zh. Van der Waals forces in liquid films. Eksp. Teor. Fiz.*, 37, 229 (1959).
15. Shishin, V. A., Zorin, Z. M., and Churaev, N. V. *Colloid J.*, USSR Academy of Sciences, 39, 520 (1977).
16. Starov, V. Nonflat equilibrium liquid shapes on flat surfaces. *J. Colloid Interface Sci.*, 269, 432 (2003).
17. Starov, V. M., and Churaev, N. V. Wetting films on locally heterogeneous surfaces: 1. Hydrophilic surface with hydrophobic inclusions. *Colloid J.*, Russian Academy of Sciences, 60 (6), 770 (1998).
18. Slavchov, R., Radoev, B., and Stöckelhuber, K. W. Equilibrium profile and rupture of wetting film on heterogeneous substrates. *Colloids Surf. A Physicochem. Eng. Asp.*, 261 (1–3), 135 (2005).
19. Kuchin, I. V., Matar, O. K., Craster, R. V., and Starov, V. M. Modeling the effect of surface forces on the equilibrium liquid profile of a capillary meniscus. *Soft Matter*, 10, 6024 (2014).

20. Young, T. An essay on the cohesion of fluids. *Philos. Trans. R. Soc.*, 95, 65–87 (1805).
21. Ahmed, G., Kalinin, V. V., Arjmandi-Tash, O., and Starov, V. M. Equilibrium of droplets on a deformable substrate: Influence of disjoining pressure. *Colloids Surf. A: Physicochem. Eng. Asp.*, 521, 3–12 (2017).
22. Koursari, N., Ahmed, G., and Starov, V. Equilibrium droplets on deformable substrates: Equilibrium conditions. *Langmuir*, 34 (19), 5672–5677 (2018).
23. Rusanov, A. Theory of wetting of elastically deformed bodies. 1. Deformation with a finite contact-angle. *Colloid J.*, 37, 614–622 (1975).
24. Rusanov, A. Theory of wetting of elastically deformed bodies. 2. Equilibrium conditions and work of deformation with a finite contact angle. *Colloid J.*, 37, 623–628 (1975).
25. Rusanov, A. Thermodynamics of deformable solid-surfaces. *J. Colloid Interface Sci.*, 63, 330–345 (1978).
26. Shanahan, M. E. R. The spreading dynamics of a liquid-drop on a viscoelastic solid. *J. Phys. D Appl. Phys.*, 21, 981–985 (1988).
27. Shanahan, M., and de Gennes, P. The ridge produced by a liquid near the triple line solid liquid fluid. *Comptes Rendus L Acad. Des Sci. Ser. II*, 302, 517–521 (1986).
28. Carre, A., and Shanahan, M. E. R. Viscoelastic braking of a running drop. *Langmuir*, 17, 2982–2985 (2001).
29. Jerison, E. R., Xu, Y., Wilen, L. A., and Dufresne, E. R. Deformation of an elastic substrate by a three-phase contact line. *Phys. Rev. Lett.*, 106, 186103 (2011).
30. Park, S. J., Weon, B. M., Lee, J. S., Lee, J., Kim, J., and Je, J. H. Visualization of asymmetric wetting ridges on soft solids with X-ray microscopy. *Nat. Commun.*, 5, 4369 (2014).
31. Schulman, R. D., and Dalnoki-Veress, K. Liquid droplets on a highly deformable membrane. *Phys. Rev. Lett.*, 115, 206101 (2015).
32. Shanahan, M., and Carre, A. Viscoelastic dissipation in wetting and adhesion phenomena. *Langmuir*, 11, 1396–1402 (1995).
33. Style, R. W., Boltyanskiy, R., Che, Y., Wettlaufer, J. S., Wilen, L. A., and Dufresne, E. R. Universal deformation of soft substrates near a contact line and the direct measurement of solid surface stresses. *Phys. Rev. Lett.*, 110, 066103 (2013).
34. Gielok, M., Lopes, M., Bonaccurso, E., and Gambaryan-Roisman, T. Droplet on an elastic substrate: Finite element method coupled with lubrication approximation. *Colloid Surf. A*, 521, 3–21 (2017).
35. Limat, L. Straight contact lines on a soft, incompressible solid. *Eur. Phys. J. E*, 35, 134 (2012).
36. Lubbers, L. A., Weijs, J. H., Botto, L., Das, S., Andreotti, B., and Snoeijer, J. H. Drops on soft solids: Free energy and double transition of contact angles. *J. Fluid Mech.*, 747, R1 (2014).
37. Style, R. W., Che, Y., Park, S. J., Weon, B. M., Je, J. H., Hyland, C. et al. Patterning droplets with durotaxis. *Proc. Natl. Acad. Sci.*, 110, 12541–12544 (2013).
38. Churaev, N., V. Starov, B. Derjaguin. The shape of the transition zone between a thin film and bulk liquid and the line tension. *J. Colloid Interface Sci.*, 89, 16–24 (1982).
39. White, L. The contact angle on an elastic substrate. 1. the role of disjoining pressure in the surface mechanics. *J. Colloid Interface Sci.*, 258, 82–96 (2003).
40. Derjaguin, B. V., Starov, V. M., and Churaev, N. V. Pressure on a wetting perimeter. *Colloid J.*, 44, 871–876 (1982).
41. Winkler, E. The theory of elasticity and strength with special reference to their application in the art for polytechnics, building academies, engineers, mechanical engineers, architects, etc., H. Dominicus, 1867.
42. Marchand, A., Das, S., Snoeijer, J. H., and Andreotti, B. Capillary pressure and contact line force on a soft solid. *Phys. Rev. Lett.*, 108, 094301 (2012).
43. Kerr, A. D. Elastic and viscoelastic foundation models. *J. Appl. Mech.*, 31, 491 (1964).
44. Starov, V. M., and Churaev, N. V. Wetting films on locally heterogeneous surfaces: Hydrophilic surface with hydrophobic spots. On the pressure applied to the wetting perimeter. *Colloids Surf. A Physicochem. Eng. Asp.*, 156, 243–248 (1999).
45. Starov, V. M., Churaev, N. V., and Derjaguin, B. V. On the pressure applied to the wetting perimeter. *Colloid J.*, USSR Academy of Sciences, 44 (5), 770 (1982).
46. Starov, V. M., and Churaev, N. V. Deformation of fluid particles in the contact zone and line tension. *Colloid J.*, USSR Academy of Sciences, 45 (5), 852 (1983).

47. Churaev, N. V., and Starov, V. M. Deformation of fluid particles in the contact zone and line tension. *J. Colloid Interface Sci.*, 103 (2), 301 (1985).

48. Lester, G. R. Contact angles of liquids at deformable solid surfaces. *J. Colloid Sci.*, 16 (4), 315 (1961).

49. Derjaguin, B. V., Muller, V. M., and Toporov, Y. P. Specific features of the condensation of liquids in narrow slits. *Colloid J.*, USSR Academy of Sciences, 37, 455 (1975); 37, 1066 (1975).

50. Derjaguin, B. V., Starov, V. M., and Churaev, N. V. Specific features of the condensation of liquids in narrow slits. *Colloid J.*, USSR Academy of Sciences, 37 (2), 219 (1975).

51. Us'yarov, O. G. Research in the field of stability of disperse systems and wetting films [in Russian]. Doctor of Sciences Thesis, SanktPetersburg (Leningrad) University, Department of Colloid Science, 1976.

52. Muller, V. M., and Yushchenko, Y. S. *Colloid J.*, USSR Academy of Sciences, 42, 500 (1980).

53. Rowlinson, J. F., and Widom, B. *Molecular Theory of Capillarity*, Clarendon, Oxford, UK, 1984.

54. Starov, V. M., and Churaev, N. V. *Colloid J.*, USSR Academy of Sciences. 42, 703 (1980).

55. de Feijter, J. A., and Vrij, A. I. Transition regions, line tensions and contact angles in soap films. *J. Electroanal. Chem.*, 37, 9 (1972).

56. Kolarov, T., and Zorin, Z. M. A measurement of line tension in a common black soap film. *Colloid Polym. Sci.*, 257, 1292 (1979).

57. Starov, V. *Emulsions: Structure, Stability and Interactions*, Edited by D. N. Petsev. Elsevier Academic Press, Amsterdam, the Netherlands, 2004, pp. 183–214.

58. Platikanov, D., Nedyalkov, M., and Nasteva, V. J. Line tension of newton black films. II. Determination by the diminishing bubble method. *Colloid Interface Sci.*, 75, 620 (1980).

59. Stevin, S., and der Weeghconst, D. B. *The Elements of the Art of Weighing*, Francois van Raphelinghen, Leyden, IL, 1586.

60. Deryagin, B. V. On the dependence of the contact angle on the microrelief or roughness of a wetted solid surface. *Dokl. Akad. Nauk SSSR*, 51, 517 (1946).

3

Hysteresis of Contact Angles Based on Derjaguin's Pressure

Introduction

It is possible, however, not easy, to measure equilibrium contact angles in thin capillaries. It is far more difficult, if possible at all, to measure equilibrium contact angles of sessile droplets because they should be at equilibrium with oversaturated vapor for a prolong period of time (see Chapters 1 and 2). That is, the experimentally measured contact angles are either static advancing or static receding contact angles.

Important: static contact angle hysteresis exists only in the case of partial wetting. There is no static contact angle hysteresis in the case of complete wetting. In Chapter 3 both static advancing and static receding contact angles were considered on the basis of the S-shaped Derjaguin's pressure isotherm (Figure 1.12, curve 2), which corresponds to the partial wetting case.

In Chapter 1, Section 1.3, a nature of contact angle hysteresis on smooth homogeneous non-deformable substrates was discussed on a phenomenological level. It was also shown in Section 1.3 that hysteresis of the contact angle on smooth homogeneous substrates is determined by surface forces action. However, the most important question remains: Is it possible to calculate both static advancing and static receding contact angles via the Derjaguin's pressure isotherm?

That is precisely what is investigated and answered in this chapter: Both static advancing and static receding contact angles are expressed via the Derjaguin's pressure isotherm. This is done in the case of both capillary menisci and sessile droplets.

It is interesting that in both cases (capillary menisci and droplets) the static receding contact angles are closer to being equilibrium contact angles than static advancing contact angles. This is in contradiction to the widely accepted point of view that the static advancing contact angle gives a reasonable estimation of the equilibrium contact angle.

The situation is even more complicated in the case of hysteresis contact angles on deformable substrates. We present a current state of theory in this area: The presented theory includes a number of very substantial simplifications and is to be developed further in the future to give a more general approach. However, even under the adopted simplifications, the main result is the same as in the case of non-deformable substrates: Both static advancing and static receding contact angles can be expressed via the Derjaguin's pressure isotherm plus the additional elasticity of the deformable substrate.

3.1 Hysteresis of Contact Angle of a Meniscus inside a Capillary with Smooth Homogeneous Non-deformable Walls

In this section we present a theory [1,2] of contact angle hysteresis of a meniscus inside a thin capillary with smooth homogeneous solid walls in terms of surface forces (Derjaguin's pressure isotherm) using a quasi-equilibrium approach. The Derjaguin's pressure isotherm includes electrostatic, intermolecular and structural components (Chapter 1). The values of the static advancing, θ_a, static receding, θ_r, and equilibrium, θ_e, contact angles in thin capillaries are calculated based on the shape of the Derjaguin's pressure isotherm. It is shown that both static advancing and static receding contact angles depend on the capillary radius. The suggested mechanism of the contact angle hysteresis has a direct experimental confirmation: The process of receding is accompanied by the formation of thick β-films on the capillary walls. The effect of the transition from partial to complete wetting in thin capillaries is predicted and analyzed. This effect takes place in very thin capillaries, when the receding contact angle decreases to zero.

It is generally believed that static hysteresis of the contact angle is determined by the surface roughness and/or heterogeneity [3]. There is no doubt that a roughness and/or a chemical heterogeneity of the solid substrate contribute substantially to the contact angle hysteresis. It is assumed in this case that, at each point of the surface, the equilibrium value of the contact angle is established depending only on the local properties of the substrate. As a result, a whole series of local thermodynamic equilibrium states can be realized, corresponding to a certain interval of contact angles. The maximum possible value corresponds to the static advancing contact angle, θ_a, and the minimum possible value corresponds to the static receding contact angle, θ_r. Hence, the dependency of the contact angle on the velocity of motion of a meniscus or a drop can be qualitatively described by the dependency presented by the solid lines in Figure 3.1.

However, roughness and/or heterogeneity of the surface are apparently not the sole reasons for contact angle hysteresis. An increasing number of publications over the last years have confirmed the presence of contact angle hysteresis even on smooth homogeneous surfaces [4–9]. However, the most convincing evidence for the presence of the contact angle hysteresis on smooth homogeneous surfaces is its presence on free liquid films [10–14], in which case there is hysteresis of a meniscus, which is located on the thin free liquid film. In this case the surfaces of free liquid films are not at all rough and are also chemically homogeneous. Hence, it is impossible to explain the hysteresis phenomenon by the presence of roughness and/or heterogeneity in the case of contact angle hysteresis on free liquid films.

A qualitative theory of contact angle hysteresis of menisci in thin capillaries [15] that was briefly described in Section 1.3 is based on an S-shape of the Derjaguin's pressure isotherm. The theory did not allow for describing the dependency of the static advancing/receding contact angles on the capillary radius. Below in this section quantitative theory of contact angle hysteresis is presented based on consideration of surface forces that act in the vicinity of the apparent three-phase contact line. This type of contact angle hysteresis exists even on a smooth homogeneous substrate. Consideration of this kind of contact angle hysteresis, that is, taking into account the surface forces action in the vicinity of the apparent three-phase contact line on rough and/or nonhomogeneous surfaces from this point of view, is to be undertaken in the future.

Evidently only a single unique value of equilibrium contact angle, θ_e, is possible on a smooth homogeneous surface. Hence, both static advancing, $\theta_a \neq \theta_e$, and static receding, $\theta_r \neq \theta_e$, contact angles as well as all contact angles in between that are observed experimentally on such surfaces, correspond only to

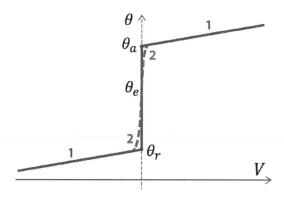

FIGURE 3.1   Solid lines (1) Idealized dependency of the contact angle on the advancing ($V > 0$) or the receding ($V < 0$) velocity of a droplet/meniscus. All contact angles, θ, between static advancing contact angle, θ_a, and static receding contact angle, θ_r, are considered as equilibrium contact angles. Curve (2) A real dependency of the contact angle on the advancing or receding velocities. At any deviation from the equilibrium contact angle, θ_e, the liquid drop/meniscus is in the state of a slow "microscopic motion," which almost abruptly transforms into "a macroscopic motion." In this case static advancing and receding contact angles, θ_a and θ_r, are extrapolations to zero velocity.

non-equilibrium states of the system. Hence, the picture presented by the solid lines in Figure 3.1 should now be replaced by a new, more realistic picture, which is represented by curve 2 in Figure 3.1.

The Derjaguin's Pressure Components

The nature of the Derjaguin's pressure was explained qualitatively earlier in Chapter 1. In this section we present more qualitative details of the Derjaguin's pressure components. It was mentioned in Chapter 1 that the properties of a liquid in the vicinity of liquid–air and solid–liquid interfaces differ from the corresponding properties in the bulk liquid because of surface forces action. The layers, where the surface forces act, are referred to as boundary layers (these boundary layers have nothing to do with hydrodynamic boundary layers). These boundary layers overlap near an apparent three-phase contact line. This overlapping of boundary layers is why the Derjaguin's pressure appears.

Contact angle hysteresis on smooth homogeneous substrates appears in the case of partial wetting, when the Derjaguin's isotherm has a special S-shape. Components contributing to the formation of the Derjaguin's pressure were discussed in Chapter 1 and see [16–19] for more details. According to the modified DLVO theory [18,19], these components are as follows:

1. The electrostatic component, which is caused by the formation of the electrical double layers and their overlapping:

$$\Pi_E = R_g T c_0 \left(\exp(\varphi) + \exp(-\varphi) \right) - 2 R_g T c_0 - \frac{(RT)^2 \varepsilon \varepsilon_0}{2F^2} \left(\frac{\partial \varphi}{\partial y} \right)^2, \tag{3.1}$$

where R_g, T, F, ε, and ε_0 are universal gas constant, temperature in K; Faraday's constant, dielectric constant of water, and dielectric constant of vacuum, respectively; c_0 is concentration of univalent electrolyte; y and φ are the coordinate normal to the liquid–air interface and dimensionless electric potential in $F/R_g T$ units, respectively.

The electric potential, φ, and the surface charge density, σ, in Eq. (3.1) are related as [18]

$$\sigma_h = \varepsilon \varepsilon_0 \frac{R_g T}{F} \left(\frac{\partial \varphi}{\partial y} \right)_{y=h} \quad \text{for the liquid/vapor interface,}$$

$$\sigma_s = -\varepsilon \varepsilon_0 \frac{R g T}{F} \left(\frac{\partial \varphi}{\partial y} \right)_{y=0} \quad \text{for the solid/liquid interface,}$$

where h is a separation between the interacting surfaces.

2. The structural component, which is caused by the orientation of water molecule dipoles in the vicinity of interfaces and the overlapping of these structured layers. This component is presented as a combination of both short-range and long-range interactions [20,21,31]:

$$\Pi_S(h) = K_1 \exp(-h/\lambda_1) + K_2 \exp(-h/\lambda_2), \tag{3.2}$$

where K_1, K_2 and λ_1, λ_2 are parameters related to the magnitude and the characteristic length of the structural forces action. The subscripts 1 and 2 correspond to the short-range and long-range structural interactions, respectively. Currently these four constants can be extracted from experimental data only. The constants K_1, K_2 usually have opposite signs (see parameters values to Figure 3.8).

3. The molecular or van der Waals component [16,18],

$$\Pi_M(h) = \frac{A}{6\pi h^3}, \tag{3.3}$$

where $A = -A_H$, A_H is the Hamaker constant. Note, the importance of the van der Waals component is usually grossly exaggerated in the literature: Other mentioned components of the Derjaguin's pressure are equally or even more important in the case of aqueous electrolyte solutions.

The resulting the Derjaguin's pressure isotherm has a characteristic S-shape [18,19]:

$$\Pi(h) = \Pi_M(h) + \Pi_E(h) + \Pi_S(h). \tag{3.4}$$

An example of possible shapes of the Derjaguin's pressure isotherms is given in Figure 1.12. The shape of the isotherm depends on the contribution of the surface forces components and corresponds to different wetting conditions.

The Derjaguin's Pressure and Wetting Phenomena

Kelvin's equation describes the change in vapor pressure over the curved liquid–vapor interface (e.g., a capillary or a droplet) [22]:

$$P_e = \frac{R_g T}{v_m} \ln \frac{p_s}{p}, \tag{3.5}$$

where $P_e = P_v - P_l$ is the excess pressure; P_l is the pressure inside the liquid; P_v is the pressure in the ambient vapor; v_m is the liquid molar volume; and p_s and p are the saturated vapor pressure (over a flat liquid surface) and the pressure over the curved interface, respectively. The excess pressure inside the droplet, P_e, should be negative (pressure inside the droplet is higher than the pressure in the ambient vapor). Thus, the right-hand side of Kelvin's equation must be negative, which is possible only if $p > p_s$; that is, droplets can only be at equilibrium with oversaturated vapor. This is why it is difficult to experimentally investigate equilibrium droplets on solid substrates: It is necessary to maintain an oversaturated vapor over the substrate under investigation for a prolonged period of time (see Chapter 1).

In contrast to a droplet, equilibrium for a meniscus according to the Kelvin's equation is possible with undersaturated vapor ($P_e > 0$ and $p < p_s$). Note, an equilibrium meniscus can exist in the case of either complete or partial wetting. This is different from equilibrium droplets, which can exist only in the partial wetting case; in the case of complete wetting, droplets spread out completely.

Schematic presentation of two possible shapes of the Derjaguin's pressure isotherms is given in Figure 1.12. These types of dependence, $\Pi(h)$, are typical for the sum of electrostatic, structural and van der Waals components of the Derjaguin's pressure (modified DLVO theory). Dependency 1 in Figure 1.12 corresponds to the complete wetting case, whereas curve 2 corresponds to the partial wetting case [26]. The structural component is included below in the Derjaguin's pressure isotherms. However, in the case of a capillary meniscus, $P_e > 0$, the secondary minimum of the structural interactions on the isotherm has no such drastic effect on the contact angle hysteresis as seen in the case of droplets (see Section 3.2).

Let us consider the conditions of equilibrium for flat wetting films in contact with a liquid meniscus or a droplet. Due to the small size of the considered systems, the gravity effect is neglected here.

In the case of complete wetting, the thickness, h_e, of the equilibrium wetting film corresponds to the intersection of a straight line, $P_e > 0$, with the Derjaguin's pressure isotherm: There is an only one intersection in case 1 in Figure 1.12.

There could be three intersections of a straight line, $P_e > 0$ (partial wetting, isotherm 2 in Figure 1.12), with the Derjaguin's pressure isotherm in the case of sufficiently "thick" capillaries (Chapter 2). However, h_β and h_u are metastable and unstable equilibrium thicknesses, respectively; only h_e corresponds to a thermodynamically stable equilibrium thickness (see Chapter 2).

The equilibrium contact angle in terms of the Derjaguin's pressure is determined by the known equation (Chapter 2):

$$\cos\theta_e \approx 1 + \frac{1}{\gamma} \int_{h_e}^{\infty} \Pi(h)dh, \tag{3.6}$$

where Π is the Derjaguin's pressure; and γ is the liquid–vapor interfacial tension.

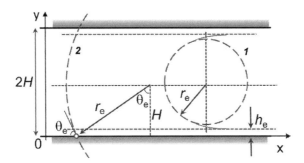

FIGURE 3.2 A schematic presentation of two possible positions of a spherical meniscus inside two- dimensional capillary. (1) complete wetting, $r_e \leq H$, $H/r_e \geq 1$; (2) partial wetting, $r_e > H$, $\cos\theta_e = H/r_e < 1$.

Let us discuss a two-dimensional capillary with a half-width H (Figure 3.2). There are two possible situations in the geometrical definition of the equilibrium contact angle, θ_e:

1. $H < r_e$ (Figure 3.2, case 2): this situation is referred to as partial wetting and the contact angle is defined as $\cos_e = H / r_e < 1$.
2. $H \geq r_e$ (Figure 3.2, case 1): this situation is referred to as complete wetting and the contact angle cannot be introduced geometrically. The case of complete wetting is characterized below by the ratio, H / r_e, which is sometimes referred to as "$\cos\theta_e > 1$."

For the complete wetting case, the integral in the right-hand side of Eq. (3.6) is always positive (Chapter 2) (see curve 1 in Figure 1.12). For partial wetting conditions ($\cos\theta_e < 1$), the integral in the right-hand side of Eq. (3.6) of the Derjaguin's pressure is negative (Figure 1.12, curve 2).

However, the integral $\int_{h_e}^{\infty} \Pi(h)dh$ can be positive even for curve 2 in Figure 1.12. This means that complete wetting conditions take place in this situation and $\theta_e = 0$. Equation (3.6) cannot be applied in the case of complete wetting case (see Section 2.4 for details).

Hysteresis of Contact Angle in Capillaries

In this section the equilibrium of a meniscus in a flat capillary is briefly considered according to Section 1.3 [4,15].

If a meniscus is at equilibrium with a reservoir under the equilibrium pressure, P_e, then there is neither flow nor evaporation in the system. However, if the pressure inside the reservoir is changed by $\Delta P \neq 0$, then the flow will start immediately. In the bulk part of the meniscus, a new local equilibrium can be achieved rather quickly. In this case, it is possible to divide the whole system (Figure 3.3) into several regions: region 1, which is the spherical meniscus with a new radius, r, in a state of a new local equilibrium, a part of the transition region 2 in a state of a local equilibrium with meniscus 1. Inside the regions 1 and 2, the pressure is constant everywhere and equal to the new excess pressure, $P = P_e + \Delta P$. Region 3 is in the flat equilibrium thin film, where the pressure equals to the initial equilibrium excess pressure, P_e; transport region 2' in which a viscous flow of liquid occurs and in which the pressure gradually changes from the value P to P_e (Figure 3.3). The largest pressure drop and the higher resistance to the flow occur in the non-equilibrium part of the transition region, 2', where the liquid film is very thin. Region 2' covers a part of the transition region of very thin films, which immediately adjoins the equilibrium thin film.

Bear in mind that the capillary is in contact with the reservoir, where the pressure, $P_a - P_e - \Delta P$, is maintained, that is, the pressure in the reservoir is lower than the atmospheric pressure, P_a.

If the pressure under the meniscus is increased, then the meniscus will not move; instead it changes its curvature to compensate for the excess pressure and, consequently, the contact angle increases accordingly. All process can be subdivided into "fast macroscopic" and "slow microscopic" processes. The change

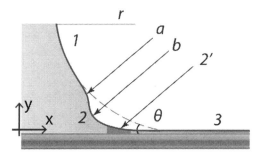

FIGURE 3.3 The liquid profile in a capillary in the case of partial wetting in the state of local equilibrium at excess pressure $P \neq P_e$; r and θ are the radius of the spherical meniscus in the central part of the capillary and the new local equilibrium contact angle, $\theta \neq \theta_e$. (1) Spherical meniscus of a new radius r, where $r \neq r_e$; (2) Profile of a part of the transition zone at local equilibrium with the meniscus; (3) Flat equilibrium liquid film of thickness h_e, with old equilibrium excess pressure P_e; 2'- a flow zone inside the transition region.

of the drop curvature over the main part of the meniscus is a "fast macroscopic" process; the flow in the relatively thick β-films is also a "macroscopic" process because the relative viscous resistance in these films is much smaller as compared with the flow in thin α-films. Hence, the idea is as follows: Until the flow takes place in relatively thin α-films, the meniscus is at a quasi-equilibrium state and in slow microscopic motion. As soon as the flow jumpwise occupies the zone of relatively thick β-films, then the meniscus starts a macroscopic motion.

Note, the thickness of β-films is almost one order of magnitude greater than the thickness of α-films. Under the applied pressure difference, $P = P - P_e$, the flow rate, Q, in the film of thickness h is $Q = \frac{P}{6\mu l} h^3$, where $l \sim \sqrt{Hh_e}$ is the characteristic length of the transition zone between the bulk meniscus and the thin films in front, and μ is the dynamic viscosity. That means, the flow inside β-films is approximately three orders of magnitude faster than in α-films because of the cubic dependency of flux Q. This is why the flow, which is located inside α-films, is referred to as "microscopic flow," while the flow inside β-films is referred to as "macroscopic flow."

In the state of "microscopic" motion the meniscus does not move macroscopically; instead, it moves microscopically. This state of microscopic motion can continue for a prolong period of time if evaporation/condensation processes can be neglected. The meniscus does not move macroscopically until some critical pressure and critical contact angle, θ_a, are reached. After a further increase in pressure, the flow zone occupies the region of the much thicker β-film and the meniscus starts to advance macroscopically. A similar phenomenon takes place if the pressure under the meniscus is decreased: The meniscus does not recede until a critical pressure and corresponding critical contact angle, θ_r, are reached. This means that in the whole range of contact angles, $\theta_r < \theta < \theta_a$ and $\theta \neq \theta_e$, the meniscus does not move macroscopically but moves microscopically.

The above qualitative explanation for the contact angle hysteresis on smooth homogeneous solid substrates is based on the S-shaped isotherm of Derjaguin's pressure in the case of partial wetting. The S-shape determines the very special shape of the transition zone in the case of an equilibrium meniscus (Figure 3.3). In the case of increasing pressure behind the meniscus (Figure 3.4a), a detailed consideration of the transition zone shows that, close to the "critical" point marked in Figure 3.4a, the slope of the profile becomes steeper with increasing pressure. In the range of very thin films (region 3 in Figure 3.4a) there is a flow zone: Viscous resistance in this region is very high, which is why the meniscus advances very slowly. After some critical pressure behind the meniscus is reached, then the slope at the "critical" point reaches $\pi/2$. After that the flow stepwise occupies the region of thick β-films, and the fast "caterpillar motion" starts, as shown in Figure 3.4a.

In the case of decreasing pressure behind the meniscus, the event proceeds according to Figure 3.4b. In this case, up to some critical pressure, the slope in the transition zone close to the "critical" marked point becomes more and more flat. In the range of very thin films (region 3 in Figure 3.4b), there is a zone of flow. As in the previous case, the viscous resistance in this region again is very high, which is

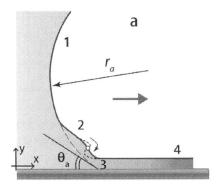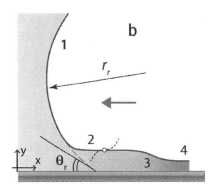

FIGURE 3.4 Contact angle hysteresis in capillaries in the case of partial wetting (S-shaped isotherm of the Derjaguin's pressure curve 2 in Figure 1.12). (a) Advancing contact angle: (1) A spherical meniscus of radius $r_a > r_e$; (2) Transition zone with a "critical" marked point (see explanation in the text); (3) Flow zone; (4) Flat film. Close to the marked "critical" point a dashed line shows the profile of the transition zone just after the contact angle reaches the critical value θ_a, the beginning of a "caterpillar motion" [38]. (b) Receding contact angle: (1) A spherical meniscus of radius $r_r < r_e < r_a$; (2) Transition zone with a "critical" marked point (see explanation in the text); (3) Flow zone, (4) Flat film. Close to the marked "critical" point the dashed line shows the profile of the transition zone just after the contact angle reaches the critical value θ_a.

why the receding of the meniscus proceeds very slowly. After some critical pressure behind the meniscus is reached, the profile near the "critical" point shows discontinuous behavior, which is obviously impossible. This means the meniscus will start to slide along the thick β-film. This phenomenon (the presence of a thick β-film behind the receding meniscus of aqueous solutions in quartz capillaries) has been discovered experimentally [23–25]. This supports the presented arguments explaining *static* contact angle hysteresis on smooth homogeneous substrates.

The suggested mechanism of contact angle hysteresis on smooth homogeneous surfaces has a direct experimental confirmation [24,25]. Quantitative calculations of advancing and receding contact angles via the Derjaguin's pressure isotherm are deduced below.

To simplify the derivation, the discussion is limited to liquids of *low volatility*, whose rates of evaporation and condensation are sufficiently low. In contrast to droplets, the effect of liquid evaporation in thin capillaries is much less significant and it may be easily neglected. The main assumption is that the liquid flow from the quasi-equilibrium meniscus to the equilibrium film in front is very slow until some critical pressure difference, ΔP_a, (in the case of *advancing* meniscus) or ΔP_r (in the case of *receding* meniscus) is reached. These conditions do not exist in the case of *complete* wetting, when the equilibrium film is sufficiently thick. However, *static* hysteresis is usually observed only in cases of *partial* wetting (at $\theta > 0$), when the surface of the solid body is covered with significantly thinner films, where the resistance to the viscous flow is much higher.

A two-dimensional (flat) capillary is under consideration below in this section, so the equations include a curvature along one direction only. The two-dimensional approach reproduces the main physical features of the system, but at the same time it makes the theoretical and numerical analysis of the problem easier. The same time the two-dimensional approach used in this section allows comparing with results on contact angle hysteresis in the case of two-dimensional droplets [26] (see Section 3.2).

The conditions for the quasi-equilibrium of the meniscus in region 1 (Figure 3.4) are listed below. Within region 1, all fluxes can be neglected, and the *excess* pressure can be considered to remain constant and equal to $P = P_e + \Delta P = \text{const.}$ It is assumed based on the previous consideration that for the description of the quasi-equilibrium profile of the liquid, $h(x)$, in region 1 in the absence of true thermodynamic equilibrium in the whole system, the known equation (Chapter 2) for liquid profile can be used:

$$\frac{\gamma h''}{\left(1 + h'^2\right)^{3/2}} + \Pi(h) = P, \tag{3.7}$$

where the equilibrium pressure, P_e, is replaced by the new non-equilibrium pressure, P.

Multiplying both sides of Eq. (3.7) by h' and integrating it with respect to x from 0 to x ($h = H$, $h' = \infty$ at $x = 0$, Figure 3.2), we obtain

$$\frac{\gamma}{\sqrt{1+h'^2}} = \psi(h,P),\qquad(3.8)$$

where

$$\psi(h,P) = P(H-h) - \int_h^\infty \Pi(\bar{h})\,d\bar{h}.\qquad(3.9)$$

The left-hand side of Eq. (3.8) ranges from 0 (at) to γ (at $h'^2 = 0$). Hence, the same should be true for the right-hand site of Eq. (3.8); this determines the region where a solution of Eq. (3.8) exists:

$$0 \le \psi(h,P) \le \gamma.\qquad(3.10)$$

There is no solution of Eq. (3.8) in the region where either $\psi > \gamma$ or $\psi < 0$. If any of these conditions is violated, then the boundary of the flow zone and the center of the meniscus cannot be connected by a continuous profile. Accordingly, quasi-equilibrium becomes impossible, that is, the meniscus cannot be at quasi-equilibrium and must start moving. The violation of one of the inequalities in Eq. (3.10) determines the static *advancing* contact angle, θ_a, and the violation of the other condition, the static *receding* contact angle, θ_r.

Although the mechanism of violation of the equilibrium is understood physically, the value of the static *advancing* contact angle, θ_a, cannot be calculated exactly because the point $h = h_1$ (Figure 3.5) belongs to a region where the liquid profile moves more abruptly and condition $h'^2 \ll 1$ is violated. This condition is required because the Derjaguin's pressure, $\Pi(h)$, was obtained with the assumption of flat (low slope) interacting surfaces [18]. Such an estimation of θ_a is given below. In the case under consideration, the value $\Delta P < 0$, and the curvature of the meniscus decreases with decreasing pressure in the reservoir. It follows from Eq. (3.9) that for $P < P_e$ the curve $\psi(h,P)$ should be below the equilibrium curve $\psi(h,P_e)$ everywhere (Figure 3.5, curve 1). At $\theta = \theta_a$ and $P = P_a$, the function Eq. (3.9) vanishes, and $\psi = 0$ at $h = h_1$ (Figure 3.5, curve 1). Hence, it follows from Eq. (3.9) that

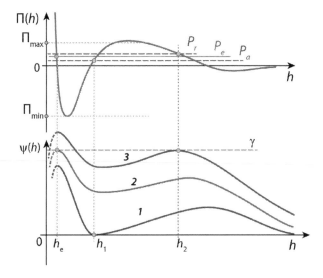

FIGURE 3.5 Schematic plot of the function $\psi(h,P)$ according to Eq. (3.9) for the following cases (1) Advancing, (2) Equilibrium and (3) Receding.

$$P_a\left(H - h_1\right) = \int_{h_1}^{\infty} \Pi\left(h\right) dh. \tag{3.11}$$

The expression for the static *advancing* contact angle, θ_a, in the case of a capillary meniscus can be found from Eq. (3.11) using the general relationship $P_a = \gamma \cos\theta_a / H = \Pi(h_1)$ obtained from the intersection of the continuation of the spherical meniscus with a capillary wall:

$$\cos\theta_a = \frac{1}{\gamma\left(1 - \dfrac{h_1}{H}\right)} \int_{h_1}^{\infty} \Pi\left(h\right) dh = \frac{\Pi(h_1)h_1}{\gamma} + \frac{1}{\gamma} \int_{h_1}^{\infty} \Pi\left(h\right) dh. \tag{3.12}$$

Equality: $P_a = \Pi(h_1)$ is a condition of equilibrium in Figure 3.5 (which corresponds to the intersection point of the isotherm with a straight-line P_a). The term $\frac{\Pi(h_1)h_1}{\gamma} = \frac{P_a h_1}{\gamma}$ can be obtained by substitution of $P_a H$ from Eq. (3.11) into the relation: $\cos\theta_a = P_a H / \gamma$.

The functional dependence Eq. (3.12) coincides with the corresponding expression for the static *advancing* contact angle in the case of drops [15,26] (see Section 3.2). However, the value h_1 in Eq. (3.12) has a different value, so the static advancing contact angle for a meniscus differs from the case of drops (see Section 3.2). The dependence $\cos\theta_a$ on the Derjaguin's pressure allows concluding that the Derjaguin's isotherm uniquely determines not only the equilibrium value of the contact angle but also the static *advancing* contact angle, θ_a.

Curve 2 in Figure 3.5 touches the dashed line γ at a single point h_e. According to Eq. (3.8) this point corresponds to the condition $h' = 0$ which is satisfied only for a flat film at $h = h_e$. For any other values of h, the liquid profile has a non-zero slope (Figure 3.5).

Now we deduce the expression for the static *receding* contact angle, θ_r. In this case, the value $\Delta P > 0$ because the curvature of the meniscus increases with decreasing pressure in the reservoir. It follows from Eq. (3.9) that, for $P > P_e$, the curve $\psi\left(h, P\right)$ should be above the equilibrium curve, $\psi\left(h, P_e\right)$, everywhere (Figure 3.5, curve 3). Violation of the conditions of quasi-equilibrium occurs in this case if $\psi\left(h, P\right) = \gamma$, that is, an increase in ΔP to such a critical value ΔP_r makes the curve $\psi(h, P_r)$ to intersect the dashed line $\gamma = \text{const}$.

For $P > P_r$, the part of the profile shown in Figure 3.4b by the dashed line will start sliding. When the meniscus is displaced from the initial position, then a thick metastable β-film should remain behind the receding meniscus. This prediction has been confirmed experimentally [24,25]. The thickness of the film h_2 belongs to the β part of the Derjaguin's isotherm.

Because the profile of the receding meniscus in the transition zone has a low slope, the value of the static *receding* contact angle, θ_r, in the case of sufficiently thick capillaries, that is, for $P_e < \Pi_{\max}$ can be determined exactly. In the case of sufficiently thick capillaries, that is, for $P_e < \Pi_{\max}$, the substitution $\psi = \gamma$, for $P = P_r$ and $h = h_2$ into Eq. (3.9) gives the following equation:

$$P_r\left(H - h_2\right) - \int_{h_2}^{\infty} \Pi\left(h\right) dh = \gamma, \tag{3.13}$$

where $P_r = \Pi\left(h_2\right)$.

The static *receding* contact angle is obtained from Eq. (3.13) as

$$\cos\theta_r = 1 + \frac{\Pi\left(h_2\right)h_2}{\gamma} + \frac{1}{\gamma} \int_{h_2}^{\infty} \Pi\left(h\right) dh \tag{3.14}$$

For comparison, the expression for the equilibrium contact angle can be obtained as before as follows [15]:

$$\cos\theta_e = 1 + \frac{\Pi\left(h_e\right)h_e}{\gamma} + \frac{1}{\gamma} \int_{h_e}^{\infty} \Pi\left(h\right) dh. \tag{3.15}$$

Calculation Procedure

The numerical calculations and analysis based on the real Derjaguin's pressure isotherms and the comparison of the calculated data for the capillary meniscus and droplet (see Section 3.2) are given below.

As explained earlier, if the contact angle θ is in between either $\theta_e < \theta < \theta_a$ or $\theta_r < \theta < \theta_e$, then the meniscus is in the state of a slow microscopic motion and, hence, this motion can be neglected from the macroscopic point of view. This is the reason to assume that continuation of the meniscus profiles for the all three cases of *advancing* (*a*), *receding* (*r*) and *equilibrium* (*e*) states intersect the capillary wall at the same macroscopic point **A** (Figure 3.6). A small microscopic shift of the point **A** during the transition between the states (*a*), (*e*), and (*r*) is neglected below.

The contact angles and a capillary width are related by the following equations (Figures 3.2 and 3.6):

$$r_a \cos\theta_a = \frac{\gamma\cos\theta_a}{\Pi(h_1)} = H, \tag{3.16}$$

$$r_r \cos\theta_r = \frac{\gamma\cos\theta_r}{\Pi(h_2)} = H, \tag{3.17}$$

$$r_e \cos\theta_e = \frac{\gamma\cos\theta_e}{\Pi(h_e)} = H. \tag{3.18}$$

The sets of Eqs. (3.12), (3.16); (3.14), (3.17); and (3.15), (3.18) were solved and the values h_e, h_1, and h_2 were found as roots of equation:

$$L(h) = \frac{\gamma\cos\theta(h)}{\Pi(h)} = H,$$

where $\theta_{r,e}(h) = \mathrm{acos}\left(1 + \frac{\Pi(h)h}{\gamma} + \frac{1}{\gamma}\int_h^\infty (h)\,dh\right)$; $\theta_a(h) = \mathrm{acos}\left(\frac{\Pi(h)h}{\gamma} + \frac{1}{\gamma}\int_h^\infty \Pi(h)\,dh\right)$, correspondingly.

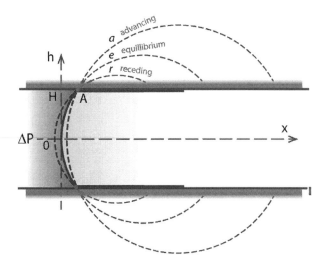

FIGURE 3.6 The deformation of the capillary meniscus profile while changing the pressure in the reservoir by ΔP.

In the case of multiple solutions, the roots, which satisfy the conditions (3.10) for (h, P), were selected.

The states of quasi-equilibrium advancing/receding or equilibrium correspond to the intersection points of the Derjaguin's pressure isotherm with different straight lines of the pressures P_a, P_r and P_e, respectively. The equilibrium and receding states correspond to positions of α- and β-films, respectively; the advancing state corresponds to intermediate (unstable) position on the isotherm.

The contact angles θ_a, θ_r and θ_e were calculated at the variation of the capillary width, H. The calculation results are presented in Figure 3.7a and Table 3.1.

The obtained data demonstrate that $\theta_r < \theta_e < \theta_a$, as expected. An agreement of the obtained values of contact angles values for droplets (Figure 3.7b) is observed. The calculation results of advancing contact angle ~57°, receding contact angle ~9° agree with calculated data for droplets (see Section 3.2) and with experimental data [20,17]. Parameters of the Derjaguin's pressure isotherm used for these calculations are like that for glass surfaces [17].

The parameters of the Derjaguin's pressure isotherm for both cases are identical and taken from Section 3.2: $\sigma_s = -150$ mC; $\sigma_h = 120$ mC; $c_0 = 1 \times 10^{-2}$ mol/m³; $A = 3.5 \times 10^{-20}$ J; $K_1 = 2.0 \times 10^7$ Pa; $K_2 = -1 \times 10^4$ Pa; $\lambda_1 = 3.6 \times 10^{-9}$ m; $\lambda_2 = 26 \times 10^{-9}$ m. A dilute aqueous solution of NaCl with the surface tension $\gamma = 72.7 \times 10^{-3}$ [N/m] was considered. The volume unit [m²] corresponds to the flat

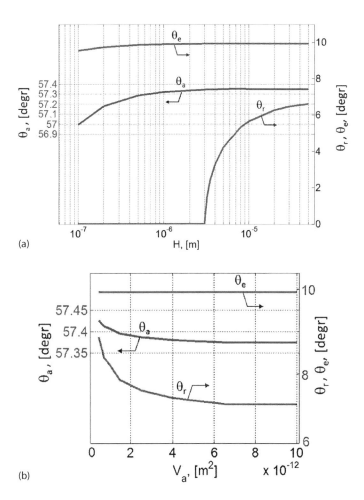

(a)

(b)

FIGURE 3.7 Dependence of the calculated values of the contact angles on the capillary width (a) (Section 3.1) and the droplet volume (b) (Section 3.2).

TABLE 3.1

The Values of the Advancing, Equilibrium and Receding Contact
Angles as Functions of the Capillary Size, H

H (m)	θ_a (degrees)	θ_e (degrees)	θ_r (degrees)
1.0×10^{-7}	56.99	9.54	0
5.0×10^{-7}	57.28	9.86	0
1.0×10^{-6}	57.32	9.89	0
5.0×10^{-6}	57.35	9.93	4.20
1.0×10^{-5}	57.35	9.93	5.65
5.0×10^{-5}	57.35	9.93	6.61

(two-dimensional) droplet per unit length of the capillary. Arrows near the curves indicate the axis (left or right) that should be seen for this curve (two vertical axes with different scales are used).

However, Figure 3.7a and b show that droplets and capillaries have completely opposite behavior as a function of droplet/capillary size: The contact angles for the meniscus grow with size of the capillary, whereas for the droplet, on the contrary, the bigger a droplet, the better wetting conditions (the lower the angles). This character of the contact angle dependence (Figure 3.7a) in capillaries has been observed experimentally in similar systems [27].

The dependences presented in Figure 3.7a and Table 3.1 demonstrate this interesting behavior: In thin capillaries with a low enough size H, a transition from partial to complete wetting occurs in the case of a receding contact angle: For capillaries with $H \approx 3 \times 10^{-6}$ m, the transition from partial to complete wetting is observed with a decrease in capillary width. This behavior of the contact angles was predicted earlier [28].

The values $\theta_r = 0$ in Table 3.1 were obtained when numerical calculations give $\cos \theta_r > 1$, meaning that $r < H$ (Figure 3.2); physically, it corresponds to the case of compete wetting, $\theta_r = 0$.

The transition from partial to complete wetting may be interpreted physically as a decrease of a spherical meniscus radius $r = \gamma/P_r$ because of increasing capillary pressure in small-sized capillaries. This leads to the receding process over thick film (β-film), which forms from the equilibrium on the Derjaguin's pressure isotherm. According to experimental observations, the process of receding frequently goes at a zero receding contact angle.

The equilibrium contact angle varies in capillaries and barely changes in droplets (Figure 3.7); however, for thick capillaries and big droplets, equilibrium contact angles coincide.

The reason for the different behavior of the equilibrium contact angles for droplets and capillaries can be clarified based on the analysis of the shape of the Derjaguin's pressure isotherm.

The value h_e [corresponding to a point of intersection between $\Pi(h)$ and P_e] changes very slightly at variations of P_e. This is because $\Pi(h)$ increases abruptly at a decreasing h. For droplets, $P_e < 0$ and the integral in Eq. (3.6) barely changes when P_e approaches zero. Therefore, the angle θ_e for droplets changes very slightly.

For capillaries, $P_e > 0$ and the integral in Eq. (3.6) starts growing with a decreasing h. As a result, a change in the contact angle, θ_e, is observed in thin capillaries.

It is important to emphasize that for both droplets and capillaries the receding contact angles are closer to the equilibrium ones than to advancing angles, that is, $\theta_e - \theta_r \ll \theta_a - \theta_e$. This is in contradiction with a well-adopted view that the static advancing contact angle is a good approximation for the equilibrium contact angle. In both cases, droplets and capillaries (Figure 3.7a and b), the receding contact angles are closer to the equilibrium contact angle values.

The constancy of the equilibrium contact angle of droplets can be explained based on the shape of the Derjaguin's pressure isotherm (e.g., Figure 1.12, curve 2): the value h_e (corresponding to a point of intersection between $\Pi(h)$ and P_e) changes very slightly at variations of P_e. This is because $\Pi(h)$ grows very abruptly at low h. Hence, according to Eq. (3.6), there is a very weak dependence of the equilibrium contact angle, θ_e, on P_e and, consequently, θ_e on the droplet volume.

In capillaries the identical Derjaguin's pressure isotherm was used as in Section 3.2 for droplets (see also [26]); however, in the case of capillaries, $P_e > 0$, and $P_e < 0$ in the case of droplets. It is necessary to increase P_e rather substantially in the region of low h values in order to have a small variation of h. This leads to a growth of the positive area under the isotherm and to a decrease of the equilibrium contact angle according to Eq. (3.6).

For droplets, the predicted dependence between the contact angles and the droplet volume was confirmed experimentally [29,30]. Prediction of thick β-films behind the receding meniscus in capillaries was experimentally confirmed by Churaev's group [16,17,21]. The presence of β-films behind the receding droplet is yet to be confirmed.

Conclusions

It is shown that both static advancing and receding contact angles in capillaries with smooth homogeneous walls can be calculated based on the isotherm of the Derjaguin's pressure.

According to the theory presented, both static advancing and receding contact angles are increasing functions of the capillary width. This is the opposite trend as compared with the calculated static advancing and receding contact angles in the case of droplets, when the contact angles decrease with an increase in droplet size (see Section 3.2). However, all three contact angles (static advancing, receding and equilibrium) coincide for big droplets and wide capillaries.

The calculation results demonstrate the effect of transition from partial to complete wetting in thin capillaries: The receding contact angle decreases to zero in this case.

3.2 Hysteresis of Contact Angle of Sessile Droplets on Non-deformable Substrates

A theory of contact angle hysteresis on smooth homogeneous solid substrates is developed in terms of the shape of the Derjaguin's pressure isotherm and quasi-equilibrium phenomena. It is shown that all contact angles, θ, in the range $\theta_r < \theta < \theta_a$, which are different from the unique equilibrium value θ_e, correspond to the state of slow "microscopic" advancing or receding motion of the liquid if $\theta_e < \theta < \theta_a$ or $\theta_r < \theta < \theta_e$, respectively. This "microscopic" motion almost abruptly becomes fast "macroscopic" advancing or receding motion after the contact angle reaches the critical values θ_a or θ_r, correspondingly. The values of the static receding, θ_r, and static advancing, θ_a, contact angles in cylindrical capillaries were calculated in Section 3.1, based on the shape of the Derjaguin's pressure isotherm. It is shown in this section that both advancing contact and receding contact angles of a droplet on a solid substrate depend on the drop volume and are not a unique characteristic of the liquid–solid system. The suggested mechanism of the contact angle hysteresis of droplets has direct experimental confirmation.

Important: This section cannot be understood without first reading Section 3.1.

Introduction

Recall some of the basic facts discussed in Chapter 1.

Why do droplets of different liquids deposited on identical solid substrates behave so differently? Why do identical droplets, for example, aqueous droplets, deposited on different substrates behave also differently?

As is well known, a mercury droplet does not spread on a glass substrate. Rather it forms a spherical cap with the contact angle bigger than $\pi/2$. An aqueous droplet deposited on the identical glass substrate spreads only partially down to some contact angle, θ, which is between 0 and $\pi/2$. However, an oil droplet (hexane or decane) deposited on the same glass substrate spreads out completely, and the contact angle decreases with time down to the zero value. These three cases—which are referred to as: non-wetting, partial wetting and complete wetting, respectively—are determined by the natures of both the liquid and the solid substrate and their interactions. The manifestation of this interaction is the Derjaguin's pressure.

It is usually believed that the generally known phenomenon of static hysteresis of the contact angle is determined by the surface roughness and/or heterogeneity. There is no doubt that a roughness and/or a chemical heterogeneity of the solid substrate contribute substantially to the contact angle hysteresis. In this case it is assumed that at each point of the surface the equilibrium value of the contact angle is established, depending only on the local properties of the substrate. As a result, a whole series of local thermodynamic equilibrium states can be realized, corresponding to a certain interval of values of the contact angle. The maximum possible value corresponds to the value of the static advancing contact angle, θ_a, and the minimum possible value corresponds to the static receding contact angle, θ_r. Hence, the dependency of the contact angle on the velocity of motion of a meniscus or a drop can be qualitatively described by the dependency presented in Figure 3.1.

However, roughness and/or heterogeneity of the surface are apparently not the only reasons for contact angle hysteresis. An increasing number of publications over the last years have confirmed the presence of contact angle hysteresis even on smooth homogeneous surfaces [4–9]. However, the most convincing evidence for the presence of the contact angle hysteresis even on smooth homogeneous surfaces is its presence on free liquid films [10–14]: In this case, there is the hysteresis of a meniscus located on thin free liquid films. In this case, the surfaces which are homogeneous and smooth of free liquid films are not rough at all and are also chemically homogeneous. Hence, in the case of contact angle hysteresis on free liquid films, it is impossible to explain the hysteresis phenomenon by the presence of roughness and/or heterogeneity.

Below we describe a mechanism of contact angle hysteresis of sessile droplets based on the consideration of surface forces that act in the vicinity of the three-phase contact line based on approach used in Section 3.1. This type of contact angle hysteresis is present even on a smooth homogeneous substrate. Consideration of the influence of surface forces on contact angle hysteresis on rough and/or nonhomogeneous surfaces from this point of view is yet to be undertaken.

Evidently, only a single unique value of equilibrium contact angle, θ_e, is possible on a smooth homogeneous surface. Hence, the hysteresis contact angles, $\theta_a \neq \theta_e$ and $\theta_r \neq \theta_e$, and all contact angles in between that are observed experimentally on such surfaces, correspond only to non-equilibrium states of the system. Hence, the picture presented by the bold lines in Figure 3.1 should be replaced by a new more realistic picture represented in the same picture by curve 2.

Thus, the following discussion of the hysteresis phenomenon is based on the analysis of non-equilibrium states of the system: menisci and/or droplets on smooth homogeneous solid substrates.

The same type of the Derjaguin's pressure isotherm is used as in Section 3.1.

An aqueous solution of a strong univalent electrolyte has been chosen with the bulk electrolyte concentration $c_0 = 1$ mol/m^3 and temperature $T = 293$ K for calculation in Figure 3.8. The electrostatic component of the Derjaguin's pressure was calculated for the case of a constant surface charge density $\sigma_{s,\,h} = $ const, with values $\sigma_s = -150$ mC and $\sigma_h = 150$ mC at the boundaries. Variable values of the van der Waals and structural interaction parameters are used.

In the case of the short-range structural repulsion, the parameter K_1 in Eq. (3.2) is positive and the structural components dependences take the shape presented in Figure 3.9a. In this situation, the structural forces change sign: They are repulsive or attractive depending on the distance. Now the Derjaguin's pressure isotherm can also include two minima (Figure 3.9b), but both minima are possible if the short-range structural repulsion is counterbalanced by attractive electrostatic or van der Waals components. In Figure 3.9b, the electrostatic force (curve 1) acts oppositely to the structural and van der Waals components (curves 3 and 2, respectively). The attractive (negative) values of the electrostatic component in Figure 3.9b were obtained for the different signs of the charged surfaces ($\sigma_s = -150$ mC; $\sigma_h = 150$ mC). The isotherms with different locations of the primary minimum (like the isotherms shown in Figure 3.8b and c) can be obtained from the data of Figure 3.9b by variation of the λ_1 parameter.

Thus, the Derjaguin's pressure isotherms with two minima are possible if they include the structural interaction consisting of two parts acting at different distances: The long-range interaction is always attractive; the short-range one can be attractive or repulsive. As shown later, these features of the Derjaguin's pressure isotherms cause the special features of the contact angle hysteresis.

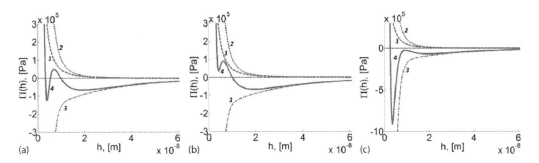

FIGURE 3.8 The Derjaguin's pressure isotherms according to Eq. (3.4). The components of the Derjaguin's pressure: (1) Electrostatic, Π_E; (2) Van der Waals, Π_W; (3) Structural, Π_S; (4) Total interaction, $\Pi = \Pi_E + \Pi_W + \Pi_S$. Parameters of the electrostatic interactions: constant surface charge, $\sigma_s = -150$ mC; $\sigma_h = -150$ mC; $c_0 = 1$ mol/m³; and $T = 293$ K. Parameters of the van der Waals and structural interactions: $A = 2 \times 10^{-18}$ J; $K_1 = -3.7 \times 10^7$ Pa; $K_2 = -2 \times 10^5$ Pa; $\lambda_2 = 25 \times 10^{-9}$ m. (a) $\lambda_1 = 1.35 \times 10^{-9}$ m; (b) 1.31×10^{-9}; (c) 1.5×10^{-9}.

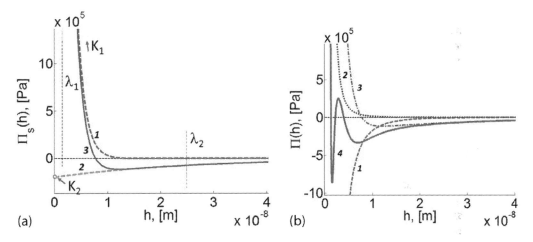

FIGURE 3.9 The components of the structural interaction (a) and the Derjaguin's pressure isotherm (b) for the case of short-range structural repulsion ($K_1 > 0$). $K_1 = 3 \times 10^7$ Pa; $\lambda_1 = 1.45 \times 10^{-9}$ m; $K_2 = -2 \times 10^5$ Pa; $\lambda_2 = 25 \times 10^{-9}$ m. (a) 1,2,3—the short-range, long-range and total structural interaction, respectively. (b) 1,2,3,4—the electrostatic, van der Waals, structural and total interaction, respectively.

Equilibrium Contact Angle and Derjaguin's Pressure for Sessile Droplets

According to Chapter 2, the equilibrium profile of a sessile two-dimensional droplet is described by the following equation:

$$\frac{\gamma h''}{(1+h'^2)^{3/2}} + \Pi(h) = P_e, \tag{3.19}$$

where γ is the liquid–vapor interfacial tension; and $h(x)$ is an equilibrium droplet profile. According to Kelvin's law

$$P_e = \frac{R_g T}{v_m} \ln \frac{p_s}{p} = -\frac{\gamma}{\Re}, \tag{3.20}$$

where v_m is the molar liquid volume; p_s and p are saturated vapor pressure and the pressure of the vapor, respectively, with which the sessile droplet is at equilibrium; and $\Re$ is the curvature of the interface.

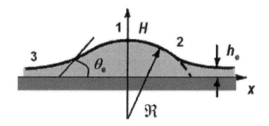

FIGURE 3.10 Equilibrium drop. (1) Spherical cap, where capillary forces dominate, (2) Transition zone, where capillary forces and the Derjaguin's pressure are equally important, and (3) Flat equilibrium film in front of the drop.

See Figure 3.10 for details. According to Eq. (3.20) the liquid can be at equilibrium only with oversaturated vapor, that is, if $p > p_s$. It is very difficult to keep oversaturated vapor over the sessile droplet for a prolonged period of time. This explains why equilibrium sessile droplets are so difficult (if possible at all) to investigate experimentally.

According to Section 2.3, the whole profile of an equilibrium drop can be subdivided into three parts (Figure 3.10): a spherical cap (using this part a *macroscopic* contact angle can be determined); a transition zone, where both capillary pressure and the Derjaguin's pressure are equally important; and a flat equilibrium liquid film region ahead of the drop.

The second-order differential Eq. (3.19) can be integrated once, and that gives

$$\frac{1}{\sqrt{1+h'^2}} = \frac{C - P_e h - \int\limits_{h}^{\infty} \Pi(h)dh}{\gamma}, \tag{3.21}$$

where C is an integration constant to be determined.

As due to the transversality condition $h'(h_e) = 0$ according to Chapter 2, the integration constant in Eq. (3.21) is $C = \gamma + P_e h_e + \int_{h_e}^{\infty} \Pi(h)dh$. Hence, the drop profile is described by the following equation:

$$\frac{1}{\sqrt{1+h'^2}} = \frac{\gamma - \varphi(h, P_e)}{\gamma} \tag{3.22}$$

where

$$\phi(h, P_e) = -P_e\left(h - h_e\right) + \int\limits_{h_e}^{h} \Pi dh. \tag{3.23}$$

Because the left-hand side of Eq. (3.22) is always positive and equal to or less than unity, it demands that

$$0 \le \phi(h, P_e) \le \gamma, \tag{3.24}$$

where the first condition corresponds to a zero derivative of h', and the second one corresponds to an infinite value of h'.

On the other hand, at the drop apex, H, the derivative vanishes, $h'(H) = 0$. This together with Eq. (3.21) results in $C = \gamma + P_e H$. Then Eq. (3.21) becomes

$$\frac{1}{\sqrt{1+h'^2}} = \frac{\gamma + P_e(H - h) - \int\limits_{h}^{\infty} \Pi(h)dh}{\gamma}. \tag{3.25}$$

Outside the range of the Derjaguin's pressure action, Eq. (3.25) reduces to

$$\frac{1}{\sqrt{1+\overline{h}'^2}} = \frac{\gamma + P_e(H - \overline{h})}{\gamma}, \tag{3.26}$$

which describes the spherical cap of the drop in Figure 3.10. The magnitudes $\overline{h}'$ and $\overline{h}$ correspond to the spherical cap in the absence of surface forces. The intersection of this profile with the thin equilibrium film of thickness, h_e, defines an apparent three-phase contact line and the macroscopic equilibrium contact angle: $\overline{h}'(\overline{h}=0) = -\tan\theta_e$. Then Eq. (3.26) can be rewritten as $\overline{h} = 0$ as $P_e = -\frac{\gamma(1-\cos\theta_e)}{H}$. Casting this expression into Eq. (3.25) at $h = h_e$ results in the following expression of the contact angle in the case of drops on a flat substrate:

$$\cos\theta_e = 1 + \frac{\dfrac{1}{\gamma}\displaystyle\int_{h_e}^{H}\Pi(h)dh}{1-h_e/H} \approx 1 + \frac{1}{\gamma}\int_{h_e}^{\infty}\Pi(h)dh, \quad \text{for} \quad t_s \ll H. \tag{3.27}$$

where $t_s \approx 100$ nm is the radius of surface forces action. Eq. (3.27) is also known as the Frumkin-Derjaguin equation (see Chapter 2). As follows from (3.27), for the *partial* wetting case,

$$\int_{h_e}^{\infty}\Pi(h)dh < 0. \tag{3.28}$$

Note that the equilibrium contact angle defined by Eq. (3.27) is not completely determined by the shape of the Derjaguin's pressure isotherm: It also depends on the lower limit of integration, h_e, which is determined by the equilibrium excess pressure, P_e. In other words, the *equilibrium* contact angle of drops depends on the *equilibrium* volume of the drop, which can vary from "infinity" (at $P_e = 0$) to a minimum value at $P_e = \Pi_{\min}$ (Figure 1.12).

Static Hysteresis of the Contact Angle of Sessile Droplets on Smooth Homogeneous Substrates

The derivation of Eq. (3.19) shows that it determines a unique equilibrium contact angle (at fixed external conditions). Experiments, however, show contact angle hysteresis with an infinite number of apparent "quasi-equilibrium positions" and "quasi-equilibrium contact angles" of a drop on a solid surface such that $\theta_r < \theta < \theta_a$, where θ_r and θ_a are denoted as (static) *receding* and *advancing* contact angles, respectively. Indeed, let us consider a liquid drop on a horizontal substrate, which is slowly growing by liquid being pumped in through an orifice in the substrate (Figure 3.11). After we stop pumping liquid and just before the droplet starts advancing, a static advancing contact angle, θ_a, is formed. Any further pumping will result in subsequent drop spreading. Let us now consider the reverse experiment. If we start from the static advancing contact angle, θ_a, reached as described above in this section and if we start sucking the liquid through the same orifice, then the contact angle will decrease without the drop base shrinking until another critical contact angle, θ_r, is reached; after further pumping the liquid out, the drop recedes and eventually disappears. For water drops on smooth homogeneous glass surfaces, $\theta_r \sim 0°$–$5°$, whereas θ_a is in the range of $40°$–$60°$.

It was shown in Chapter 2 that the height of equilibrium sessile drops, H, increases with increasing *over* saturation, when the value of P_e decreasing. However, this increase has a certain limit because, since drops can be at the equilibrium with flat films only if $P_e > \Pi_{\min}$ (Figure 1.12). It can be seen from Figure 1.12, curve 2, that $\Pi_{\min}$ is the pressure corresponding to the minimum of the isotherm $\Pi(h)$. For $P_e < \Pi_{\min}$, there is neither a film nor a drop on the surface at equilibrium. When P_e decreases and approaches $\Pi_{\min,}$ drops should be "torn" off the surface and pass into the vapor phase.

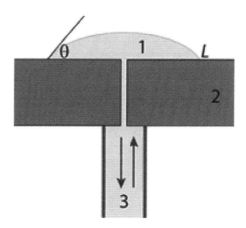

FIGURE 3.11 Schematic illustration of the formation of a drop by appropriate liquid pumping. *L*-radius of the drop base; *θ*—contact angle. (1) Liquid drop, (2) Solid substrate with a small orifice in the center, (3) Liquid source/sink (syringe).

Maintaining the external conditions, we now consider non-equilibrium profiles of drops when their volumes change by pumping liquid in or out (see Figure 3.11) and the excess pressure *P* is different from the equilibrium value. As in Section 3.1, we assume that the main part of the liquid profile is still described by Eq. (3.19), but the equilibrium pressure, P_e, is replaced now by a new non-equilibrium pressure, *P*:

$$\frac{\gamma h''}{(1+h'^2)^{3/2}} + \Pi(h) = P. \tag{3.29}$$

The qualitative reason for that assumption is presented in Section 3.1 for the case of a meniscus. However, *P* is negative now, which is different from the case of a capillary meniscus.

Expressions for the Advancing Contact Angle

Advancing contact angles are formed if $P < P_e$. Precisely as in Section 3.1, we can deduce the condition for the existence of a solution for Eq. (3.29), which can be written for a drop in the following form:

$$\gamma \geq \phi(h,P) \geq 0, \tag{3.30}$$

where now the function $\varphi(h, P)$ is given by

$$\varphi(h,P) = -P(H-h) + \int_h^\infty \Pi(h)dh. \tag{3.31}$$

Examples of the form of the function $\varphi(h, P)$ (curves 2–3) for the Derjaguin's pressure isotherm $\Pi(h)$ are shown in Figure 3.12. The extrema of $\varphi(h, P)$ are found from the condition $P = \Pi(h)$ just as in the case of equilibrium, that is, from the points of intersection of the Derjaguin's pressure isotherm with the straight line, $P =$ const. It follows from Eq. (3.31) that for the equilibrium profile of a drop, that is, for $P = P_e$, the function $\varphi(h, P_e)$ vanishes at $h = h_e$. On the other hand, $\varphi(h, P_e)$ vanishes also at $h = H$. Because $\varphi(h, P_e) > 0$, the function $\varphi(h, P_e)$ has a maximum for $h = h_2$ (Figure 3.12, curve 2).

At lower pressures, for $P < P_e$, the drop surface becomes more convex than at equilibrium; the line $\varphi(h, P)$ (Figure 3.12, curve 3) is located above the equilibrium curve 2. The condition of quasi-equilibrium is violated, and the perimeter of the drop starts to advance after the maximum of $\varphi(h, P)$ reaches the dashed line $\gamma =$ const (curve 3). This condition corresponds to the appearance of a thickness with a

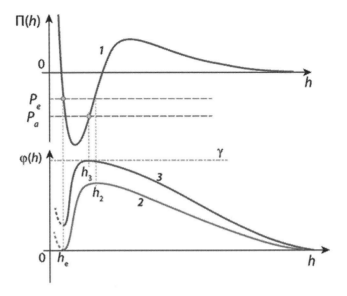

FIGURE 3.12 The Derjaguin's pressure isotherm $\Pi(h)$ upper figure: the case of *partial* wetting (curve 1); and lower figure: corresponding curves (2, 3) of the functions φ (31) determining the conditions of equilibrium (curve 2) or quasi-equilibrium of a drop before the advancing starts (curve 3).

vertical tangent $h' = \infty$ on the profile of the drop. Then liquid flows from the drop to the film by the so-called Frenkel's "caterpillar" mechanism if $\varphi(h, P) > \gamma$. This shows that the static *advancing* contact angle does not depend on the roughness of a solid substrate if the roughness size is below the value of $h_3 \sim 10$–30 nm, which is an experimentally well-known phenomenon.

Let us calculate the value of the *advancing* contact angle, θ_a, using the condition $\varphi(h, P) = \gamma$:

$$-P_a\left(H_a - h_3\right) + \int_{h_3}^{\infty} \Pi(h)\,dh = \gamma. \tag{3.32}$$

Bearing in mind that $P_a = \gamma\,(\cos\theta_a - 1)/H_a$, it follows that

$$\cos\theta_a = \frac{P_a h_3}{\gamma} + \frac{1}{\gamma}\int_{h_3}^{\infty} \Pi(h)\,dh \approx \frac{1}{\gamma}\int_{h_3}^{\infty} \Pi(h)\,dh. \tag{3.33}$$

Let us calculate the difference $\cos\theta_e - \cos\theta_a$ using Eq. (3.27) for the equilibrium contact angle, θ_e, equilibrium excess pressure, P_e, in the drop, and Eq. (3.33). We obtain

$$\cos\theta_e - \cos\theta_a = 1 + \frac{1}{\gamma}\int_{h_e}^{h_3} \Pi(h)\,dh > 0, \tag{3.34}$$

This shows that $\cos\theta_e > \cos\theta_a$ and $\theta_a > \theta_e$. Thus, in the case of a drop, the suggested theory shows, in agreement with experimental observations, that the static *advancing* contact angle is always bigger than the *equilibrium* angle.

In general, the static *advancing* contact angle can be determined in the following way. From equation $P_a = \Pi(h_3)$ we get P_a as a function of the thickness h_3. Because this equation has two solutions, the second root (not the first) should be selected (Figure 3.12, curve 3). Using this solution in Eq. (3.32) yields $-\Pi(h_3)\left(H_a - h_3\right) + \int_{h_3}^{\infty} \Pi(h)\,dh = \gamma$, which provides the unknown height of the drop, H_a, as a function of h_3:

$$H_a = h_3 + \frac{\gamma - \int_{h_3}^{\infty} \Pi(h)dh}{-\Pi(h_3)}, \tag{3.35}$$

where $P_a = \Pi(h_3) < 0$ and $\gamma - \int_{h_3}^{\infty} \Pi(h)dh > 0$.

Let R_a be the radius of the curvature of the drop at the moment of advancing. A simple geometrical consideration shows that

$$L_a = R_a \sin\theta_a = -\frac{\gamma \sin\theta_a}{\Pi(h_3)}, \tag{3.36}$$

where L_a is the radius of the droplet base at the moment when advancing contact angle is reached.

The volume of the drop at the moment of advancing, V_a, can be expressed as

$$V_a = \frac{L_a^2}{\sin^2\theta_a}(\theta_a - \sin\theta_a \cos\theta_a). \tag{3.37}$$

Note that according to experimental condition, the volume, V_a, is fixed but not the radius of the droplet base, L_a. From Eq. (3.37), we conclude

$$L_a = \sin\theta_a \sqrt{\frac{V_a}{(\theta_a - \sin\theta_a \cos\theta_a)}}. \tag{3.38}$$

Combination of Eqs. (3.36) and (3.38) results in

$$\sqrt{\frac{V_a}{(\theta_a - \sin\theta_a \cos\theta_a)}} = -\frac{\gamma}{\Pi(h_3)}, \tag{3.39}$$

where the contact angle θ_a is expressed as

$$\cos\theta_a = \frac{\Pi(h_3)h_3}{\gamma} + \frac{1}{\gamma}\int_{h_3}^{\infty}\Pi dh \tag{3.40}$$

and $\sin\theta_a = \sqrt{1 - \cos^2\theta_a}$.

The values h_3 and θ_a are found as a solution of Eqs. (3.39) and (3.40).

Thus, our theory predicts a dependency of the advancing contact angle on the droplet volume V_a. It is clearly shown now that static *advancing* contact angle in the case of drops is not a unique characteristic of solid substrates but is also determined by external conditions.

This conclusion was confirmed experimentally in [31,30]. In [32,29] the authors showed that the advancing contact angle of droplets really increases as the volume of the droplet decreases. In [31], the advancing contact angle of bubbles was considered, and the authors arrived at the same conclusion. It should be noted that experimental results on the contact angle hysteresis on smooth homogeneous surfaces are very rare in the literature. The theoretical models of this phenomenon are frequently based on various concepts [33,34], where action of surface forces is not taken into account.

At higher pressures, that is, for $P > P_e$, when the drop surface becomes less convex, the dependency $\varphi(h, P)$ (Figure 3.13, curve 4) is located below that of the equilibrium curve 2. The condition of quasi-equilibrium is violated, and the perimeter of the drop starts to recede after the minimum of $\varphi(h, P)$

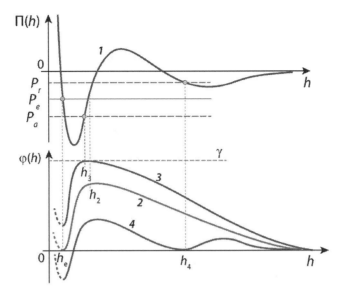

FIGURE 3.13   The Derjaguin's pressure isotherm $\Pi(h)$. The upper figure: the case of *partial* wetting (curve 1); and lower figure: corresponding curves (2, 3, 4) of the functions φ (31) determining the conditions of equilibrium (curve 2) or quasi-equilibrium of a drop before the advancing starts (curve 3) and before the receding starts (curve 4).

reaches zero (curve 4). This condition corresponds to the appearance of a thickness with a horizontal tangent $h' = 0$ on the profile of the drop. After that, the droplet starts sliding over the thick β-film. This conclusion (the presence of thick β-films behind the receding menisci) has been experimentally confirmed in [24,25]. However, experimental proof of the presence of thick β-films behind the receding droplets is yet to be made.

Expressions for the Receding Contact Angle

The expressions for the receding contact angle are found using the condition $\varphi(h, P) = 0$:

$$P_r\left(H_r - h_4\right) + \int_{h_4}^{\infty}\Pi dh = 0 \tag{3.41}$$

From Eq. (3.41), the droplet height at the moment of receding is

$$H_r = h_4 + \frac{1}{P_r}\int_{h_4}^{\infty}\Pi dh \tag{3.42}$$

For the spherical droplet

$$P_r = \gamma(\cos\theta_r - 1)/H_r \tag{3.43}$$

The receding contact angle is expressed from Eqs. (3.43) and (3.41) as

$$\cos\theta_r = 1 + \frac{P_r H_r}{\gamma} = 1 + \frac{P_r h_4}{\gamma} + \frac{1}{\gamma}\int_{h_4}^{\infty}\Pi dh \tag{3.44}$$

It is assumed that the droplet base does not shrink during the transition between the advancing and receding angles, that is, $L_a = L_r = \text{const.}$

Then the radius of the base contact line $L_r = R_r \sin\theta_r$ is expressed using Eq. (3.44) as

$$L_r = -\frac{\gamma}{\Pi(h_4)}\sqrt{1 - \left(1 + \frac{\Pi(h_4)h_4}{\gamma} + \frac{1}{\gamma}\int\limits_{h_4}^{\infty}\Pi dh\right)^2} = L_a. \qquad (3.45)$$

The value h_4 is found from Eq. (3.45), where L_a is already known from Eq. (3.38). The receding volume is expressed as

$$V_r = \frac{L_a^2}{\sin^2\theta_r}\left(\theta_r - \sin\theta_r\cos\theta_r\right) \qquad (3.46)$$

Calculated dependences of static advancing and static receding contact angles of sessile droplets on their volume are presented in Figure 3.7b.

Conclusions

Based on the developed theory for contact angle hysteresis on smooth homogeneous solid substrates in terms of the Derjaguin's pressure, it was shown that for any value of the contact angle, θ, from the range $\theta_r < \theta < \theta_a$, except the equilibrium value $\theta = \theta_e$, a slow "microscopic" advancing or receding motion of the liquid takes place.

The expressions for the advancing and receding contact angles were obtained theoretically from the analysis of the shape of the Derjaguin's pressure isotherm. At the moment when the contact angle reaches the value of θ_r and θ_a, the "microscopic" motion is replaced by fast "macroscopic" motion leading to an abrupt transformation or destruction of the droplet.

It is shown that both advancing and receding contact angles of droplets on smooth homogeneous solid substrates depend on the droplet volume. This conclusion has a direct experimental confirmation.

The evaporation, which takes place more intensively in the vicinity of the three-phase contact line [35,36], makes the measurements of both advancing and receding contact angles of drops not so straightforward and unambiguous in the case of volatile liquids. It is why a number of experiments on wetting/spreading of liquids have been performed in thin capillaries, where (i) equilibrium is reached at undersaturation and (ii) evaporation can be significantly diminished, if not ruled out.

3.3 Hysteresis of Contact Angle of Sessile Droplets on Deformable Substrates

Liquid droplet placed on a deformable/soft substrate causes the substrate to deform. The combined action of capillary pressure and surface forces, which act in the vicinity of the apparent three-phase contact line, and substrate's elasticity determine both the liquid shape and the substrate deformation. A theory of contact angle hysteresis of sessile liquid droplets on a deformable/soft substrate is developed below in this section in terms of the Derjaguin's pressure isotherm, $\Pi(h)$, which accounts for the action of surface forces in the vicinity of the contact line. Excess free energy equation for the droplet on deformable substrate results in two interconnected equations for droplet and deformed substrate shapes in terms of the Derjaguin's pressure isotherm. A simplified S-shaped Derjaguin's pressure isotherm is adopted, which allows direct calculations of static advancing/receding contact angles on deformable substrates. Elasticity of the substrate is assumed to obey a simple Winkler's model for elastic surfaces. The obtained results are in an agreement with the contact angle hysteresis theory developed in Section 3.2 for non-deformable substrates. Both the calculated advancing and receding contact angles are dependent upon the volume of the droplet and have a direct experimental confirmation (see Section 3.2). Calculated values of advancing and receding contact angles of droplets on deformable substrates are lower as compared with those on a non-deformable substrate.

Introduction

Static hysteresis of the contact angle of droplets on solid substrates is usually assumed to be determined by roughness and/or chemical heterogeneity of the substrate [3]. It is assumed in this case that at each point on the surface the equilibrium value of the contact angle is established depending only on the local properties of the substrate. As a result, a whole series of local thermodynamic equilibrium states can be realized, corresponding to a certain interval of contact angles. The maximum possible value corresponds to static advancing contact angle, θ_a, and the minimum value corresponds to static receding contact angle, θ_r.

However, roughness and chemical heterogeneity of the substrate are not the only reason behind contact angle hysteresis. We showed in Sections 3.1 and 3.2 that the contact angel hysteresis exists on smooth and homogeneous solid non-deformable substrate, which is determined by a special shape of the Derjaguin's isotherm (surface forces), which acts in the vicinity of the apparent three-phase contact line.

According to Kelvin's equation, a droplet can be at equilibrium only with oversaturated vapor and to reach the equilibrium the oversaturation should be kept for a prolonged period of time (Chapter 2). To the best of our knowledge, it is still difficult, if possible at all, to achieve the equilibrium in the case of pure liquids. As a result, only static advancing/receding non-equilibrium contact angles can be experimentally observed.

There is a regular process of deposition of droplets on any surface, such that after deposition, the droplet reaches the static advancing contact angle. It is commonly believed that this angle is close enough to the equilibrium contact angle. It was shown in Section 3.2 that this is not true, at least in the case of smooth homogeneous non-deformable substrates. Static hysteresis of the contact angle on deformable substrates is considered and compared with that on non-deformable substrates below.

The equilibrium of droplets on a deformable substrate was discussed in Section 2.7. A similar approach combined with discussion in Section 3.2 is here extended to introduce the theory of hysteresis of the contact angle on deformable substrates. The consideration is similar to that for droplets on a solid, smooth homogeneous substrate described in Section 3.2, but with the introduction of elasticity of the substrate. This is combined with the theory developed earlier for droplets at equilibrium on deformable substrates in Section 2.7, where the Derjaguin's pressure effects on the equilibrium of three-dimensional sessile liquid droplets deposited on deformable/soft substrates was considered. The theory presented here is the first attempt of development of such a theory. Therefore, substantial simplifications are unavoidable, which can be omitted in the future.

Equilibrium Contact Angle of Droplet on Deformable Substrates and the Surface Forces Action: A Simplified Model Adopted in this Section

This section deals with the equilibrium of droplets on deformable substrates. It forms the basis for the introduction of hysteresis to the problem. Apparently, there is only a single equilibrium contact angle, θ_e, on smooth homogeneous substrates regardless of whether they are deformable or non-deformable. However, according to Kelvin's equation, in order to reach this equilibrium contact angle, the droplet should be at equilibrium with oversaturated vapor (Chapter 2). Therefore, experimentally only non-equilibrium contact angles can be observed These are referred to as hysteresis contact angles: static advancing contact angle, $\theta_a > \theta_e$, and static receding contact angle $\theta_r < \theta_e$ (see Section 3.2).

That is, the consideration of equilibrium contact angles presented below should be considered from this point of view. A simple Winkler's model for the deformable solid is used. According to the Winkler's model, a linear relationship exists between the local deformation and the applied local stress [37]. According to this model, deformation of the deformable substrate, h_s, is local and is directly proportional to the applied pressure, P:

$$h_s = -KP, \tag{3.47}$$

where K is the elasticity coefficient; and h_s is the local deformation of the substrate due to the applied pressure, P, from above.

Let P_{air} be the pressure in the ambient air. Under the action of the pressure from the ambient air the solid deformation is

$$h_{se} = -KP_{air}. \tag{3.48}$$

The deformed solid substrate is covered by the thin equilibrium liquid film, which is calculated according to a combination of the well-known Kelvin's equation and the Derjaguin's pressure isotherm (Chapter 2):

$$\Pi\left(h_e\right) = P_e = \frac{R_g T}{v_m} \ln \frac{p_{sat}}{p},$$ (3.49)

where v_m is the molar volume of the liquid; T is the temperature in K; R_g is the gas constant; vapor pressure, p, is higher than the saturated pressure p_{sat}; and P_e is the equilibrium excess pressure in the liquid. Recall that at equilibrium the droplet must be at equilibrium with oversaturated vapor only according the Kelvin's equation (Eq. 3.49). The excess free energy of the equilibrium thin film on the deformed solid per unit area is given by

$$\frac{F_{e,film}}{S_{film}} = \gamma + \gamma_s + P_e h_e + \frac{h_{se}^2}{2K} + \int_{h_e}^{\infty} \Pi\left(h\right) dh,$$ (3.50)

where $P_e = P_{air} - P_{liquid}$; and γ and γ_s are liquid–vapor and solid–liquid interfacial tensions. This free energy should be subtracted from the free energy of the droplet on the deformable substrate otherwise the excess free energy of the droplet is infinite. Hence, the excess free energy of the droplet on a deformable solid substrate is as follows:

$$F - F_{e,film} = \gamma \Delta S + \gamma_s \Delta S_s + \Delta V + F_{surface\ forces} + F_{deformation},$$ (3.51)

where means "as compared with a flat equilibrium film."

Eq. (3.11) can be rewritten as

$$F - F_{e,film} = \int_0^{\infty} f\left(h, h', h_s, h_s'\right) dx,$$ (3.52)

where

$$f\left(h, h', h_s, h_s'\right) = \begin{bmatrix} \gamma\left(\sqrt{1 + h'^2\left(x\right)} - 1\right) + \gamma_s\left(\sqrt{1 + h_s'^2\left(x\right)} - 1\right) + \\[2mm] P_e\left(h - h_s\right) - P_e h_e + \frac{h_s^2}{2K} - \frac{h_{se}^2}{2K} \\[2mm] + \int_{h-h_s}^{\infty} \Pi\left(h\right) dh - \int_{h_e}^{\infty} \Pi\left(h\right) dh \end{bmatrix}.$$ (3.53)

In Eq. (3.52), x is the length along the tangential direction. The expression under the integral in Eq. (3.52) tends to zero as x tends to infinity.

Under equilibrium conditions, the excess free energy (3.52) should reach the minimum value. To satisfy this condition, the first variation of the excess free energy (3.52) should vanish, resulting in two Euler equations for profiles of both the droplet and the deformable substrate:

$$\frac{d}{dx}\left(\frac{\partial f}{\partial h'}\right) - \frac{\partial f}{\partial h} = 0,$$ (3.54)

$$\frac{d}{dx}\left(\frac{\partial f}{\partial h_s'}\right) - \frac{\partial f}{\partial h_s} = 0.$$ (3.55)

Therefore, the profile of the droplet, $h(x)$, and the profile of the substrate, $h_s(x)$, satisfy the following set of equations, *with the simplifying assumption that the substrate surface tension, γ_s, is neglected*, giving

$$\gamma \left(\frac{h''}{\left(1+h'^2\right)^{3/2}} \right) + \Pi\left(h-h_s\right) = P_e, \tag{3.56}$$

$$\Pi\left(h-h_s\right) + \frac{h_s}{K} = P_e. \tag{3.57}$$

Variable h and its derivatives, h' and h'' are all functions of the x coordinate. Eq. (3.56) is substantially different from the usual capillary equation for the droplet profile on a rigid substrate (Chapter 2) because now the Derjaguin's pressure term, $\Pi(h-h_s)$, in Eq. (3.56) depends on the profile of the deformable substrate, $h_s(x)$, which is determined by Eq. (3.57). Note, the value of P_e, is determined by the value of the oversaturated pressure in the ambient air.

Theory and Model for Hysteresis of Contact Angle on a Deformable Substrate

In this section, non-equilibrium states of a droplet on a soft substrate are considered for calculation of static advancing/receding contact angles. Consideration is based on Eqs. (3.56) and (3.57) where *a simplifying assumption is adopted that the substrate surface tension, γ_s, is neglected*. In the future this simplifying assumption should be omitted.

In the case under consideration, the droplet volume is imposed externally and can be varied externally (by pumping liquid in or out). The droplet will instantaneously start moving microscopically at $P \neq P_e$. In this scenario, excess pressure is different from its value at equilibrium (Sections 3.1 and 3.2). If the pressure inside the droplet is increased, then the droplet will start moving microscopically in the beginning. All process can be subdivided into "fast macroscopic" and "slow microscopic" processes. The change of the drop curvature over the main part of the droplet is a "fast macroscopic" process. Flow in a relatively thick β-film is a "macroscopic" process because the relative viscous resistance in these films is much smaller as compared with the flow in thin α-films. Hence, the idea is as follows: Until the flow takes place in relatively thin α-films, the droplet is at a quasi-equilibrium state and in slow microscopic motion. As soon as the flow jumpwise occupies the zone of relatively thick β-films, then the droplet starts macroscopic motion.

Note, the thickness of β-films is almost one order of magnitude greater than the thickness of α-films. Under the applied pressure difference, $P = P - P_e$, the flow rate, Q, in the film of thickness h is $Q = \frac{P}{6\mu l}h^3$, where $l \sim \sqrt{Hh_e}$ is the characteristic length of the transition zone between the bulk droplet and the thin films in front (Chapter 2), μ is the dynamic viscosity. That means, the flow inside β-films is three orders of magnitude faster than in α-films because of the cubic dependency of the flux Q on the film thickness. This is why the flow, which is located inside α-films, is referred to as "microscopic flow," while the flow inside β-films is referred to as "macroscopic flow."

In an advancing droplet, the transition from microscopic to macroscopic motion occurs when the excess pressure reaches the critical value, P_a, and, correspondingly, the macroscopic contact angle reaches the critical value, θ_a (Section 3.2). Therefore, droplet does not move macroscopically until $P = P_a$ and the advancing contact angle, θ_a, is reached. After that, the flow jumpwisely is transferred from thin α-film to much thicker β-film and becomes much faster, that is, macroscopic. Likewise, in the case of a receding droplet, the transition from microscopic to macroscopic motion occurs when the excess pressure reaches the critical value, P_r, and, correspondingly, the macroscopic contact angle reaches a critical value, θ_r (Section 3.2).

It is assumed that everywhere, except for a narrow flow zone, the droplet profile can be described by Eqs. (3.56) and (3.57), where the equilibrium excess pressure, P_e, should be replaced by a non-equilibrium

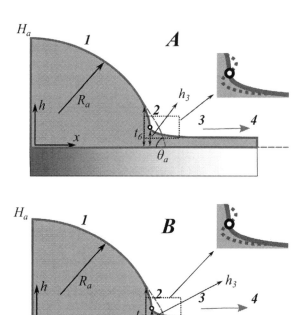

FIGURE 3.14 Schematic of a droplet on a solid non-deformable substrate (a) and soft/deformable substrate (b) at the moment before advancing begins. (1) Quasi-equilibrium part of the droplet; (2) Inflection point, where the slope becomes vertical; (3) Flow zone; (4) Equilibrium thin liquid film; h_3 is the thickness at the critical point. Close to the marked point a dashed line shows the profile of the transition zone just after the contact angle reaches the critical value θ_a and advancing starts as a caterpillar motion (see explanations in the text).

pressure, P (Figure 3.14 and Figure 3.15). Due to the increase in pressure, the curvature grows microscopically, and the slope tends to infinity at some critical point (marked in Figure 3.14) when P_a is reached. Apex, H_a, and the radius, R_a, of the advancing droplet are shown in Figure 3.14. It is assumed that, according to Sections 3.1 and 3.2, Eqs. (3.56) and (3.57) can be used to describe a quasi-equilibrium part of the droplet (Figure 3.14) even in the case of a droplet on a soft substrate. Subsequently, P_e is replaced by P_a in Eqs. (3.56) and (3.57). Similarly, when the pressure in the droplet decreases, the curvature of the droplet becomes microscopically flat and reaches a zero slope at P_r at some critical point (marked in Figure 3.15). The droplet is in this case is receding where apex, H_r, and the radius, R_r, of the receding droplet are shown in Figure 3.15.

The theory of advancing and receding contact angles was developed for the case of smooth homogeneous non-deformable substrates in Sections 3.1 and 3.2. It was shown in these section, that similar "dangerous" points exist on the profiles of the advancing and receding droplets, which are inflection points with a thickness h_3 (Figure 3.14a in the case of advancing droplet) and thickness h_4 (Figure 3.15a in the case of a receding droplet).

The slope increases with an increase in the excess pressure. At some critical pressure, P_a, the slope becomes infinite and the droplet cannot remain in the state of slow "microscopic" motion and starts "macroscopic" motion. That means at the "dangerous" point, h_3 (Figure 3.14a), which is the moment when advancing starts, two the following conditions should be satisfied:

$$h'' = 0,\ h' = -\infty,\ at\ h = h_3. \tag{3.58}$$

Similarly, the slope decreases with a decrease in excess pressure in the case of receding contact angles. At some critical pressure, P_r, the slope becomes zero and the droplet cannot remain in the state of slow

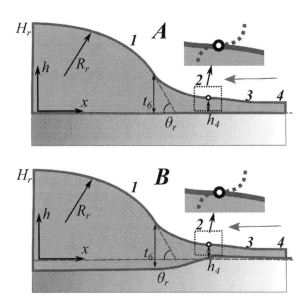

FIGURE 3.15 Schematic of a droplet on a solid non-deformable substrate (a) and soft/deformable substrate (b) at the moment before receding begins. (1) Quasi-equilibrium part of the droplet; (2) Inflection point, where the slope becomes vertical; (3) Flow zone; (4) Equilibrium thin liquid film; h_4 is the thickness at the critical point. Close to the marked point a dashed line shows the profile of the transition zone just after the contact angle reaches the critical value θ_r and receding starts (see explanations in the text).

"microscopic" motion and starts "macroscopic" motion. That means at the "dangerous" point, h_4, which is the moment when receding starts, two of the following conditions should be satisfied:

$$h'' = 0, \ h' = 0, \ at \ h = h_4. \tag{3.59}$$

Even under the simplifying assumptions adopted here, the system of Eqs. (3.56) and (3.57) is coupled in the case of deformable substrates and can be solved only numerically. This is why a simplified Derjaguin's pressure isotherm (piecewise linear function of h) is adopted below (Figure 1.12 curve 2) to simplify the system of Eqs. (3.56) and (3.57) further and to obtain an analytical solution (Figure 3.16):

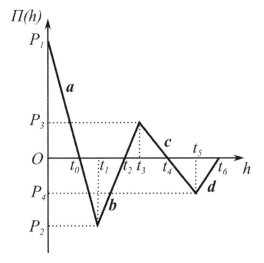

FIGURE 3.16 Simplified Derjaguin's pressure isotherms adopted for calculations below. Part **a** is referred to as thin α-films, Part **c** is referred to as thick β-films (Chapter 2).

$$\Pi(h) = \begin{cases} a(t_0 - h) & 0 \le h \le t_1 \\ b(h - t_2) & t_1 \le h \le t_3 \\ c(t_4 - h) & t_3 \le h \le t_5, \\ d(h - t_6) & t_5 \le h \le t_6 \\ 0 & h > t_6 \end{cases} \tag{3.60}$$

The Derjaguin's pressure isotherm has four regions namely, $0 \le h \le t_1$, $t_1 \le h \le t_3$, $t_3 \le h \le t_5$, and $t_5 \le h \le t_6$. A schematic plot of the adopted the Derjaguin's pressure isotherm is presented in Figure 1.16. The symbols a, b, c and d are used below for the slope of the linear parts of the isotherms inside the corresponding zones. Hence, a is the slope of the linear function in the region $h_e \le h \le t_1$ given by

$$a = \frac{P_1 - P_2}{t_1} = \frac{P_1 - P_e}{h_e}, \tag{3.61}$$

b is the slope of the linear function in the region $t_1 \le h \le t_3$ given by

$$b = \frac{P_3 - P_2}{t_3 - t_1}, \tag{3.62}$$

and c is the slope of the linear function in the region $t_3 \le h \le t_5$ given by

$$c = \frac{P_3}{t_4 - t_3}, \tag{3.63}$$

and d is the slope of the linear function in the region $t_5 \le h \le t_6$ given by

$$d = \frac{P_4}{t_5 - t_6}. \tag{3.64}$$

The selected piecewise linear dependency of the Derjaguin's pressure isotherm $\Pi(h)$ on h (see Eq. 3.60) captures the essential properties of the Derjaguin's pressure isotherm if (i) it satisfies the stability condition, $\Pi'(h) < 0$ when $0 \le h \le t_1$ (α-films and $t_3 < h < t_5$ β-films; Chapter 2); (ii) it corresponds to the partial wetting case at the proper selection of the parameters as selected below; and (iii) the influence of surface forces is short ranged and the radius of their influence is t_6.

To solve the system of Eqs. (3.56) and (3.57), subtract them, which results in

$$\gamma \left(\frac{h''}{\left(1 + h'^2\right)^{3/2}} \right) - \frac{h_s}{K} = 0. \tag{3.65}$$

Equation (3.65) must be solved for four different transition regions where the Derjaguin's pressure is applicable, that is, from h_e to t_6. According to Eq. (3.65), the condition $h'' = 0$ is satisfied when $h_s = 0$ (at $h = h_3$ and $P = P_a$ for advancing droplets and at $h = h_4$ and $P = P_r$ for receding droplets). The "dangerous points" are located on the Derjaguin's pressure isotherm in region b for advancing and in region c for receding droplets (Figure 1.16). Hence, it is possible to conclude

$$h_3 = t_2 + \frac{P_a}{b} \quad \text{and} \quad h_4 = t_4 - \frac{P_r}{c}. \tag{3.66}$$

Detailed derivation for advancing droplets are presented in the Appendix. The derivation for receding droplets is not included to avoid repetition.

Results and Discussions

Table 3.2 gives the physical properties used below for the Derjaguin's pressure isotherm shown in Figure 1.16 for calculating advancing/receding contact angles.

The values of the Derjaguin's pressure isotherm mentioned in Table 3.2 are selected to match the Derjaguin's pressure isotherm used in Sections 3.1 and 3.2. These specific values are used so that the results can be compared with those in Sections 3.1 and 3.2 for non-deformable substrates.

To begin with, the volume of the advancing droplet is varied to investigate its effect on static advancing contact angle for a non-deformable ($K = 0$ cm³/dyn) and deformable substrate ($K = 2 \times 10^{-16}$ cm³/din) (Figure 3.17). The higher the initial volume of the droplet, V_a, the lower the value of θ_a. The highest values of θ_a are found for droplets with small size, that is, those with a high excess pressure (γ/R_a). The trend of θ_a variation with V_a is in agreement with the results presented for an advancing droplet on a solid non-deformable substrate (Section 3.2), that is, a decreasing function of the volume. At $V_a \approx 5 \times 10^{-9}$ cm² and $K = 0$ cm³/dyn, the current model predicts $\theta_a \approx 57.78°$, whereas from Section 3.2 we get $\theta_a \approx 57.43°$, which are in very good agreement. The effect of increasing the elasticity of the substrate, that is, $K = 2 \times 10^{-16}$, results in an overall decrease in the values of θ_a (see red line in Figure 3.17). This shows that on a deformable substrate, the static advancing contact angle decreases as compared with the response on a non-deformable substrate.

TABLE 3.2

The Derjaguin's Pressure Isotherm Properties for Calculations of Advancing/Receding Contact Angles on Deformable Substrates

Physical Properties	Isotherm
Surface tension (γ)	72 dyn/cm
P_1	1.3×10^{13} dyn/cm²
P_2	-4.45×10^8 dyn/cm²
P_3	2.8×10^7 dyn/cm²
P_4	-6.5×10^4 dyn/cm²
t_0	$\dfrac{P_1}{a} = 2.299 \times 10^{-8}$ cm
t_1	2.3×10^{-8} cm
t_2	$t_3 - \dfrac{P_3}{b} = 2.083 \times 10^{-7}$ cm
t_3	2.2×10^{-7} cm
t_4	3×10^{-6} cm
t_5	3.007×10^{-6} cm
t_6	2×10^{-5} cm
a	$\dfrac{P_1 - P_2}{t_1} = 5.652 \times 10^{20}$ dyn/cm³
b	$\dfrac{P_3 - P_2}{t_3 - t_1} = 2.401 \times 10^{15}$ dyn/cm³
c	$\dfrac{P_3}{t_4 - t_3} = 1.007 \times 10^{13}$ dyn/cm³
d	$\dfrac{P_4}{t_5 - t_6} = 3.825 \times 10^9$ dyn/cm³
K	2×10^{-16} cm³/dyn, i.e., $bK < 1$
V_a	1×10^{-2} cm²

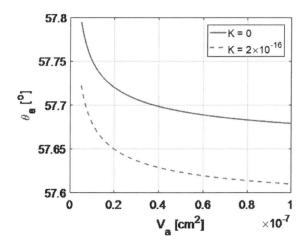

FIGURE 3.17   Effect of variation of droplet volume on static advancing contact angle.

The volume of the receding droplet is varied to investigate its effect on receding contact angle for a non-deformable ($K = 0$ cm³/dyn) and deformable substrate ($K = 2 \times 10^{-16}$ cm³/dyn) (Figure 3.18). The higher the initial volume of the droplet, V_r, the lower the value of θ_r. The highest values of θ_r are observed for droplets with small size, that is, the droplets with a high excess pressure (γ / R_r). The trend of θ_r variation with V_r is a decreasing function and agrees with the results presented for an advancing droplet on a solid substrate in Section 3.2. At $V_a \approx 5 \times 10^{-9}$ cm² and $K = 0$ cm³/dyn, the model predicts $\theta_r \approx 7.88°$, whereas from Section 3.2, the estimate for $\theta_r \approx 8.8°$. The difference is primarily due to the use of approximate values for a simple piecewise linear Derjaguin's pressure isotherm. The effect of increasing the elasticity of the substrate, that is, $K = 2 \times 10^{-16}$, results in a very slight overall decrease in the values of θ_r (see red dotted line zoomed region in Figure 3.18). Experimentally, it has been shown that the contact angles' hysteresis decreases with an increase in elasticity of the substrate [38]. Therefore, it can be assumed that with a more deformable substrate, the apparent static receding contact angle is expected to decrease, but the decrease is very small, at least according to the model used for deformable substrates.

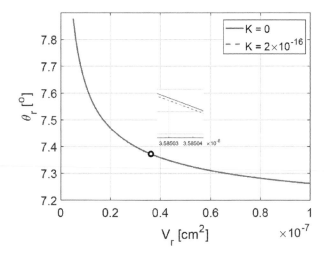

FIGURE 3.18   Effect of variation of droplet volume on apparent receding contact angle.

Conclusions

Advancing and receding contact angles on smooth and homogeneous deformable substrates are theoretically investigated. The Derjaguin's pressure action is taken into account in the vicinity of the three-phase contact line. Winkler's model is used to account for the elasticity of the deformable substrate. It is demonstrated that advancing and receding contact angles of a droplet on deformable substrate depend upon the droplet volume, the Derjaguin's pressure isotherm, and elasticity of the substrate. The result of variation of both the advancing and receding contact angles with the volume of the droplet on a non-deformable substrate are very close to the results produced in Section 3.2. For a deformable substrate, the calculated advancing contact angle is lower than the advancing contact angle on a non-deformable substrate. An increase in volume of the droplet causes the advancing contact angle to be decreased. For a deformable substrate, the estimated receding contact angle is less than the receding contact angle on a non-deformable substrate, but the decrease is very small.

APPENDIX: ADVANCING CONTACT ANGLE

To avoid repetition, detailed derivation for the Derjaguin's pressure isotherm (Figure 3.16) is presented below for the advancing contact angle only. The governing equations from Eqs. (3.56) and (3.57) can be rewritten now as

$$\gamma\left(\frac{h''}{\left(1+h'^2\right)^{3/2}}\right)+\Pi\left(h-h_s\right)=P_a, \tag{A1}$$

$$\Pi\left(h-h_s\right)+\frac{h_s}{K}=P_a. \tag{A2}$$

For the bulk of the liquid droplet, that is, the spherical region, we have, $h > t_6$. For this condition, from Figure 1.16 it can be seen that $\Pi\left(h\right)=0$, which changes Eq. (A1) into,

$$\gamma\left(\frac{h''}{\left(1+h'^2\right)^{3/2}}\right)=P_a, \tag{A3}$$

which can be integrated once to give,

$$\frac{\gamma}{\sqrt{1+h'^2}}=\gamma+P_a\left(H_a-h\right), \tag{A4}$$

where H_a is the height of the droplet at the origin, $x=0$. The solution of the differential equation in Eq. (A4) is

$$h(x)=\frac{\gamma}{P_a}\left[1+\frac{P_aH_a}{\gamma}-\sqrt{1-\left(\frac{P_ax}{\gamma}\right)^2}\right]. \tag{A5}$$

The intersection of this profile with the substrate describes the three-phase contact line, which consequently determines the macroscopic advancing contact angle, $h'(0) = -\tan\theta_a$. Using this condition and Eq. (A4) to estimate θ_a,

$$\theta_a = \cos^{-1}\left(\frac{P_a H_a}{\gamma} + 1\right) \tag{A6}$$

For the bulk of the liquid droplet, that is, the spherical region, we have $h - h_s > t_6$. For this condition, from Eq. (3.60) it can be seen that $\Pi(h) = 0$, which changes Eq. (A2) into

$$h_s = KP_a. \tag{A7}$$

From Eqs. (3.60) and (A2), h_s is evaluated for the different regions of the Derjaguin's pressure isotherm,

$$h_s = \frac{K(P_a + bt_2 - bh)}{1 - bK} \quad \text{for} \quad t_1 \le h \le t_3, \tag{A8}$$

and

$$h_s = \frac{K(P_a - ct_4 + ch)}{1 + cK} \quad \text{for} \quad t_3 \le h \le t_5. \tag{A9}$$

and

$$h_s = \frac{K(P_a + dt_6 - dh)}{1 - dK} \quad \text{for} \quad t_5 \le h \le t_6. \tag{A10}$$

It is necessary to note here that regions of interest for the advancing contact angle are reduced to three regions: $t_1 \le h \le t_3$, $t_3 \le h \le t_5$ and $t_5 \le h \le t_6$. In a similar method of solution, integrating Eq. (3.65) with respect to h for the Derjaguin's pressure isotherm region $t_1 \le h \le t_3$ results in

$$\frac{1}{\sqrt{1 + h'^2}} = C_{2,a} + \frac{b}{2\gamma(1 - bK)}h^2 - \frac{(P_a + bt_2)}{\gamma(1 - bK)}h, \tag{A11}$$

where $C_{2,a}$ is the integration constant. For the Derjaguin's pressure isotherm region $t_3 \le h \le t_5$, integrate Eq. (3.65) with respect to h,

$$\frac{1}{\sqrt{1 + h'^2}} = C_{3,a} - \frac{c}{2\gamma(1 + cK)}h^2 - \frac{(P_a - ct_4)}{\gamma(1 + cK)}h, \tag{A12}$$

where $C_{3,a}$ is the integration constant. For the Derjaguin's pressure isotherm region $t_5 \le h \le t_6$, integrate Eq. (3.65) with respect to h,

$$\frac{1}{\sqrt{1 + h'^2}} = C_{4,a} + \frac{d}{2\gamma(1 - dK)}h^2 - \frac{(P_a + dt_6)}{\gamma(1 - dK)}h, \tag{A13}$$

where $C_{4,a}$ is the integration constant. Let $x = L_{6,a}$ be the position, where the total thickness of the liquid film is equal to the radius of the Derjaguin's pressure action, that is, at $x = L_{6,a}$, $h(L_{6,a}) - h_s(L_{6,a}) = t_6$. Then from Eq. (A7) and this boundary condition

$$h(L_{6,a}) = t_6 + KP_a. \tag{A14}$$

From Eq. (A5), $h(L_{6,a})$ is equal to

$$h(L_{6,a}) = \frac{\gamma}{P_a}\left[1 + \frac{P_a H_a}{\gamma} - \sqrt{1 - \left(\frac{P_a L_{6,a}}{\gamma}\right)^2}\right].$$
(A15)

According to Eqs. (A14) and (A15), the thickness of both sides must be equal. This allows determining

$$L_{6,a} = \sqrt{\left(\frac{\gamma}{P_a}\right)^2 - \left(t_6 + KP_a - \frac{\gamma}{P_a} - H_a\right)^2}.$$
(A16)

Let $x = L_a$ be the position where the spherical profile intersects the initial non-deformed surface; hence, at $x = L_a$, $h(L_a) = 0$. For this condition, evaluating for L_a gives

$$L_a = \sqrt{\left(\frac{\gamma}{P_a}\right)^2 - \left(\frac{\gamma}{P_a} + H_a\right)^2}.$$
(A17)

Boundary conditions are

At $h = h_3$, $h' = -\infty$, from Section 3.2,

At $h = t_3$, $h'_c = h'_b$ and $h_c = h_b$,

At $h = t_5$, $h'_c = h'_d$ and $h_c = h'_d$,

where subscripts only point to different regions on the Derjaguin's pressure isotherm plotted in Figure 1.16. Using the above boundary conditions, unknowns $L_{4,a}$, L_a, $C_{2,a}$, $C_{3,a}$ and $C_{4,a}$ can be evaluated. The volume of the advancing droplet, V_a, is fixed. Therefore,

$$V_a = 2\left\{L_a\left(H_a + \frac{\gamma}{P_a} - KP_a\right) - \frac{1}{2}\left[-L_a\sqrt{\left(\frac{\gamma}{P_a}\right)^2 - L_a^2} + \left(\frac{\gamma}{P_a}\right)^2 \sin^{-1}\left(\frac{L_a P_a}{\gamma}\right)\right]\right\}.$$
(A18)

REFERENCES

1. Arjmandi-Tash, O., Kovalchuk, N. M., Trybala, A., Kuchin, I. V., and Starov, V. Kinetics of wetting and spreading of droplets over various substrates. *Langmuir* (Feature Article), 33, 4367–4385 (2017).
2. Kuchin, I., and Starov, V. Hysteresis of the contact angle of a meniscus inside a capillary with smooth, homogeneous solid walls. *Langmuir*, 32, 5333–5340 (2016).
3. Joanny, F., and de Gennes, P. G. A model for contact angle hysteresis. *J. Chem. Phys.*, 81, 552–562 (1984).
4. Starov, V. Static contact angle hysteresis on smooth, homogeneous solid substrates. *Colloid Polym. Sci.*, 291, 261–270 (2013).
5. Chibowski, E. Surface free energy of a solid from contact angle hysteresis. *Adv. Colloid Interface Sci.*, 103, 149–172 (2003).
6. Extrand, C. W., and Kumagai, Y. An experimental study of contact angle hysteresis. *J. Colloid Interface Sci.*, 191, 378–383 (1997).

7. Extrand, C. W. Water contact angle and hysteresis of polyamid surfaces. *J. Colloid Interface Sci.*, 248, 136–142 (2002).

8. Zorin, Z. M., Sobolev, V. D., and Churaev, N. V. *Surface Forces in Thin Films and Disperse Systems.* Nauka, Moscow, Russia [in Russian], 1972, p. 214.

9. Romanov, E. A., Kokorev, D. T., and Churaev, N. V. Effect of wetting hysteresis on state of gas trapped by liquid in a capillary. *Int. J. Heat Mass Transfer*, 16, 549–554 (1973).

10. Rangelova, N., and Platikanov, D. Contact angles in black foam films of albumin. *Ann. Univ. Sofia Fac. Chim.*, 71/72, 109 (1976/1977); Platikanov, D., Yampolskaya, G. P., Rangelova, N., Angarska, Z., Bobrova, L. E., and Izmailova, V. N. Free black films of proteins. Thermodynamic parameters. *Colloid J.USSR*, 43, 177–180 (1981).

11. Rangelova, N., and Platikanov, D. *Ann. Univ. Sofia Fac. Chim.*, 78, 126 (1984).

12. Rangelova, N. I., Izmailova, V. N., Platikanov, D. N., Yampol'skaya, G. P., and Tulovskaya, S. D. Free black films of proteins: Dynamic hysteresis of the contact angle (film-bulk liquid) and the rheological properties of adsorption layers. *Colloid J. USSR*, 52, 442–447 (1990).

13. Platikanov, D., Nedyalkov, M., and Petkova, V. Phospholipid black foam films: Dynamic contact angles and gas permeability of DMPC bilayer films. *Adv. Colloid Interface Sci.*, 101–102, 185–203 (2003).

14. Petkova, V., Platikanov, D., and Nedyalkov, M. Phospholipid black foam films: Dynamic contact angles and gas permeability of DMPC+DMPG black films. *Adv. Colloid Interface Sci.*, 104, 37–51 (2003).

15. Starov, V. M., and Velarde, M. G. Surface forces and wetting phenomena. *J. Phys. Condens. Matter*, 21, 464121 (11pp) (2009).

16. Churaev, N. V., Sobolev, V. D., and Starov, V. M. Disjoining pressure of thin nonfreezing interlayers. *J Colloid Interface Sci.*, 247, 80–83 (2002).

17. Churaev, N. V., and Sobolev, V. D. Prediction of contact angles on the basis of the Frumkin-Derjaguin approach. *Adv Colloid Interface Sci.*, 61, 1–16 (1995).

18. Derjaguin, B. V., Churaev, N. V., and Muller, V. M. *Surface Forces.* Springer, New York, 1987.

19. Kuchin, I. V., Matar, O. K., Craster, R. V., and Starov, V. M. Influence of the disjoining pressure on the equilibrium interfacial profile in transition zone between a thin film and a capillary meniscus. *Coll. Int. Sci. Comm.*, 1, 18–22 (2014).

20. Djikaev, S., and Ruckenstein, E. The variation of the number of hydrogen bonds per water molecule in the vicinity of a hydrophobic surface and its effect on hydrophobic interactions. *Curr. Opin. Colloid Interface Sci.*, 16 (4), 272–284 (2011).

21. Churaev, N. V., and Sobolev, V. D. Wetting of low energy surfaces. *Adv. Colloid Interf. Sci.*, 134–135, 15–23 (2007).

22. Rowlinson, J. S., and Widom, B. *Molecular Theory of Capillarity.* Clarendon Press, Oxford, UK, 1982.

23. Derjaguin, B. V., and Zorin, Z. M. Investigation of the surface condensation and vapour adsorption close to saturation conditions with micropolarization method. *Zh. Fiz. Khim.*, 29, 1755–1770 (1955).

24. Zorin, Z. M., Novikova, A. V., Petrov, A. K., and Churaev, N. V. *Surface Forces in Thin Films and Stability of Colloids.* Nauka, Moscow, Russia [in Russian], 1974, p. 94.

25. Zorin, Z. M., Romanov, V. P., and Churaev, N. V. The contact angles of surfactant solution on the quartz surface. *Colloid Polym.Sci.*, 257, 968–972 (1979).

26. Kuchin, I., and Starov, V. Hysteresis of contact angle of sessile droplets on smooth homogeneous solid substrates via disjoining/conjoining pressure. *Langmuir*, 31, 5345–5352 (2015).

27. Gu, Y. Drop size dependence of contact angles of oil drops on a solid surface in water. *Colloids Surf. A*, 181, 215–224 (2001).

28. Churaev, N. V., Setzer, M. J., and Adolphs, J. Influence of surface wettability on adsorption isotherms of water vapor. *J. Colloid Interface Sci.*, 197, 327–333 (1998).

29. Mack, G. L. The determination of contact angles from measurements of the dimensions of small bubbles and drops. I. The spheroidal segment method for acute angles. *J. Phys. Chem.*, 40 (2), 159–167 (1936).

30. Veselovsky, V. S., and Pertsev, V. N. Adhesion of the bubbles to solid surfaces. *J. Phys Chem (USSR Academy of Sciences)*, 8 (2), 245–259 (1936).

31. Drelich, J. The effect of drop (Bubble) size on contact angle at solid surfaces. *J. Adhesion*, 63, 31–51 (1997).

32. Good, R. J., and Koo, M. N. The effect of drop size on contact angle. *J. Colloid Interface Sci.*, 71 (2), 283–292 (1979).

33. Timmons, C. O., and Zisman, W. A. The effect of liquid structure on contact angle hysteresis. *J. Colloid Interface Sci.*, 22, 165–171 (1966).

34. Schwartz, A. M. Contact angle hysteresis: A molecular interpretation. *J. Colloid Interface Sci.*, 75 (2), 404–408 (1980).

35. Lee, K. S., Cheah, C. Y., Copleston, R. J., Starov, V. M., and Sefiane, K. Spreading and evaporation of sessile droplets: Universal behavior in the case of complete wetting. *Colloids and Surf. A: Physicochem. Eng. Asp.*, 323, 63–72 (2008).

36. Semenov, S., Trybala, A., Rubio, R., Kovalchuk, N., Starov, V., and Velarde, M. Simultaneous spreading and evaporation: Recent developments.*Adv Colloid Interface Sci.*, 206, 382–398 (2014).

37. Winkler, E. The theory of elasticity and strength with special reference to their application in the art for polytechnics, building academies, engineers, mechanical engineers, architects, etc., H. Dominicus, 1867.

38. Extrand, C. W., and Kumagai, Y. Contact angles and hysteresis on soft surfaces. *J. Colloid Interface Sci.*, 184, 191–200 (1996).

4

Kinetics of Wetting

Introduction

Consideration in Chapter 3 showed that the mechanism behind both static advancing and static receding contact angles are rather sophisticated. It is probably why kinetics of spreading based on consideration of the Derjaguin's pressure in the case of particle wetting has not been developed yet. In this Chapter we consider kinetics of wetting processes in the case of complete wetting, whereas the Derjaguin's pressure includes only one component: the molecular part.

In Chapters 1 through 3 the importance of surface forces in the vicinity of the apparent three-phase contact line has been shown and investigated in the case of equilibrium and static hysteresis of the contact angle. It is obvious that surface forces are equally important in the kinetics of spreading. In this Chapter we consider the complete wetting case only. Investigation of kinetics of spreading using the same approach as in this Chapter 4 should be undertaken in the future.

The consideration in Chapter 4 shows that in the case of spreading the whole drop profile should be divided in several regions, which are briefly discussed below in this section.

We should decide which dimensionless parameters are important in the case of spreading and which drops can be considered as "small drops" and which as "big drops." This is determined by the action of gravity.

Let us consider the simplest possible example of spreading of two-dimensional (cylindrical) droplets (Figure 4.1a).

The capillary regime is an initial stage of spreading of small drops, whereas the gravitational regime is the final stage of spreading of small drops or the regime of spreading of big drops. The transition from the capillary regime of spreading to the gravitational regime takes place at the moment t_c, when

$$R(t_c) \sim a = \sqrt{\frac{\gamma}{\rho g}},$$

where $R(t)$ is the radius of the drop base at time t, which is referred to here as the radius of spreading; γ and ρ are the interfacial tension and density of the liquid, respectively; g is the gravity acceleration.

In the case of water ($\gamma = 72.5$ dyn/cm, $\rho = 1$ g/cm^3), hence, $a \sim 0.27$ cm.

The next two important parameters are the Reynolds number and the capillary number. The Reynolds number characterizes the importance of inertial forces as compared with viscous forces. To deduce the relevant expression for the Reynolds number, let us consider a spreading of a two-dimensional droplet over a solid surface (gravity action is neglected). In this case, the Navier-Stokes equation with the incompressibility condition takes the following form:

$$\rho\left(u\frac{\partial u}{\partial x} + v\frac{\partial u}{\partial y}\right) = -\frac{\partial p}{\partial x} + \eta\left(\frac{\partial^2 u}{\partial x^2} + \frac{\partial^2 u}{\partial y^2}\right)$$

$$\rho\left(u\frac{\partial v}{\partial x} + v\frac{\partial v}{\partial y}\right) = -\frac{\partial p}{\partial y} + \eta\left(\frac{\partial^2 v}{\partial x^2} + \frac{\partial^2 v}{\partial y^2}\right),$$

$$\frac{\partial u}{\partial x} + \frac{\partial v}{\partial y} = 0$$

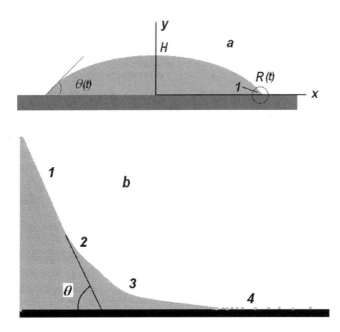

FIGURE 4.1 (a) Spreading of a liquid droplet over a flat solid substrate. $R(t)$ is the radius of spreading, which is the position of the apparent three-phase contact line; $\theta(t)$ is the dynamic contact angle, (1) vicinity of the apparent three-phase contact line. (b) A magnification of the vicinity of the moving apparent three-phase contact line in the case of complete wetting: (1) spherical part of the drop, which forms a dynamic contact angle, θ, with the solid substrate; (2) a region, where a spherical shape is distorted by the hydrodynamic force; and (3) a region where the Derjaguin's pressure comes into play and becomes increasingly important toward the end of the region 3; and (4) a region, where a macroscopic description is no longer valid, and either surface diffusion takes place in the case of nonvolatile liquid or the surface is covered by an adsorbed liquid layer, which is in the thermodynamic equilibrium with the ambient vapor pressure in the surrounding air.

where $\vec{v} = (u, v)$ is the velocity vector; and the gravity action is neglected. Note, in Chapter 4, the symbol "p" is used for hydrodynamic pressure in the liquid. Let U_* and v_* are scales of the velocity components in the tangential and the vertical directions, respectively. Using the incompressibility condition, we conclude

$$\frac{U_*}{r_*} = \frac{v_*}{h_*},$$

or

$$v_* = \varepsilon U_*, \quad \varepsilon = \frac{h_*}{r_*}.$$

If the droplet has a low slope, then $\varepsilon \ll 1$ and, hence, the velocity scale in the vertical direction is much smaller than the velocity scale in the tangential direction. Now using the first Navier-Stokes equation we can estimate

$$\rho u \frac{\partial u}{\partial x} \sim \rho v \frac{\partial u}{\partial y} \sim \frac{\rho U_*^2}{r_*}, \quad \eta \frac{\partial^2 u}{\partial x^2} \sim \varepsilon^2 \eta \frac{\partial^2 u}{\partial y^2} \ll \eta \frac{\partial^2 u}{\partial y^2}, \quad \eta \frac{\partial^2 u}{\partial y^2} \sim \frac{\eta U_*}{h_*^2},$$

hence, all derivatives in the low slope approximation in the direction x can be neglected as compared with derivatives in the axial direction y. Now we can estimate the Reynolds number

$$Re \sim \frac{\rho u \dfrac{\partial u}{\partial x}}{\eta \dfrac{\partial^2 u}{\partial y^2}} \sim \frac{\dfrac{\rho U_*^2}{r_*}}{\dfrac{\eta U_*}{h_*^2}} = \frac{\rho U_* h_*^2}{\eta r_*} = \varepsilon^2 \frac{\rho U_* r_*}{\eta}$$

or

$$Re = \varepsilon^2 \frac{\rho U r_*}{\eta}. \tag{4.1}$$

This expression shows that the Reynolds number under the low slope approximation is proportional to ε^2. Hence, during the initial stage of spreading, when $\varepsilon \sim 1$, the Reynolds number is not small, but as soon as the low slope approximation is valid, Re becomes small even if

$$\frac{\rho U r_*}{\eta}$$

is not small enough. This means, during the short initial stage of spreading, both the low slope approximation and low Reynolds number approximations are not valid. However, we are interested only in the main part of the spreading process when the short initial stage is over. Below we see that Re number should be calculated only in the close vicinity of the moving contact line, where the low slope approximation is valid (see below) because in the main part of the spreading droplet the liquid is moving much slower than close to the edges. Hence, the inertial terms in Navier-Stokes equations can be safely omitted and only Stokes equations should be used

$$0 = -\frac{\partial p}{\partial x} + \eta \left(\frac{\partial^2 u}{\partial x^2} + \frac{\partial^2 u}{\partial y^2} \right)$$

$$0 = -\frac{\partial p}{\partial y} + \eta \left(\frac{\partial^2 v}{\partial x^2} + \frac{\partial^2 v}{\partial y^2} \right).$$

$$\frac{\partial u}{\partial x} + \frac{\partial v}{\partial y} = 0$$

The tangential stress on the free drop surface at $y = h(x)$ is

$$\eta \left[h'^2 \left\{ h' \left(\frac{\partial u}{\partial y} + \frac{\partial v}{\partial x} \right) \right\} + 2h'^2 \left(\frac{\partial u}{\partial x} - \frac{\partial v}{\partial y} \right) + h' \left(\frac{\partial u}{\partial y} + \frac{\partial v}{\partial x} \right) \right] = 0.$$

Under the low slow approximation, $\varepsilon \ll 1$, we can easily check that $h'^2 \ll 1$. Using this estimation, this condition can be rewritten as

$$\eta \left(\frac{\partial u}{\partial y} + \frac{\partial v}{\partial x} \right) = 0$$

Using the previous estimations, we conclude

$$\frac{\partial u}{\partial y} \sim \frac{U}{h_*}; \quad \frac{\partial v}{\partial x} \sim \frac{\varepsilon U}{r_*} = \varepsilon^2 \frac{U}{h_*} \ll \frac{\partial u}{\partial y}.$$

This estimation shows that under the low slope approximation the tangential stress on the free liquid interface is

$$\eta \frac{\partial u}{\partial y} = 0.$$

This boundary condition is used hereafter in Chapter 4.

The capillary number,

$$Ca = \frac{U\eta}{\gamma},$$

characterizes the relative influence of the viscous forces as compared with the capillary force. Let us estimate possible values of the Ca. Let $r_* \sim 0.1$ cm, $\gamma \sim 30$ dyn/cm, and $\eta \sim 10^{-2} P$ (oils). Let the droplet edge on the distance be equal to its radius over 1 s, which can be considered as a very high velocity of spreading. This gives the following estimation $Ca \sim 3 \cdot 10^{-5} \ll 1$. That means, we should expect Ca to be even less than 10^{-5} over the duration of spreading. According to the previous, we assume below that both the capillary and Reynolds numbers are very small except for a very short initial stage of spreading. We estimate the duration of the initial stage of spreading in Chapter 5 (Section 5.1) immediately after the drop is deposited on the solid substrate.

Let us consider the consequence of the smallness of the capillary number, $Ca \ll 1$, using the simplest possible example of spreading of two-dimensional (cylindrical) droplets (Figure 4.1a). Let the length scales in both x and y directions in the main part of the spreading drop (Figure 4.1a) be r_* (i.e., it is not a low slope case), then the pressure inside the main part of the droplet has the order of magnitude of the capillary pressure, that is

$$p \sim \frac{\gamma}{r_*}.$$

Using the incompressibility condition, we immediately conclude that velocity in both directions, u and v, have the same order of magnitude U_*. Let us introduce the following dimensionless variables, which are marked by an overbar

$$\bar{p} = \frac{p}{\gamma/r_*}, \quad \bar{x} = \frac{x}{r_*}, \quad \bar{y} = \frac{y}{r_*}, \quad \bar{u} = \frac{u}{U_*}, \quad \bar{v} = \frac{v}{U_*}.$$

Using these variables, the Stokes equations can be rewritten as

$$\frac{\partial \bar{p}}{\partial \bar{x}} = Ca \left(\frac{\partial^2 \bar{u}}{\partial \bar{x}^2} + \frac{\partial^2 \bar{u}}{\partial \bar{y}^2} \right)$$

$$\frac{\partial \bar{p}}{\partial \bar{y}} = Ca \left(\frac{\partial^2 \bar{v}}{\partial \bar{x}^2} + \frac{\partial^2 \bar{v}}{\partial \bar{y}^2} \right).$$

We already concluded that $Ca \ll 1$, which means the right-hand side of both latter equations in very small. Hence, these equations can be rewritten as

$$\frac{\partial \bar{p}}{\partial \bar{x}} = 0$$

$$\frac{\partial \bar{p}}{\partial \bar{y}} = 0,$$

which means that the pressure remains constant inside the main part of the spreading droplet. If we now write down the normal stress balance on the main part of the spreading droplet, we get

$$\bar{p} = \frac{\bar{h}''}{(1+\bar{h}'^2)^{3/2}} + Ca\left[-\frac{2}{\left(1+\bar{h}'^2\right)}\left\{-\bar{h}'\left(\frac{\partial \bar{u}}{\partial y} + \frac{\partial \bar{v}}{\partial x}\right) - \frac{\partial \bar{v}}{\partial y} - \bar{h}'^2\frac{\partial \bar{u}}{\partial x}\right\}\right].$$

Using the condition $Ca \ll 1$, we conclude from this equation

$$\bar{p} = \frac{\bar{h}''}{(1+\bar{h}'^2)^{3/2}} = \text{const},$$

even in the case when the droplet profile does not have the low slope approximation, that is, even if $\bar{h}'^2 \sim 1$ not small. This shows that the spreading droplet retains its spherical shape over the main part of the droplet. Note, that the radius of the droplet base, $R(t)$, changes over time, and this change results in a quasi-steady-state change of the droplet profile.

In the case of a moving meniscus in a capillary, similar estimations show that $Ca \ll 1$ results in a spherical shape of the meniscus in the main part of the capillary.

Once again, the smallness of the Ca means that the surface tension is much more powerful over the most part of the droplet/meniscus and, hence, the droplet/meniscus has a spherical shape everywhere except for the vicinity of the apparent three-phase contact line. A size of this region, l_*, is estimated in Section 4.2. It will be shown that the following inequality is satisfied:

$$h_* \ll l_* \ll r_*.$$

Hence,

$$\delta = \frac{h_*}{l_*} \ll 1$$

is a small parameter inside the vicinity of the moving contact line. This means that the curvature of the liquid interface inside the vicinity of the moving contact line can be estimated as

$$\frac{\gamma h''}{\left(1+\bar{h}'^2\right)^{3/2}} \sim \frac{\gamma \frac{h_*}{l_*^2}\bar{h}''}{\left(1+\frac{h_*^2}{l_*^2}\bar{h}'^2\right)^{3/2}} = \frac{\gamma \frac{h_*}{l_*^2}\bar{h}''}{\left(1+\delta^2\bar{h}'^2\right)^{3/2}} \approx \gamma \frac{h_*}{l_*^2}\bar{h}''.$$

This gives a very important conclusion: the low slow approximation is valid inside the vicinity of the moving contact line even if the drop profile is not low sloped, that is, even if $\bar{h}'^2 \sim 1$ is not small. Hence, we can always use the low slope approximation inside the vicinity of the moving contact line except for the case when the slope is close to $\pi/2$ (see Chapter 3).

Let us estimate a possible range of capillary numbers. If $Ca \sim 1$, then in the case of water, we conclude

$$Ca \sim \frac{U_*\eta}{\gamma} \sim \frac{U \cdot 10^{-2}P}{72\,\text{dyn/cm}} \sim 1.$$

This results in $U_* \sim 72$ m/s. This velocity is so high that it probably can be achieved only under very special conditions. In the case of $r_* \sim 0.1-1$ cm, this velocity results in $Re \sim 7.2 \cdot 10^4 - 7.2 \cdot 10^5$, which is a turbulent flow and is beyond the scope of this book. This means that the possible range of Ca is between 0 and 1. Low $Ca \ll 1$ means a relatively low rate of spreading, while $Ca \sim 1$ means a very high velocity of motion. In turn, the case $Ca << 1$ include "a high capillary number limit" (see Sections 4.5 through 4.7). This means that the case of low capillary numbers, $Ca << 1$, and intermediate capillary numbers, $Ca \sim 1$, should be considered in a completely different way. The situation is similar to the case of the Reynolds number: consideration of flows at low Reynolds numbers is very much different from the consideration of flows at high Reynolds numbers.

Now we are prepared to consider in more detail the vicinity of the moving contact line (region 1 in Figure 4.1a), which is magnified in Figure 4.1b.

The whole vicinity of the apparent three-phase contact line can be subdivided into four regions (Figure 4.1b). Region 1 is a spherical meniscus in the main part of the spreading droplet. This region is included to show the dynamic contact angle, $\theta(t)$, which is defined at the intersection of the tangent to the spherical part of the droplet with the solid substrate. The dynamic contact angle is unknown and should be determined by matching of all regions presented in Figure 4.1b. Inside the next region, 2, the spherical shape is distorted by the hydrodynamic flow. This region is followed by region 3, where the Derjaguin's pressure comes into play. Over region 3, the Derjaguin's pressure action becomes increasingly important as compared with the capillary force. Toward the end of region 3, the Derjaguin's pressure overcomes the capillary force and becomes the only driving force of the spreading process. Region 3 is followed by region 4. In region 4, we should distinguish two possible options: (i) volatile liquid and (ii) nonvolatile liquid. In case (i), the solid surface is covered by a thin adsorbed layer, which is in equilibrium with the vapor pressure in the surrounding air. In case (ii), a macroscopic description of the spreading process becomes impossible because the characteristic scale in the vertical direction is of the order of the molecular size and surface diffusion take place.

The picture of a spreading drop profile in the vicinity of the three-phase contact line, presented in Figure 4.1b has been understood only recently. A number of a simplified physical mechanisms has been introduced previously based on a simplification of the above picture.

For a long time, the so-called "singularity on a three-phase contact line" [1] has been considered as a major problem in the consideration of kinetics of spreading. We explain below the source of this singularity and why it is removed by the Derjaguin's pressure acting in the vicinity of the apparent three-phase contact line.

It is easy to see that the viscous stress in the tangential direction close to the three-phase contact line (Figure 4.1b) is

$$\eta \frac{\partial v_r}{\partial y} \sim \frac{\eta U_*}{h} \to \infty,$$

as $h \to 0$. This means that the drop cannot spread out because the friction force at the moving front becomes infinite. The way to overcome this problem has been suggested in [2]. The idea is as follows. The very first layer of the liquid molecules on the liquid–solid interface is attached to the solid substrate by a force of adhesion. However, the adhesion force is not infinite, but finite. If the tangential stress is becoming big enough, then the first layer of the liquid molecules will be swept away by the tangential stress. The result is "a slippage velocity." The slippage velocity is introduced as follows:

$$\left[\eta \frac{\partial v_r}{\partial y} \right]_{y=0} = \left[\alpha v_r \right]_{y=0},$$

where α is a proportionality coefficient. That is, the slippage velocity is proportional to the applied shear stress on the solid substrate. This definition can be rewritten as

$$\left[\frac{\partial v_r}{\partial y} \right]_{y=0} = \left[\frac{v_r}{\lambda} \right]_{y=0}, \quad \lambda = \frac{\eta}{\alpha}, \tag{4.2}$$

where λ has a dimension of a length and can be referred to as a slippage length. However, it turns out [3] that $\lambda \sim 10^{-6}$ cm, which is located just in the range where surface forces are the most powerful. This means that condition (4.2) cannot be used as a macroscopic condition because it should be used in the region of surface forces action. Note, that in the case of complete wetting, the Derjaguin's pressure is equal to

$$\Pi(h) = \frac{A}{h^3} \to \infty$$

as $h \to 0$, that is, it grows even faster than the tangential stress. Hence, the Derjaguin's pressure is the driving force of spreading in the vicinity of a three-phase contact line. Let us estimate the thickness where the Derjaguin's pressure overcomes the increasing tangential stress:

$$\frac{\eta U_*}{h} < \frac{A}{h^3},$$

or

$$h < \left(\frac{A}{\eta U_*}\right)^{1/2} = h_t.$$

That is, the lower the velocity of spreading, U, the higher the thickness below which the Derjaguin's pressure overcomes the tangential stress. If, as before, we adopt at the initial stage of spreading a very high spreading velocity, $U_* \sim 10^{-1}$ cm/s, $A \sim 10^{-14}$ erg, $\eta \sim 10^{-2}$ P, then $h_t \sim 0.3 \cdot 10^{-6}$ cm. That is, it is in the range where the Derjaguin's pressure is the most powerful.

A simple way to overcome the problem of "singularity on the moving contact line" has been suggested in [4]: a cutting length was introduced in the vicinity of the three-phase moving contact line. However, the introduction of the cutting length is similar to the introduction of the slippage velocity and is not considered in this book.

A simplifying approach was suggested in [5]. According to this approach, the hydrodynamic flow in regions 2 and 3 is ignored as well as the Derjaguin's pressure action in region 3 (Figure 4.1b). According to this approach, a spherical meniscus is followed directly by region 4, where surface diffusion takes place. This approach results in the following equation for the velocity of spreading:

$$\dot{R} = \text{const} \cdot \left[\cos\theta_e - \cos\theta(t)\right]$$

that is, the velocity of spreading is proportional to the difference between the $\cos\theta_e$, where θ_e is the equilibrium contact angle (θ_e is a fitting parameter in the theory [5]) and $\cos\theta(t)$, where $\theta(t)$ is the instantaneous dynamic contact angle. This equation results in the case of complete wetting, that is, if $\cos\theta_e = 1$ in,

$$\dot{R} = \text{const} \cdot \theta^2(t)$$

It is well established that in the case of complete wetting (see Sections 4.1 and 4.2), this law is

$$\dot{R} = \text{const} \cdot \theta^3(t)$$

Comparison of these two equations shows that the approach suggested in [5] does not agree with both the well-established theoretical predictions and experimental data.

The next approach to be mentioned was tried long ago and based on the consideration of the surface tension of in the vicinity of the apparent moving three-phase contact line. It has been assumed that the surface tension of "the fresh interface" (which appears close to the apparent three-phase contact line) is higher than the surface tension behind the apparent three-phase contact line on "the old interface." This surface difference could be the driving force of spreading. However, both experimental investigations [6] and theoretical estimations [7] showed that the relaxation time of the surface tension on "a fresh" liquid–air interface of pure liquids is too small and, hence, cannot influence the spreading process, which proceeds on much large time scales. However, recently an attempt was made to revive the idea of a high surface tension on "a fresh liquid–air" interface [8]. The approach suggested in [8] also completely ignores the Derjaguin's pressure action in the vicinity of a moving three-phase contact line. This approach was properly criticized in [7].

In the case of nonvolatile liquid surface, diffusion takes place inside region 4 in Figure 4.1b, which results in an effective slippage [9]. The first attempt to introduce the surface slippage base on the consideration of surface diffusion was undertaken in [9]. This approach is to be developed further.

4.1 Spreading of Nonvolatile Liquid Drops over Flat Solid Substrates: Qualitative Analysis

In this section, we discuss the viscous spreading of drops over solid surfaces in the case of *complete wetting*. In this case, the spreading proceeds completely down to the molecular level. We disregard on this stage the influence of surface forces in the vicinity of the *apparent* three-phase contact line. Thus, we know there is a singularity at the moving three-phase contact line (as shown in the introduction to this chapter). Despite that, we find the correct time dependence of spreading (spreading laws in the case of capillary and gravitational spreading). Subsequently, we consider the spreading of microdrops, that is, very small drops, entirely subject to the action of the surface forces. As expected, we see that the Derjaguin's pressure action removes the singularity on the moving three-phase contact line. Also, to be expected is that with due account of the Derjaguin's pressure, the liquid profile does not end at any particular point but tents asymptotically to zero thickness. Hence, only an *apparent* moving front can be identified.

Considerable experimental and theoretical material dealing with the spreading of droplets over a solid, non-deformable, dry surface has been accumulated over decades [10–13]. Below we present a derivation of equations which describe kinetics of spreading in different situations.

Note, all consideration below is undertaken in the case of complete wetting. Consideration in the case of partial wetting is undertaken in Chapter 3, which shows that the partial wetting case is far more complicated as compared with the complete wetting. This explains why the spreading law in the case of partial wetting is to be deduced.

Let us consider an axisymmetric spreading of liquid droplet over a dry solid substrate (Figure 4.1). Let $R(t)$ be the radius of spreading, $\theta(t)$ the dynamic contact angle, and $h(t, r)$ is the unknown liquid profile. Our objective is to deduce the spreading law, $R(t)$ (Figure 4.1). As it was discussed in the introduction to Chapter 4, the Reynolds number is small over main duration of the spreading process, hence, $Re \ll 1$. We also assume that the low slope profile approximation

$$\varepsilon \sim \left|\frac{\partial h}{\partial r}\right| \ll 1$$

is valid. This means, the scale in the vertical direction, h_*, is much smaller than the scale in the radial direction, r_*.

The cylindrical coordinate system (r, φ, z) should be used in the case of axisymmetric spreading. Because of symmetry: $v_\varphi = 0$ and all unknown values are independent of the angel, φ.

Let U_* and v_* be scales of the velocity components in the tangential and the vertical directions, respectively. Using the incompressibility condition, we conclude

$$\frac{U_*}{r_*} = \frac{v_*}{h_*}, \text{ or } v_* = \varepsilon U_* \ll U, \quad \varepsilon = \frac{h_*}{r_*} \ll 1. \tag{4.3}$$

Hence, the velocity scale in the vertical direction is much smaller than the velocity scale in the tangential direction. This means that

$$|v_z| \ll |v_r|.$$

Using the same small parameter ε, we can conclude that all derivatives in radial direction, r, are much smaller than derivatives in a vertical direction z

$$\left| \frac{\partial f}{\partial r} \right| << \left| \frac{\partial f}{\partial z} \right|, \tag{4.4}$$

where $f(r, z)$ is any function.

Now we can easily conclude from the Stokes equation for the vertical velocity component

$$\frac{\partial p}{\partial z} = 0,$$

hence, the pressure depends only on the radial coordinate, r, that is $p = p(r)$ and remains constant through the cross-section of the spreading drop. This means that the pressure can be presented as

$$p = p_a - \gamma K - \Pi(h) + \rho gh,$$

where p_a is the pressure in the ambient air, $-\gamma K$ is the capillary pressure, K is the mean curvature of the interface (negative in the case of droplets of a flat solid substrate and positive in the case of menisci in capillaries), g is the gravity acceleration, ρ is the liquid density and $\Pi(h)$ is the isotherm of the Derjaguin's pressure.

In the low slope case (see Eq. 4.3) we conclude

$$K = \frac{1}{r} \frac{\partial}{\partial r} \left(r \frac{\partial h}{\partial r} \right).$$

Hence,

$$p = p_a - \gamma \frac{1}{r} \frac{\partial}{\partial r} \left(r \frac{\partial h}{\partial r} \right) - \Pi(h) + \rho gh. \tag{4.5}$$

After that the only equation left is the equation for the radial velocity component, v_r, which now can be written as

$$\frac{\partial p}{\partial r} = \eta \frac{\partial^2 v_r}{\partial z^2}, \tag{4.6}$$

with non-slip condition boundary conditions at $z = 0$

$$v_r(t, z) = 0, \tag{4.7}$$

and non-tangential stress condition on the liquid–air interface

$$\eta \frac{\partial v_r}{\partial z} = 0 \quad \text{at} \quad z = h(t, r). \tag{4.8}$$

To deduce this condition, we should take into account both (4.3) and (4.4), after that all terms except for one shown in Eq. (4.8) disappear (see Introduction to Chapter 4).

Integration of Eq. (4.6) with boundary conditions (4.7) and (4.8) results in the following expression for the velocity profile:

$$v_r = -\frac{1}{\eta} \frac{p}{\partial r} \left(hz - \frac{z^2}{2} \right). \tag{4.9}$$

This equation allows calculating the flow rate, Q:

$$Q = 2\pi \int_0^h r v_r dz = -\frac{2\pi}{3\eta} r h^3 \frac{\partial p}{\partial r}. \tag{4.10}$$

Conservation of mass reads

$$2\pi r \frac{\partial h}{\partial t} + \frac{\partial Q}{\partial r} = 0.$$

Substitution of Eq. (4.10) into this equation results in

$$\frac{\partial h}{\partial t} = \frac{1}{3\eta r} \frac{\partial}{\partial r}\left(r h^3 \frac{\partial p}{\partial r} \right), \tag{4.11}$$

which is referred below as "the equation of spreading."

Nonvolatility of the liquid results in the conservation of the total liquid volume, V:

$$2\pi \int_0^{R(t)} r \, h \, dr = V = \text{const}, \tag{4.12}$$

where $R(t)$ is the radius of the drop base, which is referred below as a microscopic radius of spreading (Figure 4.2). Finally, we have two equations (4.11) and (4.12) with two unknowns: the liquid profile, $h(t, r)$, and the radius of spreading, $r_0(t)$. Note, that the pressure, $p(t, r)$, in Eq. (4.9) is specified according to Eq. (4.5).

All the thicknesses of a spreading droplet should be divided into three parts: the big thickness, the intermediate thickness, and the small thickness.

Big thickness: $h \geq 10^{-5}$ cm $\sim t_s$. The effects of the Derjaguin's pressure may be neglected. According to Eq. (4.5) the hydrodynamic pressure within this region is equal to

$$p = p_a - \gamma \frac{1}{r} \frac{\partial}{\partial r}\left(r \frac{\partial h}{\partial r} \right) + \rho g h. \tag{4.13}$$

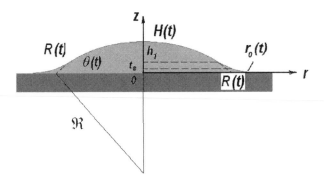

FIGURE 4.2 Spreading of a spherical droplet. At $h > h_1$, the spherical droplet profile is not distorted by the hydrodynamic flow, $t_s < h_1$ is radius of action of the Derjaguin's pressure. $R(t)$ is the macroscopic wetting perimeter (the apparent three-phase contact line); $r_0(t)$ is the true microscopic wetting perimeter; $\theta(t)$ is the dynamic contact angle; and $H(t)$ is the drop apex.

Let us introduce the following dimensionless variable, which is marked with an overbar:

$$\bar{r} = \frac{r}{r_*}, \quad \bar{h} = \frac{h}{h_*} \quad \text{or} \quad r = \bar{r}r_*, \quad h = \bar{h}h_*.$$

Note, all dimensionless values have the same order of magnitude of 1. Using the dimensionless variable, we conclude from Eq. (4.13)

$$p = p_a - \frac{\gamma h_*}{r_*^2} \frac{1}{\bar{r}} \frac{\partial}{\partial \bar{r}} \left(\bar{r} \frac{\partial \bar{h}}{\partial \bar{r}} \right) + \rho g h_* \bar{h}. \tag{4.14}$$

This expression includes two dimensionless parameters:

$$\frac{\gamma h_*}{r_*^2} \quad \text{and} \quad \rho g h_*.$$

The first one gives the intensity of capillary forces, and the second one the intensity of gravitational force.

If capillary forces prevail, then the capillary regime of spreading takes place, that is, if

$$\frac{\gamma h_*}{r_*^2} \gg \rho g h_* \quad \text{or} \quad r_* \ll \sqrt{\frac{\gamma}{\rho g}} = a,$$

where a is the capillary length.

If gravity forces prevail, then the gravitational regime of spreading takes place, that is, if

$$\frac{\gamma h_*}{r_*^2} \ll \rho g h_* \quad \text{or} \quad r_* \gg \sqrt{\frac{\gamma}{\rho g}} = a_c.$$

In the case of water, the capillary length is equal to $\bar{a} \approx 0.27$ cm. The capillary regime is an initial stage of spreading of small drops, whereas the gravitational regime is the final stage of spreading of small drops or the regime of spreading of big drops. The transition from the capillary regime of spreading to the gravitational regime should takes place at the moment t_c, when $R(t_c) \sim a$, where $R(t)$ is the radius of the drop base at time t, which is referred to below as the radius of spreading.

Intermediate thickness: 10^{-6} cm $\leq h \leq 10^{-5}$ cm. In this region, both the capillary and the Derjaguin's pressures act simultaneously, that is, the hydrodynamic pressure takes on the form

$$p = p_a - \frac{\gamma}{r} \frac{\partial}{\partial r} \left(r \frac{\partial h(t,r)}{\partial r} \right) - \Pi(h), \tag{4.15}$$

Small thickness: $h \leq 10^{-6}$ cm. Here the value of the capillary pressure is negligible in comparison to the Derjaguin's pressure:

$$p = p_a - \Pi(h). \tag{4.16}$$

Accordingly, the following stages of the spreading of a drop can be identified: (i) inertial, when Re is not small; however, Ca can be still small. We do not consider this stage in this book; however, an estimation of the duration of this stage is given in Section 5.1; this stage is followed by (ii) the gravitational stage in the case of big droplets; however, in the case of small droplets (which is considered below), the gravitational stage is preceded by (iii) a capillary stage and only after that does the gravitational stage starts; the last stage considered in this section is the stage when (iv) the complete droplet is in the range of the Derjaguin's pressure action.

All stages of spreading are determined, to different degrees, by the Derjaguin's pressure.

The apparent macroscopic wetting perimeter, $R(t)$ (Figure 4.2), and the "true" microscopic wetting perimeter, $r_0(t)$, differ because of the formation of the precursor film, which is caused by the action of both hydrodynamic and surface forces. In the vicinity of the point $R(t)$, there is a transitional region where both the capillary and the Derjaguin's pressures act. Below, we consider so-called similarity solutions of the equation of spreading (4.11) in the cases of capillary and gravitation regimes of spreading.

Capillary Regime of Spreading

Small drops with an initial characteristic size smaller than the capillary length, a_c, are considered below in this part. That means, after a short inertial period, the capillary regime of spreading begins. In this case, the pressure is given by

$$p = p_a - \gamma \frac{1}{r}\frac{\partial}{\partial r}\left(r\frac{\partial h}{\partial r}\right),$$

and Eq. (4.11) takes the following form:

$$\frac{\partial h}{\partial t} = -\frac{\gamma}{3\eta r}\frac{\partial}{\partial r}\left\{r h^3 \frac{\partial}{\partial r}\left[\frac{1}{r}\frac{\partial}{\partial r}\left(r\frac{\partial h}{\partial r}\right)\right]\right\}, \tag{4.17}$$

with conservation law (4.12) and the boundary conditions imposed by the symmetry at the center of the droplet

$$\left.\frac{\partial h}{\partial r}\right|_{r=0} = \left.\frac{\partial^3 h}{\partial r^3}\right|_{r=0} = 0. \tag{4.17'}$$

First, let us introduce dimensionless values using the following characteristic scales: h_*, r_*, and t_*. From Eq. (4.17), we conclude

$$\frac{h_*}{t_*}\frac{\partial \bar{h}}{\partial \bar{t}} = -\frac{\gamma h_*^4}{3\eta r_*^4}\frac{1}{\bar{r}}\frac{\partial}{\partial \bar{r}}\left\{\bar{r}\,\bar{h}^3 \frac{\partial}{\partial \bar{r}}\left[\frac{1}{\bar{r}}\frac{\partial}{\partial \bar{r}}\left(\bar{r}\frac{\partial \bar{h}}{\partial \bar{r}}\right)\right]\right\}.$$

Hence,

$$\frac{h_*}{t_*} = \frac{\gamma h_*^4}{3\eta r_*^4}, \quad \text{or} \quad t_* = \frac{3\eta r_*^4}{\gamma h_*^3}.$$

In the same way from Eq. (4.12),

$$2\pi r_*^2 h_* \int_0^{r_0(t)} \bar{r}\,\bar{h}\,d\bar{r} = V.$$

Hence,

$$2\pi r_*^2 h_* = V, \text{ or } h_* = \frac{V}{2\pi r_*^2}, t_* = \frac{3(2\pi)^3 \eta r_*^{10}}{\gamma V^3}.$$

In Section 4.2 we will see that this estimation of the characteristic time scale is unrealistically small. However, for the moment, we ignore this because (as we will see below) the spreading problem cannot be solved precisely in this way.

Now Eqs. (4.17) and (4.12) can be rewritten as

$$\frac{\partial \bar{h}}{\partial \bar{t}} = -\frac{1}{\bar{r}}\frac{\partial}{\partial \bar{r}}\left\{ \bar{r}\,\bar{h}^3\frac{\partial}{\partial \bar{r}}\left[\frac{1}{\bar{r}}\frac{\partial}{\partial \bar{r}}\left(\bar{r}\frac{\partial \bar{h}}{\partial \bar{r}} \right) \right] \right\},$$

(4.18)

$$\int_0^R \bar{r}\bar{h}d\bar{r} = 1.$$

(4.19)

Similarity solution of Eqs. (4.18) and (4.19).

Let ξ be a new variable, which is a combination of $\bar{r}$ and $\bar{t}$. The dimensionless thickness, $\bar{h}(\bar{t},\bar{r})$, should depend on this new single variable. This type of solution is referred to as a similarity solution.

Let $\xi = \bar{r}f(\bar{t})$, where $f(\bar{t})$ is a new unknown function. From Eq. (4.19), we conclude

$$\int_0^R \bar{r}f(\bar{t})\frac{1}{f(\bar{t})}\,\bar{h}\,d\bar{r}f(\bar{t})\frac{1}{f(\bar{t})} = 1,$$

or

$$\int_0^{R\,f(t)} \xi\frac{\bar{h}}{f^2(t)}\,d\xi = 1$$

(4.20)

This equation must be independent of $\bar{t}$. This gives two conditions

$$\bar{R}(\bar{t})\,f(\bar{t}) = \lambda_c = \text{const}, \quad r_0(\bar{t}) = \frac{\lambda_c}{f(\bar{t})}$$

(4.21)

where λ_c is an unknown constant. Equations (4.18) and (4.19) do not include any dimensionless parameters, so this means that $\lambda_c \sim 1$. As we see below, this constant cannot be determined without consideration of the Derjaguin's pressure and the constant will be determined only in Section 4.2.

The second condition, which immediately follows from Eq. (4.20) is

$$\bar{h} = A(\xi)f^2(\bar{t}).$$

(4.22)

According to Eq. (4.21) the spreading law $\bar{R}(\bar{t})$ is known if the function, $f(\bar{t})$, is determined.

Similarity solution (4.22) should be substituted in Eq. (4.18).

Let us calculate

$$\frac{\partial \bar{h}}{\partial \bar{t}} = A'(\xi)\left[\bar{r}f'(\bar{t}) \right]f^2(\bar{t}) + A(\xi)2f(\bar{t})f'(\bar{t})$$

$$= A'(\xi)\left[\bar{r}f(\bar{t}) \right]f'(\bar{t})f(\bar{t}) + 2A(\xi)f(\bar{t})f'(\bar{t})$$

$$= f(\bar{t})f'(\bar{t})\left[A'(\xi)\xi + 2A(\xi) \right]$$

Using this equation and $\bar{r} = \dfrac{\xi}{f(\bar{t})}$, we conclude from Eq. (4.18),

$$f(\bar{t})f'(\bar{t})\left[A'(\xi)\xi + 2A(\xi) \right] = -f^8(\bar{t})f^4(\bar{t})\frac{1}{\xi}\frac{d}{d\xi}\left\{ \xi A^3\frac{d}{d\xi}\left[\frac{1}{\xi}\frac{d}{d\xi}\left(\xi\frac{dA}{d\xi} \right) \right] \right\}.$$

(4.23)

This equation should not depend on time but on variable ξ, hence

$$f(\bar{t})f'(\bar{t}) = -f^8(\bar{t})f^4(\bar{t}),$$

or

$$f'(\bar{t}) = -f^{11}(\bar{t}). \tag{4.24}$$

The solution of this equation is

$$f(\bar{t}) = \frac{1}{\left[10(\bar{t}+C)\right]^{0.1}},$$

where C is an integration constant.

From this equation and Eq. (4.21), we get

$$\bar{R}(\bar{t}) = \lambda_c \left[10(\bar{t}+C)\right]^{0.1}.$$

Using the initial condition $R(0) = R_0$, where R_0 is different from the initial radius of the droplet but equal to the radius of the droplet at the end of the inertial period of spreading (see further discussion in Section 5.1). This means that we should choose the scale of the radius of the droplet as follows: $r_* = R(0)$. According to this choice, $\bar{R}(0) = 1$. Hence,

$$C = \frac{1}{10\lambda_c^{10}}.$$

It is reasonable to assume (it will be justified partially in Section 4.1), that this constant is equal to the duration of the inertial stage of spreading. Hence, we adopt

$$C = \frac{t_{in}}{t_*} = \bar{t}_{in},$$

where t_{in} is the duration of the inertial stage of spreading. Using this notation, we can write the spreading law as

$$\bar{R}(\bar{t}) = (\lambda_c 10^{0.1})(\bar{t}+\bar{t}_{in})^{0.1}.$$

From hereafter, we do not distinguish between $R(t)$ and $r_0(t)$ because of the small difference between these two values. Now back to dimensional variables,

$$\frac{R(t)}{R_0} = (\lambda_c 10^{0.1})\left(\frac{t+t_{in}}{t_*}\right)^{0.1}$$

$$R(t) = \frac{(\lambda_c 10^{0.1})}{\left[3(2\pi)^3\right]^{0.1}}\left(\frac{\gamma V^3}{\eta}\right)^{0.1}(t+t_{in})^{0.1} = 0.65\lambda_c\left(\frac{\gamma V^3}{\eta}\right)^{0.1}(t+t_{in})^{0.1}.$$

Finally

$$R(t) = 0.65\lambda_c\left(\frac{\gamma V^3}{\eta}\right)^{0.1}(t+t_{in})^{0.1}. \tag{4.25}$$

Using Eq. (4.25) and the definition (4.22), we conclude

$$H(t) \approx \left(\frac{\eta V^2}{\gamma}\right)^{1/5} (t + t_{in})^{-1/5}$$

and

$$\theta(t) \approx \left(\frac{\eta^3 V}{\gamma^3}\right)^{1/10} (t + t_{in})^{-0.3},$$

where the dynamic contact angle is determined as

$$\theta(t) \sim \frac{H(t)}{R(t)}.$$

The spreading rate according to Eq. (4.25) is

$$U = \frac{dR(t)}{dt} = 0.065\lambda_c \left(\frac{\gamma V^3}{\eta}\right)^{0.1} (t + t_{in})^{-0.9}.$$

If we express time in this equation via the dynamic contact angle, $\theta(t)$, we arrive to

$$U \sim 0.065\lambda_c \frac{\gamma}{\eta} \theta^3,$$

which is the well-known and well-verified Tanner's law of dynamics of spreading in the case of complete spreading. This law was mentioned in the introduction to Chapter 4.

As mentioned, the unknown constant, λ_c, is close to 1.

According to Eq. (4.25), the exponent 0.1 and $R(t)/V^{0.3}$ are independent of the drop volume. These two conclusions have been well confirmed in the case of complete wetting by numerous experimental data.

Similar calculations can be carried out for the dynamics of spreading of a one-dimensional droplet (cylinder), which results in

$$R(t) \approx \left(\frac{\gamma}{\eta} V^3\right)^{1/7} (t + t_{in})^{1/7}.$$

However, there is one very substantial drawback of the obtained solution (see below).

Let us return to the Eq. (4.23), which takes the following form after selection (4.24):

$$[A'(\xi)\xi + 2A(\xi)] = \frac{1}{\xi} \frac{d}{d\xi} \left\{ \xi A^3 \frac{d}{d\xi} \left[\frac{1}{\xi} \frac{d}{d\xi} \left(\xi \frac{dA}{d\xi} \right) \right] \right\}.$$

This equation can be transformed as follows:

$$\left[A'(\xi)\xi^2 + 2\xi A(\xi) \right] = \frac{d}{d\xi}\left\{ \xi A^3 \frac{d}{d\xi}\left[\frac{1}{\xi}\frac{d}{d\xi}\left(\xi\frac{dA}{d\xi} \right) \right] \right\}$$

$$\left[A(\xi)\xi^2 \right]' = \frac{d}{d\xi}\left\{ \xi A^3 \frac{d}{d\xi}\left[\frac{1}{\xi}\frac{d}{d\xi}\left(\xi\frac{dA}{d\xi} \right) \right] \right\}.$$

$$A(\xi)\xi^2 = \xi A^3 \frac{d}{d\xi}\left[\frac{1}{\xi}\frac{d}{d\xi}\left(\xi\frac{dA}{d\xi} \right) \right]$$

$$A\left\{ \xi - A^2 \frac{d}{d\xi}\left[\frac{1}{\xi}\frac{d}{d\xi}\left(\xi\frac{dA}{d\xi} \right) \right] \right\} = 0.$$

This equation shows that either

$$A(\xi) = 0$$

or

$$\left\{ \xi - A^2 \frac{d}{d\xi}\left[\frac{1}{\xi}\frac{d}{d\xi}\left(\xi\frac{dA}{d\xi} \right) \right] \right\} = 0.$$

That is, this equation describes the spreading droplet profile up to the point of intersection with the axis ξ, that is, to the point where $A(\xi_0) = 0$. Hence, the following problem should be solved:

$$\begin{cases} \dfrac{\xi}{A^2} = A''' + \dfrac{A''}{\xi} - \dfrac{A'}{\xi^2} \\[2mm] A(0) = a \\[2mm] A'(0) = 0 \\[2mm] A''(0) = -b \end{cases}.$$

This equation should be solved with two additional conditions:

$$\int_0^{\xi_0} \xi A(\xi)d\xi = 1, \quad \text{where} \quad A(\xi_0) = 0,$$

which determine the unknown constants a and b. Note the first condition is simply the conservation law (4.21). Hence, we can now determine that $\lambda_C = \xi_0$. It could be the complete solution of the problem of the capillary regime of spreading.

Unfortunately, there is no such solution of the above problem. At any choice of constants a and b, the solution behaves as shown in Figure 4.3: The similarity drop profile never intersects the axis (curve 1; calculated according to the above system). Curve 2 in Figure 4.3 presents the parabolic profile

$$a - \frac{b}{2}\xi^2$$

for comparison. This is the manifestation of the so-called "singularity" at the moving apparent three-phase contact line.

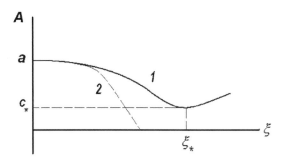

FIGURE 4.3 Dependency of a similarity profile of the spreading droplet, $A(\xi)$: it never intersects the axis. (1) Calculated according to Eq. (4.23); and (2) the spherical (parabolic profile).

This contradiction is resolved in Section 4.2. Surprisingly, the spreading law (4.25) remains almost untouched except for the determination of the unknown constant λ_c.

Gravitational Spreading as a Continuation of the Capillary Spreading Regime

In this case, the radius of spreading is bigger than the capillary length. Hence, according to Eq. (4.13), we can omit the capillary pressure and this results in the following expression for the pressure inside $p = p_a + \rho g h$.

Substitution of this expression into the equation of spreading (4.11) results in

$$\frac{\partial h}{\partial t} = \frac{\rho g}{3 \eta r} \frac{\partial}{\partial r} \left\{ r h^3 \frac{\partial h}{\partial r} \right\},$$ (4.26)

with the same conservation law (4.12) and the boundary condition

$$\left. \frac{\partial h}{\partial r} \right|_{r=0} = 0.$$

Let us introduce the following dimensionless values using characteristic scales h_*, a, t_*. The end of the capillary stage of spreading is when the radius of spreading reaches a. This is why this length is selected as the characteristic length scale. From Eq. (4.26), we conclude

$$\frac{h_*}{t_*} \frac{\partial \bar{h}}{\partial \bar{t}} = \frac{\rho g h_*^4}{3 \eta a^2} \frac{1}{\bar{r}} \frac{\partial}{\partial \bar{r}} \left\{ \bar{r} \, \bar{h}^3 \frac{\partial \bar{h}}{\partial \bar{r}} \right\}.$$

Hence,

$$\frac{h_*}{t_*} = \frac{\rho g h_*^4}{3 \eta a^2}, \quad \text{or} \quad t_* = \frac{3 \eta a^2}{\rho g h_*^3}.$$

Now from the conservation law (4.12), we conclude:

$$2 \pi a^2 h_* \int_0^{R(\bar{t})} \bar{r} \, \bar{h} \, d\bar{r} = V.$$

Hence,

$$2 \pi a^2 h_* = V, \quad \text{or} \quad h_* = \frac{V}{2 \pi a^2}, \quad t_* = \frac{3(2\pi)^3 \eta a^8}{\rho g V^3}.$$

Using these notations, we can conclude from Eq. (4.26) and conservation law (4.12)

$$\frac{\partial \bar{h}}{\partial \bar{t}} = \frac{1}{\bar{r}} \frac{\partial}{\partial \bar{r}} \left\{ \bar{r} \, \bar{h}^3 \frac{\partial \bar{h}}{\partial \bar{r}} \right\},$$ (4.27)

$$\int_0^{R(\bar{t})} \bar{r} \, \bar{h} \, d\bar{r} = 1.$$ (4.28)

Similarity Solution

Let ξ be a new similarity variable which is a combination of r and t; as in the previous case of capillary spreading, $h(t, r)$ should depend on this new single variable. Such a solution, as before, is referred to as a similarity solution.

Let $\xi = r f(t)$, where $f(t)$ is a new unknown function. Substituting this definition into Eq. (4.28), we find

$$\int_0^{\bar{R}(\bar{t})f(\bar{t})} \xi \frac{\bar{h}}{f^2(\bar{t})} \, d\xi = 1.$$

This equation must be independent of t, this gives two conditions

$$\bar{R}(\bar{t}) f(\bar{t}) = \lambda_g = \text{const}, \quad \bar{R}(\bar{t}) = \frac{\lambda_g}{f(t)},$$ (4.29)

and

$$\bar{h} = A(\xi) f^2(\bar{t}),$$ (4.30)

where λ_g should be close to 1. Equation (4.29) shows that the spreading law, $R(t)$, is known if the unknown function, $f(t)$, is determined.

The similarity solution (4.30) should be substituted into Eq. (4.27). The calculations, like the previous case, result in the identical expression for the time derivative. Using the definition

$$\bar{r} = \frac{\xi}{f(\bar{t})},$$

we conclude from Eq. (4.27)

$$f(\bar{t}) f'(\bar{t}) \left[A'(\xi)\xi + 2A(\xi) \right] = f^8(\bar{t}) f^2(\bar{t}) \frac{1}{\xi} \frac{d}{d\xi} \left\{ \xi A^3 \frac{dA}{d\xi} \right\}.$$ (4.31)

This equation should not depend on time but on variable ξ only, hence

$$f(\bar{t}) f'(\bar{t}) = -f^8(\bar{t}) f^2(\bar{t})$$

or

$$f'(\bar{t}) = -f^9(\bar{t}).$$ (4.32)

Note "minus" in this equation because the radius of spreading, $R(t)$, should be an increasing function of time. The solution of Eq. (4.32) is

$$f(\bar{t}) = \frac{1}{\left[8(\bar{t} + C) \right]^{1/8}},$$

where C is an integration constant.

From this equation and Eq. (4.29), we conclude

$$\bar{R}(\bar{t}) = \lambda_g \left[8(\bar{t} + C) \right]^{1/8}. \tag{4.33}$$

The initial condition can be specified as $R(0) = a$ or, in dimensionless units, as $\bar{R}(0) = 1$. Using this initial condition, we conclude from Eq. (4.33) that

$$C = \frac{1}{8\lambda_g^8},$$

which we can refer to as the duration of the capillary stage of spreading, t_c. This means that Eq. (4.33) can be rewritten as

$$\bar{R}(\bar{t}) = (\lambda_g 8^{1/8})(\bar{t} + \bar{t}_c)^{1/8}.$$

After we return to dimensional variables, this equation becomes

$$\frac{\bar{R}(t)}{a} = (\lambda_g 8^{1/8}) \left(\frac{t + t_c}{t_*} \right)^{1/8},$$

or

$$R(t) = 0.57 \lambda_g \left(\frac{\rho g V^3}{\eta} \right)^{1/8} (t + t_c)^{1/8}, \tag{4.34}$$

where λ_g close to 1. Exponent 1/8 agrees very well with a number of experimental observations.

Now we can consider the shape of the spreading droplet over the duration of the gravitational regime of spreading. Equation (4.31), taking into account Eq. (4.32), which now takes the following form:

$$\left[A'(\xi)\xi + 2A(\xi) \right] = -\frac{1}{\xi} \frac{d}{d\xi} \left\{ \xi A^3 \frac{dA}{d\xi} \right\}$$

After simple rearrangements (like the capillary case) we arrive to

$$A' = -\frac{\xi}{A^2}.$$

Solution of this equation is

$$A(\xi) = \left(\frac{3}{2} \right)^{1/3} \left(\lambda_g^2 - \xi^2 \right)^{1/3}, \tag{4.35}$$

where the integration constant ξ_0 should be determined from the conservation law

$$\int_0^{\lambda_g} \xi A(\xi) d\xi = 1,$$

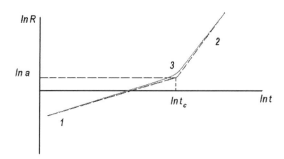

FIGURE 4.4 Time evolution of the radius of spreading in log–log coordinate system. (1) Capillary spreading; (2) gravitational regime of spreading; and (3) is a "real" experimental dependency, which is shown to guide the eye.

which results in $\lambda_g \approx 1.37$, which is close to 1, as predicted earlier. Using this value, we can rewrite Eq. (4.34) as

$$R(t) = 0.78 \left(\frac{\rho g V^3}{\eta} \right)^{1/8} (t + t_c)^{1/8}. \qquad (4.36)$$

That is, in the case of gravitational spreading, the spreading law can be determined completely according to Eq. (4.36).

However, we still have one substantial problem in the case of gravitational spreading. The profile of the droplet in this case is presented by Eq. (4.35) and shown in Figure 6.25 (the case $n = 1$). This figure shows that the low slope approximation used in our consideration is severely violated in the vicinity of the edge of the spreading droplet: It is necessary to introduce the transition zone from the gravitational part toward the contact with the solid substrate, which will include both capillary and surface forces.

In Figure 4.4, the time evolution of the radius of spreading is schematically presented in log–log coordinates. Capillary and gravitational regimes are shown by lines 1 [according to Eq. (4.25) with λ_c still unknown] and 2 [according to Eq. (4.36)], respectively.

The capillary regime of spreading switches to the gravitational regime of spreading at the moment, marked as t_c when the radius of spreading reaches the value of the capillary length, a. The "real-time" evolution is shown by curve 3 in Figure 4.4. The experimental evidence of such dependency was obtained in [14], where a transition from the capillary regime of spreading to the gravitation regime has been demonstrated to be similar to that shown in Figure 4.4.

The highlighted text below can be omitted at the first reading.

Spreading of Very Thin Droplets

In the previous part of this section, we showed that the pure capillary spreading results in the inconsistency of the mathematical treatment in the vicinity of the apparent moving contact line of the spreading droplet. This inconsistency is usually referred to as "a singularity" at the three-phase contact line. As mentioned in the introduction to Chapter 4, this inconsistency is a result of neglect of the Derjaguin's pressure action in the vicinity of the moving apparent three-phase contact line. Recall that as the drop thickness tends to zero, the Derjaguin's pressure overcomes the capillary pressure and dominates the spreading process.

It is why in this part we consider the spreading of microdroplets, whose apex is located in the range of action of the surface forces. Below we neglect the effect of the capillary pressure; that is, the hydrodynamic pressure in the liquid is described by Eq. (4.16).

Thus, the case under consideration refers to the final stage of the spreading of droplets that completely wet the substrate. The equation describing the droplet profile, $h(t, r)$, of a spreading droplet has the following form

$$\frac{\partial h}{\partial t} = -\frac{1}{3\eta r}\frac{\partial}{\partial r}\left\{r\,h^3\Pi'(h)\frac{\partial h}{\partial r}\right\}, \tag{4.37}$$

which is obtained by a substitution of Eq. (4.16) into the equation of spreading (4.11). The boundary conditions for the second-order differential equation (4.37) are as follows

$$\frac{\partial h}{\partial r} = 0, \quad r = 0, \tag{4.38}$$

$$h(t,\infty) = 0, \tag{4.39}$$

and the conservation law (4.12). Note, Eq. (4.37) has the form of a nonlinear equation of thermal conductivity. As we know, an equation of thermal conductivity often (for example, in the linear case) results in an infinite rate of propagation of perturbations. Therefore, features similar to those of Eq. (4.37) should be expected in the case under consideration.

Let us first consider the simplest form of the isotherm of the Derjaguin's pressure, $\Pi(h)$, in the case of complete wetting:

$$\Pi(h) = \begin{cases} b(h - t_s), & 0 \le h \le t_s \\ 0, & h \ge t_s \end{cases}. \tag{4.40}$$

In this case, Eq. (4.37) takes the form

$$\frac{\partial h}{\partial t} = \frac{b}{3\eta r}\frac{\partial}{\partial r}\left\{r\,h^3\frac{\partial h}{\partial r}\right\}. \tag{4.41}$$

Introducing dimensionless values, as before, in this equation we conclude:

$$2\pi r_*^2 h_* = V, \quad \text{or} \quad h_* = \frac{V}{2\pi r_*^2}, \quad t_* = \frac{3(2\pi)^3\eta r_*^8}{bV^3}.$$

Using these notations, we can conclude from Eq. (4.41) and the conservation law (4.12):

$$\frac{\partial \bar{h}}{\partial \bar{t}} = \frac{1}{\bar{r}}\frac{\partial}{\partial \bar{r}}\left\{\bar{r}\,\bar{h}^3\frac{\partial \bar{h}}{\partial \bar{r}}\right\}, \tag{4.42}$$

$$\int_0^{\bar{R}(\bar{t})} \bar{r}\,\bar{h}\,d\bar{r} = 1. \tag{4.43}$$

Surprisingly, Eqs. (4.42) and (4.43) are identical to Eqs. (4.27) and (4.28) in the case of gravitational spreading. Hence, we can use the already deduced similarity solution (4.35) for the droplet profile with $\lambda_g \approx 1.37$. After that, the spreading law becomes as (4.36)

$$R(t) = 0.78\left(\frac{aV^3}{\eta}\right)^{1/8}(t + t_0)^{1/8}, \tag{4.44}$$

where t_0 is the time when this stage of spreading starts. The only difference of this expression from (4.36) is that ρg in Eq. (4.36) is replaced by b in Eq. (4.44). At any rate, the Derjaguin's pressure action, even in the simplest possible form (4.40), removed the artificial singularity on the moving apparent contact line.

Let us consider an isotherm of the Derjaguin's pressure of the form

$$\Pi(h) = A/h^n, \tag{4.45}$$

which as we already know is relevant for the case of complete wetting (e.g., spreading of oils over glass, metals).

Equation (4.37) now becomes

$$\frac{\partial h}{\partial t} = \frac{nA}{3\eta r}\frac{\partial}{\partial r}\left\{r h^{2-n}\frac{\partial h}{\partial r}\right\}. \tag{4.46}$$

Introducing dimensionless values using the following characteristic scales h_*, r_*, t_* in Eq. (4.46) we conclude

$$\frac{h_*}{t_*}\frac{\partial \bar{h}}{\partial \bar{t}} = -\frac{nAh_*^{3-n}}{3\eta r_*^2}\frac{1}{\bar{r}}\frac{\partial}{\partial \bar{r}}\left\{\bar{r}\,\bar{h}^{2-n}\frac{\partial \bar{h}}{\partial \bar{r}}\right\}.$$

Hence, we can choose

$$t_* = \frac{3\eta r_*^2}{nAh_*^{2-n}}.$$

Using the conservation law (4.12), we conclude as before:

$$h_* = \frac{V}{2\pi r_*^2}, \quad t_* = \frac{3\eta V^{n-2}}{(2\pi)^{n-2} r_*^{2n-6}}.$$

Now Eqs. (4.46) and (4.12) can be rewritten as

$$\frac{\partial \bar{h}}{\partial \bar{t}} = \frac{1}{\bar{r}}\frac{\partial}{\partial \bar{r}}\left\{\bar{r}\,\bar{h}^{2-n}\frac{\partial \bar{h}}{\partial \bar{r}}\right\}, \tag{4.47}$$

and (4.19).

Let $\xi = \bar{r}f(\bar{t})$, where $f(\bar{t})$ is a new unknown function. Substitution of this definition into Eq. (4.19) we find

$$\int_0^{\bar{R}(\bar{t})f(\bar{t})}\xi\frac{\bar{h}}{f^2(\bar{t})}\,d\xi = 1.$$

This equation must be independent of t and gives two conditions

$$\bar{R}(\bar{t})f(\bar{t}) = \lambda_n = \text{const}, \quad \bar{R}(\bar{t}) = \frac{\lambda_n}{f(t)}, \tag{4.48}$$

and

$$\bar{h} = A(\xi)f^2(\bar{t}), \tag{4.49}$$

where λ_n should be close to 1 and depends on the exponent n of the Derjaguin's pressure isotherm. Eq. (4.48) shows that the spreading law, $R(t)$, is known if the unknown function, $f(t)$, is determined.

The similarity solution (4.49) should be substituted into Eq. (4.47). The calculations similar to the previous cases result in the identical expression for the time derivative. Using the definition

$$\bar{r} = \frac{\xi}{f(\bar{t})}$$

we conclude from Eq. (4.47)

$$f(\bar{t})f'(\bar{t})\left[A'(\xi)\xi + 2A(\xi)\right] = f^{8-2n}(\bar{t})\frac{1}{\xi}\frac{d}{d\xi}\left\{\xi A^{2-n}\frac{dA}{d\xi}\right\}. \tag{4.50}$$

This equation should not depend on time but only on variable ξ, hence

$$f(\bar{t})f'(\bar{t}) = -f^{8-2n}(\bar{t})$$

or

$$f'(\bar{t}) = -f^{7-2n}(\bar{t}) \tag{4.51}$$

Note, the "minus" sign in this equation is because the radius of spreading, $R(t)$, should be an increasing function of time. Now Eq. (4.50) takes the following form:

$$\left[A'(\xi)\xi + 2A(\xi)\right] = -\frac{1}{\xi}\frac{d}{d\xi}\left\{\xi A^{2-n}\frac{dA}{d\xi}\right\},$$

or after rearrangement similar to the previous cases,

$$\xi = -A^{1-n}\frac{dA}{d\xi}. \tag{4.52}$$

Let us consider two cases, $n = 3$ and $n = 2$, in these two equations. In the case $n = 2$, we conclude

$$f'(\bar{t}) = -f^3(\bar{t}) \text{ and } \xi = -A^{-1}\frac{dA}{d\xi}.$$

Taking into account Eq. (4.48), we conclude

$$\bar{R}(\bar{t}) = \lambda_2\sqrt{2(\bar{t} + \bar{t}_2)}, \tag{4.53}$$

and

$$A(\xi) = C\exp\left(-\frac{\xi^2}{2}\right), \tag{4.54}$$

where $\bar{t}_2$ is the dimensionless time of the beginning of this stage of spreading and C is the integration constant to be determined from the conservation law. According to Eq. (4.54), the droplet profile does not vanish anywhere, but tends asymptotically to zero. That is, we can determine only the effective apparent three-phase contact line as follows. According to the conservation law,

$$\int_0^{\lambda_2} \xi C\exp\left(-\frac{\xi^2}{2}\right)d\xi = 1,$$

hence, after integration,

$$C\left(1-\exp\left(-\frac{\lambda_2^2}{2}\right)\right)=1.$$

If we select the apparent contact line as the point where the profile is close to zero, that is,

$$\exp\left(-\frac{\lambda_s^s}{2}\right)=0.1,$$

then $\lambda_2=\sqrt{2\ln10}\approx2.15$. In this case $C\approx0.9$, and we conclude

$$\bar{R}(\bar{t})=3.04\sqrt{(\bar{t}+\bar{t}_2)},\tag{4.55}$$

$$A(\xi)=0.9\exp\left(-\frac{\xi^2}{2}\right).$$

Note, the spreading law (4.55) with exponent 0.5 is the fastest we considered yet.

Let us emphasize again, there is no distinct three-phase contact line in the case under consideration, the droplet profile tends only to zero asymptotically. This is a consequence of neglecting the surface diffusion in front of the moving contact line. How to connect the macroscopic description based on the consideration of the Derjaguin's pressure with surface diffusion, which is a microscopic description, is still an open problem.

In the case $n=3$, we conclude from Eqs. (4.51) and (4.52)

$$f'(\bar{t})=-f(\bar{t}) \text{ and } \xi=-A^{-2}\frac{dA}{d\xi}.$$

Solution of these equations is

$$f(\bar{t})=\exp\left(-(\bar{t}+\bar{t}_3)\right) \text{ and } A(\xi)=\frac{1}{C+\dfrac{\xi^2}{2}},$$

where $\bar{t}_3$ is the time when this stage of spreading begins and C is an integration constant. As in the previous case, there is no distinct three-phase contact line because the droplet profile tends to zero asymptotically only. We select an apparent contact line as follows from the conservation law:

$$\int_0^{\lambda_3}\frac{\xi d\xi}{C+\dfrac{\xi^2}{2}}=1,$$

which gives

$$\ln\left(1+\frac{\lambda_3^2}{2C}\right)=1.$$

This equation has the following solution:

$$1+\frac{\lambda_3^2}{2C}=e,$$

or

$$\frac{\lambda_3^2}{2C} = e - 1.$$

According to the previous consideration, λ_3 should be around 1. Hence, if we select

$$C \sim \frac{1}{2(e-1)} = 0.29,$$

then

$$\bar{R}(\bar{t}) = \lambda_3 \exp(\bar{t} + \bar{t}_3), \qquad (4.56)$$

and

$$A(\xi) \approx \frac{1}{0.29 + \dfrac{\xi^2}{2}}.$$

This Eq. (4.56) gives the highest possible rate of spreading.

4.2 Spreading of Nonvolatile Liquid Drops over Dry Surfaces: Influence of Surface Forces

In this section we consider the spreading of an axisymmetric liquid drop on a plane solid substrate in the case of complete wetting. Both capillary and the Derjaguin's pressure are taken into account [15]. As we concluded in Section 4.1 and in the introduction to Chapter 4, neglecting the Derjaguin's pressure in the vicinity of the moving apparent three-phase contact line results in a singularity and a contradiction. On the other hand, we have already showed at the end of Section 4.1 that the Derjaguin's pressure action removes the singularity from the moving apparent contact line.

The cylindrical coordinate system is used below. The initial size of the droplet is assumed to be smaller than the capillary length

$$a = \sqrt{\frac{\gamma}{\rho g}},$$

that is, the gravity action is neglected.

We already estimated in the Introduction to Chapter 4 that

1. The capillary number $Ca \sim 10^{-5} \ll 1$, that is, the main part of the spreading droplet retains the spherical shape.
2. A low slope approximation is valid in the vicinity of the moving apparent contact line. However, we should still estimate the scale of the narrow zone, where this approximation is satisfied.

This means that the liquid profile in the main part of the spreading droplet is

$$h(r) = \sqrt{\Re^2 - r^2} - (\Re - H), \qquad (4.57)$$

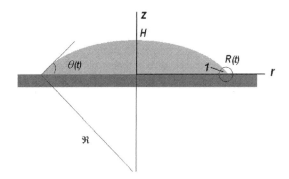

FIGURE 4.5 Spreading of axisymmetric droplet. (1) Vicinity of the moving contact line.

where r, z are cylindrical coordinates (Figure 4.5). Using simple geometrical consideration, we conclude from Figure 4.2 or 4.5 that H and $\Re$ are expressed via the dynamic contact and the radius of spreading as

$$\Re = \frac{R}{\sin\theta}, \quad H = \frac{R(1-\cos\theta)}{\sin\theta}.$$

If we assume that the liquid is located mostly in the spherical part of the droplet, then the volume, V, is

$$V = \frac{\pi}{6} R^3 \tan\frac{\theta}{2}\left(3 + \tan^2\frac{\theta}{2}\right), \tag{4.58}$$

or

$$R(t) = \left(\frac{6V}{\pi}\right)^{1/3} \frac{1}{\left[\tan\dfrac{\theta(t)}{2}\left(3 + \tan^2\dfrac{\theta(t)}{2}\right)\right]^{1/3}}. \tag{4.59}$$

In the case of a small contact angle, this equation results in

$$R(t) = \left(\frac{4V}{\pi\theta}\right)^{1/3}. \tag{4.60}$$

This equation gives a good approximation of the right-hand side of Eq. (4.59) in the range of dynamic contact angles $0 < \theta(t) < \pi/4$.

Using the low slope approximation, we already deduced the equation of spreading, which describes the shape of the liquid profile (4.11). Now this equation is used only in the vicinity of the moving contact line and will be matched with the solution (4.57).

Inside the vicinity of the moving contact line, the pressure in given by expression (4.15), which should be substituted into Eq. (4.11). In the case of complete wetting, which is under consideration in this section, we adopt the Derjaguin's pressure isotherm, $\Pi(h)$, as

$$\Pi(h) = \frac{A}{h^n}, \tag{4.61}$$

with $A > 0$ is the Hamaker constant; and $n = 2$ or 3. Substituting these equations into Eq. (4.11) results in

$$\frac{\partial h}{\partial t} = -\frac{\gamma}{3\eta}\frac{1}{r}\frac{\partial}{\partial r}\left[rh^3\left[\frac{\partial}{\partial r}\left(\frac{1}{r}\frac{\partial}{\partial r}\left(r\frac{\partial h}{\partial r}\right)\right)\right] - \frac{nA}{\gamma}\frac{\partial h}{h^{n+1}}\frac{\partial h}{\partial r}\right]. \tag{4.62}$$

The first and the second terms in the brackets in the right-hand side of Eq. (4.62) describe the effects of capillarity and the Derjaguin's pressure, respectively. The drop thickness decreases from the center of the drop to its spreading edge. In this section, we are interested in the three regions of the spreading drop (Figure 4.1b). The spherical part of the spreading droplet, that is, the region from the center of the drop to a point where the film thickness is still large such that in this region where both viscous and the Derjaguin's pressure effects are negligible compared to capillary effects is referred to as the "outer region." Further radially outward there is a region where both capillary and viscous effects dominate (region 2 in Figure 4.1b). The next region, the "inner region," is further outward where the film is thin and the Derjaguin's pressure effects are of a comparable importance to viscous and capillary effects (see Figure 4.1b).

Equation (4.62) can be made dimensionless by introducing appropriate scales for thickness, radius, and time

$$-\varepsilon \frac{\partial h}{\partial t} = \frac{1}{r} \frac{\partial}{\partial r}\left[r h^3 \left[\frac{\partial}{\partial r}\left(\frac{1}{r} \frac{\partial}{\partial r}\left(r \frac{\partial h}{\partial r}\right)\right)\right] - \frac{\lambda}{h^{n+1}} \frac{\partial h}{\partial r}\right],$$ (4.63)

where we use the same notations for dimensionless values as for dimensional: $h \to h/h_*$, $r \to r/r_*$, $t \to t/t_*$, $\varepsilon = 3\eta\, r_*^4 / \gamma\, h_*^3 t_*$, $\lambda = nA r_*^2 / h_*^{n+1}$. The scales, h_*, r_* and t_* are determined below. According to Eq. (4.12) $2\pi\, r_*^2 h_* = V$, where V is the drop volume. Hence, from this equation h_* is determined as $h_* = V / \left(2\pi r_*^2\right)$, and there are now only two unknown scales, r_* and t_*.

The volume of the liquid drop is constant during spreading, that is, Eq. (4.12) should be used.

If the whole liquid profile satisfies the condition $h'^2 \ll 1$, then Eq. (4.63) describes the liquid profile in all regions. In this case, the following symmetry conditions are valid at $r = 0$:

$$\frac{\partial h}{\partial r} = \frac{\partial^3 h}{\partial r^3} = 0.$$ (4.64)

This case does not differ substantially from the more general case (then the dynamic contact angle is not sufficiently small). This is why only the low slope approximation, which is valid over the whole droplet, is under consideration here.

The initial solid surface is dry in front of the spreading droplet, that is,

$$h \to 0, \text{ at } r \to \infty$$ (4.65)

It is necessary to comment on this boundary condition. As we already established in Section 4.1, the liquid profile may tend asymptotically to zero, in this case the boundary condition is specified below.

It is shown below that both dimensionless constants in Eq. (4.63) are small, that is,

$$\varepsilon \ll 1,$$ (4.66)

and

$$\lambda \ll 1.$$ (4.67)

Condition (4.66) expresses the smallness of the viscous forces, F_η, as compared with capillary forces, F_γ: $F_\eta \sim \eta\, \partial^2 v_r / \partial z^2 \sim \eta\, r_* / t_* h_*^2$, and

$$F_\gamma \sim \frac{\partial}{\partial r}\left(\frac{\gamma}{r} \frac{\partial}{\partial r}\left(r \frac{\partial h}{\partial r}\right)\right) \sim \gamma \frac{h_*}{r_*^3},$$

where z is the vertical coordinate and v_r the radial component of velocity. Hence, $F_\eta / F_\gamma \sim \eta r_*^4 / \left(\gamma t_* h_*^3 \right) = \varepsilon / 3 \ll 1$. This consideration shows that according to its physical meaning, ε is the modified capillary number, Ca. Condition (4.67) expresses the smallness of the Derjaguin's pressure as compared with capillary forces at large thicknesses.

Letting $\varepsilon = \lambda = 0$, we obtain the "outer" solution of Eq. (4.63) in the region of the spherical part of the droplet as

$$h(r,t) = 4f^2(t)\left(1 - \xi^2\right), \qquad 0 < \xi < 1, \tag{4.68}$$

where $\xi = rf(t)$ and an unknown function $f(t)$ is determined below. Eq. (4.68) determines the parabolic profile of the drop away from the drop edge. The apparent contact line corresponds to the condition $\xi = 1$, that is,

$$R(t) = 1/f(t), \tag{4.69}$$

where $R(t)$ is the macroscopic apparent radius of the spreading drop (Figure 4.2). In deriving Eq. (4.68), the conservation law (4.12) was used (see Appendix 1). It was supposed that almost the whole volume of the spreading droplet is located in the spherical part.

To derive the "inner" solution of Eq. (4.63) in the region where the drop profile is distorted by the action hydrodynamic flow and the Derjaguin's pressure, we introduce new variables,

$$\mu = (\xi - 1)/\chi(t), \tag{4.70}$$

and

$$h = h_0(t)\psi(\mu), \tag{4.71}$$

where $\chi(t)$, $\psi(\mu)$ and $h_0(t)$ are new unknown functions. Letting $\chi \ll 1$, $h_0 \ll 1$ and focusing only on the largest terms (see Appendix 1), from Eq. (4.63) we obtain

$$-\frac{\varepsilon h_0 \dot{f}}{f \chi} \frac{d\psi}{d\mu} = \frac{h_0^4 f^4}{\chi^4} \frac{d}{d\mu}\left[\psi^3\left[\frac{d^3\psi}{d\mu^3} - \frac{\lambda \chi^2}{f^2 h_0^{n+1} \psi^{n+1}} \frac{d\psi}{d\mu}\right]\right],$$

or

$$-\frac{\varepsilon \chi^3 \dot{f}}{h_0^3 f^5} \frac{d\psi}{d\mu} = \frac{d}{d\mu}\left[\psi^3\left[\frac{d^3\psi}{d\mu^3} - \frac{\lambda \chi^2}{f^2 h_0^{n+1} \psi^{n+1}} \frac{d\psi}{d\mu}\right]\right], \tag{4.72}$$

where the overdot indicates the first derivative with respect to dimensionless time t. For both sides of Eq. (4.72) to be explicit functions of μ but not time, we require

$$\frac{\varepsilon \chi^3 \dot{f}}{h_0^3 f^5} = -1, \tag{4.73}$$

$$\frac{\lambda \chi^2}{f^2 h_0^{n+1}} = 1. \tag{4.74}$$

Equations (4.73) and (4.74) are two ordinary differential equations for three unknown functions $f(t)$, $\chi(t)$ and $h_0(t)$.

Integrating Eq. (4.72) and using condition (4.65), which in a dimensionless form becomes $\psi(\mu) \to 0$, as $\eta \to \infty$, we obtain

$$\psi^2 \left[\psi''' - \frac{\psi'}{\psi^{n+1}} \right] = 1, \tag{4.75}$$

where the symbols $'$ and $'''$ indicate first and third derivatives with respect to μ, respectively. The first and second terms in the brackets in the left-hand side of this equation are due to the capillary and the Derjaguin's pressure effects, respectively.

The condition for matching the "inner" and "outer" solutions is

$$\frac{d\psi}{d\mu} = -8 \frac{f^2 \chi}{h_0} = -B, \qquad \mu = -\infty, \tag{4.76}$$

where B is a positive constant to be determined. The matching condition (4.76) is the third equation, which is required to determine the three unknown functions $f(t)$, $\chi(t)$ and $h_0(t)$. From Eqs. (4.73), (4.74), and (4.76), we find (see Appendix 2)

$$f(t) = 0.5 \left(\frac{10E^3 t}{2^{10} \varepsilon} + 1 \right)^{-0.1}, \tag{4.77}$$

$$\chi(t) = \left(\frac{2^{2(n+2)} \lambda}{E^{n+1}} \right)^{\frac{1}{n-1}} \left(\frac{10E^3 t}{2^{10} \varepsilon} + 1 \right)^{\frac{n+1}{5(n-1)}}, \tag{4.78}$$

$$h_0(t) = \left(\frac{2^6 \lambda}{E^2} \right)^{\frac{1}{n-1}} \left(\frac{10E^3 t}{2^{10} \varepsilon} + 1 \right)^{\frac{3}{5(n-1)}}, \tag{4.79}$$

where $E = 8/B$.

It follows from Eqs. (4.69) and (4.77) that the dimensionless macroscopic radius of the spreading drop $R(t)$ has the following time dependence

$$R(t) = 2 \left(\frac{10E^3 t}{2^{10} \varepsilon} + 1 \right)^{0.1}. \tag{4.80}$$

The constant B in Eqs. (4.77) and (4.79) can be determined using the matching condition according to Eq. (4.76)

$$\frac{d\psi}{d\mu} = -B, \qquad \mu = -\infty. \tag{4.81}$$

Unfortunately, Eq. (4.75) (see Appendix 3) does not have the required asymptotic behavior at $\mu \to -\infty$ to satisfy the condition (4.81). This suggests that in order for both the "outer," Eq. (4.68), and the "inner," Eq. (4.75), solutions to be meaningful in some common range of variables ξ and μ, it is necessary to replace the matching condition (4.81) with an approximate condition

$$\frac{d\psi}{d\mu} = -B, \qquad \mu = -\mu_*, \tag{4.82}$$

where $\mu_* > 0$ is an unknown constant. According to the physical meaning of the point μ_* the requirement $\mu_* \gg 1$ should be satisfied, which is checked below.

Let us find an expression for μ_*. The "outer" solution, Eq. (4.68), obtained from Eq. (4.63) at $\varepsilon = \lambda = 0$, becomes meaningless when $h^3 \sim \varepsilon$. Changing the variable according to Eq. (4.71) gives $h_0^3 \psi_*^3 \sim \varepsilon$, where $\psi_* = \psi(-\mu_*) \gg 1$. Substituting Eq. (4.22) into this expression and omitting the time-dependent term, we have

$$\left(2^6 \lambda / E^2\right)^{\frac{3}{n-1}} \psi_*^3 \sim \varepsilon.$$

The aforementioned approximate condition is chosen from this equation, where $\sim$ is replaced by equality, that is,

$$\left(2^6 \lambda / E^2\right)^{\frac{3}{n-1}} \psi_*^3 = \varepsilon,$$

and, at last, the necessary equation for μ_* is

$$\psi_* = \varepsilon^{1/3} \left(E^2/64\lambda\right)^{\frac{1}{n-1}}. \tag{4.83}$$

Note, the condition $\psi_* \gg 1$ should be satisfied.

We are now in a position to calculate the parameter ε. In terms of dimensional variables, Eq. (4.80) becomes

$$R(t) = 2r_* \left(\frac{10E^3 t}{2^{10} \varepsilon t_*} + 1\right)^{0.1}.$$

Differentiating this equation, we conclude

$$\frac{dR(t)}{dt} = \frac{E^3 r_*}{2^9 \varepsilon t_*} \left(\frac{10E^3 t}{2^{10} \varepsilon t_*} + 1\right)^{-0.9}.$$

Based on these two equations, we choose $r_* = R(0)/2$ and

$$\frac{r_*}{t_*} = -0.5 \frac{dR(0)}{dt}.$$

These two equations determine the unknown scales, r_* and t_*, and, hence, ε.

$$\varepsilon = \frac{E^3}{2^{10}} = \frac{1}{2B^3}. \tag{4.84}$$

From the definitions of ε, h_*, r_* and Eq. (4.84), we conclude

$$t_* = \frac{3\pi^3 \eta r_{00}^{10} B^3}{2^6 \gamma V^3} = 1.45 \frac{\eta\, r_{00}^{10} B^3}{\gamma V^3},$$

where r_{00} is the initial macroscopic drop radius,

$$r_{00} = R(0). \tag{4.85}$$

We show in Appendix 3 that as $\mu \to -\infty$,

$$\psi' \approx -3^{1/3} \ln^{1/3}|\mu|, \tag{4.86}$$

$$\psi \approx 3^{1/3} \ln^{1/3}|\mu|. \tag{4.87}$$

From Eqs. (4.82), (4.83), (4.86) and (4.87), we deduce

$$\psi_* = \varepsilon^{1/3}\left(\frac{1}{\lambda B^2}\right)^{\frac{1}{n-1}} = 3^{1/3}\mu_* \ln^{1/3}\mu_* = B\mu_* = B\exp\left(B^3/3\right). \tag{4.88}$$

From Eqs. (4.84) and (4.88), we have an equation for parameter B

$$\left(\frac{1}{\lambda B^2}\right)^{\frac{1}{n-1}} = 2^{1/3}B^2\exp(B^3/3). \tag{4.89}$$

Now from Eqs. (4.84), (4.88), and (4.89), we can determine B, μ_* and ε. Neglecting the time-dependent term in deriving Eqs. (4.83) and (4.89) is justified, as discussed in Appendix 4, where the dependence of B on time is shown to be weak.

Let us consider the solution of Eq. (4.89) for the two particular cases of the Derjaguin's pressure isotherm mentioned above.

$n = 2$ Case

From Eq. (4.89), we conclude in this case

$$B^4\exp\left(B^3/3\right) = \frac{1}{2^{1/3}\lambda}. \tag{4.90}$$

Let us examine the magnitude of the parameter λ for typical values of h_*, r_*, A and γ. For $h_* = 0.01$ cm, $r_* = 0.05$ cm, $A = 10^{-7}$dyn and $\gamma = 20$ dyn/cm, we conclude from the λ definition: $\lambda = 2.5 \cdot 10^{-5} \ll 1$. Substituting the value of λ into Eq. (4.90), we have $B^4\exp(B^3/3) = 3.18 \cdot 10^4$. This gives $B = 2.68$. Substituting this value of B into Eqs. (4.84) and (4.88) we find $\varepsilon = 0.026 \ll 1$, $\psi_* = 1648 \gg 1$, and $\mu_* = 615 \gg 1$. As λ changes from $1.25 \cdot 10^{-5}$ to $5 \cdot 10^{-5}$, B and ε change from 2.76 to 2.60 and from 0.024 to 0.029, respectively. For the same range of λ, ψ_*, and μ_* change from 3025 to 904 and from 1096 to 348, respectively. This means that the approximate conditions (4.82) and (4.83) are practically equivalent to the matching condition (4.81) over some range containing the point $\mu = -\mu_*$. Also note that the values of parameters satisfy the conditions mentioned before: $\varepsilon \ll 1$, $\lambda \ll 1$, $\mu_* \gg 1$, and $\psi_* \gg 1$.

$n = 3$ Case

From Eq. (4.89), we conclude that in this case

$$B^3\exp\left(B^3/3\right) = \frac{1}{2^{1/3}\sqrt{\lambda}}. \tag{4.91}$$

The magnitude of λ for typical values of h_*, r_*, A and γ are as follows: $h_* = 0.01$ cm, $r_* = 0.05$ cm, $A = 10^{-14}$erg and $\gamma = 20$ dyn/cm: $\lambda = 3.75 \cdot 10^{-10} \ll 1$. Substituting the value of λ into Eq. (4.91) we have $B^3\exp(B^3/3) = 4.10 \cdot 10^4$. This gives $B = 2.82$. Substituting this value of B into Eqs. (4.84) and (4.88), we find $\varepsilon = 0.022 \ll 1$, $\psi_* = 5138 \gg 1$, and $\mu_* = 1819 \gg 1$. As λ changes from $1.875 \cdot 10^{-10}$ to $7.5 \cdot 10^{-10}$, B and ε change from 2.86 to 2.79 and from 0.0213 to 0.0231, respectively. For the same range of λ, ψ_*, and μ_* change from 7073 to 3735 and from 2472 to 1341, respectively. Like the previous case, this means that the approximate conditions (4.82) and (4.83) are practically equivalent to the matching condition (4.81) over some range containing the match point $\mu = -\mu_*$. Also note that the values of parameters satisfy the conditions mentioned before: $\varepsilon \ll 1$, $\lambda \ll 1$, $\mu_* \gg 1$, and $\psi_* \gg 1$.

Using Eqs. (4.84) and (4.85), we can simplify the equation for $r_0(t)$ dependency as

$$R(t) = r_{00} \left(\frac{10t}{t_*} + 1 \right)^{0.1}. \tag{4.92}$$

From Eqs. (4.68), (4.77), and (4.84), we can derive the droplet profile as a function of radial position and time $h(r,t) = 4f^2(t)\left(1 - r^2 f^2(t)\right)$ or in dimensional variables

$$h(r,\,t) = h_* \left(10t/t_* + 1\right)^{-0.2} \left[1 - \left(r^2/4r_*^2\right)\left(10t/t_* + 1\right)^{-0.2} \right]. \tag{4.93}$$

The apex height of the drop $H(t) = h(0,t)$ is

$$H(t) = h(0,\,t) = h_* \left(10t/t_* + 1\right)^{-0.2} = h_{00} \left(10t/t_* + 1\right)^{-0.2}, \tag{4.94}$$

where h_{00} is the initial apex height of the drop. The error of assigning $h_* = h_{00}$ is negligible, as discussed in Appendix 2.

For a small angle, the advancing dynamic contact angle, $\theta(t)$, can be derived from Eq. (4.93) as $\theta(t) = -\partial\, h/\partial r R$, at $r = r_0(t)$, or

$$\theta(t) = \left(h_* / r_*\right)\left(10\,t/t_* + 1\right)^{-0.3} = \left(2\,h_{00}/r_{00}\right)\left(10\,t/t_* + 1\right)^{-0.3}. \tag{4.95}$$

Assuming the spreading drop has the shape of a spherical cap and the angle is small, another advancing dynamic contact angle, $\theta_{RH}(t)$, can be derived from Eqs. (4.92) and (4.94)

$$\theta_{RH}(t) = 2\tan^{-1}\left(H(t)/r_0(t)\right) = 2H(t)/R(t)\left(2\,r_{00}/r_{00}\right)\left(10\,t/t_* + 1\right)^{-0.3}. \tag{4.95'}$$

The expression in Eq. (4.95') is the same as that in Eq. (4.95), as it should be in the case of low slops.

We now derive the relationship between the advancing dynamic contact angle and the capillary number, Ca, which is defined as $Ca = \eta\, U\, /\gamma$, where the spreading speed at the drop edge, U, is defined as

$$U = dR/dt = \left(r_{00}/t_*\right)\left(10\,t/t_* + 1\right)^{-0.9}. \tag{4.96}$$

From Eq. (4.95) we conclude

$$\theta(t) = 2^{1/3} B\, Ca^{1/3} \quad \text{(in radians)}$$

$$= 82.64\, B\, Ca^{1/3} \quad \text{(in degrees)}. \tag{4.97}$$

We can rewrite the power laws in $\chi \ll 1$, $h_0 \ll 1\,(10t/t_* + 1)$ of Eqs. (4.92), (4.94), and (4.95) to power laws in t_e, where t_e is the experimentally measured time. Then

$$R(t_e) = \left(640\gamma V^3/3\pi^2\eta B^3\right)^{0.1} t_e^{0.1} = 1.21\left(\gamma V^3/\eta B^3\right)^{0.1} t_e^{0.1},$$

$$= K\left(\gamma V^3/\eta\right)^{0.1} t_e^{0.1}, \tag{4.98}$$

where $K = 1.21 \cdot B^{0.3}$. Using definitions of h_*, r_*, t_*, and Eq. (4.85) we can reduce Eq. (4.94) to

$$H(t_e) = \left(2B^3\right)^{0.2}\left(3\eta V^3/40\pi^2\gamma\right)^{0.1} t_e^{-0.2},\qquad(4.99)$$

and Eq. (4.95) to

$$\theta(t_e) = \left(2B^3\right)^{0.3}\left(27\eta^3 V/2000\pi\gamma^3\right)^{0.1} t_e^{-0.3} \ (\text{in radians}).\qquad(4.100)$$

The Eqs. (4.98), (4.99), and (4.100) are the same as those derived in Section 4.1 except for the pre-factor, which could not be deduced in Section 4.1.

From Eq. (4.89) we know that B is a function of λ

$$\lambda = nAr_{00}^{2n+4}\pi^{n+1}/2^{n+3}\gamma V^{n+1},$$

and in turn depends on the parameters, n and A, of the Derjaguin's pressure isotherm. For the $n=2$ case with typical values of h_*, r_*, A and γ given above, we have $K = 0.900$ and Eq. (4.98) becomes

$$R(t_e) = 0.900\left(\gamma V^3/\eta\right)^{0.1} t_e^{0.1}.\qquad(4.101)$$

Similarly, for the $n=3$ case, we have $K = 0.887$ and

$$R(t_e) = 0.887\left(\gamma V^3/\eta\right)^{0.1} t_e^{0.1}.\qquad(4.102)$$

Let us now examine the influence of the constant A (Hamaker constant in the $n=3$ case) on constant K in the spreading laws (4.102) and (4.103). For the $n=2$ case, the value of B changes from 2.76 to 2.60 and, hence, K changes from 0.892 to 0.909. For the $n=3$ case, the value of B changes from 2.86 to 2.79 and, hence, K changes from 0.883 to 0.890. It is obvious that although the value of K is somewhat larger in the $n=2$ case than in the $n=3$ case, the difference is not significant between the two cases. Thus, K is only slightly dependent on n and the constant A of the Derjaguin's pressure isotherm in the case of complete wetting.

From Eqs. (4.92) and (4.98), the wetted area predicted by our theory, S_t, is

$$S_t = \pi R^2(t) = \pi r_{00}^{0.2}\left(10t/t_* + 1\right)^{0.2} = \left(640\gamma V^3\pi^2/3\eta \ B^3\right)^{0.2} t_e^{0.2}$$

$$= K^2\left(\gamma V^3/\eta\right)^{0.2} t_e^{0.2}$$

We now briefly discuss the applicable conditions of the results:

1. The solid surface must be flat and smooth, and the liquid a Newtonian liquid and nonvolatile.
2. The gravity and inertia effects must be negligible compared to capillary effects. That means the Bond number,

$$Bo = \frac{\rho g H^2}{\gamma} \ll 1,$$

and Weber number,

$$We = \frac{\rho U^2 H}{\gamma} \ll 1,$$

where ρ is the liquid density.

3. From the solution, it follows that $H \to 0$, $as \, t/t_* \to \infty$. However, because the droplet apex is in the range of the surface forces action, the influence of the Derjaguin's pressure in the drop center and the even a weak volatility of the liquid become significant. In the case of a volatile liquid, the equilibrium film thickness, h_e, is determined by the vapor pressure in the surrounding air and according to Chapter 2 is

$$\frac{A}{h_e^n} = \left(RT/v_m \right) \ln \left(p_S/p \right).$$

This means that $H \to h_e$, as $t/t_* \to \infty$ and, hence, our results are valid only when the drop apex is much bigger than the final equilibrium film thickness, $H \gg h_e$;

4. Given that we only discuss the macroscopic aspect of the spreading problem, the condition $h(r,t) \to 0$ at $r \to \infty$ is the macroscopic condition. Macroscopic description (including the Derjaguin's pressure) is meaningless if the scale of the thickness is comparable with the molecular length scale m. Thus, our results in the case of nonvolatile liquid are valid in the wetted area within the radius r_m, where $h(r_m, t) > m$.

In conclusion, we compared the calculated radius of time dependency according to Eq. (4.92) and Eq. (4.25) deduced in Section 4.1. We rewrite Eq. (4.92) as

$$R(t) = 0.89 \left(\frac{\gamma V^3}{\eta} \right)^{0.1} \left(t + t_{in} \right)^{0.1},$$

where $t_{in} = t_* / 10$. Comparison of this spreading law and the similar deduced in Section 4.1 (Eq. 4.25) results in $\lambda_C = 1.34$, that is, $\lambda_C \sim 1$, as predicted in Section 4.1.

Comparison with Experiments

Chen [16] and [15] reported a series of experiments on spreading drops of polydimethylsiloxane. His data allow us to compare with the deduced Eqs. (4.92), (4.94), (4.95), and (4.97). In those experiments the conditions described in the theory are satisfied. The experiments involve depositing a liquid drop on a glass surface and monitoring the silhouette of the spreading drop for the drop radius, R(t), the drop apex height, $H(t)$, and the advancing dynamic contact angle, $\theta(t)$. All experiments are run at room temperature. The errors of measurement are within 0.001 cm, 0.0002 cm, and 0.75°, respectively. These errors are estimated from three repeated measurements for each quantity.

The liquid used is a silicone liquid (a polydimethylsiloxane, Dow Corning 200 fluid, Dow Corning Corporation). It has a number-average molecular weight of 7500. At the room temperature of $22.5°C - 24.0°C$, its viscosity ranges from 1.98 to 1.93 poises, its density from 0.970 to 0.969 g/cm^3, and its surface tension against air from 20.9 to 20.8 dyn/cm. The glass samples used are soda-lime glass plates and a borosilicate microscope slide.

An example of the comparison is shown in Figure 4.6. The straight lines are the results of least-square-fit to data in power laws. We first compare the least-square-fit results in power laws of

$$\left(10 \, t/t_* + 1 \right), R_0, H_0 \theta_0 \text{ and } \theta_{RH} = M \cdot \left(10 \, t/t_* + 1 \right)^N, \tag{4.103}$$

with the theory, according to Eqs. (4.92), (4.94), and (4.95). The values of M and N obtained from least-square-fit of experimental data show an excellent agreement with the theory prediction not only in exponents, N, but also in a pre-factor, M.

Figure 4.7 shows a typical comparison between our theory predictions and experimental data. These comparisons show that the predicted pre-factors by the theory agree well with the experimental values.

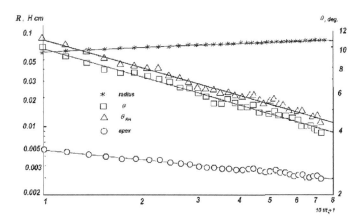

FIGURE 4.6 Experimental dependencies of radius of spreading, dynamic contact angle, and the drop apex height on time fitted according to Eq. (4.103).

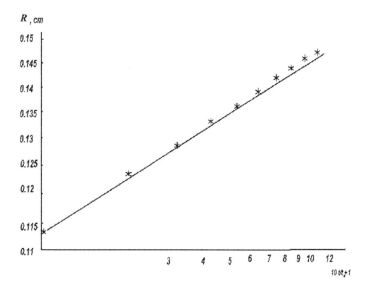

FIGURE 4.7 Comparison between the theoretical prediction (according to Eq. 4.92) and the same experimental data as in Figure 4.6. Comparison of all other dependences (contact angle, the drop apex) show the similar excellent agreement between the theoretical equations and experimental data.

The least-square fit results for data points of $\theta(Ca)$ dependency and for the same number of data points of $\theta_{RH}(Ca)$ from all experiments results in

$$\theta = 222 \cdot Ca^{0.353} \quad (\text{in degrees}), \tag{4.104}$$

$$\theta_{RH} = 242 \cdot Ca^{0.353} \quad (\text{in degrees}). \tag{4.105}$$

The value of Ca ranges from $1.8 \cdot 10^{-6}$ to $3.3 \cdot 10^{-4}$ for experiments. The pre-factors and exponents in Eqs. (4.104) and (4.105) are close to those predicted by Eq. (4.97), which can be written as

$$\theta = \theta_{RH} = 238 \cdot Ca^{0.333} \quad (\text{in degrees}). \tag{4.106}$$

Ausserre et al. [17,18] measured the time-dependence of the radius and advancing dynamic contact angle for a number of spreading drops of polydimethylsiloxane with different molecular weight. They found that the radius and contact angle show a power law dependence on time with an exponent of 0.100 ± 0.010 and -0.3 ± 0.015, respectively. These values agree with our predictions, 0.1 and -0.3. With similar experiments Tanner [19] found that the dynamic contact angle follows a power law in time with an exponent ranging from -0.317 to -0.335, which is close to our predicted value, -0.3.

Conclusions

The combined effect of viscosity, surface tension, and the Derjaguin's pressure is included in the theory for an axisymmetric, nonvolatile, Newtonian liquid drop spreading over a horizontal, dry, smooth, flat solid surface in the case of complete wetting. It is shown that the Derjaguin's pressure action removes the "singularity" on the moving three-phase contact line. The drop profile is calculated as a function of time and radial position. The spreading radius, the apex height of the drop, and the advancing dynamic contact angle are found to follow different power laws in time. The dynamic contact angle is found to follow a power law in capillary number. Both the pre-factors and exponents in the power laws are predicted and the predicted power laws agree with known experimental data.

APPENDIX 1

This appendix shows the derivation of Eqs. (4.68) and (4.72).

Let $\varepsilon = \lambda = 0$, Eq. (4.63) becomes in this case

$$\frac{\partial}{\partial r}\left[rh^3 \left[\frac{\partial}{\partial r}\left(\frac{1}{r}\frac{\partial}{\partial r}\left(r\frac{\partial h}{\partial r} \right) \right) \right] \right] = 0.$$

Integrating once with respect to r results in

$$h^3\left[\frac{\partial}{\partial r}\left(\frac{1}{r}\frac{\partial}{\partial r}\left(r\frac{\partial h}{\partial r} \right) \right) \right] = \frac{C_1}{r}.$$

Given that h finite at $r = 0$, C_1 must be 0. Integrating with respect to r three times, we have $h = C_2 r^2/4 + C_3 \ln r + C_4$. For the same reason, C_3 must be 0. Recognizing $h(r,t) = 0$ at $r = R$, we have $h = (C_2/4)(r^2 - R^2)$. Now taking into account the conservation law (4.12) and defining $R = 1/f(t)$, we arrive to Eq. (4.68).

From Eq. (4.71), we conclude

$$\frac{\partial h}{\partial t} = \dot{h}_0(t)\psi(\mu) + h_0(t)\frac{d\psi}{d\mu}\frac{\partial \mu}{\partial t} = \dot{h}_0\psi + h_0\psi' \frac{r\dot{f}\chi - (rf - 1)\dot{\chi}}{\chi^2}$$

$$= \dot{h}_0\psi + h_0\psi' \frac{\left(\mu + \dfrac{1}{\chi}\right)\dfrac{\dot{f}}{f}\chi - \mu\dot{\chi}}{\chi},$$

where the overhead dot and the superscript indicate the first derivative with respect to dimensionless time, t, and coordinate, μ, respectively.

As $\mu = rf \to 1$ and $\chi \ll 1$, then $\mu \ll 1/\chi$, hence,

$$\left(\mu + \frac{1}{\chi}\right)\frac{\dot{f}}{f}\chi - \mu\dot{\chi} \approx \frac{\dot{f}}{f} - \mu\dot{\chi} \approx \frac{\dot{f}}{f}.$$

In view of this, we conclude

$$\frac{\partial h}{\partial t} = \dot{h}_0 \psi + h_0 \psi' \frac{\dot{f}}{f\chi}.$$

Here we use the following estimations

$$\dot{h}_0 \approx \frac{h_0}{t}, \psi' \approx \frac{\psi}{\mu}, \dot{f} \approx \frac{f}{t}.$$

These estimations give

$$\frac{h_0 \psi}{h_0 \psi' \frac{f}{f\chi}} \approx \frac{h_0 \psi \eta}{h_0 \psi \frac{f}{f\chi}} = \eta\chi \ll 1.$$

Hence, combining all previous estimations, we get the following

$$\frac{\partial h}{\partial t} = h_0 \psi' \frac{f}{f\chi}. \tag{A1.1}$$

Now for $\mu\chi \ll 1$, the right-hand side of Eq. (4.63) can be simplified as

$$\frac{1}{r}\frac{\partial}{\partial r}\left[rh^3\left[\frac{\partial}{\partial r}\left(\frac{1}{r}\frac{\partial}{\partial r}\left(r\frac{\partial h}{\partial r}\right)\right) - \frac{\lambda}{h^{n+1}}\frac{\partial h}{\partial r}\right]\right]$$

$$= h_0^4 \frac{f^4}{\chi^4}\frac{1}{1+\mu\chi}\frac{d}{d\mu}\left[(1+\mu\chi)\psi^3\left[\frac{d}{d\mu}\left(\frac{1}{1+\mu\chi}\frac{d}{d\mu}\left((1+\mu\chi)\frac{d\psi}{d\mu}\right)\right) - \frac{\lambda\chi^2}{f^2 h_0^{n+1}\psi^{n+1}}\frac{d\psi}{d\mu}\right]\right] = \tag{A1.2}$$

$$= h_0^4 \frac{f^4}{\chi^4}\frac{d}{d\mu}\left[\psi^3\left[\frac{d^3\psi}{d\mu^3} - \frac{\lambda\chi^2}{f^2 h_0^{n+1}\psi^{n+1}}\frac{d\psi}{d\mu}\right]\right].$$

Now from Eqs. (A1.1) and (A1.2), we arrive to Eq. (4.72).

APPENDIX 2

This appendix shows the derivation of Eqs. (4.77) through (4.79), and discusses the negligible error in assigning $H(0,0) = H_*$ in the derivation.

From Eqs. (4.73) and (4.76), we have

$$\frac{\dot{f}}{f^{11}} = -\frac{E^3}{\varepsilon}. \tag{A2.1}$$

Here the overhead dot indicates the first derivative with respect to the dimensionless time, t. Integrating Eq. (A2.1) with respect to t once, we have

$$f^{-10} 10 E^3 t / \varepsilon + \text{const.} \tag{A2.2}$$

To determine the integration constant, we make use of the following initial condition at $R_0 = R(0)$ and $t = 0$: $H(0,0) = H_*$, which in a dimensionless form is

$$h(0,0) = 1, \text{ at } r = 0 \text{ and } t = 0, \tag{A2.3}$$

where $H(0,0)$ is the initial apex height of the drop. Substituting Eq. (A2.3) into Eq. (4.68) we find

$$f(0) = 1/2. \tag{A2.4}$$

Substituting Eq. (A2.4) into Eq. (A2.2), we find constant $= 2^{10}$ and

$$f(t) = 0.5\left(10E^3 t / 2^{10}\varepsilon + 1\right)^{-0.1},$$

which coincides with Eq. (4.77). From Eqs. (4.74), (4.76) and (4.77) we can derive Eq. (4.78) and, at last, from Eq. (4.76), (4.77), and (4.78), we can derive Eq. (4.79).

We now discuss the error in assigning $H_* = H(0,0)$ instead of using $H_* = V/(2\lambda R^2_*)$. In view of $r_* = r_0(0)/2$ and Eq. (4.85), H_* can be written as

$$H_* = V / \left(2\pi R_*^2\right) = 2V / \left(\pi R_0^2\right). \tag{A2.5}$$

If the drop shape is a spherical cap then,

$$V = \left(\pi H_{00} / 6\right)\left(3r_{00}^2 + H_{00}^2\right), \tag{A2.6}$$

where $H_{00} = H(0,0)$ is the initial apex height of drop and $r_{00} = R_0(0)$ is the initial macroscopic drop radius. From Eq. (A2.5) and (A2.6), we conclude

$$H_* = \left(H_{00} / 3r_{00}^2\right)\left(3r_{00}^2 + H_{00}^2\right) = H_{00}\left(1 + H_{00}^2 / 3r_{00}^2\right). \tag{A2.7}$$

From Eq. (A2.7), we estimate the error in assigning $H_* = H(0,0)$ is 0.75%, when H_{00}/r_{00} is 0.15. Thus, the error is negligibly small.

APPENDIX 3

This appendix shows the derivation for Eqs. (4.86) and (4.87).

As $\mu \to -\infty$, $\psi \to \infty$, and, hence, Eq. (4.75) becomes

$$\psi^2 \psi''' = 1. \tag{A3.1}$$

Let us define $\varphi(\psi) = \psi' = d\psi / d\mu$. According to the chain rule, we have

$$\psi'' = d^2\psi / d\mu^2 = d\varphi / d\mu = d\varphi / d\psi \cdot d\psi / d\mu = \varphi'\varphi$$

$$\psi''' = d\left(\varphi'\varphi\right) / d\mu = d\varphi' / d\mu\, \varphi + \varphi' d\varphi / d\mu = \varphi''\varphi^2 + \left(\varphi'\right)^2 \varphi.$$

Equation (A3.1) now becomes

$$\psi^2\left(\varphi''\varphi^2 + \left(\varphi'\right)^2 \varphi\right) = 1. \tag{A3.2}$$

Let $y = \ln \varphi$ and apply the chain rule once again, we have

$$\varphi' = d\varphi / d\psi = d\varphi / dy \cdot dy / d\psi = \varphi_y / \psi$$

$$\varphi'' = d^2\varphi / d\psi^2 = dy / d\psi\, d\left(\varphi_y / \psi\right) / dy = \left[d\left(\varphi_y e^{-y}\right) / dy\right] / \psi.$$

$$= e^{-y}\left(\varphi_{yy} e^{-y} - \varphi_y e^{-y}\right) = e^{-2y}\left(\varphi_{yy} - \varphi_y\right).$$

Substituting these results into Eq. (A2.2), we have

$$\varphi_y^2 \varphi + \left(\varphi_{yy} - \varphi_y\right)\varphi^2 = 1. \tag{A3.3}$$

Assume the solution to Eq. (A3.3) has the form $\varphi = Dy^G$, where D and G are constants to be determined, then

$$\varphi_y^2 \varphi + \varphi_{yy}\varphi^2 = (2G-1)GD^3 y^{3G-2}$$

$$\varphi_y \varphi^2 = GD^3 y^{3G-1}.$$

If $G = 2/3$, then $\left|\varphi_y \varphi^2\right| = |D|^3 y^{1/3} 2/3 \to \infty$, at $y \to \infty$. Hence, $G = 2/3$ cannot satisfy Eq. (A3.3) at $y \to \infty$. However, if $G = 1/3$, then

$$\left|\varphi_y^2 \varphi + \varphi_{yy}\varphi^2\right| = |D|^3 / (9y) \to 0, \text{ at } y \to \infty.$$

Hence, at $y \to \infty$, for $G = 1/3$ we have from Eq. (A3.3) $D = -3^{1/3}$ and

$$\varphi = Dy^{1/3} = \psi' = -3^{1/3} y^{1/3} = -3^{1/3} \ln^{1/3}\psi. \tag{A3.4}$$

From this equation, we have Eq. (4.86) and $d\psi / \ln^{1/3}\psi = -3^{1/3} d\mu$.

Introducing $z = \ln\psi$ and integrating the above equation by parts, we conclude

$$\int \left(\ln^{-1/3}\psi\right) d\psi = \int e^z z^{-1/3} dz = \left[e^z z^{-1/3}\right] + (1/3)\int e^z z^{-4/3} dz$$

$$= \left[\psi \ln^{-1/3}\psi\right] + (1/3)\int \ln^{-4/3}\psi \, d\psi \cong \psi \ln^{-1/3}\psi,$$

given that $\ln^{4/3}\psi \gg \ln^{1/3}\psi'$, at $\psi \to \infty$. Hence,

$$\psi / \ln^{1/3}\psi = -3^{1/3}\mu. \tag{A3.5}$$

Solving Eq. (A3.5) in the limits $|\mu| \gg 1$ and $\psi \gg 1$, we conclude

$$\psi = 3^{1/3}|\mu|\ln^{1/3}\psi = 3^{1/3}|\mu|\left(\ln\left(3^{1/3}|\mu|\ln^{1/3}\psi\right)\right) \approx 3^{1/3}|\mu|\ln^{1/3}|\mu|$$

This expression coincides with Eq. (4.87).

APPENDIX 4

This appendix discusses the weak dependence of B on the dimensionless time, t.

In deriving Eqs. (4.83) and (4.89) we have neglected the time-dependence term in h_0^3. Here we keep the time-dependent term for am estimation. We determine match point $\mu = -\mu_\bullet$ as the point where the following condition is satisfied

$$h_0^3 \psi_\bullet^3 = \varepsilon, \tag{A4.1}$$

where $\psi_\bullet = \psi\left(-\mu_\bullet\right) \gg 1$, and, in view of Eq. (4.79), Eq. (A4.1) can be rewritten as

$$\left(2^6 \lambda / E^2\right)^{3/(n-1)} \left(10E^3 \tau / 2^{10}\varepsilon + 1\right)^{9/5(n-1)} \psi_\bullet^3 = \varepsilon. \tag{A4.2}$$

From Eqs. (4.84) and (4.2) we conclude

$$\psi_\bullet = \varepsilon^{1/3}\left(E^2/64\lambda\right)^{1/(n-1)}\left(10t+1\right)^{-3/5(n-1)} = \varepsilon^{1/3}\left(E^2/64\Lambda\right)^{1/(n-1)},$$

where $\Lambda = \lambda\left(10t+1\right)^{3/5}$. Following the same derivation for Eq. (4.89), we have

$$\left(\Lambda B^2\right)^{-1(n-1)} = 2^{1/3}B^2\exp\left(B^3/3\right).$$

Let us define $F(B) = 1/\Lambda = 2^{(n-1)/3}B^{2n}\exp[(n-1)B^3/3]$, then

$$dF/dB = \left(2n/B+(n-1)B^2\right)/\Lambda \text{ and } dF/dB\cdot dB/dt = -\left(d\Lambda/dt\right)/\Lambda^2.$$

Given that $\left(d\Lambda/dt\right)/\Lambda = 6/\left(10t+1\right)$, we have $dB/dt = -6B/\left[\left(10t+1\right)\left(2n+(n-1)B^3\right)\right]$
For $n = 3$ and $B \approx 3$, we conclude from this equation

$$dB/dt = -0.03/\left(t+0.1\right)$$

Hence,

$$B = 3 - 0.03\ln\left(t+0.1\right).$$

This equation proves that $B(t)$ weakly dependent upon time and justifies the omission of the time-dependent term in deriving Eqs. (4.83) and (4.89).

4.3 Spreading of Drops over a Surface Covered with a Thin Layer of the Same Liquid

In this section a solution is obtained for the problem of viscous spreading of a liquid drop on a plane solid surface that has been prewetted with a film of the same liquid. The film thickness is assumed "thick enough," that is, the thickness is bigger than the radius of the Derjaguin's pressure action. Appearing in the spreading equation is a universal small parameter, which is independent of the nature of the liquid/substrate system and which is a characteristic of the viscous spreading regime. By an expansion in terms of this small parameter, a solution has been obtained for the problem, through which the profile of the spreading drop and the velocity of motion of the drop boundary can be determined [11].

Let us examine a drop of a viscous liquid on a planar horizontal solid surface covered with a layer of the same liquid with thickness, h_0 (Figure 4.8). The thickness h_0 is assumed to be outside the range of the Derjaguin's pressure action. However, the droplet is still small enough to neglect the gravity action. Because the spreading process is axisymmetric in the case under consideration, the height of the drop, $h(r,t)$, is a function of both the distance r from the coordinate origin, which is located on the plane of the

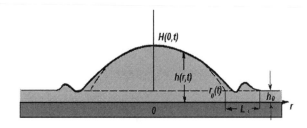

FIGURE 4.8 Spreading of liquid droplets over the solid surface covered with a film of the same liquid of thickness, h_0.

solid surface in the center of the drop (Figure 4.8), and the time, t. For a sufficiently low sloped drop, in the case in which gravitational and inertial forces can be neglected, the equation describing the process of viscous spreading of the drop is deduced in Section 4.1 (Eq. 4.17). This assumption means: $h_* \ll r_*$, that is, the length scale in the vertical direction is much smaller than in the horizontal direction. In view of the symmetry of the drop, the boundary conditions in the droplet center (4.17′) are satisfied.

Because flow occurs only where the liquid surface is curved, so that there is no flow going out into the film to infinity, the excess volume of liquid above the film remains constant:

$$2\pi \int_0^\infty r\left[h(r,t) - h_0\right] drR = V = \text{const.} \tag{4.107}$$

Far from the droplet its profile tends to that of the film, hence:

$$h(r,t) = h_0, \quad r \to \infty. \tag{4.108}$$

In addition to the four boundary conditions (4.17′), (4.107), (4.108) Eq. (4.17), which is the fourth order partial differential equation, requires the assignment of an initial condition, which is formulated below. Let us introduce the dimensionless quantities using the characteristic scales r_*, h_*, t_*, and U_* for the horizontal dimension, the height of the drop, the time, and the spreading velocity, respectively, with $U_* = r_*/t_*$ and $V = 2\pi r_*^2 h_*$, this relationship determining the horizontal scale of r_* with a given drop volume and initial height follows from the conservation law (4.107) in the same way as in the previous Sections 4.1 and 4.2. Using the following dimensionless variables, $r \to r/r_*$, $t \to t/t_*$, $h \to h/H_*$, and $h_0 \to h_0/h_*$, and keeping the same symbols for dimensionless values as for dimensional, we arrive to the equation of spreading:

$$\frac{1}{r}\frac{\partial}{\partial r}\left[rh^3\frac{\partial}{\partial r}\left(\frac{1}{r}\frac{\partial}{\partial r}\left(r\frac{\partial h}{\partial r}\right)\right)\right] = -\varepsilon\frac{\partial h}{\partial t}, \tag{4.109}$$

where the dimensionless parameter ε,

$$\varepsilon = \frac{3\eta}{\gamma}\frac{U_* r_*^3}{h_*^3}, \tag{4.110}$$

which as before represents the ratio of characteristic values of the force of viscous friction in the drop $F_\mu = \eta U_*/h_*^2$ and the horizontal component of the gradient of capillary pressure $F_\gamma = \gamma h_* r_*^3$.

The conservation law (4.107) now becomes

$$\int_0^\infty r\left[h(r,t) - h_0\right] dr = 1, \tag{4.111}$$

and the condition (4.108) has the identical form.

It is easy to check that:

$$\varepsilon \ll 1. \tag{4.112}$$

We also assume that the film thickness is much smaller than the characteristic size of the droplet, that is:

$$h_0 \ll 1. \tag{4.113}$$

In solving the above problem, we use the method of matching asymptotic expansions, which we used in Section 4.2. In view of the smallness of the parameter ε, the entire drop can be subdivided into two zones: the outer zone, $0 \le r < r_0(t)$, which determines the macroscopic boundary of the drop, that is, the coordinate at which $h(r_0(t),t) \approx h_0$, and the inner zone, $r > r_0(t)$, with the characteristic scale, L_*, much smaller than the scale of the outer region (Figure 4.8): $L_* \ll r_*$. Thus, the integral in Eq. (4.111) can be rewritten as follows:

$$\int_0^\infty r(h - h_0)\,dr = \int_0^{r_0} r(h - h_0)\,dr - \int_0^\infty r(h - h_0)\,dr.$$

These integrals can be estimated using dimensional variables as

$$\int_0^{r_0} r(h - h_0)\,dr \approx h_* r_* \quad \text{and} \quad \int_{r_0}^\infty r(h - h_0)\,dr \approx h_0 L_*.$$

This estimation shows that

$$\int_{r_0}^\infty r(h - h_0)\,dr \Big/ \int_0^{r_0} r(h - h_0)\,dr \approx \frac{h_0}{h_*} \frac{L_*}{r_*} \ll 1,$$

hence, the entire volume of the drop, with the accuracy with which the problem is being solved, is actually concentrated in the outer zone; that is, we can set

$$\int_0^{r_{00}(t)} r(h - h_0)\,dr \approx 1. \tag{4.114}$$

In zeroth approximation, we conclude from Eq. (4.109), setting $\varepsilon = 0$, the equation for the outer solution:

$$\frac{\partial}{\partial r}\left(\frac{1}{r}\frac{\partial}{\partial r}\left(r\frac{\partial h}{\partial r}\right)\right) = 0. \tag{4.115}$$

The boundary condition (4.108) should now be written as

$$h(r_0,t) \approx h_0. \tag{4.116}$$

We obtain the solution of the problem (4.115), (4.111), (4.114), (4.116) in the following form

$$h(r,t) = 4f^2(t)\left(1 - \xi^2\right) + h_0, \tag{4.117}$$

where

$$\xi = rf(t), \tag{4.118}$$

and $f(t)$ is a new unknown function of time, which is determined below; $0 \le \xi < 1$. It follows from Eq. (4.117) that the drop has the parabolic shape; the apparent radius of the spreading drop corresponds to $\xi = 1$, or

$$r_0(t) = 1/f(t). \tag{4.119}$$

To obtain the inner solution of the problem (4.109) and (4.108), which is valid close to the boundary of the spreading drop, we introduce new scales for the space variable and a new form of the unknown $h(r,t)$:

$$\xi - 1 = \chi(t)\mu, \tag{4.120}$$

$$h = h_0 \psi(\mu),$$

where $\chi(t)$ and $\psi(\mu)$ are new unknown functions subject to determination; μ is the new local variable inside the inner zone; and $\chi(t) \ll 1$. Below we present a slightly modified way of deducing the inner equation for the droplet profile as compared with the previous section. This method is more appropriate for the problem under consideration. To do this, we integrate Eq. (4.109) with respect to the variable r:

$$rh^3 \frac{\partial}{\partial r}\left(\frac{1}{r}\frac{\partial}{\partial r}\left(r\frac{\partial h}{\partial r}\right)\right) = -\varepsilon \int_0^r r \frac{\partial h}{\partial t} dr. \tag{4.121}$$

We now calculate the integral in the right side of Eq. (4.121), using the outer solution, because the integration takes place mostly across this region:

$$-\varepsilon \int_0^r r \frac{\partial h}{\partial t} dr = -\varepsilon \int_0^\xi \frac{\xi}{f^2(t)}\frac{\partial h}{\partial t} d\xi = -\frac{4\varepsilon}{f^2(t)}\int_0^\xi \left[2ff'\left(1-\xi^2\right)-2f^2\xi rf'\right]\xi d\xi$$

$$= -8\varepsilon \frac{f'}{f}\int_0^\xi \xi\left(1-2\xi^2\right)d\xi = -4\varepsilon\frac{f'}{f}\left(\xi^2-\xi^4\right) = -\varepsilon\frac{f'}{f^3}\xi^2\left(h-h_0\right)$$

Substituting the resulting expression, together with the transformation (4.120), into Eq. (4.121), we obtain:

$$\psi^3 \frac{d^3\psi}{d\mu^3} = -\frac{\varepsilon}{h_0^3}\frac{\chi^3(t)f'(t)}{f^5(t)}(\psi-1). \tag{4.122}$$

Equation (4.122) should include only the variable μ, this gives the following requirement:

$$\frac{\varepsilon}{h_0^3}\frac{\chi^2 f'}{f^5} = 1. \tag{4.123}$$

Hence, Eq. (4.122) can be rewritten as

$$\psi^3 \frac{d^3\psi}{d\mu^3} = \psi - 1. \tag{4.124}$$

The joining of the inner (4.120) and outer (4.118) solutions is determined by the condition

$$\frac{d\psi}{d\mu} = -\frac{8f^2(t)\chi(t)}{h_0}. \tag{4.125}$$

This means that the following condition must be fulfilled:

$$8f^2(t)\chi(t)/h_0 = B = \text{const}, \tag{4.126}$$

where B can no longer be selected arbitrarily, but is determined by the matching condition.

The two Eqs. (4.123) and (4.126) determine the two unknown functions $f(t)$ and $\chi(t)$ that have been introduced above. The expressions (4.125) and (4.126), together with the boundary condition (4.108), give the following boundary conditions for Eq. (4.124):

$$\frac{d\psi}{d\mu} = -B, \tag{4.127}$$

$$\psi\big|_{\mu \to \infty} = 1. \tag{4.128}$$

The problems (4.124), (4.127), (4.128) are investigated in Section 4.5. It is shown in Section 4.5 that the function $\psi(\mu)$ has the form of damped oscillations as $\mu \to +\infty$,

$$\psi(\mu) = 1 + C \exp(-\mu/2) \sin\left(\frac{\sqrt{3}}{2}\mu\right). \tag{4.129}$$

When we take conditions (4.123) and (4.126) into account, we obtain

$$\left(\frac{B}{8}\right)^3 \varepsilon \frac{f'}{f^{11}} = -1. \tag{4.130}$$

Now we can return to the question of the initial condition for Eq. (4.109). The analysis that has been performed is valid only after the passage of a certain initial period of time t_{in} at which the spreading regime described in this section is established. As already mentioned, the duration of the initial stage of spreading when both Re and Ca numbers are not small is estimated in Section 5.1. The moment $t = t_{in}$ is taken as the initial moment of time. At this time, a parabolic profile has been formed in the center of the drop; for assignment of this profile at the initial moment, all that must be known [apart from the condition of constancy of volume and initial radius $r_0(0)$] is the height at the center, $h(0,0)$. Selecting $h(0,0) - h_0 = h_*$ at the scale on the z axis, we obtain the missing initial condition for Eq. (4.109):

$$h(0,0) \approx 1 \tag{4.131}$$

This condition determines the initial condition for the function $f(t)$ that is the solution of Eq. (4.130): $f(0) = 1/2$. This makes it possible to determine from Eqs. (4.130) to (4.126) the functions $f(t)$ and $\chi(t)$:

$$f(t) = \frac{1}{2}\left(\frac{5}{B^3} t/\varepsilon + 1\right)^{-1/10}, \tag{4.132}$$

$$\chi(t) = \frac{B}{2} h_0 \left(\frac{5}{B^3} t/\varepsilon + 1\right)^{1/5}. \tag{4.133}$$

This expression for $\chi(t)$ gives the scale of the inner zone, which should be small, that is, $\chi(t) \ll 1$. It follows from Eq. (4.133) that this condition is satisfied if

$$\frac{B}{2} h_0 \left(\frac{5}{B^3} t/\varepsilon + 1\right)^{1/5} \ll 1, \quad \text{or} \quad t \ll \frac{2^5 \varepsilon}{h_0^5 B^2 5}. \tag{4.134}$$

This gives the required restriction on the duration of the spreading process. After that the droplet is becomes too small and indistinguishable from the film.

From Eqs. (4.119) and (4.132), we find an expression for the radius of the spreading drop: $r_0(t) = 2\left(\frac{5}{B^3} t/\varepsilon + 1\right)^{1/10}$ or in dimensional variables

$$r_0(t) = 2r_* \left(\frac{5}{B^3} \frac{\gamma h_*^3}{3\eta r_*^4} t + 1 \right)^{0.1}.$$ (4.135)

From these expressions, we obtain the time dependence of the dynamic contact angle at the drop boundary as

$$tg\theta = -\left. \frac{\partial h}{\partial r} \right|_{r=r_0} = \frac{h_*}{r_*} \left(\frac{5}{B^3} t/\varepsilon + 1 \right)^{-3/10}.$$

Expressing the right side of the last equation in terms of the spreading velocity, $U(t) = dr_0/dt$, we arrive to

$$tg\theta = B\left(\frac{3\eta U}{\gamma} \right)^{1/3} = B(3Ca)^{1/3}.$$ (4.136)

Let us check that the dimensional combination appearing in Eq. (4.110) for the parameter, ε, does not depend upon the selection of the initial moment of time. As we established above, the characteristic values of the quantities appearing in this combination vary with time in accordance with the laws $r_* \approx r_0(t) \approx t^{1/10}$, $h_* \approx h(0,t) \approx t^{-1/5}$, $t_* \approx t$, that is, it remains constant with time: $r_*^4 / t_* h_*^3 = \text{const}$.

Differentiating both sides of Eq. (4.135) with respect to time, we find that at the initial moment, $t = 0$,

$$\frac{dr_0}{dt} = \frac{1}{3B^3} \frac{\gamma h_*^3}{\eta r_*^3}.$$ (4.137)

Given that $r_0(0) = 2r_*$, then, taking as the characteristic value of velocity $2U_* = \frac{dr_0}{dt} = \frac{2r_*}{t_*}$, from Eq. (4.137) we find the characteristic value of spreading time: $t_* = (6B\eta r_*^4/(\gamma h_*^3))$.

In Section 4.2 we showed that Eq. (4.124) does not have any solutions satisfying the matching condition (4.127); therefore, in place of the matching of the outer (4.118) and inner (4.120) solutions, we require, as in Section 4.2, that they must be patched at a certain point $\mu = -\mu_* : d\psi/d\mu = -B$.

We determine the patching point, μ_*, in precisely the same way as in the previous Section 4.2. The outer solution (4.118) obtained from Eq. (4.109) with $\varepsilon = 0$ loses its meaning if its left-hand side becomes of the same order of magnitude as the right-hand side, that is, when $h^3 = \varepsilon$. We set as the patching point $h^3 = \varepsilon$ or, in view of (4.120), $h_0^3 \psi^3(-\mu_*) = \varepsilon$.

Equation (4.124) as $\mu \to -\infty$ has the following asymptotic representation:

$$\psi(\mu) \approx \sqrt[3]{3}|\mu|\ln^{1/3}|\mu|, \frac{d\psi}{d\mu} \approx -\sqrt[3]{3}\ln^{1/3}|\mu|$$

(see Section 4.2 for details), which gives the following equation for determining of the quantity B. All details are identical to those presented in Section 4.2. This results in

$$B^3 \exp(B^3) = 1/2h_0^3 \text{ and } \varepsilon = 1/2B^3.$$ (4.138)

If we now set $h_* \approx 0.1$ cm and $h_0 \approx 10^{-5}$ cm, then the dimensionless value $h_0 \approx 10^{-4}$; and for B and ε, we obtain the following values: $B \approx 2.753$, $\varepsilon \approx 0.024$; and correspondingly, $\mu_* \approx 1048$ and $\psi(-\mu_*) \approx 2886$.

The calculated values of B and ε are very close to those found in the previous section for the spreading over dry substrate. This means that in the case of complete wetting, the pre-exponential constant is almost insensitive to the conditions in front of the spreading droplet. We will see a further evidence of this insensitivity in Section 5.1.

4.4 Quasi-Steady-State Approach in the Kinetics of Spreading

A simplified approach is suggested below to the solution of problems of liquid drop spreading. The approach consists essentially of assuming constancy of the drop spreading velocity, U, at each fixed moment of time, t. This gives the possibility of determining the relationship $U = f(r_0)$, where r_0 is the radius of the drop base. Given that $U = dr_0/dt$, this equation can be used to obtain an expression for the spreading radius as a function of time. The applicability of the method has been demonstrated using examples of spreading on a solid substrate (with the Derjaguin's pressure acting in the vicinity of the apparent three-phase moving contact line), spreading on a prewetted solid substrate, spreading under the influence of gravity and under the influence of applied temperature gradient [20].

A number of spreading problems were solved in Sections 4.1 through 4.3. In the case of spreading of small drops over dry solid substrate (Section 4.2), the explicit expressions obtained for the drop radius and height and the dynamic contact angle as functions of time. These expressions do not include any fitting parameters and depend solely on the hydrodynamic characteristics of the liquid, including the Derjaguin's pressure isotherm, proved to be in good agreement with experimental data. This solution was limited to the case of complete wetting for the Derjaguin's pressure isotherms of the type $\Pi(h) = A/h^n$, where $A > 0$ and $n \geq 2$. The similarity solution of gravitational regime of spreading and spreading of very small liquid drops were obtained in Section 4.1 and spreading over pre-wetted solid substrate was solved in Section 4.3. In all cases, the solution was based on more or less sophisticated mathematical treatment. From an analysis of all the relationships found between the drop radius and time, we arrive to a hypothesis of a "quasi-steady-state nature" of the spreading process. According to this hypothesis, the characteristics of spreading are determined for the most part only by the instantaneous value of the spreading velocity.

Let us first examine the problem of spreading of a low sloped drop of a viscous liquid on a horizontal surface covered with a layer of the same liquid with a thickness h_0, which is the same problem as in Section 4.3. The same notations as in Section 4.3 are used here. Let $h(r,t)$ be the equation of the drop profile; r is the radial coordinate; and t is the time. We use the following characteristic scales: h_*, r_* and t_*. The following relationships are satisfied: $h_* \gg h_0$; $r_* = r_0(0)/2$, where $r_0(t)$ is the radius of drop spreading; and $h_* \ll r_*$. Then, as shown in Section 4.3, in dimensionless variables $h \rightarrow h/h_*$, $r \rightarrow r/r_*$, $t \rightarrow t/t_*$, and $h_0 \rightarrow h_0/h_*$, the spreading process is described by the differential equation (4.109), with conservation law (4.107) and boundary conditions (4.17') and (4.108).

As shown in Section 4.3, the condition $\varepsilon \ll 1$ is usually met; that is, viscous forces are small in comparison with capillary forces. Now, in Eq. (4.109), we use a new quasi-steady-state variable,

$$\xi = r - r_0(t),\qquad(4.139)$$

where $r_0(t)$ is the spreading radius. Let $U = \dot{r}_0(t)$ be the spreading velocity; the overdot denotes the derivative with respect to time t. Using the above notations and assuming that the liquid profile depends on the new variable only, that is, $h = h(\xi)$, we obtain from Eq. (4.109)

$$\varepsilon U \frac{dh}{d\xi} = \frac{1}{\xi}\frac{d}{d\xi}\left[\xi h^3 \frac{d}{d\xi}\left(\frac{1}{\xi}\frac{d}{d\xi}\left(\xi \frac{dh}{d\xi}\right)\right)\right].\qquad(4.140)$$

Setting $\varepsilon = 0$ in this equation, we obtain in the same way as in Section 4.3, the outer solution of the problem

$$h = C\left(1 - \xi^2/r_0^2\right) + h_0,$$

where C is the integration constant, determined from the conservation law (4.107). This gives $C = 4/r_0^2$. Where, the outer solution is

$$h = \frac{4}{r_0^2}\left(1 - \xi^2/r_0^2\right) + h_0,\qquad(4.141)$$

which coincides with the outer solution deduced in Section 4.3 (Eq. 4.117). Eq. (4.141) describes a parabolic profile of the drop and is valid far from the apparent moving three-phase contact line.

In the vicinity of the moving apparent three-phase contact lien, $\xi = 0$, where the profile of the outer solution (4.141) intersects the surface of the liquid film, we introduce the inner variable as before:

$$\mu = \frac{\xi - r_0}{\chi}, \quad \chi << 1$$

$$h = h_0 \psi(\mu)$$

(4.142)

Using the new inner variable and retaining in Eq. (4.140) only the leading terms, we obtain

$$\frac{\varepsilon U r_0^3 \chi^3}{h_0^3} \frac{d\psi}{d\mu} = \frac{d}{d\mu}\left[\psi^3 \frac{d^3\psi}{d\mu^3}\right].$$

(4.143)

The requirement of "self-similarity" with the "frozen" time t results in

$$\frac{\varepsilon U r_0^3 \chi^3}{h_0^3} = 1$$

$$\frac{d\psi}{d\mu} = \frac{d}{d\mu}\left[\psi^3 \frac{d^3\psi}{d\mu^3}\right].$$

(4.144)

Integration of this equation with the condition $\psi \to 1$, $\mu \to \infty$ yields Eq. (4.124).

The condition of "matching" the outer solution (4.141) and the inner solution (4.124) at a certain point $\mu = -\mu_*$ has the form similar to that deduced previously (4.127):

$$\frac{d\psi}{d\eta}\Big|_{\eta=-\eta_*} = -\frac{8\chi}{r_0^2 h_0} = -B = \text{const.}$$

(4.145)

Now using the first equation (4.144) and (4.145), we obtain

$$\varepsilon U = \left(\frac{8}{B}\right)^3 r_0^{-9}.$$

(4.146)

We now take into account the fact that $U = dr_0/dt$. Then, from Eq. (4.146) with the initial condition $r_0(0) = 2$, we obtain

$$r_0 = 2\left(\frac{5}{B^3}\frac{t}{\varepsilon} + 1\right)^{0,1}.$$

(4.147)

The law of spreading (4.147) coincides with that obtained in Section 4.3 by a more rigorous method. However, the constant B that appears in the spreading law (4.147) and determines the point of matching of the outer and inner solutions differs from that found in Section 4.3. We assume that, as in Section 4.3, at the point of matching, $h^3 = \varepsilon$. In this case, the right and left sides of Eq. (4.140) have identical orders. Using this condition, we conclude

$$h_0^3 \psi^3(-\mu_*) = h_0^3 \psi_*^3 = \varepsilon.$$

(4.148)

Taking as the initial condition for the velocity $U(0) = 2$ as in Section 4.3, we conclude from this Eqs. (4.146)–(4.148) the identical to Section relation (4.138): $\varepsilon = 1/(2B^3)$.

The solution of Eq. (4.124) has the asymptotic form (see Section 4.2)

$$\psi^3 \approx 3|\eta|^3 \ln|\eta|$$

$$\left(\frac{d\psi}{d\eta}\right)^3 \approx -3\ln|\eta|, \eta \to -\infty. \tag{4.149}$$

Now using Eqs. (4.145), (4.148) and (4.149), we conclude

$$\psi_*^3 = \varepsilon U / h_0^3 = 3\mu_*^3 \ln\eta_* = B^3\mu_*^3 = B^3 \exp(B^3). \tag{}$$

For the determination of the constant B, we can limit ourselves in Eq. (4.148) to the initial moment of time; that is, we can set $U = U(0) = 2$. Then the value of B satisfies the relationship

$$B^6 \exp\left(B^3\right) = 1/h_0^3. \tag{4.150}$$

Both Eqs. (4.150) and (4.138) determine the constant B in a similar fashion. However, Eq. (4.138) was deduced using more rigorous procedure as compared with Eq. (4.150) in Section 4.3. Note, the right hand site of Eq. (4.150) is exactly twice as large as the right hand side in Eq. (4.138).

Let us pass on now to the case of spreading of a liquid drop over a dry surface in the case of complete wetting, that is, with the Derjaguin's pressure in the following form: $\Pi(h) = A/h^n$ ($A > 0$, $n \geq 2$). In this case, as was shown in Section 4.2, the spreading equation is given by Eq. (4.63).

We now change over, the same as previously, to a quasi-steady-state variable (4.139). This is substituted into Eq. (4.63), which gives

$$\varepsilon U \frac{dh}{d\xi} = \frac{1}{\xi}\frac{d}{d\xi}\left[\xi h^3\left(\frac{d}{d\xi}\left(\frac{1}{\xi}\frac{d}{d\xi}\left(\xi\frac{dh}{d\xi}\right)\right) - \frac{\lambda}{h^{n+1}}\frac{dh}{d\xi}\right)\right]. \tag{4.151}$$

At $\varepsilon = \lambda = 0$, we obtain the outer solution of Eq. (4.151) similar to Eq. (4.68), which is valid far from the moving three-phase contact line

$$h = \frac{4}{r_0^2}\left(1 - \xi^2/r_0^2\right). \tag{4.152}$$

In the vicinity of the point $\xi = 0$, which corresponds to the apparent three-phase contact line, we carry out a replacement of variables in (4.142), in which h_0 is now a new unknown function to be determined.

The requirement of self-similarity of Eq. (4.151) results in the following conditions:

$$\frac{\varepsilon U r_0^3 \chi}{h_0^3} = 1,$$

$$\frac{\lambda r_0^2 \chi}{h_0^{n+1}} = 1. \tag{4.153}$$

and the equation itself for the inner solution $\psi(\mu)$ describing the drop surface profile in the vicinity of the moving apparent three-phase contact line takes the form

$$\frac{d\psi}{d\mu} = \frac{d}{d\mu}\left(\psi^3\left(\frac{d^3\psi}{d\mu^3} - \frac{1}{\psi^{n+1}}\frac{d\psi}{d\mu}\right)\right)$$

or, after integration, considering that $\psi \to 0$, $\mu \to +\infty$,

$$\psi^2 \left(\frac{d^3\psi}{d\mu^3} - \frac{1}{\psi^{n+1}} \frac{d\psi}{d\mu} \right) = 1. \tag{4.154}$$

Equation (4.154) was investigated in details in Section 4.2 (see also the end of Section 4.1).

The condition of matching of the inner and outer solutions, as above, has the form (4.145) and it leads, after taking into account that $U = dr_0/dt$, to the expression (4.147) for the spreading radius r_0. Expressions for the unknown functions $h_0(t)$ and $\chi(t)$, determining the scale of the zone of the inner solution, can be found easily using Eqs. (4.153) and (4.147) (see Section 4.2 for details).

We focus below on determining the constant B, which according to Eq. (4.148), determines the point of matching of the solutions, and, in view of Eq. (4.138), determines the magnitude of the small parameter ε.

Equation (4.154), as $\mu \to -\infty$, has the previous asymptotic behavior as Eq. (4.149). Hence, in the same way as previously, we deduce

$$\psi_* = (\varepsilon U)^{1/3}/h_0 = B \exp(B^3/3). \tag{4.155}$$

Substituting into Eq. (4.153) $r_0 = 2$ and $U = 2$ (values corresponding to the initial moment of spreading), we now obtain

$$h_0^{n-1} = \lambda/(\varepsilon U)^{2/3} = \lambda B^2.$$

After that, using Eq. (4.155) and taking into account that $\varepsilon = 1/B^3$, we obtain

$$\frac{1}{B} \left(\lambda B^2 \right)^{-\frac{1}{n-1}} = B \exp(B^3/3). \tag{4.156}$$

Equation (4.156) for the determination of the constant B differs from Eq. (4.89), which was obtained by a more rigorous method in Section 4.2, only in the absence of the factor $1/^3\sqrt{2}$ in the left-hand side. Comparison of numerical results obtained using the quasi-steady-state approach, Eq. (4.156), with those obtained in Section 4.2, Eq. (4.89), for the examples that were considered in Section 4.2 gives almost identical numerical values for both $n = 2$ and $n = 3$ for both B and ε.

Thus, in this particular problem as well, the proposed quasi-steady-state method of investigation of spreading has led to results that are essentially not different from those obtained previously by a more rigorous method.

Now let us apply the proposed method to the solution of the problem of liquid drop spreading on a horizontal substrate under the influence of gravitational forces. Neglecting capillary forces and considering the spreading of a two-dimensional (cylindrical) drop (along the OX axis), we write the spreading equation in dimensionless form, according to Section 4.1, as

$$\frac{\partial h}{\partial t} = \beta \frac{\partial}{\partial x} \left(h^3 \frac{\partial h}{\partial x} \right), \tag{4.157}$$

where $\beta = \rho g h_*^3 t_*/(\eta r_*^2)$; ρ is the density of the liquid; and g is the acceleration of gravity. The conservation of volume, V, in this case has the form

$$2\int_0^{x_0} h\,dx = V, \tag{4.158}$$

where x_0 is the spreading radius; and V is the volume per unit length of the spreading droplet. Selection

$$h_* = \frac{V}{2r_*},$$

results in

$$\int_0^{x_0} h\,dx = 1,$$ (4.159)

and $\beta = \rho g V^3 t_* / (8\eta r_*^8)$, $x_0(t)$ is now the dimensionless radius of spreading.

Now performing the replacement of variables according to (4.139) we obtain from Eq. (4.157):

$$-U\frac{dh}{d\xi} = \beta\frac{d}{d\xi}\left(h^3\frac{dh}{d\xi}\right),$$

or, after integration

$$h^2\frac{dh}{d\xi} = -\frac{U}{\beta}.$$

where

$$h = -\left(\frac{3U}{\beta}\right)^{1/3}\xi^{1/3}.$$

Using the condition of conservation of volume, we obtain

$$1 = \int_{-x_0}^0 h\,d\xi = \frac{3}{4}\left(\frac{3U}{\beta}\right)^{1/3} x_0^{4/3}.$$

Hence, from this equation we conclude

$$U = \dot{x}_0 = \frac{3}{4}\beta/\left(3x_0^4\right),$$

and the drop spreading law has the following form:

$$x_0 = k\left(\beta t\right)^{1/s}, k = \left(\frac{320}{81}\right)^{1/s} \approx 1.316.$$ (4.160)

The shape of the drop surface is determined here by the equation

$$h = 4\left(x_0 - x\right)^{1/3}\left(3x_0^{4/3}\right).$$ (4.161)

The spreading law (4.160) differs from the self-similar solution obtained in ref. [21] only in the value of the constant coefficient, $k = 1.411$ instead of $k = 1.316$.

Now let the drop spreading take place under the influence of a temperature gradient only. We consider the spreading of a liquid droplet using a low slope approximation, that is $h_* \ll r_*$ and only spreading of a two-dimensional droplet is considered.

The Navier-Stokes equations in the case under consideration take the following form:

$$\frac{\partial P}{\partial x} = \eta\frac{\partial^2 v_x}{\partial z^2}$$

$$\frac{\partial P}{\partial z} = 0$$

These equations give

$$P = P(x), \quad v_x = \frac{1}{2\eta}\frac{\partial P}{\partial x}z^2 + C_1 z + C_2, \tag{4.162}$$

with the following two boundary conditions: the non-slip condition on the solid substrate,

$$v_x(0) = 0, \tag{4.163}$$

and the condition of the applied tangential stress on the droplet surface, caused by the surface tension gradient, which in turn is caused by the applied temperature gradient:

$$\eta \left.\frac{\partial v_x}{\partial z}\right|_{z=h} = \frac{\partial \gamma}{\partial x} = \frac{d\gamma}{dT}\frac{\partial T}{\partial x} = \Lambda. \tag{4.164}$$

Note, that in this condition we assume that the thickness of the droplet is small enough and the temperature is constant through the vertical cross section of the droplet and equal to the temperature of the solid support. Note, usually

$$\frac{d\gamma}{dT} < 0,$$

that is, the liquid–air interfacial tension decreases with temperature. Below we consider the spreading of the droplet from the cold side in the center of the droplet to the hotter part of the solid substrate under the action of the constant temperature gradient. It is also assumed that in the range of temperature under consideration the derivative, $d\gamma/dT$, is a negative constant. Hence, under above assumption $\Lambda < 0$ and remains constant.

Using these two boundary conditions, we conclude from Eq. (4.162)

$$v_x = \frac{1}{\eta}\frac{\partial P}{\partial x}\left(\frac{z^2}{2} - zh\right) + \frac{\Lambda}{\eta}z. \tag{4.165}$$

The governing equation results from the integration of the continuity equation is

$$\frac{\partial h}{\partial t} = -\frac{\partial}{\partial x}\int_0^h v_x dz. \tag{4.166}$$

Substitution of expression for the velocity according to (4.165) into the governing equation (4.166) yields

$$\frac{\partial h}{\partial t} = \frac{1}{\eta}\frac{\partial}{\partial x}\left(\frac{h^3}{3}\frac{\partial P}{\partial x} - \frac{\Lambda h^2}{2}\right) \tag{4.167}$$

With the conservation law according to Eq. (4.158). The pressure inside the spreading droplet is given by Eq. (4.5). However, below we concentrate on the spreading under the action of the temperature gradient only. That is, we assume that

$$\frac{2h_*}{r_*}P_* \ll \Lambda.$$

Below we consider the negative values of the constant Λ.

In this case, Eq. (4.167) becomes

$$\frac{\partial h}{\partial t} = \alpha \frac{\partial h^2}{\partial x},$$

(4.168)

where

$$\alpha = \frac{|\Lambda| h_* t_*}{2\eta r_*}.$$

After the replacement of the variable according to Eq. (4.139), we obtain from this equation

$$U \frac{dh}{d\xi} = \alpha \frac{dh^2}{dx},$$

(4.169)

or after integration

$$Uh = \alpha h^2 + C.$$

(4.170)

In the situation under consideration, the drop surface has an abrupt change at the point $x = x_0 (\xi = 0)$ corresponding to the liquid propagation front. Following the method described in Section 5.6, we obtain the condition that must be satisfied by the solution of Eq. (4.168) at the moving front.

We integrate Eq. (4.168) over x from $x_- = x_0 - \delta$ to $x_+ = x_0 + \delta$, where δ is a small value:

$$\int_{x_-}^{x_+} \frac{\partial h}{\partial t} dx = \alpha \left(h_+^2 - h_-^2 \right).$$

Let us calculate

$$\frac{d}{dt} \int_{x_-}^{x_+} h\, dx = \dot{x}_0 \left(h_+ - h_- \right) + \int_{x_-}^{x_+} \frac{\partial h}{\partial t} dx.$$

Hence,

$$\int_{x_-}^{x_+} \frac{\partial h}{\partial t} dx = -\frac{d}{dt} \int_{x_-}^{x_+} h\, dx + \dot{x}_0 (h_+ - h_-),$$

where $h_+ = h(x_+)$ and $h_- = h(x_-)$ are thickness of the droplet just above and behind the moving front.

Taking the limit in this equation at $x_- \to x_0$, $x_+ \to x_0$ (i.e., $\delta \to 0$), we obtain

$$\dot{x}_0 \left(h_+ - h_- \right) = \alpha \left(h_+^2 - h_-^2 \right).$$

In front of the moving droplet, the solid substrate is dry, that is, $h_+ = 0$. Hence, from this equation we conclude

$$\dot{x}_0 = \alpha h_-.$$

(4.171)

Equation (4.171) upon substitution into Eq. (4.170), gives $C = 0$, and thus the drop has the form of a flat step ("pancake") with a height

$$h = U / \alpha,$$

(4.172)

and a radius of the moving front $x_0(t)$. Equation (4.172) shows that the thickness of the flat drop is a function of time only.

From the conservation law (4.159), we conclude now

$$1 = \int_0^{x_0} h\,dx = \int_{-x_0}^0 h\,d\xi = Ux_0/\alpha.$$

Taking into account $U = \dot{x}_0 = \alpha/x_0$ and $h = 1/x_0$, we get finally the spreading law as

$$x_0 = \sqrt{2\alpha t} \tag{4.173}$$

Summarizing the results obtained in this section, we can say that the proposed approximate quasi-steady-state method for solving spreading problems gives a satisfactory accuracy of solution in all of the cases examined in this section. The method itself is mathematically simple, and it offers a means for reducing a problem of complex nonlinear partial differential equations to the solution of ordinary differential equations. The quasi-steady-state approach that we have examined in this section can be used for the investigation of more complex spreading problems.

4.5 Dynamic Advancing Contact Angle and the Form of the Moving Meniscus in Flat Capillaries in the Case of Complete Wetting

In this section we analyze how both the Derjaguin's pressure and the width of the capillary influence the *dynamic* advancing contact angle of the meniscus of a liquid completely wetting the capillary walls. For simplicity, we consider that the meniscus moves from an equilibrium position in a *flat* capillary. The effect of both the Derjaguin's pressure and the width of the capillary on the dynamic contact angle is investigated following [22]. The problem of the form of the advancing meniscus and the dynamic contact angles has been considered earlier [23–25], neglecting the Derjaguin's pressure of the thin layer of liquids and limiting the discussion only to capillary forces. However, in the close vicinity of the moving apparent three-phase contact line the thickness on the liquid is so thin that the influence of the Derjaguin's pressure becomes significant.

The thickness of the film on the walls of the capillary ahead of the advancing meniscus was assumed to be arbitrary in [23–25]. This means that, up to the start of the flow, the system is not in equilibrium and that, consequently, there is flow from the meniscus to the film or the contrary, not connected with the dynamics of the meniscus. These flows were also neglected in solution of the problems in [23–25].

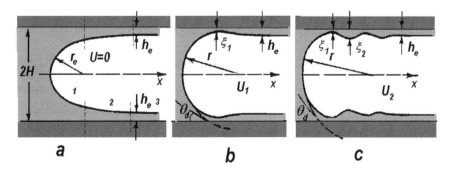

FIGURE 4.9 Schematic presentation of the profile of the meniscus in a flat capillary. (a) at equilibrium: (1) the spherical meniscus, (2) the transition zone, and (3) flat equilibrium film; (b) velocity of motion below the critical velocity, ξ_1—the only minimum on the liquid profile; and (c) velocity of motion is above the critical velocity, ξ_1 and ξ_2 are the thickness of the first minimum and maximum on the liquid profile.

As shown in Section 2.2, in a state of equilibrium (Figure 4.9a), the capillary pressure of the meniscus, P_e, relates to the isotherm of the Derjaguin's pressure of flat wetting films, $\Pi(h)$, by the following relationship:

$$P_e = \left(\gamma + \int_{h_e}^{\infty} \Pi dh \right) / (H - h_e) = \Pi_e, \tag{4.174}$$

where γ is the surface tension of the liquid; H is the half-width of the capillary; h_e is the thickness of the equilibrium film on the solid surfaces corresponding to $\Pi(h_e) = P_e$.

The relative pressure of the vapor, p/p_s, above the films and the meniscus is related to the equilibrium pressure P_e according to Eq. (4.7) in Chapter 1. In sufficiently wide capillaries ($H \gg h_e$), the meniscus in a central region has a spherical shape with a constant radius of curvature $r_e = \gamma/P_e$.

The further solution is carried out for the case of complete wetting, that is, for isotherms of the Derjaguin's pressure of the type

$$\Pi = A/h^n > 0, \tag{4.175}$$

where $n \geq 2$ and A is the constant of the surface forces (Hamaker constant in the case $n = 3$). In Figure 4.9 schemes illustrating the profiles of the meniscus in a flat capillary in a state of equilibrium (*a*) and with different rates of motion (*b*, *c*) are presented.

In Section 2.4 we showed that the radius of curvature of an equilibrium meniscus $r_e < H$, and in this case of complete wetting, is equal to

$$r_e = \frac{\gamma}{P_e} = \left[H - \left(\frac{n}{h-1} \right) h_e \right].$$

The form of the equilibrium profile of the liquid in the transitional zone between the meniscus and the film for isotherms of the type (4.175) was investigated in Section 2.4.

Let us consider the motion of the advancing meniscus in a flat capillary (Figure 4.9b and c) from a state of equilibrium (Figure 4.9a). The zone of the flow in which the main hydrodynamic resistance is exerted takes in a region of the thicknesses of the layer $h(x)$ above the surface of the substrate on the order of h_e. Axis x is directed along the capillary axis (Figure 4.9).

According to the introduction to Chapter 4, the profile of the moving meniscus (Figure 4.9b and c) can be subdivided at a small capillary number, $Ca \ll 1$, into two regions: an outer region, where the meniscus has a spherical shape of an unknown radius of the curvature, r, which is the function of the velocity of motion (or the same, capillary number, Ca); and an inner region between the spherical meniscus and the initial equilibrium flat film. In this region, according to our estimations in the introduction to Chapter 4, the curvature of the meniscus is small and low slope approximation can be used.

In the case under consideration, the equation of spreading (4.62) from Section 4.2 should be rewritten as

$$\frac{\partial h}{\partial t} = -\frac{\gamma}{3\eta} \frac{\partial}{\partial x} \left[h^3 \frac{\partial h}{\partial x} - \frac{nA}{\gamma h^{n-2}} \frac{\partial h}{\partial r} \right],$$

with the boundary condition

$$h \to h_e, \quad x \to \infty.$$

At steady-state motion, we can introduce a new coordinate system moving with the meniscus $y = x - Ut$, which results in

$$U \frac{dh}{dy} = \frac{\gamma}{3\eta} \frac{d}{dy} \left[h^3 \frac{dh}{dy} - \frac{nA}{\gamma h^{n-2}} \frac{dh}{dy} \right], \tag{4.176}$$

with the boundary condition

$$h \to h_e, \quad y \to \infty. \tag{4.177}$$

Integration of Eq. (4.176) with boundary condition (4.177) yields

$$\frac{\gamma h^3}{3\eta}\left(h''' - \frac{nA}{\gamma h^{n+1}}h'\right) = U\left(h - h_e\right), \tag{4.178}$$

where $'$ means differentiation with y.

Let us introduce the following dimensionless variables $\xi = h/h_e$; $z = y/y_*$, ξ' and ξ''' means differentiation with z,

$$y_* = h_e\left(\gamma/3\eta U\right)^{1/3} = \frac{h_e}{(3Ca)^{1/3}},$$

and the only one dimensionless parameter, which characterizes the intensity of the Derjaguin's pressure action:

$$\alpha = \frac{nA^{1/n}P_e^{(n-1)/n}}{\gamma^{1/3}\left(3\eta U\right)^{2/3}}. \tag{4.179}$$

It is important to note that using these scales, we get the following estimation of the derivative in the flow zone:

$$h' \sim \frac{h_e}{y_*} = (3Ca)^{1/3} \ll 1,$$

and the low slope condition is satisfied in the flow zone in the vicinity of the apparent moving three-phase contact line as we predicted in the introduction to Chapter 4. We should remind the reader that in the case under consideration, $Ca \sim 10^{-6}$, hence, $(Ca)^{1/3} \sim 0.01 \ll 1$.

Substituting this dimensionless variables into Eq. (4.178) results in

$$\xi^3\xi''' - a\xi'\xi^{2-n} = \xi - 1. \tag{4.180}$$

In Appendix 5 we show (see Eq. A5.9 in Appendix 5) that Eq. (4.180) has the proper asymptotic behavior at $z \to -\infty$. In this sense, the problem under consideration is completely mathematically correct: It is possible to match asymptotic expansions in the outer region (meniscus) and inner region in the vicinity of the moving apparent three-phase contact line. Below in this sections we undertake the matching procedure numerically.

Note an unusual feature of Eq. (4.180): It does not have a proper asymptotic behavior in the case of a droplet (see Section 4.2) but it does in the case of a moving meniscus.

Before presenting the results of the numerical calculations according to Eq. (4.180), let us analyze this equation. Let us consider the behavior of the dimensionless thickness, ξ, far from the meniscus (i.e., at $1 - \xi \ll 1$), introducing $\upsilon = \xi - 1$ and the linearization of Eq. (4.180) results in:

$$\upsilon''' - \alpha\upsilon' = 0. \tag{4.181}$$

We seek the solution of Eq. (4.181) in the form $\upsilon = \exp(\lambda x)$. Substitution of this expression into Eq. (4.181) yields

$$\lambda^3 - \alpha\lambda - 1 = 0. \tag{4.182}$$

The discriminant of this equation $Q = (1/4) - (\alpha/3)^3$ can be either positive or negative. If $Q > 0$ (i.e., with $\alpha < \alpha_c \cong 1.89$), Eq. (4.182) has one real positive root and two conjugate complex roots, whose real parts are negative. This corresponds to the presence of damped waves ahead of the moving meniscus. Such a situation (Figure 4.9c) is possible with $U > U_c$ (with A and H = const), or with $A < A_c$ (with U and H = const), or, finally, with $H > H_c$ (with U and A = const). With a decrease in α (with $\alpha < \alpha_c$), the imaginary part of the roots rises monotonically, and the real part decreases, which corresponds to an increase in the amplitude and a decrease in the length of the surface waves.

At $Q < 0$ (i.e., with $\alpha > \alpha_c$), Eq. (4.182) has two negative real roots: $-\lambda_1$ and $-\lambda_2$ ($\lambda_1 > \lambda_2 > 0$) and one positive real root, $\lambda_3 > 0$. These three roots, as functions of α, have the following properties: $\lambda_1 < \sqrt{\alpha}$, $\lambda_2 < \sqrt{\alpha}$. At $\alpha \to \infty$, $\lambda_2 \to \sqrt{\alpha}$ and $\lambda_2 \approx 1/\alpha$. It follows from this that, at $\alpha < \alpha_c$, the function $\xi(x)$ has a single minimum at the point $z \cong (2/\sqrt{\alpha})\ln\sqrt{\alpha}$, which corresponds to the situation illustrated in Figure 4.9b. Thus, profiles of this type, as in Figure 4.9b, are realized with $\alpha > \alpha_c$, that is, with rather large values of A, and/or at a small velocity of motion of the meniscus U, or in narrow capillaries.

With an increase in the velocity of the flow or a decrease in the value of A such that α becomes less than α_c, wavy films are formed ahead of the moving meniscus, as in Figure 4.9c. Thus, if the effect of surface forces is neglected ($A = 0$), that is, $\alpha = 0$, the solution of Eq. (4.8) can only have wavy profiles of the film in front of the moving meniscus, as has been obtained earlier in [24]. The effect of surface forces results (at $\alpha < \alpha_c$) in damping of the waves and to an increase in the wavelength (still at $\alpha < \alpha_c$) ahead of the moving meniscus. This consideration shows the qualitative effect of the influence of the Derjaguin's pressure of thin layers of liquids on hydrodynamics of moving of menisci.

Figure 4.10a gives the numerically calculated values of H/r [calculated according to Eq. (4.180)] as a function of the parameter $(Ca)^{2/3}$ for different values of the width of the capillaries H. The values of r are located along a section of constant curvature, adjacent to the zone of the flow, appearing with a rise in the value of the thickness, h, but still in the region of the low sloped profile. The transition to this section corresponds to the condition $d^2h/dy^2 = \gamma/r = $ const. At $H/r < 1$, the values of $H/r = \cos\theta_d$, where θ_d is the dynamic contact angle (Figure 4.9b and c). Calculations were made for the following parameters $A = 10^{-7}$ dyn and $n = 2$, characteristic for the value of β-films of water on the surface of quartz [26]. The corresponding value of γ was taken equal to 72 dyn/cm and the viscosity $\eta = 0.01 P$.

Figure 4.10b gives similar results for another isotherm $\Pi(h)$, characteristic for wetting films of nonpolar liquids on a solid dielectric [26]: $A = 10^{-4}$ erg, $n = 3$, $\gamma = 30$ dyn/cm, $\eta = 0.01 P$.

Figure 4.10 shows that the calculated dependences of $\cos\theta_d$ on Ca (curves 1–3) are similar to those calculated using the approximate Fritz' equation [24]: $\tan\theta_d \cong 2.36\sqrt{Ca}$ (curve 4). However, in distinction

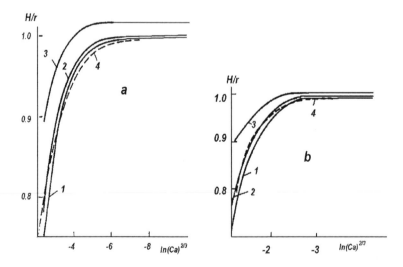

FIGURE 4.10 Dependence of the ratio H/r on the capillary number, $Ca = \frac{U\eta}{\gamma}$. (a) The Derjaguin's pressure isotherm $\Pi(h) = 10^{-7}/h^2$, (1) $H = 10^{-2}$ cm, $h_e = 360$ Å; (2) $H = 1.25.10^{-3}$, $h_e = 125$ Å; and (3) $H = 10^{-5}$ cm, $h_e = 11$ Å; (b) The Derjaguin's pressure isotherm $\Pi(h) = 10^{-14}/h^3$, (1) $H = 10^{-2}$ cm, $h_e = 150$ Å; (2) $H = 1.25.10^{-3}$ cm, $h_e = 74$ Å; and (3) $H = 1.25.10^{-5}$ cm, $h_e = 16$ Å. The dotted line 4 corresponds to the Fritz's equation. (From Friz, G., *Angew Z. Phys.*, 19, 374, 1965.)

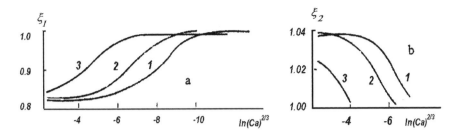

FIGURE 4.11 Dependences of the thickness of the first minimum, $\xi_1 = h_1 / h_e$ (Figure 4.9c), (a) and the first maximum, $\xi_2 = h_2 / h_e$ (Figure 4.9c); (b) on Ca, obtained for values of (1) $H = 10^{-2}$ cm, (2) 10^{-3} cm, and (3) 10^{-5} cm. The parameters of the isotherm: $A = 10^{-5}$ dyn, $n = 2$.

from the Fritz' calculations, our calculations show a dependence of θ_d on the width of the capillary. This becomes particularly noticeable "in narrow capillaries," when $H < 10^{-3}$ cm, where the thickness of the surface film becomes less than 100 Å. In the general case, an increase in H, as can be seen from Figure 4.10, leads to a decrease in $\cos\theta_d$, that is, to a rise in θ_d.

Figure 4.11 presents the dependence on Ca of the characteristic thickness inside the flowing zone: $\xi_1 = h_1/h_e$ is the point of the first minimum and $\xi_2 = h_2/h_e$ is the point of the first maximum. As can be seen from these curves, the values of ξ_1 fall with an increase in the rate of motion of the meniscus. Under these circumstances a lowering of the thickness of the film near the moving meniscus is observed at lower rates of motion the wider the capillary. Thus, in a capillary $H = 10^{-2}$ cm, the thinning of the film near the meniscus starts with $Ca > 10^{-6}$ (for water, at $U > 10^{-2}$ cm/s).

The relative height of the first maximum of ξ_2 on the contrary, rises with an increase in Ca, that is, with a rise in the rate of motion of the meniscus. However, the appearance of convexity of the film is observed with considerably greater velocities U than the appearance of concavity.

It is important to note that, with a further rise in the value of the velocity of the meniscus, U, the values of ξ_1 and ξ_2 are stabilized, that is, the first wave barely changes its form; an increase in U is accompanied by a change in the form of the second, third, etc. waves, propagating ahead of the meniscus. Under these circumstances, the value of α, corresponding to a transition to a wavy profile, found from the values of $(3Ca)^{2/3}$, with $\xi_2 = 0$ by extrapolation of the curve of $\xi_2 (Ca)$ in Figure 4.11b, were found to lie in the interval from 1.5 to 2.0, that is, they were close to the values of $\alpha_c \cong 1.89$, obtained with a linearized preliminary analysis of Eq. (4.180).

As can be seen from Figure 4.11, the form of the wavy film ahead of a moving meniscus depends more strongly than $\cos\theta_d$ on the width of the capillary, H, and the thickness of the equilibrium film, h_e in front of the moving meniscus. An increase in the value of the constant A, signifying an increase in the effect of the surface forces, leads (with $H = $ const) to a rise in ξ_1 and to a lowering of ξ_2, with identical values of Ca. Thus, the effect of surface forces, as followed from the preliminary analysis of Eq. (4.180), leads to damping of the waves and makes the profile of the film of the moving meniscus smoother.

The calculations made offer the possibility of evaluating the effect of surface forces on the dynamic contact angles θ_d and the profile of the film ahead of the moving meniscus for isotherms of the Derjaguin's pressure, corresponding to complete wetting.

APPENDIX 5

Asymptotic Behavior of Solution of $y^3 \dfrac{d^3y}{dx^3} = y - 1$ *at* $y \to \infty$

At a sufficiently big thickness, we can neglect the Derjaguin's pressure action in Eq. (4.180). Hence, we come to the following differential equation:

$$y^3 \frac{d^3y}{dx^3} = y - 1, \qquad (A5.1)$$

with the following boundary conditions:

$$y \rightarrow +\infty, \; x \rightarrow -\infty. \tag{A5.2}$$

According to the boundary condition (A5.2) this equation asymptotically can be written as

$$y^2 \frac{d^3 y}{dx^3} = 1. \tag{A5.3}$$

Let us introduce a new unknown function $w(y)$ in Eq. (A5.3)

$$w(y) = \frac{1}{2} \left(\frac{dy}{dx} \right)^2. \tag{A5.4}$$

Hence,

$$\frac{dw}{dy} = \frac{d^2 y}{dx^2}, \quad \frac{d^2 w}{dy^2} = -\frac{1}{\sqrt{2w}} \frac{d^3 y}{dx^3}.$$

Substitution of this expression into Eq. (A5.3) results in

$$y^2 \frac{d^2 w}{dy^2} = -\frac{1}{\sqrt{2w}}. \tag{A5.5}$$

Let us introduce a new variable in this equation:

$$\frac{dw}{d\xi} = p(w), \quad \xi = \ln y. \tag{A5.6}$$

Using the new variable in Eq. (A5.6), we arrive to

$$p \frac{dp}{dw} - p = -\frac{1}{\sqrt{2w}}. \tag{A5.7}$$

Solution of Eq. (A5.7) can be expressed via cylindrical functions. However, for the investigation of the asymptotic behavior of Eq. (A5.2), we limit ourselves by considering only main power terms in asymptotic expansion:

$$p = Cw^k. \tag{A5.8}$$

Substitution of this expression into Eq. (A5.7) results in

$$C^2 w^{2k-1} = Ckw^k - \frac{w^{-1/2}}{\sqrt{2}}.$$

Equalizing different exponents in this equation, we conclude that only three options are available for the exponent k: (i) $k = 1$, (ii) $k = 1/4$, (iii) $k = -1/2$.

At $k = 1$, we conclude from Eq. (A5.7): $C = 1$, hence, $p = w$. Substitution of this expression into Eq. (A5.6) gives $w' = w$, or $w = 2C_1 y$, where $C_1 > 0$ is the integration constant. Next step is the substitution into Eq. (A5.4), which results in $y' = \sqrt{4C_1 y}$. Solution of this equation is

$$y = C_1 (x + C_2)^2, \tag{A5.9}$$

where C_2 is a new integration constant. Note, the constant C_1 can be only positive. We use the asymptotic solution (A5.9) in the current section (Section 4.5) and in Section 4.6.

The second solution (at $k = 1/4$) does not satisfy the requirement (A5.2).

Let us consider now the last, third exponent $k = -1/2$, which results, according to Eq. (A5.8), in

$$p = \frac{1}{2} w^{-1/2}. \tag{A5.10}$$

Substitution of this expression into Eqs. (A5.10) and (A5.6) results in

$$w = \frac{1}{2} (3 \ln y)^{2/3}.$$

Casting this expression into Eq. (A5.4) yields

$$\frac{dy}{dx} = -3^{1/3} \ln^{1/3} y \tag{A5.11}$$

Integration of this equation results in

$$\int \frac{dy}{\ln^{1/3} y} = -3^{1/3} x.$$

Integration by parts gives

$$\int \frac{dy}{\ln^{1/3} y} = \frac{y}{\ln^{1/3} y} + \frac{1}{3} \int \frac{dy}{\ln^{4/3} y}.$$

At $y \to +\infty$ the following inequality holds: $\ln^{4/3} y \gg \ln^{1/3} y$. Using this equation can be simplified as

$$\frac{y}{\ln^{1/3} y} = -3^{1/3} x. \tag{A5.12}$$

Eq. (A5.12) gives a possibility of deducing an explicit asymptotic behavior of function $y(x)$:

$$y = -3^{1/3} x \ln^{1/3} y = 3^{1/3} |x| \ln^{1/3} \left(3^{1/3} |x| \ln^{1/3} y \right).$$

Considering that $|x| \gg \ln^{1/3} y$, we arrive to

$$y = 3^{1/3} |x| \ln^{1/3} |x|, \qquad x \to -\infty. \tag{A5.13}$$

Direct differentiation of Eq. (A5.13) gives

$$\frac{dy}{dx} = 3^{1/3} \ln^{1/3} |x|, \qquad x \to -\infty. \tag{A5.14}$$

We used this asymptotic behavior in Section 4.2.

4.6 Motion of Long Drops in Thin Capillaries in the Case of Complete Wetting

By now the reader is expected to be familiar with the fact that for nonpolar liquids we can use the Derjaguin's pressure isotherms $\Pi(h) = A/h^n$, $A > 0$, $n \geq 2$ (h is the thickness of the film). Such isotherms pertain to the case of complete wetting. In this section, we consider the motion of long oil drops or air bubbles in thin capillaries [27–29].

Let us consider the motion of a long drop/bubble in a thin capillary (gravity action is neglected) (Figure 4.12). Under the action of applied pressure difference $p^- - p^+ > 0$ the drop/bubble moves from left to right with velocity U to be determined as a function of the applied pressure difference. Note, the velocity U is different from the average Poiseuille velocity because the drag force in the system presented in Figure 4.12 is different from the drag force in the same capillary completely filled with the liquid 1.

We consider below a relatively slow motion, when the capillary number,

$$Ca = \frac{U\eta_1}{\gamma} \ll 1,$$

where η_1 is the viscosity of the liquid 1 in the capillary, γ is an interfacial tension, which can be substantially different from liquid–air interfacial tension. This interfacial tension can be used only in the case of the motion of an air bubble. We will show that the viscosity of liquid 2 inside the drop/bubble, η_2, does not play any significant role and can be usually omitted. However, the interfacial tension in the case liquid bubble is still very important as well as its difference from the liquid–air interfacial tension.

Let us estimate the velocity of the motion in the case $Ca \sim 1$: if $\eta_1 \approx 10^{-2} P$, $\gamma \approx 72$ dyn/cm, then the corresponding velocity $U \approx 7200$ cm/sec = 72 m/sec. Only under very special conditions such huge velocity can be achieved in a sufficiently thin capillary with radius less than 0.1 cm. That means we can safely consider $Ca \ll 1$.

As we discussed in the introduction to Chapter 4, in this case,

1. Advancing meniscus EE′ has a constant curvature, up to the zone of the flow, ED, where the low slope approximation is valid, the curvature of the advancing meniscus EE′ is a function of the capillary number Ca to be determined,
2. The same is valid for the receding meniscus BB′ and the zone of flow CB. It is obvious that the curvature of the receding meniscus BB′ is smaller than the curvature of the advancing meniscus EE′.

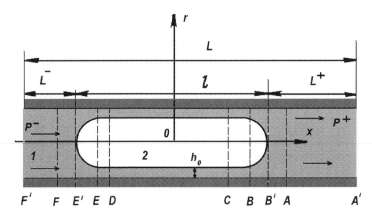

FIGURE 4.12 Schematic presentation of a motion of a drop/bubble in a capillary under the action of applied pressure difference $p^- - p^+ > 0$, that is, the motion from left to right. (1) Liquid in a capillary of length L, and (2) drop/bubble of the length ℓ. L^- and L^+ are parts of the capillary without drop/bubble; F′F and AA′ are parts of the capillary, where Poiseuille flow takes place; E and B positions of the end of spherical menisci (curvature of meniscus EE′ is bigger than the curvature of the meniscus BB′); and ED and CB transition zones from menisci to the region of DC of the film of constant thickness, h_0.

The author of [23] had studied the motion of a long drop/bubble in a capillary without allowance for the effect of the Derjaguin's pressure. The following dependence of film thickness, h_0, on drop velocity U has been deduced:

$$h_0 = 1.337 R (Ca)^{2/3}, \tag{4.183}$$

where R is the radius of the capillary. Relation (4.183) yields a zero film thickness at $Ca \to 0$. However, as understood from Chapter 2, this is impossible, because the film thickness should tends to the equilibrium thickness, h_e, at $Ca \to 0$. This equilibrium value of the film thickness should be found from the following condition (see Chapter 2) $\Pi (h_e) = P_e$, where P_e is the excess pressure under meniscus. Note, the excess pressure, P_e, in the case of capillary meniscus is positive. Hence, ignoring the action of the Derjaguin's pressure results in a wrong prediction of the film thickness at low capillary numbers. According, to Eq. (4.183) the film thickness increases unboundedly as capillary number growth. Hence, if $h_0 > t_s$, where t_s is the radius of surface forces action, then Eq. (4.183) should become valid. Hence, at low capillary numbers we should expect a substantial deviation from the prediction according to Eq. (4.183) and at high capillary numbers, Eq. (4.183) should hold asymptotically. Our calculations (see Figure 4.13) confirm the above-suggested dependency of the film thickness on the capillary number.

In the present section, we obtain the main characteristics of the motion of a drop under an applied pressure gradient taking into account the Derjaguin's pressure action.

We examine a motion of a long drop of an immiscible fluid 2 with a length ℓ (Figure 4.12) inside the cylindrical capillary of radius R filled with a fluid 1. The motion is axisymmetric, and we introduce a coordinate system connected with the moving drop. The x axis coincides with the axis of the capillary. Let ℓ be the length of the drop, L^+ and L^- the lengths of the drop free sections of the capillary, and L its overall length of the capillary, $(L = L^+ + L^- + \ell)$. The pressure difference, $p^+ - p^-$ is applied at the ends of the capillary, where $p^- > p^+$. We examine the steady motion of the drop along the capillary in the positive direction of the x axis at the velocity U to be determined.

As before, we restrict ourselves to the cases of low Reynolds and capillary numbers. We also assume that the drop is long, that is, $R/\ell \ll 1$. Let η_1 and η_2 be the viscosities of fluids 1 and 2, respectively. At $\eta_1 \approx 10^{-2}$ P, $\gamma \approx 30$ dyn/cm, $U \approx 1$ cm/sec, we have $Ca \approx 3 \cdot 10^{-4} \ll 1$. Because the value of $U \sim 1$ cm/sec can be taken as the upper bound of the velocity of the drop, we assume that the condition $Ca \ll 1$ is always satisfied.

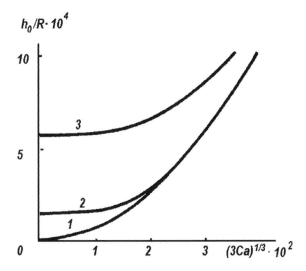

FIGURE 4.13 Dependency of the constant film thickness, h_0, inside the zone DC (Figure 4.12) on capillary number, Ca, (1) according to [23], when the Derjaguin's pressure action is ignored, (2) the Derjaguin's pressure $\frac{A}{h^3}$, and (3) the Derjaguin's pressure $\frac{A}{h^2}$.

We divide the flow field in the capillary into following regions ($A'A$ and FF'), those parts of the capillary not containing the drop, where Poiseuille flow is realized; AB and EF are spherical menisci at the ends of the drop; CD is the region with a constant film thickness, h_0, to be determined; and BC and DE are transitional regions from the constant-thickness film to menisci.

In Appendix 6, we show that flow in zone EB (both transition zones, ED and CB, and the zone of the constant film thickness, DC) nearly coincides with Eq. (4.178) deduced in the previous Section 4.5. It is assumed that the viscosity ratio of liquids 1 and 2 is of the order of 1, that is, $\bar{\eta} = (\eta_2 / \eta_1) \sim 1$. Equation (4.178), however, does not include the viscosity of the liquid inside the drop/bubble, η_2. This means that the viscosity of the inner fluid η_2 can be neglected. The proof of that is given in Appendix 6. Here we give a qualitative explanation.

As we showed in the introduction to Chapter 4, in zone EB the low slope approximation is valid. Hence, the equality of tangential stress at the liquid–liquid interface, $h(x)$, has the following form

$$\eta_1 \frac{\partial v_1}{\partial x}\bigg|_{h(x)} = \eta_2 \frac{\partial v_2}{\partial x}\bigg|_{h(x)},$$

where $v_1 = v_2 \sim U$ are velocities at the liquid–liquid interface, $h(x)$. Let us estimate the ratio of the tangential stress from the drop/bubble side to the tangential stress from the film side

$$\eta_2 \frac{\partial v_2}{\partial x}\bigg|_{h(x)} \bigg/ \eta_1 \frac{\partial v_1}{\partial x}\bigg|_{h(x)} \sim \left(\eta_2 \frac{U}{R}\right) \bigg/ \left(\eta_1 \frac{U}{h}\right) = \bar{\eta} \frac{h}{R} \ll 1.$$

This estimation shows that the flow inside the drop/bubble can be safely ignored.

Using the low slope approximation, the curvature in of the interface, $h(x)$, inside zone EB (Figure 4.12) can be written as

$$K(x) = \frac{d^2 h}{dx^2} + \frac{1}{R - h}. \tag{4.184}$$

Using this expression, we conclude from Eq. (4.178)

$$h^3 \left(\frac{d^3 h}{dx^3} + \frac{1}{(R-h)^2} \frac{dh}{dx} + \frac{\Pi'(h)}{\gamma} \frac{dh}{dx}\right) = 3Ca(h - h_0).$$

In this equation we neglect the small difference between the actual effective Derjaguin's pressure,

$$\Pi_{ef} = \frac{R}{R - h} \Pi(h),$$

where $\Pi(h)$ is the Derjaguin's pressure of the corresponding flat films. See the justification below.

In the flow zone EB (Figure 4.12), the thickness of the film is much smaller than the capillary radius, hence,

$$(R - h)^2 = R^2 \left(1 - \frac{h}{R}\right)^2 \approx R^2$$

and this equation can be rewritten as

$$h^3 \left(\frac{d^3 h}{dx^3} + \frac{1}{R^2} \frac{dh}{dx} + \frac{\Pi'(h)}{\gamma} \frac{dh}{dx}\right) = 3Ca(h - h_0). \tag{4.185}$$

Here we consider only the case of complete wetting, and we use the following isotherms of the Derjaguin's pressure:

$$\Pi(h) = \frac{A}{h^n}, \quad n = 2, 3, 4,$$

Let us introduce the following dimensionless values in Eq. (4.185): $z = x / x_*, \xi = h / h_0$, where the characteristic scale x_* is the characteristic dimension of the regions BC and DE. In this section, our choice of this scale is different from the choice in the previous section because we are going to concentrate on a transitional regime of flow from equilibrium to a relatively high velocity. That is why we select this scale equal to the length of the transition zone at the equilibrium [see Chapter 2, Section 2.3, Eq. (4.49)]: $x_* = \sqrt{h_e R}$. It is important to note that using these scales we get the following estimation of the derivative in the flow zone,

$$h' \sim \frac{h_e}{x_*} = \frac{h_e}{\sqrt{h_e R}} = \sqrt{\frac{h_e}{R}} \ll 1,$$

and the low slope condition is really satisfied in the flow zone in the vicinity of both advancing and receding menisci.

This choice shows also that

$$h''' \sim \frac{h_e}{(h_e R)^{3/2}} = \frac{1}{\sqrt{h_e R^3}} \gg \frac{1}{R^2} h' \sim \frac{h_e}{R^2 \sqrt{h_e R}} = \frac{h_e}{R} \frac{1}{\sqrt{h_e R^3}},$$

and hence, the second term in the right-hand side of Eq. (4.185) can be omitted. This means that Eq. (4.185) can be rewritten as

$$\frac{d^3 \xi}{dz^3} - \beta \frac{1}{\xi^{n+1}} \frac{d\xi}{dz} = \varepsilon \frac{\xi - 1}{\xi^3}, \tag{4.186}$$

where

$$\beta = \frac{nR}{\gamma} \frac{A}{h_0^n}, \quad \varepsilon = 3Ca(R / h_0)^{3/2}.$$

The case $\beta = 0$ corresponds to the case of filtrative motion at a "high" velocity examined in [23], where the effect of the Derjaguin's pressure was ignored.

Equation (4.186) has the solution $\xi \equiv 1$ corresponding to a film of constant thickness in the zone CD. This solution should ensure matching of this zone with the surfaces of the spherical menisci in regions AB and EF at the ends of the drop.

The matching conditions for the leading end of the drop have the form

$$\frac{d^2 \xi}{dz^2} \to \frac{R}{R^+}, \quad z \to +\infty,$$
$$\xi \to 1, \quad z \to -\infty \tag{4.187}$$

whereas for the trailing end,

$$\frac{d^2 \xi}{dz^2} \to \frac{R}{R^-}, \quad z \to -\infty.$$
$$\xi \to 1, \quad z \to +\infty \tag{4.188}$$

Satisfaction of conditions (4.187) and (4.188) is assured by the existence of asymptotic solutions of Eq. (4.186) having the form (see Appendix 5)

$$\eta \approx \frac{A^{\pm}(Ca)}{2}z^{2} + B^{\pm}(Ca), \qquad z \to \pm\infty, \qquad (4.189)$$

(the "+" sign denotes the leading meniscus, while the "−" sign denotes the trailing meniscus). The parameters $A^{\pm}$ and $B^{\pm}$, dependent on the capillary number, Ca, are determined below using the numerical solution of Eq. (4.186).

Comparing Eq. (4.189) with Eqs. (4.187) and (4.188), we conclude

$$R^{\pm} = R / A^{\pm}$$

$$h_{0} = \frac{R(A^{+}-1)}{A^{+}B^{+}} = \frac{R(A^{-}-1)}{A^{-}B^{-}}. \qquad (4.190)$$

Assuming in Eq. (4.186) that $\varepsilon = 0$ (or $Ca = 0$), we obtained an equation for the transitional zone between the film and the meniscus for the equilibrium case, which was examined in Section 2.4. For the case of complete wetting, the meniscus has the radius $R_{e}^{\pm} = R - (n/n-1)h_{e}$, where h_{e} is the thickness of the equilibrium film determined from the condition

$$\frac{A}{h_{e}^{n}} = 2\gamma / R_{e}.$$

Comparing these relations with those in (4.190), we conclude

$$\lim_{\varepsilon \to 0} A^{\pm}(\varepsilon) = 1 + \frac{n}{n-1}\frac{h_{e}}{R}$$

$$\lim_{\varepsilon \to 0} B^{\pm}(\varepsilon) = \frac{n}{n-1}. \qquad (4.191)$$

Figure 4.13 shows the dependence of film thickness h_{0} on the dimensionless velocity $(3Ca)^{1/3}$, whereas Figure 4.14 shows the dependences of the parameters $A^{\pm}, B^{\pm}$, and the radii of both the leading R^{+} and trailing R^{-} menisci on dimensionless velocity $(3Ca)^{1/3}$. In each case, we adopted the radius of the capillary and the interfacial tension $R = 10^{-2}$ cm, $\gamma = 30$ dyn/cm, respectively.

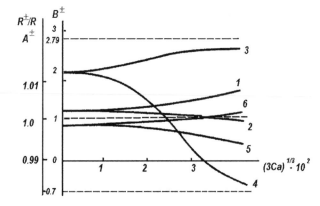

FIGURE 4.14 Dependency of parameters A^{+} (curve 1), A^{-} (curve 2), B^{+} (curve 3), B^{-} (curve 4) and the radii of the leading meniscus, R^{+} (curve 5) and the trailing meniscus, R^{-} (curve 6) on the capillary number, Ca.

Let us study the solution of Eq. (4.186) at $\varepsilon \ll 1$, having represented it in the form

$$\xi = \xi_e + 3\xi_1. \tag{4.192}$$

Using this representation, we arrive to the following equation for the zeroth approximation

$$\frac{d^3\xi_e}{dz^3} - \frac{\beta}{\xi_e^{n+1}} \frac{d\xi_e}{dz} = 0. \tag{4.193}$$

Having integrated Eq. (4.193), we obtain

$$\frac{d^2\xi_e}{dz^2} + \frac{\beta}{n\xi_e^n} = \text{const} = \frac{R}{R_e}. \tag{4.194}$$

Equation (4.194) describes the profile of the transitional zone between the film and the equilibrium meniscus and, as was shown in Section 2.4, has the asymptotic solution

$$\eta_e \sim A_e \frac{\xi^2}{2} + B_e, \tag{4.195}$$

where

$$A_e = \frac{R}{R_e} = A_e^+ = A_e^-, \quad B_e = \frac{n}{n-1} = B_e^+ = B_e^-. \tag{4.196}$$

The equation of the first approximation for (4.186) using the representation (4.192) takes the following form:

$$\xi_e^3 \frac{d}{dz}\left(\frac{d^2\xi_1}{dz^2} - \frac{\beta\xi_1}{n\eta_e^{n+1}}\right) = \xi_e - 1. \tag{4.197}$$

Having integrated this equation, we conclude

$$\frac{d^2\xi_1}{dz^2} - \frac{\beta\xi_1}{n\xi_e^{n+1}} = C^\pm - \int_\xi^{\pm\infty} \frac{\xi_e - 1}{\xi_e^3} dz, \tag{4.198}$$

where $C^\pm = \text{const}$.

At $z \to -\infty$ for the transitional region and $z \to +\infty$ for the trailing meniscus, we conclude $\xi_1 = 0$. After that we conclude from Eq. (4.198)

$$C^\pm = \int_{\mp\infty}^{\pm\infty} \frac{\xi_e - 1}{\xi_e^3} dz. \tag{4.199}$$

Given that the solution of equation (4.198) satisfies the asymptotic conditions,

$$\xi_1 = \frac{A_1^+}{2} z^2 + B_1^+, \qquad z \to +\infty, \tag{4.200}$$

for the leading meniscus and

$$\xi_1 = \frac{A_1^-}{2} z^2 + B_1^-, \qquad z \to -\infty, \tag{4.201}$$

for the trailing meniscus, we find from Eq. (4.198) that

$$A^\pm = C^\pm. \tag{4.202}$$

Equation (4.190) can be rewritten in the following form:

$$h_0 B^\pm = R\left(1 - \frac{1}{A^\pm}\right)$$

or, in a first approximation with respect to the parameter ε

$$h_e B_1^\pm = \frac{R A_1^\pm}{A_e^2}. \tag{4.203}$$

Hence,

$$B_1^\pm = \frac{R A_1^\pm}{h_e A_e^2} = \frac{R C^\pm}{h_e A_e^2}. \tag{4.204}$$

As shown in Section 2.4,

$$A_e = 1 + \frac{n}{n-1} \frac{h_e}{R}. \tag{4.205}$$

To within the leading terms, we obtain the following relation using Eqs. (4.204) and (4.205) in ε:

$$B_1^\pm = \frac{R}{h_e} C^\pm. \tag{4.206}$$

The final asymptotic representation of the parameters $A^\pm$ and $B^\pm$ takes the form

$$A^\pm = A_e + \varepsilon A_1^\pm = 1 + \frac{n}{n-1} \frac{h_e}{R} + \varepsilon C^\pm, \tag{4.207}$$

$$B^\pm = B_e + \varepsilon B^\pm = \frac{n}{n-1} + \varepsilon \frac{R}{h_e} C^\pm. \tag{4.208}$$

The constants, $C^\pm$, were found from the numerical solution of Eq. (4.194). It follows from Eqs. (4.207) and (4.208) that at $\varepsilon \ll 1$ the parameters $A^\pm$ and $B^\pm$ are linear functions of ε (or on Ca). At $\varepsilon = 0$ ($Ca = 0$), we recover the equilibrium value. At "high" velocities (Figure 4.13), the values of the parameters $B^\pm$ are close to the values of the constants obtained in [23] as expected: $B^+ \sim 2.79$, $B^- \sim -0.7$.

The thickness of the film in the region CD (Figure 4.13) can be considered equal to the equilibrium thickness h_e at $(3Ca)^{1/3} \le 8 \cdot 10^{-3}$ for a Derjaguin's pressure isotherm of the form $\Pi(h) = A/h^3$ and at $(3Ca)^{1/3} \le 10^{-2}$ for $\Pi(h) = A/h^2$. In the case of "high" velocities, the plot of the relation $h_0(Ca)$ tends asymptotically to that given by Eq. (4.183), which is valid at a sufficiently high velocity of the drop/bubble.

To close the problem, we need to establish the relationship between the applied pressure gradient, $\Delta p = p^- - p^+$, at the ends of the capillary and the velocity of the drop, U. This relation can be represented in the following form (see Appendix 6 for details):

$$\Delta p = \frac{2\gamma}{R} \left[\frac{R}{R^+} - \frac{R}{R^-} + 4Ca \frac{L^+ + L^- + \bar{\eta}\, l}{R} \right]. \tag{4.209}$$

With allowance for Eq. (4.190), we conclude from Eq. (4.209)

$$\Delta p = \frac{2\gamma}{R} \left[A^+ (Ca) - A^- (Ca) + 4Ca \frac{L^+ + L^- + \bar{\eta}\, l}{R} \right]. \tag{4.210}$$

The plot of velocity, U, according to of Eq. (4.210) obtained using the previously numerically calculated dependencies are shown in Figure 4.15, where we used:

$$R = 10^{-2}\,\text{cm}, \quad \gamma = 30\ \text{dyn/cm}, \quad \bar{\eta} = 0, \ L^- + L^+ = 4\ \text{cm}, \quad l = 1\,\text{cm}, \eta_1 = 10^{-3} P.$$

For a small ε, Eq. (4.210) takes the following form

$$\Delta p = \frac{2\gamma}{R} \left[\varepsilon\, (C^+ - C^-) + 4Ca \frac{L^+ + L^- + \bar{\eta}\, l}{R} \right], \tag{4.211}$$

or

$$\Delta p = \frac{2\gamma}{R} Ca \left[3 \left(\frac{R}{h_0} \right)^{3/2} (C^+ - C^-) + 4 \frac{L^+ + L^- + \bar{\eta}\, l}{R} \right]. \tag{4.212}$$

Equation (4.212) shows that at a small ε, the velocity U of the drop/bubble motion depends linearly on the applied pressure difference, Δp.

Let us examine the motion of a sequence k of drops/bubbles in the capillary. Let $\ell_i, i = 1, 2, \ldots, k$ be the lengths of individual drop. We assume that the influence of drops on one another can be neglected if the distance between their ends exceeds $2R$. Then for a chain of k drops, Eq. (4.210) can be rewritten as:

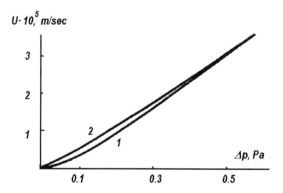

FIGURE 4.15 Dependence of velocity of motion of a single drop/bubble, U, on the applied pressure difference, Δp. (1) $\Pi(h) \equiv 0$; (2) $\Pi(h) = A/h^2$ according to Eq. (4.210).

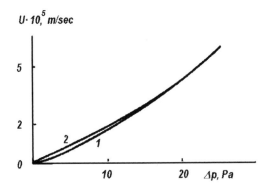

FIGURE 4.16 Dependence of velocity of motion of a chain of drops on the applied pressure difference, Δp: (1) $\Pi(h) \equiv 0$; (2) $\Pi(h) = A/h^3$ according to Eq. (4.213).

$$\Delta p = \frac{2\gamma}{R}\left[k\left(A^+ - A^-\right) + 4Ca \frac{L^+ + L^- + \bar{\mu}\sum\limits_{i=1}^{k}l_i}{R}\right].$$ (4.213)

Figure 4.16 shows the plot of the velocity of motion, U, of the chain of bubbles according to Eq. (4.213) at

$$k = 10, \quad \sum_{i=1}^{k}l_i = 9\,\text{cm},$$

and $L^+ + L^- = 1\,\text{cm}$, $\gamma = 30\,\text{dyn/cm}$, $R = 10^{-2}\,\text{cm}$.

In numerical calculations for the isotherm of the Derjaguin's pressure $\Pi(h) = A/h^3$, we used $A = 10^{-14}$ J/m at $n = 2$ and $A = 10^{-21}$ J/m at $n = 3$.

APPENDIX 6

The low slope approximation is valid inside the zone EB (two flow zones ED and CB, and the zone of the flat film DC, Figure 4.12). In this Appendix, the symbol P is used for hydrodynamic pressure. At a low Reynolds number and taking into account that all unknown function does not depend on the angle, we get the following system of equation, which describes the flow in both liquids 1 and 2:

$$0 = -\frac{\partial P}{\partial r} + \eta\left(\frac{\partial^2 v_r}{\partial r^2} + \frac{\partial^2 v_r}{\partial x^2} + \frac{1}{r}\frac{\partial v_r}{\partial r} - \frac{v_r}{r^2}\right),$$ (A6.1)

$$0 = -\frac{\partial P}{\partial x} + \eta\left(\frac{\partial^2 v_x}{\partial r^2} + \frac{\partial^2 v_x}{\partial x^2} + \frac{1}{r}\frac{\partial v_x}{\partial r}\right),$$ (A6.2)

and the continuity equation

$$\frac{\partial v_r}{\partial r} + \frac{\partial v_x}{\partial x} + \frac{v_r}{r} = 0.$$

This equation can be rewritten as

$$\frac{\partial r v_r}{\partial r} + \frac{\partial r v_x}{\partial x} = 0. \tag{A6.3}$$

The low slope in the zone EB results in $v_r \ll v_x$, $\dfrac{\partial}{\partial x} \ll \dfrac{\partial}{\partial r}$. Using that, we conclude from Eqs. (A6.1) and (A6.2)

$$0 = -\frac{\partial P}{\partial r} \tag{A6.4}$$

$$0 = -\frac{\partial P}{\partial x} + \eta \left(\frac{\partial^2 v_x}{\partial r^2} + \frac{1}{r} \frac{\partial v_x}{\partial r} \right). \tag{A6.5}$$

Equation (A6.4) means that the pressure in both liquids 1 and 2 is a function of the axial coordinate x only. Integrating Eq. (A6.3) over the radius in liquids 1 and 2, we conclude

$$\int_0^{R-h} \frac{\partial r v_x}{\partial x} dr + (R-h) v_r^h = 0, \tag{A6.6}$$

$$\int_{R-h}^{R} \frac{\partial r v_x}{\partial x} dr - (R-h) v_r^h = 0, \tag{A6.7}$$

where v_r^h, v_x^h are velocities on the liquid–liquid interface. The integral in the right-hand side of these equations can be transformed as

$$\frac{\partial}{\partial x} \int_0^{R-h} r v_x dr = -\frac{\partial h}{\partial x} (R-h) v_x^h + \int_0^{R-h} \frac{\partial r v_x}{\partial x} dr,$$

or

$$\int_0^{R-h} \frac{\partial r v_x}{\partial x} dr = \frac{\partial}{\partial x} \int_0^{R-h} r v_x dr + \frac{\partial h}{\partial x} (R-h) v_x^h.$$

Substitution of this expression into Eq. (A6.6) and considering that

$$\frac{\partial h}{\partial t} = v_x^h \frac{\partial h}{\partial x} + v_r^h$$

results in

$$(R-h) \frac{\partial h}{\partial t} = -\frac{\partial}{\partial x} \int_0^{R-h} r v_x dr,$$

or

$$\frac{\partial (R-h)^2}{\partial t} = \frac{\partial}{\partial x} \int_0^{R-h} 2 r v_x dr. \tag{A6.8}$$

Similar transformations of Eq. (A6.7) result in

$$\frac{\partial (R-h)^2}{\partial t} = -\frac{\partial}{\partial x}\int_{R-h}^{R} 2rv_x dr. \tag{A6.9}$$

Integration of Eqs. (A6.4) and (A6.5) using (i) non-slip boundary condition on the capillary surface at $r = R$, (ii) symmetry condition in the capillary center at $r = 0$, and (iii) equality of velocities and tangential stresses at the liquid–liquid interface at $r = R - h$ we can deduce expressions for axial velocities, v_x, in both liquids 1 and 2. Substitution of the deduced expressions for the axial velocities in both liquids 1 and 2 in Eqs. (A6.8) and (A6.9) results in

$$\frac{\partial (R-h)^2}{\partial t} = \frac{\partial}{\partial x}\left\{\frac{P_2'(R-h)^4}{8\eta_2} - \frac{P_1'}{4\eta_1}(R-h)^2\left[(R-h)^2 - R^2\right] - \frac{(R-h)^4(P_2'-P_1')}{2\eta_1}\ln\left(1 - \frac{h}{R}\right)\right\} \tag{A6.10}$$

$$\frac{\partial (R-h)^2}{\partial t} = \frac{\partial}{\partial x}\left\{\frac{P_2'(R-h)^4}{8\eta_1} - \frac{P_1}{8\eta_1}\left[(R-h)^2 - R^2\right] - \frac{(R-h)^2(P_2'-P_1')}{8\eta_1}\left[\frac{(R-h)^2 - R^2}{2}\right.\right.$$
$$\left.\left. -(R-h)^2\ln\left(1 - \frac{h}{R}\right)\right]\right\}, \tag{A6.11}$$

where $P_1(x)$ and $P_2(x)$ are pressures in liquids 1 and 2, respectively; and ′ means "a derivation with *x*."

Subtracting Eq. (A6.11) from Eq. (A6.10) and integrating over resulting equation *x* results in

$$-\frac{P_2'}{8\eta_2}(R-h)^4 - \frac{P_1'}{8\eta_1}\left[R^4 - (R-h)^4\right] + \frac{(P_2'-P_1')}{4\eta_1}(R-h)^2\left[(R-h)^2 - R^2\right] = 2A(t), \tag{A6.12}$$

where $A(t)$ is an integration function, which can dependent only on time.

We consider the steady motion of the drop with the velocity *U*. Changing over to the variable $z = x - Ut$ in equations (A6.10) and (A6.11), we conclude

$$-\frac{P_2'}{8\eta_2}(R-h)^2 - \frac{P_1'}{4\eta_1}\left[R^2 - (R-h)^2\right] + \frac{(P_2'-P_1')}{2\eta_1}(R-h)^2\ln\left(1 - \frac{h}{R}\right) = U. \tag{A6.13}$$

$$-\frac{P_1'}{8\eta_1}\left[R^2 - (R-h)^2\right]^2 + \frac{(P_2'-P_1')}{2\eta_1}(R-h)^2\left[(R-h)^2\ln\left(1 + \frac{h}{R}\right) + \frac{R^2 - (R-h)^2}{2}\right] = 2A - U(R-h)^2. \tag{A6.14}$$

In the steady-state case being examined, the integration constant *A* should be independent of time. The film has a constant thickness h_0 far from the ends of the drop, hence, from Eqs. (A6.13) and (A6.14), we obtain the solutions

$$P_1' = P_2' = -\frac{8\eta_2 U}{(R-h_0)^2 + 2\bar{\eta}\left[R^2 - (R-h_0)^2\right]}, \tag{A6.15}$$

$$A = \frac{U}{2} \frac{(R-h_0)^4 + \bar{\eta}[R^4 - (R-h_0)^4]}{(R-h_0)^4 + 2\bar{\eta}[R^2 - (R-h_0)^2]}, \tag{A6.16}$$

where we introduced $\bar{\eta} = \eta_1 / \eta_2$.

Equation (A6.16) gives the dependence of the flow rate in the capillary on the thickness of the film, h_0, and the velocity of the drop, U.

Equations (A6.13) and (A6.16) are used to obtain an equation for the thickness of a film of liquid under a drop. Limiting ourselves to the leading terms of the expansion in $h/R \ll 1$, we arrive to the following from Eq. (A6.16):

$$h^3 \left(\frac{d\aleph}{dx} + \frac{1}{\gamma} \Pi'(h) \frac{dh}{dx} \right) = 3Ca(h-h_0)F \tag{A6.17}$$

where $\aleph = h'' + \dfrac{h}{R^2}$ is the curvature and

$$F = \frac{\left(1 + 2\bar{\eta}\, \dfrac{h-h_0}{R} + 8\bar{\eta}\, \dfrac{hh_0}{R^2}\right)\left(1 + 4\bar{\eta}\, \dfrac{h}{R}\right)}{\left(1 + \bar{\eta}\, \dfrac{h}{R}\right)\left(1 + 4\bar{\eta}\, \dfrac{h-h_0}{R} + 16\bar{\eta}^2\, \dfrac{hh_0}{R^2}\right)}$$

If we assume that $\bar{\eta} \sim 1$, then $F = 1$ and Eq. (A6.17) takes the form (4.185).

4.7 Liquid Film Coating of a Moving Thin Cylindrical Fiber

In this section, we study how the Derjaguin's pressure affects the thin film adhering on the surface of a fiber being pulled out of a liquid. The problem of coating by liquid films of a moving support is frequently encountered in technological processes. The problem of coating by a liquid film of a flat vertical support pulled out of the liquid in a gravity field has been investigated earlier in [30,31], and the film thickness versus support velocity was deduced for a relatively high velocity of coating. In Section 4.6, we found that the effect of the Derjaguin's pressure is predominant at low velocities of support. The problem of coating from this point of view is considered in this section. The effect of the Derjaguin's pressure in the thin film deposited on the surface of a fiber pulled out of a liquid is taken into consideration in this section.

Let us examine the problem of pulling out of a thin cylindrical fiber of radius a through an interface between two immiscible liquids with a constant velocity U. The schematic presentation of the experimental system is shown in Figure 4.17. The fiber moves along its own axis perpendicular to the interface of the liquids. We neglect the forces of gravity in comparison to viscous and capillary forces.

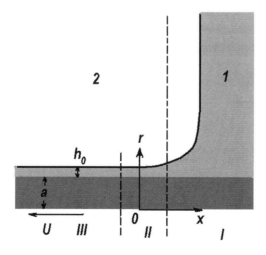

FIGURE 4.17 System under consideration: moving cylinder of radius a with velocity U. (I) Zone of the meniscus, (II) the flow zone, (III) zone of the film of the uniform thickness h_0, to be determined as a function of Ca.

Statement of the Problem

Let us introduce a cylindrical system of coordinates, whose axis x coincides with the axis of the fiber and axis r is perpendicular to it. In this system of coordinates, the fiber moves with velocity U in the negative direction of axis x (Figure 4.17). It is assumed that the liquid 1 completely wets the fiber surface.

Both the Reynolds number, $Re = Ua/\eta$, and the capillary number, $Ca = \eta U/\gamma$, are supposed to be small, $Re \ll 1$, $Ca \ll 1$, and the film thickness, h_0, is much smaller than the radius of the fiber, a, $h_0/a \ll 1$, where η is the dynamic viscosity of the liquid 1, γ is the liquid–liquid interfacial tension on the interface of liquids 1 and 2.

The field of flow is subdivided as in Section 4.6 into three zones (Figure 4.17): zone I, the steady-state meniscus of the liquid, where the movement of the liquid in this zone can be neglected; zone II, the flow zone between the film of constant thickness and the meniscus of the liquid; and zone III, a film of constant thickness, h_0, to be determined as a function of the Ca.

According to the previous consideration in zones II and III, the low slope approximation is valid. Hence, the equations of flow considering the previous assumptions in zones II and III can be written in the following form:

$$\frac{\partial P_1}{\partial r} = 0, \tag{4.214}$$

$$\frac{\partial P_1}{\partial x} = \eta \frac{1}{r}\frac{\partial}{\partial r}\left(r\frac{\partial u}{\partial r}\right). \tag{4.215}$$

$$P_2 = \text{const.} \tag{4.216}$$

The continuity equation is

$$\frac{\partial}{\partial x}u + \frac{1}{r}\frac{\partial}{\partial r}(rv) = 0. \tag{4.217}$$

The boundary conditions for Eqs. (4.214) through (4.217) are
Non-slip boundary condition on the fiber surface

$$u = -U, \quad r = a, \tag{4.128}$$

absence of tangential stress on the liquid–liquid interface (see the discussion in Section 4.6):

$$\eta\frac{\partial u}{\partial r} = 0, \quad r = a+h, \tag{4.219}$$

normal stress jump on the same interface results in

$$P_2 - P_1 = \gamma\aleph(x) + \frac{a}{a+h}\Pi(h), \quad r = a+h, \tag{4.220}$$

where u and v are the velocity along axes x and r, respectively; P_1 and P_2 are the pressures in liquids 1 and 2; $\aleph(x)$ is the curvature of the surface of the film; $\Pi(h)$ is the isotherm of the Derjaguin's pressure of flat films. Note the presence of the term $\frac{a}{a+h}$ in the boundary condition (4.220) (see Section 2.12 for the derivation).

Derivation of the Equation for the Liquid–Liquid Interface Profile

Integration of Eqs. (4.214) and (4.215) with consideration of boundary conditions (4.218) and (4.219) gives the following equation for the velocity of the liquid:

$$u = (P_1'/4\eta)\left[r^2 - a^2 - 2(a+h)^2 \ln r/a\right] - U, \tag{4.221}$$

where $dP_1/dx = P_1'$. The flow rate of the liquid in the liquid layer is

$$Q = 2\pi \int\limits_{a}^{a+h} urdr. \tag{4.222}$$

Substituting Eq. (4.221) into Eq. (4.222), we obtain

$$Q/2\pi = \left\{ \left[(a+h)^2 - a^2 \right]^2 / 4 - \left((a+h)^2 / 2 \right) \left[(a+h)^2 \left(2\ln\left[(a+h)/a \right] - 1 \right) + a^2 \right] \right\}$$

$$P_1' / 4\eta - U \left[(a+h)^2 - a^2 \right] / 2. \tag{4.223}$$

The condition of constancy of flow in any section of the film, $Q = $ const, follows from the equation of continuity (4.217). For a film of constant thickness h_0 in zone III, we deduce from boundary condition (4.220) and Eq. (4.216) that $P_1' = 0$. This allows rewriting Eq. (4.223) as

$$Q = -V \left[(a+h_0)^2 - a^2 \right] / 2. \tag{4.224}$$

Neglecting terms higher than the first order in zones II and III with respect to $h/a \ll 1$, we obtain from Eqs. (4.223) and (4.224)

$$-P_1' h^3 / 3\eta = U (h - h_0), \tag{4.225}$$

Boundary condition (4.220) with consideration of (4.216) can be used to obtain the equation for the profile of the liquid–liquid profile in zones II and III from Eq. (4.225),

$$\frac{d}{dx} \left[\gamma \aleph(x) + \frac{a}{a+h} \Pi(h) \right] = (3\eta U) \frac{h - h_0}{h^3}. \tag{4.226}$$

The curvature of the surface of the liquid–liquid interface in the cylindrical system of coordinates is $\aleph(x) = h''/(1+h'^2)^{3/2} - 1/[(a+h)(1+h'^2)^{1/2}]$.

Let us introduce as before the following dimensionless variables $z = x/\ell$ and $H = h/h_0$, where ℓ is the scale of the flow zone. We are interested in low and intermediate capillary numbers, that is, the thickness of the film, h_0, does not differ very much from the equilibrium thickness, h_e. This allows us, as in Section 4.6, to select this scale as $\ell = \sqrt{ah_0}$. According to that choice, $\frac{h_0}{\ell} = \sqrt{\frac{h_0}{a}} \ll 1$. Using the new variables, we can rewrite the expression for the curvature as

$$\aleph(z) = (1/a)[H'' - 1]. \tag{4.227}$$

Using the same dimensionless variables in Eq. (4.226) in combination with Eq. (4.227), we obtain

$$\frac{d}{dz} \left[H'' - 1 + \frac{a}{\gamma \left(1 + \frac{h_0}{a} H \right)} \Pi(h_0 H) \right] = 3Ca \left(\frac{a}{h_0} \right)^{3/2} \frac{H - 1}{H^3}. \tag{4.228}$$

According to Section 4.6, the dependency of the film thickness, h_0, on the capillary number is given by Eq. (4.183). This means that our choice of the scale of the transition zone $\ell = \sqrt{ah_0}$ tends to $L = \sqrt{ah_e}$ at Ca tends to zero, that is to the length of the transition zone at the equilibrium and $\ell = \sqrt{ah_0} \sim \sqrt{a \cdot a \cdot Ca^{2/3}} = aCa^{1/3}$ at sufficiently high Ca. This corresponds to the choice adopted in [23], which is valid only at sufficiently high Ca. Note, in this section a means radius of the fiber, not a capillary length.

Equilibrium Configuration

In zone 1 (Figure 4.17), the curvature of the interface should be constant given that the movement of the liquid in this region can be neglected (for details, see the introduction to Chapter 4). However, away from the fiber, the interface is flat, and the curvature vanishes. It is evident that $h \approx a$, and the scale of this zone has the same order of magnitude, a. Hence, in this zone $h'^2 \approx 1$. From the expression for the curvature, we conclude

$$h'' / \left(1 + h'^2\right)^{3/2} - 1 / \left[(a+h)\sqrt{1+h'^2} \right] = 0. \tag{4.229}$$

Integrating Eq. (4.229) once results in

$$h' = \sqrt{C^2 (a+h)^2 - 1}, \tag{4.230}$$

where C is the integration constant. The solution of Eq. (4.230) is

$$h = (1/C)\cosh\left[C(x+C_1) \right] - a, \tag{4.231}$$

where C_1 is the new integration constant. This equation has a stationary point $h = h_s$, $h'_s = 0$. Then we can determine using Eq. (4.231) that

$$C = 1 / (a + h_s). \tag{4.232}$$

Let us shift the origin for axis x to the stationary point, that is, we select $C_1 = 0$ and expand the solution (4.231) in Taylor series, which results in

$$h = h_s + \frac{x^2}{2(a+h_s)} + \cdots \tag{4.233}$$

Matching of Asymptotic Solutions in Zones I and II (Figure 4.17)

Now we should match the solution of the stationary meniscus according to Eq. (4.233) with the solution of Eq. (4.228), which describes the liquid profile in zones II and III. Let us examine the asymptotic solution of Eq. (4.228) for $z \to +\infty$ (or $H \to \infty$) then we obtain from Eq. (4.228). As mentioned above, (see Eq. (4.189))this equation has the following asymptotic solution:

$$H = A_0 z^2 / 2 + B_0, \tag{4.234}$$

where the origin is transferred to the stationary point selected above to get rid of the term proportional to z; A_0 and B_0 are some functions of the capillary number Ca determined from the solution of Eq. (4.228) with the boundary condition $H \to 1$ for $z \to -\infty$ and from the matching conditions, which is obtained below.

Let us rewrite solution (4.233) using the "inner variables" used in Eq. (4.228):

$$H = \frac{h_s}{h_0} + \frac{1}{1 + h_s / a} \frac{z^2}{2} + \cdots. \tag{4.235}$$

Comparing Eqs. (4.234) and (4.235) and because these expansions should coincide, we obtain two matching conditions for the solutions of the film profile equations in zones I and II:

$$h_s / h_0 = B_0, \tag{4.236}$$

$$1/(1 + h_s / a) = A_0, \tag{4.237}$$

where both thickness h_0 and h_s are still unknown. Solution of these two equations is

$$h_0 = a(1 - A_0) / A_0 B_0, \tag{4.238}$$

$$h_s = a(1 - A_0) / A_0. \tag{4.239}$$

The constants A_0 and B_0 and their dependency on capillary number, Ca, can be determined numerically only. The results of the numerical calculations are presented in Figure 4.18. Only the case of complete wetting is under consideration in this section, which is why the isotherm of the Derjaguin's pressure used in the problem was selected as

$$\Pi(h) = \frac{A}{h^3}, \tag{4.240}$$

where A is the Hamaker constant. If $Ca \to 0$, then $h_0 \to h_e$ and $B_0 \to 3/2$ (as we show below). Hence,

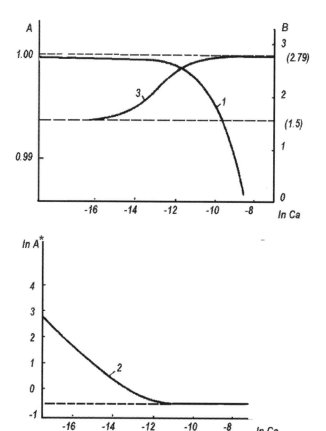

FIGURE 4.18 Calculated parameters (1) A_0, (2) A^*, and (3) B_0 on capillary number, Ca.

$$A_0 \rightarrow \frac{1}{1+\dfrac{3h_e}{2a}} \approx 1 - \frac{3h_e}{2a}, \quad Ca \rightarrow 0,$$

where h_e is the equilibrium thickness of the film.

Equilibrium Case ($Ca = 0$)

In the equilibrium case ($Ca = 0$), the film profile equation will have the following form according to Eq. (4.226):

$$\frac{d}{dx} \left[\frac{h''}{\left(1+h'^2\right)^{3/2}} - \frac{1}{(a+h)\sqrt{1+h'^2}} + \frac{a}{a+h} \Pi(h) \right] = 0.$$

Integrating once, we obtain

$$\frac{h''}{\left(1+h'^2\right)^{3/2}} - \frac{1}{(a+h)\sqrt{1+h'^2}} + \frac{a}{a+h} \Pi(h) = \text{const} = 0. \tag{4.241}$$

The integration constant is equal to zero because the interface is flat away from the fibber.

In zone III, we obtain the equation for determining the film thickness from (4.241), assuming $h = h_e$, $h' = 0$, and $h'' = 0$:

$$\gamma/a = \Pi(h_e), \tag{4.242}$$

or using the Derjaguin's pressure isotherm in the case of complete wetting (4.240), Eq. (4.242) is rewritten as follows: $\gamma/a = A/h_e^3$. This equation results in

$$h_e = \left(\frac{Aa}{\gamma} \right)^{1/3}. \tag{4.243}$$

Equation (4.241) in the case of complete wetting takes the following form:

$$\frac{h''}{\left(1+h'^2\right)^{3/2}} - \frac{1}{(a+h)\sqrt{1+h'^2}} + \frac{Aa}{\gamma(a+h)h^3} = 0. \tag{4.244}$$

In zone III, the film profile according to Eq. (4.244) becomes flat and coincides with Eq. (4.243), in zone I, where the Derjaguin's pressure can be neglected, we obtain the meniscus profile, which is like that given by Eq. (4.231). All terms of Eq. (4.244) are important in zone II. Integrating Eq. (4.244) once, we obtain

$$1/\sqrt{1+h'^2} = \left[C_e - \left(\frac{Aa}{2\gamma h^2} \right) \right] / (a+h),$$

where C_e is the integration constant. At $h \rightarrow h_e$, $h' \rightarrow 0$, and we conclude from this equation that $C_e = a + h_e + \left(\frac{Aa}{2\gamma h_e^2} \right)$. Hence, the liquid profile is

$$1/\sqrt{1+h'^2} = \left[a + h_e + \left(\frac{Aa}{2\gamma h_e^2} \right) - \left(\frac{Aa}{2\gamma h^2} \right) \right] / (a+h). \tag{4.245}$$

The matching conditions of solutions of Eqs. (4.245) in zone II and the equation for the steady-state meniscus in zone I can be written as follows:

$$\lim_{h_{II} \to \infty} h'_{II} = \lim_{h_{II} \to 0} h'_{I},$$ (4.246)

where h'_{I} and h'_{II} refer to zones I and II, respectively.

The solution of Eq. (4.229), which describes the film profile in zone I, is

$$1/\sqrt{1 + h'^2_{I}} = C_2/(a + h_{I}),$$ (4.247)

where C_2 is the integration constant. Using Eqs. (4.245) and (4.247), we find using the matching condition (4.247) that

$$C_2 = C_e.$$ (4.248)

At the stationary point $h_{I} = h_{se}$, where $h'_{I} = 0$, we conclude from Eqs. (4.247) and (4.248) that

$$h_{se} = C_2 - a = h_e + \left[\frac{Aa}{2\gamma h_e^2}\right],$$

or

$$h_{se} = 1.5 h_e,$$ (4.249)

which agrees with the conclusion obtained in Section 2.4 (see Figure 2.20).

Hence, we obtain the following expression for coefficient B_e in the equilibrium case (see Eq. 4.234):

$$B_e = \lim_{Ca \to 0} B_0 = 1.5.$$ (4.250)

Numerical Results

The coefficients $A^* = (1-A_0)/B_0 (3\,Ca)^{2/3}$, A_0, and B_0 as functions of the capillary number according to Eqs. (4.237) and (4.236) obtained from the numerical calculation and are presented in Figure 4.18. The isotherm of the Derjaguin's pressure is defined by Eq. (4.240), that is, in the case of complete wetting. Figure 4.18 shows that at $Ca \to 0, B \to 3/2$, which corresponds to the prediction (4.249). At the same time, $A_0 \to 1-(3/2)H_e$ as shown above, and $A^* \to \infty$.

For relatively "large" capillary numbers, Ca, the film thickness h_0 becomes thick enough such that the effect of the Derjaguin's pressure can be neglected. At the same time, $A^* \to 0.643, B \to 2.79$ (these values were obtained in [23]) and

$$A_0 \approx 1 - 0,643(3Ca)^{2/3} \cdot 3/2.$$

The dimensionless thickness H_0 of the film remaining on the fibber as a function of the capillary number Ca (dimensionless velocity) is presented in Figure 4.19. At $Ca \to 0$, the graph has a plateau, $h_0 \to h_e$. For "large" Ca

$$H_0 \approx 1.33(Ca)^{2/3},$$ (4.251)

as was first deduced in [23] earlier.

Let us determine the critical velocity, Ca_*, characteristic of the transitional velocity from "small" to "large" capillary numbers as the point of intersection of the lines $\ln h_e = $ const and $\ln 1.33 + 2/3 \ln Ca$. In our case, $\ln Ca_* = -12.85$. The critical velocity obtained from the experimental data in [32] is equal to $\ln Ca_* \approx -13.26$, which is close to the above theoretical prediction.

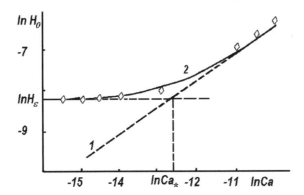

FIGURE 4.19 Dimensionless film thickness $H_0 = h_0/a$ on the capillary number, Ca. (1) without the Derjaguin's pressure according to Eq. (4.251), and (2) with consideration of the Derjaguin's pressure according to Eq. (4.238). (Experimental point from Quere, D. et al., *Science*, 249, 1256, 1990.)

4.8 Blow-off Method for Investigating Boundary Viscosity of Volatile Liquids

The highlighted text below can be omitted at the first reading.

A blow-off method allows determining the boundary viscosity as a function of the distance to a solid substrate. A theory is suggested that includes consideration of both the flow of a liquid film and film evaporation with gas being blown through a plane-parallel channel over the film. The theory allows finding the dependence of the dynamic viscosity, η, on the distance to the substrate, h, with η being supposed to be a continuous function of h. A procedure is outlined for calculating the viscosity, η, on h dependency based on available experimental data. The theory is applied to the calculation of the boundary viscosity of hexadecane. It turns out that the viscosity in thin layers (4.40–200 Å) is lower than that in the bulk [33,34].

Boundary Viscosity

The measurements of the boundary viscosity are of a substantial interest. Attempts have been made to apply the conventional methods adapted to the measurement of the boundary viscosity. These methods may be divided into three groups:

- Methods based on measurement of the rate of liquid flow through capillaries [35]
- Methods based on the rotation of coaxial cylinders or discs
- Methods based on measurement of the velocity of a falling sphere

These methods afford the possibility of determining only the mean value of viscosity for a sufficiently thick layer.

The blow-off method [36] consists of the following. One of the walls of a channel formed by two plane-parallel surfaces is coated with a layer of the tested liquid. If a current of air (or an inert gas) is then passed through the channel, a flow is generated in the film due to a tangential force induced by the flowing gas. In this case, the film profile becomes wedge shaped, the slope of this wedge being dependent on the viscosity of the liquid.

If the liquid viscosity over the whole distance to the wall is constant, the slope remains constant until the wetting boundary is reached. In the case of an increased viscosity close to the wall, the slope increases; in the case of a decreased viscosity, the slope decreases. The viscosity is calculated from the slope at a given point of the profile.

The blow-off method was used to examine the boundary viscosity of some organic liquids [37]. It has been established that the profile of a film of very pure vaseline oil is rectilinear. In this case, no anomaly of viscosity is observed, and vaseline oil preserves its bulk viscosity value up to the layer thickness of about 10^{-7} cm. The viscosity of incompletely hydrated benzontron was found to decrease in the boundary layer, whereas in a layer of 300 Å, the viscosity of chloroderivatives of saturated hydrocarbons was found to increase abruptly (jumpwise). An increase in the viscosity of some organic liquids and some vinyl polymers and their solutions in films up to 50,000 Å thick as well as the instability of the films of the solutions of vinyl polymers at certain thicknesses were observed in Refs. [38–40].

Up to now, a limitation of the blow-off method has been its inapplicability to the measurement of the boundary viscosity of volatile liquids. In this section, the blow-off method has been modified to be applicable to studies of the boundary viscosity of volatile liquids. It is necessary to consider the fact that when gas is blown through a plane-parallel channel, the flow of a film of liquid occurs (as in the usual variant of the blow-off method) simultaneously with its evaporation.

The theory of the blow-off method, allowing the determination of the boundary viscosity of volatile liquids, is developed below.

Theory of the Method

Let us consider the profile of a film at instant of time t, which is characterized by the relationship

$$x = x(h,t), \tag{4.252}$$

where x is the distance from the wetting boundary to the point at the surface of the film area having local thickness h. An initial condition is adopted as follows

$$x(h,0) = 0. \tag{4.253}$$

An increment, dx, per elementary increment of time, dt, is the sum of an increment, dx_1, resulting from the flow under the action of the tangential shear stress per unit area of the blow-off current, τ, and an increment, dx_2, because of evaporation. It is obvious that the first increment due to the gas flow is

$$dx_1 = \tau \int_0^h \frac{dh}{\eta(h)} \cdot dt, \tag{4.254}$$

where $\eta(h)$ is the local viscosity in the boundary layer at distance h from the substrate. Evaporation results in

$$dx_2 = i \frac{dt}{\sin\alpha}, \tag{4.255}$$

where i is the linear evaporation rate and $\sin\alpha$ is the slope of the film surface at a given point.

Except for a very short initial period, the film has a very low slope profile, which allows one to substitute,

$$\left(\frac{dx}{dh} \right)^{-1}$$

for $\sin\alpha$.

Summing Eqs. (4.254) and (4.255) and transforming them, we eventually obtain

$$\frac{dx}{\partial t} - i\frac{\partial x}{\partial h} = \tau \int_0^h \frac{dh}{\eta(h)}.$$

(4.256)

The general solution of partial differential Eq. (4.256) is

$$x = f(h+it) - \frac{\tau}{i}\int_0^h dh_1 \int_0^{h_1} \frac{dh_2}{\eta(h_2)}.$$

(4.257)

The initial condition (4.253) results in

$$x = \frac{\tau}{i}\left[\int_0^{h+it} dh_1 \int_0^{h_1} \frac{dh_2}{\eta(h_2)} - \int_0^h dh_1 \int_0^{h_1} \frac{dh_2}{\eta(h_2)}\right] = \frac{\tau}{i}\times\int_h^{h+it} dh_1 \int_0^{h_1} \frac{dh_2}{\eta(h_2)}.$$

(4.258)

Consider the consequences of the solution obtained. If there is no special boundary viscosity, that is, $\eta(h) = \eta_\infty = \text{const}$, then

$$x = \frac{\tau t}{2\eta_\infty}(2h+it).$$

(4.259)

Thus, having experimentally determined the relationship $x = x(t,h)$ at constant thickness, h, and having revealed its strictly parabolic form on time, one may conclude that the viscosity has a constant value. Therefore, using the parameters of the parabolic relationship $x(t,h)$, at any $h = \text{const} \neq 0$, one may determine both η_∞ and the evaporation rate i.

It is more convenient to attain the same objective by experimentally determining the dependence of h on t for a given value x and by checking its hyperbolic form.

If one determines only the velocity, at which the wetting boundary recedes (which is the simplest of all), then Eq. (4.259) becomes

$$\frac{i}{2\eta_\infty} = \frac{x}{\tau t^2},$$

(4.260)

which is of methodical interest, thus allow one to determine the ratio $\frac{i}{\eta_\infty}$.

From Eq. (4.259), we conclude

$$h = \frac{x\eta_\infty}{\tau t} - \frac{it}{2}.$$

If $\eta(h) = \eta_\infty = \text{const}$, this equation gives η_∞ and i at any constant x.

From Eq. (4.258) we conclude

$$\left(\frac{\partial x}{\partial t}\right)_h = \tau \int_0^{h+it} \frac{dt}{\eta(h)},$$

(4.261)

$$\left(\frac{\partial^2 x}{\partial t^2}\right)_h = \tau i \frac{1}{\eta(h+it)},$$

(4.262)

$$\frac{\partial^2 x}{\partial h \partial t} = \tau \frac{1}{\eta(h+it)}. \tag{4.263}$$

Using Eqs. (4.262) and (4.263) as the basis, it is difficult to obtain precise data on the viscosity of boundary layers because it is difficult to measure x at a given h as a function of t. If we confine ourselves to the case where $h = \text{const} = 0$, then the problem is rendered somewhat easier. Yet, we have to determine $(\partial x / \partial h)_t$ and to beyond the time interval, in which $it < H_N$, where H_N is the thickness of the boundary layer, where the viscosity is different from its bulk value.

Measuring the values of h as a function of t at the fixed coordinate x, it is possible to derive the experimental dependence $h(t)$ or, conversely, $t(h)$. Let us explore the possibility of obtaining $\eta(h)$ using such an approach.

For convenience, hereafter we use the following notation:

$$C = \frac{xi}{\tau}, \quad \varphi(h) = it(h), \tag{4.264}$$

$$f(h) = \int_0^h dh_1 \int_0^{h_1} \frac{dh_2}{\eta(h_2)}, \tag{4.265}$$

$$f(0) = f'(0) = 0, \quad h \geq 0$$

It follows from Eq. (4.258) that $f(h)$ satisfies the following functional equation

$$f[h + \varphi(h)] = f(h) + C. \tag{4.266}$$

Because the viscosity is always positive, the definition (4.265) gives for all thicknesses h

$$f(h) > 0, \quad f'(h) > 0, \quad f''(h) > 0.$$

According to the definition, $\varphi(h) = it(h)$ does not vanish anywhere, except for at $h = \infty$. Note, $\varphi(h)$ has the dimension of length.

Using Eq. (4.266) we determine the dependence $\eta(h)$. We continue according to the following procedure:

i. Let us define a sequence of film thicknesses, H_m, $m = 0,1,2,3, \ldots$ such that

$$\int_{H_{m-1}}^{H_m} dh_1 \int_0^{h_1} \frac{dh_2}{\eta(h_2)} = C, \quad m = 0, 1, 2, \ldots$$

Having determined the dependence $\eta(h)$ at $h = H_m$, we find the value of i.

ii. We show that the function $f(h)$ can be determined at $h \geq H_N$, where $h = H_N$ is a film thickness such that $\eta(h) = \eta_\infty = \text{const}$ for $h \geq H_N$, that is,

$$f''(h) = \frac{1}{\eta_\infty} = \text{const}.$$

iii. The function $f(h)$ for $h \geq H_N$, constructed according to (ii), will be extended to include values $h < H_N$ up to $h = 0$, which allows one to find the dependence of viscosity on thickness for all $h > 0$, because

$$\eta(h) = \frac{1}{f''(h)}.$$

iv. We develop a method for determining the H_N thickness of the boundary layer of liquid.

Everywhere in steps (i)–(iv), the viscosity is considered as a continuous function of thickness, h.

The dependency $t(h)$ is determined experimentally with an accuracy sufficient for further consideration.

We now turn to the realization of the program indicated above.

1. Let us introduce the following notations:

$$\varphi_{-1} = \varphi(0), \quad \varphi_0 = \varphi(\varphi_{-1}), \quad \ldots, \quad \varphi_m = \varphi(\varphi_{-1} + \varphi_0 + \ldots + \varphi_{m-1}), \quad \ldots$$

$$H_{-1} = 0, \quad H_0 = \varphi_{-1}, \quad \ldots, \quad H_m = \varphi_{-1} + \varphi_0 + \ldots + \varphi_{m-1}, \quad \ldots \tag{4.267}$$

$$H_{m+1} = H_m + \varphi_m,$$

where $m = 0, 1, \ldots$.

At $h = 0$ from Eq. (4.267), we conclude $f(\varphi_{-1}) = f(0) + C = C$, or $f(H_0) = C$. Let us assume that $h = H_m$

$$f(H_m) = (m+1) \cdot C, \tag{4.268}$$

and prove Eq. (4.268) by induction. We assume that it is valid at $m = l$ and show then that Eq. (4.268) holds identically for $m = l+1$.

Equation (4.256) at $h = H_l$ gives $f(H_l + \varphi_l) = f(H_l) + C = (l+1)C + C = (l+2) \cdot C$, or

$$f(H_{l+1}) = (l+2)C.$$

This relation proves the validity of Eq. (4.268). Because Eq. (4.268) is valid at $m = 0, 1$, it is valid for any m.

It follows from definition (4.267) that $H_m \le H_{m+1}$, that is, H_m is an increasing sequence of thicknesses; hence, it has a finite or an infinite limit $H = \lim_{m \to \infty} H_m$. Passing over to the limit in the relationship $H_{m+1} = H_m + \varphi(H_m)$, we obtain $H = H + \varphi(H)$. Hence, if $H < \infty$ then $\varphi(H) = 0$; however, as stated above, this is impossible; $\varphi(h) = it(h)$ does not vanish anywhere, except for at $h = \infty$.

Below we use the following notation

$$f(H_m) = f_m, \quad f'(H_m) = f'_m, \quad f''(H_m) = f''_m,$$

$$\varphi'(H_m) = \varphi'_m, \quad \varphi''(H_m) = \varphi''_m.$$

By differentiating Eq. (4.266) with respect to h, we conclude

$$f'\left[h + \varphi(h)\right] \cdot \left[1 + \varphi'(h)\right] = f'(h). \tag{4.269}$$

At $h = 0$, we obtain from Eq. (4.269) the following expression

$$f'(\varphi_0)(1 + \varphi'_{-1}) = f'(0) = 0. \tag{4.270}$$

Given that $f'(h) > 0$ at $h > 0$, then $\varphi'_{-1} = -1$. We show below that the point at which $\varphi'(h) = -1$ is unique. Let us assume that $\varphi'(h_0) = -1$, when $h_0 > 0$. In this case, from Eq. (4.269) we have

$$f'(h_0) = f'\left[h_0 + \varphi(h_0)\right] \cdot \left[1 + \varphi'(h_0)\right] = 0,$$

that is, $f'(h_0) = 0$, which is a contradiction because $f'(h_0) > 0$.

Bearing in mind Eq. (4.265), the definition $\varphi(h)$ gives $it'(0) = -1$, or

$$i = -\frac{1}{t'(0)}. \tag{4.271}$$

We show below that derivatives of the function $f(h)$ at $h = H_m < H_N$ satisfy the following relation

$$f'_m = f'_N \cdot \prod_{i=m}^{N-1}(1 + \varphi'_i). \tag{4.272}$$

Equation (4.269) at $h = H_{N-1}$ yields

$$f'(H_{N-1}) = f'_N \cdot (1 + \varphi'_{N-1}). \tag{4.273}$$

If Eq. (4.272) is valid at $m = l$, we show its validity at $m = l - 1$. From Eq. (4.269) at $h = H_{l-1}$ we have

$$f'_{l-1} = f' \cdot (1 + \varphi'_{l-1}) = \left[f'_N \cdot \prod_{i=l}^{N-1}(1 + \varphi'_i)\right] \cdot (1 + \varphi'_{-i}) = \left[f'_N \cdot \prod_{i=l-1}^{N-1}(1 + \varphi'_i)\right] \cdot (1 + \varphi'_{l-1})$$

$$= f'_N \cdot \prod_{i=l-1}^{N-1}(1 + \varphi'_i), \tag{4.274}$$

which proves the relation (4.272).

Now we show by induction that the second derivative with respect to h of the function $f(h)$ at $H_m < H_N$ satisfies the following relation:

$$f''_{N-m} = f''_N \cdot \prod_{i=N-m}^{N-1}(1 + \varphi'_i)^2 + f'_{N-m+1} \cdot \varphi''_{N-m} + \sum_{k=0}^{m-2} f'_{N-k} \cdot \varphi''_{N-k+1}$$

$$\times \prod_{i=k+1}^{m-1}(1 + \varphi'_{N-i-1}). \tag{4.275}$$

By differentiating Eq. (4.266) twice with respect to h, we obtain

$$f''\left[h + \varphi(h)\right] \cdot \left[1 + \varphi'(h)\right]^2 + f'\left[h + \varphi(h)\right] \cdot \varphi''(h) = \varphi''(h). \tag{4.276}$$

At $h = H_{N-1}$, Eq. (4.276) changes to Eq. (4.275). Assuming Eq. (4.275) to hold at $m = l$, we show its validity at $m = l + 1$. From Eq. (4.275) at H_{N-l-1}, we get

$$f''_{N-l-1} = f''_{N-l} \cdot \left(1 + \varphi'_{N-l-1}\right)^2 + f'_{N-l} \cdot \varphi''_{N-l-1}$$

$$= \left[f''_N \cdot \prod_{i=N-l}^{N-1} \left(1 + \varphi'_i\right)^2 + f'_{N-l+1} \cdot \varphi''_{N-l} + \sum_{k=0}^{l-2} f'_{N-k} \cdot \varphi''_{N-k-1} \cdot \prod_{i=k+1}^{l-1} \left(1 + \varphi'_{N-i-1}\right)^2 \right]$$

$$\times \left(1 + \varphi'_{N-l-1}\right)^2 + f'_{N-l} \cdot \varphi''_{N-l-1'} = f'_N \cdot \prod_{i=N-l-1}^{N-1} \left(1 + \varphi'_i\right)^2 + f'_{N-l} \cdot \varphi''_{N-l-1}$$

$$+ \sum_{k=0}^{l-1} f'_{N-k} \cdot \varphi''_{N-k-1} \cdot \prod_{i=k+1}^{l} \left(1 + \varphi'_{N-i-1}\right)^2$$

It follows from Eq. (4.275) that when $h = 0$,

$$f''(0) = f'(\varphi_{-1}) \cdot \varphi''_{-1}, \qquad (4.277)$$

here $f'(\varphi_{-1})$ being determined from Eq. (4.272) at $m = 0$. By differentiating Eq. (4.266) three times with respect to h and putting $h = 0$, we obtain

$$f'''(0) = f'(\varphi_{-1}) \cdot \varphi'''_{-1}.$$

However, on the other hand,

$$f'''(0) = -\frac{\eta'(0)}{\eta^2(0)} = -\eta'(0) \cdot f''^2(0).$$

Consequently,

$$\eta'(0) = -\frac{\varphi'''(0)}{f'(\varphi_{-1}) \cdot \varphi''^2(0)}. \qquad (4.278)$$

Thus, the sign of $\eta'(0)$ is opposite to that of the third derivative, $\varphi'''(h)$, at $h = 0$. It is shown below that in the case $\eta(h) = \eta_\infty = $ const.

$$\varphi'''(0) = 0, \text{ that is, } \varphi'(0) = \eta'_\infty = 0.$$

The relations (4.268), (4.272), and (4.275) allow the determination of the dependence $\eta(h)$ at thicknesses $h = H_m$

$$\eta(H_m) = \frac{1}{f''(H_m)}, \qquad (4.279)$$

while the relation (4.278) gives a criterion for the behavior of viscosity at small h.

Thus, we have fully completed the first part of the program of determining the viscosity as a function of h.

2. Given that $f''(h) = \frac{1}{\eta_\infty} = \text{const}$ at $h \geq H_N$, then at such h

$$f(h) = \frac{h^2}{2\eta_\infty} + Ah + B, \tag{4.280}$$

where A and B are integration constants, which are as yet unknown.

Let us set

$$\lambda_N = \int_0^{H_N} dh / \eta(h);$$

then at $h \geq H_N$, we derive from Eq. (4.266)

$$f\left[h + \varphi(h)\right] - f(h) = \int_h^{h+\varphi} dh_1 \int_0^{h_1} \frac{dh_2}{\eta(h_2)} = \int_h^{h+\varphi} \left(\lambda_N + \frac{h_1 - H_N}{\eta_\infty}\right) dh_1$$

$$= \frac{(h + \varphi)^2}{2\eta_\infty} - \frac{h^2}{2_\infty} + \left(\lambda_N - \frac{H_N}{\eta_\infty}\right)\varphi = C. \tag{4.281}$$

On the other hand, from Eq. (4.280) we find

$$f(h + \varphi) - f(h) = \frac{(h + \varphi)^2}{2\eta_\infty} - \frac{h^2}{2\eta_\infty} + A \cdot \varphi = C. \tag{4.282}$$

Comparing Eqs. (4.281) and (4.282), we infer

$$A = \lambda_N - \frac{H_N}{\eta_\infty}. \tag{4.283}$$

Thus, A specifies an integral deviation of $\eta(h)$ from η_∞ at $h < H_N$. At $\eta(h) \equiv \eta_\infty$ for all h, we have from Eq. (4.283): $\lambda_N = H_N / \eta_\infty$ and $A = 0$. At this point, we assume λ_N and H_N to be known that they are defined in step (iv).

From the second part of equality (4.281), we get

$$\frac{\varphi^2}{2\eta_\infty} + \varphi\left[\lambda_N - \frac{h - H_N}{\eta_\infty}\right] - C = 0, \tag{4.284}$$

hence,

$$\lambda_N = \frac{C}{\varphi(h)} - \frac{\varphi(h)}{2\eta_\infty} - \frac{h - H_N}{\eta_\infty}. \tag{4.285}$$

Given that, $\lambda_N = \text{const}$, in the case $h \geq H_N$ that impose some restriction on the function $\varphi(h)$. From Eq. (4.284), $\varphi(h)$ is determined by the following relation:

$$\varphi(h) = \sqrt{\left(\lambda_N \cdot \eta_\infty + h - H_N\right)^2 + 2C\eta_\infty} - \left(\lambda_N \cdot \eta_\infty + h - H_N\right). \tag{4.286}$$

For $\left(\lambda_N \cdot \eta_\infty + h - H_N\right)^2 \gg 2C\eta_\infty$, we get from Eq. (4.286)

$$\varphi(h) = \frac{C\eta_\infty}{\lambda_N \eta_\infty + h - H_N}. \tag{4.287}$$

This expression at $h \gg \left[H_N - \lambda_N \cdot \eta_\infty\right]$ results in

$$\varphi(h) = \frac{C\eta_\infty}{h}. \tag{4.288}$$

Multiplying Eq. (4.259) by $i\eta_\infty$ and dividing it by τ yields $2\eta_\infty \frac{x \cdot i}{\tau} = 2hit + \left(it\right)^2$ or $2\eta_\infty C = 2h\varphi + \varphi^2$. Hence,

$$\varphi(h) = \sqrt{h^2 + 2C\eta_\infty} - h, \quad h \geq 0, \tag{4.289}$$

$$\varphi(h) = \frac{C \cdot \eta_{V\infty}}{h}, \quad h \gg \sqrt{2C\eta_\infty}. \tag{4.290}$$

As has been pointed out previously, at $\eta(h) \equiv \eta_\infty, \lambda_N = \frac{H_N}{\eta_\infty}$ and, hence, Eq. (4.286) transforms to Eq. (4.289). From Eq. (4.289), we conclude

$$0 > \varphi'(h) = -1 + \frac{h}{\sqrt{h^2 + 2C\eta_\infty}} > -1,$$

$$\varphi''(h) = \frac{2C\eta_\infty}{\left(h^2 + 2C\eta_\infty\right)^{3/2}} > 0,$$

$$\varphi'''(h) = -\frac{6C\eta_\infty h}{\left(h^2 + 2C\eta_\infty\right)^{5/2}} < 0,$$

$$\varphi'''(0) = 0$$

These properties mean that the inverse to the $\varphi(h) = it(h)$ function, which is $h(t)$, decreases monotonically and is concave.

Using Eq. (4.268), from Eq. (4.280) at $h = H_N$, we obtain

$$\frac{H_{N^2}}{2\eta_\infty} + AH_N + B = (N+1)C. \tag{4.291}$$

As a result, we find

$$B = (N+1)C - \frac{H_{N^2}}{2\eta_\infty} - A \cdot H_N. \tag{4.292}$$

Thus, Eq. (4.280) determines $f(h)$ at $h \geq H_N$ (with λ_N and H_N assumed to be known constants).

3. Let us set

$$f(h) = f_m(h) \tag{4.293}$$

at $H_m \le h \le H_{m+1}$, and

$$f(h) = f_N(h) = \frac{h^2}{2\eta_\infty} + Ah + B, \tag{4.294}$$

at $h \ge H_N$. Let now h vary from H_{N-1} to H_N. In this case, by definition, $h + \varphi(h)$ varies from H_N to H_{N+1}. However, at $h \ge H_N$, the unknown function, $f(h)$, is already defined and given by Eq. (4.294). Hence, in accordance with Eq. (4.266), we can determine $f(h)$ at a smaller thickness in the interval, $H_{N-1} \le h \le H_N$, as

$$f_{N-1}(h) = f_N\left[h + \varphi(h)\right] - C = \frac{\left[h + \varphi(h)\right]^2}{2\eta_\infty} + A\left[h + \varphi(h)\right] + B - C. \tag{4.295}$$

Let us show, that $f(h)$ constructed according to Eq. (4.295) is continuous at $h = H_N$. Indeed, $f_{N-1}(H_N) = f_N(H_{N+1}) - C$; however, by definition of H_{N+1}: $f_N(H_{N+1}) = f_N(H_N) + C$.

From the two last equalities, we deduce $f_{N-1}(H_N) = f_N(H_N)$. That is, the constructed function is continuous at $h = H_N$.

Let us assume now that we have already determined the function $f_{m+1}(h)$, that is, the value of $f(h)$ in the interval of thickness $H_{m+1} \le h \le H_{m+2}$. Now, let h vary from H_m to H_{m+1}. According to the definition (4.267), $h + \varphi(h)$ varies within H_{m+1} to H_{m+2}. In this range, however, $f(h)$ has already been determined and is equal to $f_{m+1}(h)$.

According to Eq. (4.266), we define

$$f_m(h) = f_{m+1}\left[h + \varphi(h)\right] - C. \tag{4.296}$$

By construction, the function $f(h)$ satisfies the functional Eq. (4.266). The uniqueness of the function obtained follows from the additional condition: at $h \ge H_N$:

$$f''(h) = \frac{1}{\eta_\infty} = \text{const.}$$

Eqs. (4.295) and (4.296) result in

$$f_m(H_m) = f_{m+1}(H_{m+1}) - C = \cdots = f_N(H_N) - (N-m)\cdot C = (N+1)\cdot C - (N-m)\cdot C = (m+1)\cdot C,$$

which coincides with Eq. (4.268).

Let us show that the deduced function is continuous at the joining thicknesses, that is, $f_m(H_{m+1}) = f_{m+1}(H_{m+1})$. We do this for the case of $f_{N-2}(h)$ as an example, because other joining thickness can be considered in the same way:

$$f_{N-1}(H_{N-1}) = f_N(H_N) - C = (N+1)\cdot C - C = N\cdot C,$$

$$f_{N-2}(H_{N-1}) = f_{N-1}(H_N) - C = f_N(H_N) - C = (N+1)\cdot C - C = N\cdot C.$$

In this derivation, we used the continuity of $f(h)$ at the thickness $h = H_N$. The two last equalities give

$$f_{N-1}(H_{N-1}) = f_{N-2}(H_{N-1}) = N\cdot C.$$

Similarly, assuming the continuity of $f_{m+1}(h)$, we can show the continuity of $f_m(h)$ at thickness $h = H_{m+1}$.

At all other thickness (different from the joining thicknesses), the continuity of $f(h)$ follows from the continuity $\varphi(h)$ and relation (4.296).

Differentiating relation (4.296) with respect to h yields

$$f'_m(h) = f'_{m+1}\left[h + \varphi(h)\right] \cdot \left[1 + \varphi'(h)\right] > 0.$$

Given that $0 > \varphi' > -1$ and $f'_N(h) > 0$ at $h = H_m$, we derive from the last equality

$$f'_m(H_m) = f'_{m+1}(H_{m+1})\left[1 + \varphi'_m\right] = f'_{m+2}(H_{m+2}) \cdot \left(1 + \varphi'_m\right) \cdot \left(1 + \varphi'_{m+1}\right)$$

$$= \ldots = f'_N(H_N) \cdot \prod_{i=m}^{N-1}\left(1 + \varphi'_i\right),$$

which coincides with Eq. (4.272).

Differentiating relation (4.296) twice with respect to h, we obtain

$$f''_m(h) = f''_{m+1}(h + \varphi)\left(1 + \varphi'\right)^2 + f'_{m+1}(h + \varphi) \cdot \varphi''. \tag{4.297}$$

From Eq. (4.297) at $h = H_m$, it is likewise possible to derive formula (4.275) by induction.

Thus, the function $f(h)$ obtained is a unique solution of Eq. (4.266) and satisfies all the previous relations (4.268), (4.272), and (4.275). At any value of h, the viscosity is now given by

$$\eta(h) = \frac{1}{f''_m(h)}, H_m \leq h \leq H_{m+1}. \tag{4.298}$$

We now turn to the determination of the quantities

$$\lambda_N = \int_0^{H_N} \frac{dh}{\eta(h)}$$

and H_N, which are as yet unknown.

Let N_0 be some value of the number that we provisionally take as the N sought. In this case, according to Eq. (4.285), at $h = H_{N_0}$ we conclude

$$\lambda_{N_0} = \frac{C}{\varphi_{N_0}} - \frac{\varphi_{N_0}}{2\eta_\infty}. \tag{4.299}$$

Let $\varphi^{(N_0)}(h)$ denote function (4.286) at λ_{N_0}, that is,

$$\varphi^{(N_0)}(h) = \sqrt{\left(\lambda_{N_0} \cdot \eta_\infty + h - H_{N_0}\right)^2 + 2C\eta_\infty} - \left(\lambda_{N_0}\eta_\infty + h - H_{N_0}\right),$$

at $h \geq H_{N_0}$. Determine N as the smallest number N_0, at which

$$\left|\frac{\varphi(h) - \varphi^{(N_0)}(h)}{\varphi(h)}\right| \leq 0.01, \tag{4.300}$$

at all $h \geq H_{N_0}$.

Relations (4.299) and (4.300) determine the values of the thickness of a boundary layer and λ_N.

Experimental Part [33]

The experimental device used for studying the boundary viscosity of volatile substances is substantially the same as that for nonvolatile liquids (Figure 4.20).

The body of device 1, made of brass, has a rectangular cutout, which receives a steel insert 3, with its upper surface polished to the finish quality 14-b and then subjected to additional optical polishing. Electron microscopy examination revealed that the surface of the insert was fairly smooth. There are separate randomly oriented scratches with rounded off edges. Their depth is so small that shading failed to be obtained. Their horizontal size varies from 20–160 Å, and their number is small. In the horizontal scale, the main working surface exhibits roughness of the size less than 80 Å; hence, their depth is of the order of 40 Å. At the top, the instrument is covered with prism 5. A plane parallel channel is formed between the base of the prism and the upper surface of the insert. A current of gas (air or nitrogen) is passed through the channel to blow off the liquid being tested.

The thickness of the channel is adjusted by putting calibrated shims (4) between insert (3) and the base (2). For measuring the pressure gradient, two narrow openings (6) with connections to connect a pressure gauge are provided in the channel (the connections are not shown in Figure 4.20).

In the device, a blow-off current is jetted up by a water jet pump installed in an exhaust cabinet. This is done to prevent the vapor of volatile liquids from getting into the air of laboratory rooms. The pressure differential is checked against a liquid differential pressure gauge, using a reading microscope.

The accuracy in the determination of differential pressure is 1%. Such accuracy is ensured because an additional capillary tube is connected to the system in series with the instrument channel, the pressure drop being measured at the ends of the capillary tube.

The capillary tube is chosen such that the pressure drop in the channel equal to 0.25 mmH$_2$O corresponds to the pressure drop of 200 mmH$_2$O in the capillary tube. This facilitates measurement and ensures the measurement accuracy required. In the experiments, the pressure drop was equal to 0.25 mm of water over the length of the working portion of the film, which was equal to 3 cm. This corresponds to the shear stress of about $\tau = 0.4$ dyn/cm^2. Experiments were also conducted at other pressure differentials: namely, at 0.06, 0.125, and 0.5 mmH$_2$O.

A saturated hydrocarbon, hexadecane, was used as a volatile liquid to be tested. Hexadecane of various degrees of purification was used: industrial grades qualified as "chemically pure," a grade which has been additionally purified on silica gel using the displacement chromatography method, and a "chemically pure" grade which has been purified in a chromatography unit. The content of organic admixtures in hexadecane, which has been qualified as "chemically pure," did not exceed 0.5%; that in the chromatographically pure grade was below 0.3%. The admixtures in such amounts, certainly, could not affect the measurement results.

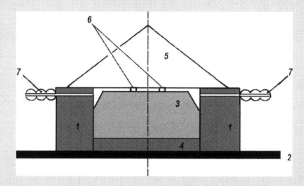

FIGURE 4.20 Schematic diagram of the experimental device for investigations of the boundary viscosity of liquids.

As before, interference bands were found to occur on the insert because of testing hexadecane. The movement of the wetting boundary was recoded. The gap width was chosen to be 1 mm to ensure complete evaporation. It is difficult to blow off substances of very high volatility. In this case, evaporation is so prevalent over flow that during an experiment the spacing between the interference bands would remain extremely narrow, and measurements would become difficult.

The experiment is carried out as follows. The insert 3 (Figure 4.20) was coated with a film of the liquid to be tested. Then the insert coated with the film is placed in the device, and the film thickness is measured in the course of blowing off at different distances l from the wetting boundary ($l = 10$ mm, $l = 20$ mm, and $l = 30$ mm) and at different shear stresses (0.1, 0.2, and 0.4 dyn/cm^2). The thickness of films was measured by the ellipsometry method.

Figure 4.21 gives experimental curves representing a variation in the film thickness versus the blow-off time. The curves were obtained at $\Delta p = 0.25$ mm of water ($\tau = 0.4$ dyn/cm^2 and $l = 2$ cm). The curves that have been obtained at other pressure differentials differ from the given curves in their slope because an increase or a decrease in the pressure differential is equivalent to a decrease or an increase in the blow-off time.

The different curves in Figure 4.21 correspond to the virtually identical experimental conditions. However, the curves are shifted from each other along the time axis. It may be supposed that this shift depends on the difference in deposition of the original film. The initial condition (4.253) means that the initial contact angle of the film is $90° = \pi/2$. Because the real initial state of the film being blown off differed from the aforementioned one, this determines certain scattering in the effective blow-off times. At considerable blow-off times, the influence of the initial state $x(h)$ results only in a shift of experimental curves from each other, the curves themselves remaining similar.

In the case under consideration, the dependences $\eta(h)$ calculated by the above method prove to be practically the same. Certain minor differences are observed for larger thicknesses. This is

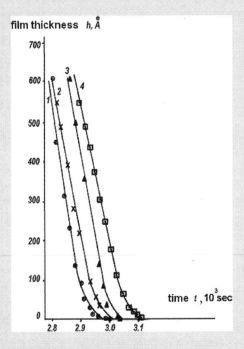

FIGURE 4.21 Thickness of the hexadecane film versus the blow-off time, *t*.

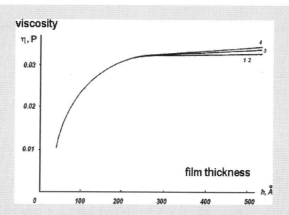

FIGURE 4.22 Calculated dependencies of the boundary viscosity of hexadecane on the film thickness. Curves 1–4 correspond to curves 1–4 in Figure 4.21.

quite understandable because the influence of the deposition of the initial film prior to blowing off can influence only the initial stage of the process (at high enough thicknesses).

The experimental data were processed according to Eq. (4.298), giving the dependence of the boundary viscosity on the film thickness.

Figure 4.22 represents the deduced dependences of the boundary viscosity on the film thickness. Figure 4.22 shows that in the thickness range of 50–200 Å, the viscosity of hexadecane is lower than the bulk viscosity of liquid.

It should be noted that the same decreased value of the hexadecane viscosity in the boundary layer was obtained for different pressure differences $\Delta p = 0.25$ mm, and $\Delta p = 0.06, 0.125$, and 0.5 mm of water when the evaporation rates differed. This shows the independence of the procedure from the evaporation rates. Measurements of viscosity at distances $l = 1$ cm and $l = 3$ cm from the wetting boundary also yielded also the same decreased viscosity values within the indicated thickness range. This means that the phenomenon of the film spreading at the solid substrate (in the opposite direction to the blow off gas) can be neglected. Otherwise, the spreading phenomenon would be most pronounced at $l = 1$ cm.

A hexadecane molecule is composed of 14 methylene groups that interact with both the substrate and each other, and of two methyl end groups. The number of methyl groups being small, they do not play an important role. In the layer closest to the wall, one may expect a horizontal orientation of molecules owing to their sufficient rigidity and length.

The observed decrease in the viscosity of hexadecane seems to be attributable to the horizontal orientation of molecules.

Conclusions

The theory thus developed allows determining the boundary viscosity as a function of the distance to the substrate, $\eta(h)$. The dependency $\eta(h)$ is determined according to Eq. (4.298) using the extension procedure set forth in (iii).

In the case of constant viscosity, the thickness of a layer at a given point varies with time according to the hyperbolic law,

$$h_\infty(t) = -\frac{it}{2} + \frac{x\eta_\infty}{\tau \cdot t}.$$

4.9 Combined Heat and Mass Transfer in Tapered Capillaries with Bubbles under the Action of a Temperature Gradient

Simultaneous flows of vapor and a liquid in a thin liquid film under the action of both temperature and pressure gradients are investigated in this section as a function of the radius and taper of capillaries for decane and hexane. The regions of the greatest effect of film flow are established [41].

Combined heat and mass transfer in porous media is of great interest in a number of areas. The space inside any porous material has a sophisticated structure and is difficult to model even using computer simulations. It is even more difficult to model a combined heat and mass transfer in such complicated structures. Below we investigate a mass transfer of oils in a model porous system, which is a tapered capillary. The mass transfer takes place between two menisci of oil at $x = 0$ and $x = L$ under the action of an imposed constant temperature gradient (Figure 4.23).

We assume that the characteristic scale in the axial direction is much bigger than the capillary radius, that is $r(x) \ll L$.

Because of vapor adsorption, a film of adsorbed liquid, $h(x)$, forms on the walls and the liquid flows in this film on the capillary walls. It is assumed that in each cross section there is a local equilibrium between the vapor and the adsorbed liquid film. According to Chapter 2, the hydrodynamic pressure in the adsorbed liquid film, P_l, is equal to

$$P_l(x) = P_a - \frac{\gamma[T(x)]}{r(x) - h(x)} - \frac{r(x)}{r(x) - h(x)}\Pi[h(x)], \tag{4.301}$$

where Pa is the pressure in the ambient air, $\gamma(T)$ and $\Pi(h)$ are the liquid–vapor interfacial tension, which is temperature dependent, and the Derjaguin's pressure of the flat oil films, respectively.

A bubble of air separates two menisci of the wetting liquid, each having a different curvature. The surface of the capillary between the two menisci is covered by a wetting film, whose thickness, $h(x)$, is a function of the coordinate x; boundary values of the film thickness are $h(0) = h_0$ and $h(L) = h_1$.

The excess pressure in the liquid film as before is as follows:

$$P = P_a - P_l. \tag{4.302}$$

Note, now the excess pressure is a function of the position. Equality of chemical potentials of the liquid molecules in the vapor phase and in the liquid films results in the same equation as in the Chapter 2, which applied however locally:

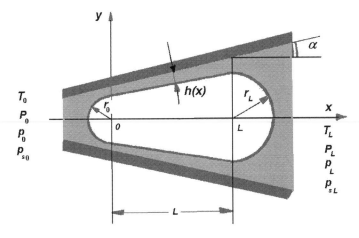

FIGURE 4.23 Combined heat and mass transfer in a tapered capillary of the length L. Radius of the capillary at cross section x is $r(x) = r_0 + x \cdot \tan\alpha$; and imposed temperature gradient is constant, $\frac{dT}{dx} = \frac{T_L - T_0}{L} = \text{const}$.

$$P = \frac{RT}{v_m} \ln \frac{p_s(T)}{p(x)}, \tag{4.303}$$

where $T(x)$, $p_s(T)$, and $p(x)$ are the local temperature, the local pressure of the saturated vapor, and the local vapor pressure, respectively; R is the gas constant; and v_m is the molar volume of the liquid. Below we assume that the liquid under investigation is an oil, which completely wets the capillary walls. Note, in this section we are using R for gas constant, not R_g as before. We are considering a flow and evaporation of oil, that is, the Derjaguin's pressure in this case is determined by dispersion forces only,

$$\Pi(h) = \frac{A}{h^3},$$

where A is the Hamaker constant.

The combination of Eqs. (4.301) and (4.303) results in

$$P_l(x) = P_a - \frac{\gamma[T(x)]}{r(x) - h(x)} - \frac{r(x)}{[r(x) - h(h)]} \frac{A}{h^3(x)} = \frac{RT(x)}{v_m} \ln \frac{p_s[T(x)]}{p(x)}, \tag{4.304}$$

see Section 2.9 for the explanation why the modified Derjaguin's pressure

$$\frac{r(x)}{[r(x) - h(h)]} \frac{A}{h^3(x)}$$

should be used.

Now we should deduce equation for a flow in thin liquid films under the action of both pressure gradient and surface tension gradient.

We consider a flow in thin films at both a low Reynolds number, $Re \ll 1$, and low capillary numbers, $Ca \ll 1$ because below we are considering only a slow, steady-state flow. This means that both front and rear menisci (Figure 4.23) have an equilibrium profile and we ignore any deviation from the equilibrium shapes.

Taking into account the condition $r(x) \ll L$, we can conclude as in Chapter 4 that (i) the velocity in the radial direction is much smaller than the velocity in the axial direction; (ii) the velocity in the axial direction, v, depends only on the radial position r; and (iii) the pressure in the films depends only on the axial component x. After that, the remaining Stokes equations taking into account the definition (4.302) becomes

$$-P' = \eta \left(\frac{d^2 v}{dr^2} + \frac{1}{r} \frac{dv}{dr} \right),$$

with a non-slip condition on the capillary walls

$$v(r(x)) = 0,$$

and the condition on the free film surface

$$\eta \frac{dv}{dr} \bigg|_{r=r(x)-h} = \frac{d\gamma}{dx} = \frac{d\gamma}{dT} T'.$$

This condition shows the tangential stress on the free film surface caused by the presence of the surface tension gradient, which is in turn caused by the temperature gradient. The symbol ′ indicates the derivative with respect to x.

Solution of this equation with the two boundary conditions and Eq. (4.304) results in the following expression for the mass flow rate, Q:

$$Q = Q_1 + Q_2 + Q_3, \tag{4.305}$$

where the following expressions for mass flow rates in vapor, Q_1, in the thin liquid film under the action of the temperature gradient (thermocapillary flow), Q_2, and under the action of the pressure gradient, Q_3, are

$$Q_1 = -\lambda_1 \frac{\pi (r-h)^2 D\mu}{RT} p'$$

$$Q_2 = \lambda_2 \frac{\pi r \rho h^2}{\eta} \frac{d\gamma}{dT} T' \tag{4.306}$$

$$Q_3 = -\lambda_3 \frac{2\pi r \rho R T h^3}{3\eta v_m} \left(\frac{p'}{p} - \frac{p_s'}{p_s} + \frac{T'}{T} \ln \frac{p_s}{p} \right),$$

where D is the diffusion coefficient of the vapor; and ρ, η, and μ are density, viscosity, and mass of the molecule of the liquid, respectively. We introduced the additional coefficients λ_1, λ_2, and λ_3. Each of these coefficients can be 1 or zero to "switch" on or off the corresponding part of the mass flow rate.

The dependency of the film thickness on x, $h(x)$, is determined by Eq. (4.304).

The boundary conditions for the first Eq. (4.305) are as follows:

$$p(0) = p_s \left[T(0) \right] \exp\left(-\frac{2\gamma \left[T(0) \right] v_m}{r(0) RT(0)} \right)$$

$$p(L) = p_s \left[T(L) \right] \exp\left(-\frac{2\gamma \left[T(L) \right] v_m}{r(L) RT(L)} \right) \tag{4.307}$$

where we considered that the saturated pressure depends only on the temperature, T.

Equation (4.305), differentiated with respect to x, and after that was solved numerically using the quasi-linearization method [41] combined with the method of iterations:

$$\left\| \left(Q_n - Q_{n+1} \right) / Q_n \right\| < \varepsilon, \tag{4.308}$$

where n is the number of iterations, and $\varepsilon = 0.01$ is the given relative error. Even in the case where condition (4.308) is satisfied, not less than 10 iterations were made.

The calculations were made using mean values of ρ, η, and D inside the bubble length, L, corresponding to the mean temperature $T_m = (T_1 + T_2)/2$. It was assumed that a constant temperature gradient is imposed $\frac{dT}{dx} = $ const, that is, that the distribution of the temperature, T, was linear inside the bubble. The distribution of the vapor pressure, $p(x)$, corresponding to the imposed value of $\frac{dT}{dx} = $ const and the value of the total mass flow rate, Q, were calculated. Tabulated values of the liquid–air interfacial tension, $\gamma(T)$, and the saturated vapor pressure, $p_s(T)$, were used in calculations. The intermediate values were found by linear (for the interfacial tension, γ) and logarithmic (for the saturated pressure, p_s) interpolations.

Calculations were made for two nonpolar liquids with a different volatility: decane and hexane. The well-known equation of the molecular component of the Derjaguin's pressure, $\Pi(h) = A/h^3$, where $A \sim 10^{-13}$ erg [41], was used as the isotherm $\Pi(h)$. The value of the Hamaker constant, A, was regarded as independent of temperature [42]. Dependences of the total mass flow rate, Q, and of the individual components of the flow rates on the radius and the taper of the capillaries were obtained for $\frac{dT}{dx} = $ const from $-1°$ to $-100°$/cm with a constant mean temperature $T_m = 300$ K. The total mass flow rate was calculated using Eq. (4.305), setting $\lambda_1 = \lambda_2 = \lambda_3 = 1$. The vapor component, Q_1, corresponding only to the diffusion of the vapor (in the presence of fixed films, only decreasing the cross section of the capillary), was calculated with $\lambda_1 = 1$ and $\lambda_2 = \lambda_3 = 0$. An analogous method was used to calculate the components Q_2 (with $\lambda_2 = 1$ and $\lambda_1 = \lambda_3 = 0$) and Q_3 (with $\lambda_3 = 1$ and $\lambda_1 = \lambda_2 = 0$), respectively, to the thermocapillary flow of the film and to film flow under the action of the pressure gradient dP/dx, due to the gradients of the Derjaguin's and capillary pressures (see Eq. 4.304).

The instantaneous radius of the conical capillary is given by the following equation:

$$r = r_1 + \alpha x, \tag{4.309}$$

where $\alpha = [r(L) - r(0)]/L = \Delta r/L$ is the angle of taper of the capillary. Note, we are using the small angle approximation.

For hexane, the following values were used: $D = 0.075$ cm^2/s; $\mu = 86.17$ g/mol; $\rho = 0.6534$ g/cm^3; $\eta = 2.95 \cdot 10^{-3}$ P; $v = 132$ cm^3/mol. For decane, the corresponding values are: $D = 0.046$ cm^2/s; $\mu = 142.28$ g/mol; $\rho = 0.7245$ g/cm^3; $\eta = 8.2 \cdot 10^{-3}$ P, $v = v196$ cm^3/mol. The ratio L/r varied from 10 (for large capillary radii, r) to 1000 (for small capillary radii, r). The absolute values of L were ranged from 10^{-1} to 10^{-3} cm. In view of this choice, with high a temperature gradient, $\frac{dT}{dx} =$ const, the temperature difference at the boundaries of a bubble was very small. This justified the possibility of using constant mean values of ρ, η, and D.

Cylindrical Capillaries

We consider first the effect of the radius of the capillaries on film flow. To this end, the values of the ratio Q/Q_1 were calculated for cylindrical capillaries of equal radius r (Figure 4.24). At $Q/Q_1 = 1$, the principal mechanism of mass transfer is the diffusion of vapor: $Q_1 \gg Q_2 + Q_3$. Figure 4.24 shows that, with an increase in the radius of the capillaries, the effect of film flow decreases. However, in narrow capillaries ($r \sim 10^{-6}$ cm), film flow is the principal mechanism of transfer. The flow of vapor, Q_1, in such thin capillaries contributes to the total flow, which is an order of magnitude less, although the thickness of the films remains considerably less than the radius of the capillaries

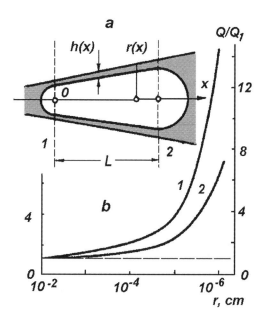

FIGURE 4.24 (a) Calculating scheme, which was used both in the case of tapered [$r(0) < r(L)$] and cylindrical capillaries [$r(0) = r(L)$]; and (b) dependences of the ratio of the total mass flow rate, Q, to the mass flow rate of vapor, Q_1, on the radius of cylindrical capillaries for (1) decane and (2) hexane: $r = 10^{-2}$ cm, $L = 10^{-1}$ cm; $r = 10^{-3} - 10^{-5}$, $L = 10^{-2}$; and $r = 10^{-6}$, $L = 10^{-3}$.

($h/r = 0.14$). The effect of film-transfer is more pronounced for the less volatile decane: curve 1 passes above curve 2 for hexane. The results of the calculations do not disclose a dependence of Q/Q_1 on the value of the temperature gradient, $\frac{dT}{dx} = $ const, which is explained by the smallness of the absolute values of $T_1 - T_2$, leading to a practically linear dependence of all the flows on the temperature gradient, $\frac{dT}{dx} = $ const.

The flow in thin films in cylindrical capillaries is mainly determined by thermocapillary flow. The pressure-driven flow, Q_3, is directed to the opposite side as compared with the flows of vapor, Q_1, and thermocapillary flow, Q_2, and is always much less as compared with the thermocapillary flow, Q_2. This can also be shown by obtaining analytical dependences for the individual components of the flow. Thus, in a cylindrical capillary, the ratio Q_3/Q_2 is equal to

$$Q_3 / Q_2 \approx 4/3\left(A/\gamma r^2\right)^{1/3}. \tag{4.310}$$

At $r \to \infty$, $(Q_3/Q_2) \to 0$. Therefore, we evaluate the ratio (4.310) for the thinnest capillaries, setting $A = 10^{-13}$ erg and $\gamma = 30$ dyn/cm. Calculations show that, for $r \geq 10^{-6}$ cm, the contribution of the flow Q_3 does not exceed 3% in comparison with the flow Q_2.

Tapered Capillaries

The picture of the flows in conical (tapered) capillaries has quite different features as compared with the flow in cylindrical capillaries. With an increase in the angle of taper of the capillary, α, the contribution of the flow, Q_3, due to the differences in the capillary pressures of the menisci, rises and, at sufficiently large values of α, becomes dominating. As an example, Figure 4.25 shows dependences of Q/Q_1 on α for $r_1 = 10^{-4}$ cm. As can be seen from Figure 4.25, at $\alpha \leq 10^{-5}$, the ratio of Q/Q_1 remains constant, and it has the same value as in a cylindrical capillary of radius $r = r(0)$. The dashed lines show the course of the dependences of $(Q_1 + Q_2)/Q_1$ on α. The flow rates Q_1 and Q_2 depend only weakly on α. Thus, the deviation of curves 1 and 2 downward with $\alpha \geq 10^{-5}$ is connected only with the effect of the flow Q_3 directed

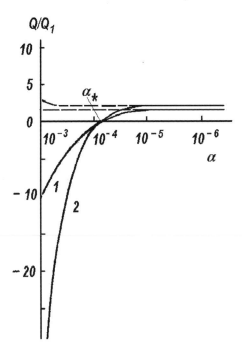

FIGURE 4.25 Dependences of Q/Q_1 on the taper of the capillaries, α, for (1) decane and (2) hexane: $r_1 = 10^{-4}$ cm, $L = 10 = ^{-2}$ cm, and $\frac{dT}{dx} = -1°$/cm.

toward the side of the narrowing of the capillary. At $\alpha \geq 10^{-4}$, its effect becomes dominating, and because of which the total flow rate, Q, changes sign.

At some value of the taper $\alpha = \alpha^*$, the total flow rate vanishes, $Q = 0$: the flow of vapor, Q_1, and the thermocapillary flow, Q_2, equalize out in a direction opposite to the direction of the film flow, Q_3. Such circulating flows are realized, for example, in heat pipes, as well as in porous media completely saturated by a liquid. The local taper of a pore, corresponding to the condition $Q = 0$, depends on the temperature gradient and the radius of the capillaries. It can be found using the calculating procedure suggested in this section.

In porous bodies, where the capillaries have a variable radius, bubbles of air, displaced toward the hot side, can be held in expanded pores if the values of r, α, and the imposed temperature gradient, ΔT, are such that the condition $Q = 0$ is satisfied. This hold-up means that the air- and, consequently, the moisture-content cannot vary with time, despite the existence of the imposed temperature gradient, ΔT, because the flow under the action of the temperature gradient, ΔT, is compensated by a reverse flow under the action of the difference arising in the capillary pressures.

With a transition to thinner capillaries, the picture of the mass transfer is becoming even more complicated. Figure 4.26 gives the results of calculations for $r(0) = 10^{-6}$ cm with the same values of L and ΔT as in Figure 4.25. Whereas, at a small taper ($\alpha < 10^{-7}$), the dependences of Q/Q_1 have qualitatively the same shape as in Figure 4.25; in the region of larger values of α, the dependency of Q/Q_1 on α undergoes a discontinuity and changes the sign. The reasons for such a course of the curves can be established by analyzing the dependences of the individual components of the flow on α. It follows from the calculations that the thermocapillary flow, Q_2, is relatively small and almost independent of α. The flow of vapor, Q_1, directed at small angles α toward the cold side, starts to be retarded with a rise in α because of an increase in the pressure of the vapor above the meniscus $r(L)$. Note, that in the calculations it was assumed that $r(0) = $ const, $r(L) = r(0) + \alpha L$, that is, $r(0)$ increases with α. At $\alpha = \alpha_0 = 5 \cdot 10^{-7}$, the difference Δp, due to the different temperatures of the menisci, is compensated by the Kelvin's difference in the pressures Δp, connected with the different curvature of the menisci. At $\alpha > 5 \cdot 10^{-7}$, the flow of vapor changes direction: the Kelvin's difference in the pressures of the vapor exceeds the thermal difference. An increase in the taper, α (at $L = $ const) leads to a sharp increase in the rate of the backflow of vapor ($Q_1 < 0$).

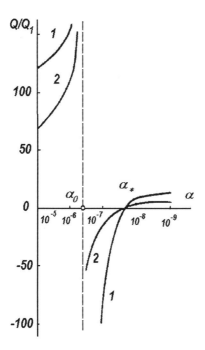

FIGURE 4.26 Dependences of Q/Q_1 on the taper of the capillaries, α, for (1) decane and (2) hexane: $r = 10^{-6}$ cm, $L = 10^{-2}$ cm, and $\frac{dT}{dx} = -1$°/cm.

In distinction from the vapor flow, Q_1, the pressure flow in thin liquid films, Q_3, is always directed toward the side of the narrower part of the capillary and $Q_3 < 0$. With an increase in the taper, α, the absolute values of Q_3 rises. At $\alpha > 5 \cdot 10^{-8}$, Q_3 makes the principal contribution to the total flow; hence, here, it can be assumed that $Q \sim Q_3$. As the flow Q_1 approaches zero, the ratio $Q/Q_1 = Q_3/Q_1 \to \pm\infty$, where $Q/Q_1 \to -\infty$ at $Q_1 > 0$ and $Q_3/Q_1 \to +\infty$ at $Q_1 < 0$, which explains the results given in Figure 4.26.

As can be seen from Figure 4.26, at $\alpha > \alpha_0$, the values of Q/Q_1 fall with an increase in the value of α. This is explained by the decreasing contribution of the film flow, Q_3, with an increase in the mean radius of the capillary $r_m = r(0) + (\alpha/2)$. In thin pores ($r \sim 10^{-6}$ cm) and for small bubbles ($L = 10^{-2}$ cm), even a small taper leads to a sharp increase in the rate of mass transfer due to the flow of the liquid films. In the region of values of the taper $\alpha \sim \alpha_0$, the flow $Q_3 \sim Q$ can exceed the flow of vapor. At $\alpha = \alpha_0$, the flow in the liquid phase is the sole mass transfer mechanism in the system. For the less volatile decane, the effect of film transfer is more strongly expressed (curves 1) than for hexane (curves 2) (Figure 4.26).

The smaller the radius of the capillary, the greater the role of the taper, even the absolute value of the taper is insignificant. Whereas, at $r(0) = 10^{-4}$ cm, the effect of the taper starts to be appreciable at $\alpha > 10^{-5}$, with $r = 10^{-6}$ cm, it is already appreciable at $\alpha > 10^{-8}$.

4.10 Spreading of Non-Newtonian Liquids over Solid Substrates

In this section, the spreading of drops of a non-Newtonian liquid (Ostwald–de Waele liquid) over horizontal solid substrates is theoretically investigated in the case of complete wetting and small dynamic contact angles. Both gravitational and capillary regimes of spreading are considered. The evolution equation deduced for the shape of the spreading drops has self-similar solutions, which allows obtaining spreading laws for both gravitational and capillary regimes of spreading. In the gravitational regime case of spreading, the profile of the spreading drop is provided [43].

Note, in this section the action of the Derjaguin's pressure in the vicinity of an apparent three-phase contact line is not considered at all because very little (if anything) is known about the Derjaguin's pressure in the case of non-Newtonian liquids. This is why we are unable to fit the obtained below solutions within the transition zone and, as a result, all solutions include an unknown parameter, which can be determined by matching the obtained solution and within the transition zone in the same way as in Section 4.2. However, this matching can be made only in the case when the Derjaguin's pressure is known.

The spreading of liquids over solid surfaces has been studied from both theoretical and experimental points of view in Sections 4.1 and 4.2, where investigations dealt with the kinetics of spreading of Newtonian liquids. Both gravitational and capillary spreading regimes were considered and the spreading laws were established. It was shown in Sections 4.1 and 4.2 that the singularity at the three-phase contact line is removed by the action of the surface forces (Derjaguin's pressure).

The theoretically predicted spreading laws for gravitational and capillary regimes have been deduced as $R(t) \sim t^{1/8}$ and $R(t) \sim t^{1/10}$, respectively, where $R(t)$ is the radius of the base of the spreading drop and t is the time (see Section 4.1). Comparison of the predicted spreading laws with experimental data in Section 4.2 has shown the excellent agreement.

However, a number of liquids (polymer liquids and suspensions [44,45]) show a non-Newtonian behavior. The aim of this section is to extend the similarity solution method used in Section 4.1 to the case of spreading of non-Newtonian liquids over solid surfaces and to deduce the corresponding spreading laws for both gravitational and capillary regimes of spreading.

Governing Equation for the Evolution of the Profile of the Spreading Drop

The problem below is solved under the following assumptions:

   (i) It is a complete wetting case.
   (ii) The dynamic contact angle is low.

(iii) Reynolds number is low, $Re \ll 1$.

(iv) The rheological properties of the liquid are determined by the viscosity dependency, $\eta(S)$, on the shear deformation rate, S [44].

We do not consider the flow in the vicinity of the three-phase contact line, where the influence of the surface forces become important. The influence of these forces determines only a pre-exponential factor in the spreading law according to Sections 4.1 and 4.2. In the case of a Newtonian liquid, the pre-exponential factor has been found to be almost insensitive to the details of the surface forces (Section 4.2).

The second assumption means that $R_* \gg H_*$, where R_*, H_* are characteristic scales in the radial and axial directions, respectively. In the case of the complete wetting assumptions (ii) and (iii) are always satisfied at the final stage of spreading.

Let us consider a drop of an incompressible non-Newtonian liquid with density ρ and surface tension γ and which spreads over a horizontal solid substrate. The density, viscosity and the pressure gradient in the surrounding air are neglected. The solid substrate is assumed rigid and non-deformable.

Both the axisymmetric and cylindrical problems of spreading (Figure 4.27) are considered below.

The liquid flow inside the spreading drop obeys the following equations:

The incompressibility condition,

$$\frac{1}{r}\frac{\partial}{\partial r}(rv_r) + \frac{\partial v_z}{\partial z} = 0, \tag{4.311}$$

and Navier-Stokes equations, which are considerably simplified using assumptions (i)–(iv):

$$-\frac{\partial p}{\partial z} - \rho g = 0, \tag{4.312}$$

$$-\frac{\partial p}{\partial r} + \frac{\partial}{\partial z}\left[\eta(S)\frac{\partial v_r}{\partial z}\right] = 0, \tag{4.313}$$

where (r, φ, z) is the cylindrical coordinate system, the z-axis coincides with the axis of symmetry, and $z = 0$ corresponds to the solid substrate; all functions are independent of the angle, φ, because of symmetry; g is the gravity acceleration; p is the pressure; v_r and v_z are radial and axial components of the velocity vector, respectively.

Let $D_{rr} = \frac{\partial v_r}{\partial r}$, $D_{\varphi\varphi} = \frac{v_r}{r}$, $D_{rz} = D_{zr} = \frac{1}{2}\left(\frac{\partial v_r}{\partial z} + \frac{\partial v_z}{\partial r}\right)$, $D_{zz} = \frac{\partial v_z}{\partial z}$ be components of the deformation rate tensor. The parameter S is expressed in terms of the components of the deformation rate tensor as $S = 2\left(D_{rr}^2 + D_{\varphi\varphi}^2 + D_{zz}^2 + 2D_{rz}^2\right)$. Under the assumption (ii) this expression becomes

$$S = \left(\frac{\partial v_r}{\partial z}\right)^2, \tag{4.314}$$

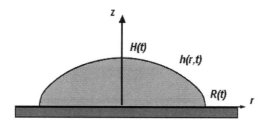

FIGURE 4.27 Cross section of the spreading drop.

No-slip conditions are adopted on the solid substrate:

$$v_z = 0, \quad z = 0, \tag{4.315}$$

$$v_r = 0, \quad z = 0. \tag{4.316}$$

Let the profile of the spreading drop be $z = h(r,t)$, which is to be determined. Boundary conditions on the free liquid–air interface include the kinematic condition

$$\frac{\partial h}{\partial t} = v_z - v_r \frac{\partial h}{\partial r}, \quad z = h(r,t); \tag{4.317}$$

and the conditions for the normal and tangential components of the stress tensor

$$p = p_a - \gamma \frac{1}{r} \frac{\partial}{\partial r}\left(r \frac{\partial h}{\partial r}\right), \quad z = h, \tag{4.318}$$

$$\eta(S)\frac{\partial v_r}{\partial z} = 0, \quad z = h, \tag{4.319}$$

where p_a is the pressure in the surrounding air and small terms are omitted based on the assumption $R_* \gg H_*$. This assumption means $h'^2 \ll 1$, that is, the low slope approximation is valid.

Using the continuity Eq. (4.311) and the kinematic condition (4.317), an equation describing the evolution of the drop profile becomes

$$\frac{\partial h}{\partial t} + \frac{1}{r} \frac{\partial}{\partial r}\left(r \int_0^h v_r dz\right) = 0. \tag{4.320}$$

Integration of Eq. (4.312) over z with boundary condition (4.318) results in the following expression for the pressure distribution:

$$p = p_a + \rho g(h - z) - \gamma \frac{1}{r} \frac{\partial}{\partial r}\left(r \frac{\partial h}{\partial r}\right), \tag{4.321}$$

This equation shows that $\frac{\partial p}{\partial r}$ is independent of z. Integration of Eq. (4.313) over z with boundary condition (4.319) gives

$$\eta\left(\left(\frac{\partial v_r}{\partial z}\right)^2\right)\frac{\partial v_r}{\partial z} = -\frac{\partial p}{\partial r}(h - z). \tag{4.322}$$

Integration over $f(h)$ of this equation results in the following expression for the radial component of the velocity

$$v_r = \left\{ G\left(\frac{\partial p}{\partial r}(h - z)\right) - G\left(\frac{\partial p}{\partial r}h\right)\right\}\bigg/ \frac{\partial p}{\partial r}, \tag{4.323}$$

where the function $G(x)$ is determined as

$$G(x) = \int_0^x F(y)dy, \quad \eta\left(F^2(x)\right)F(x) \equiv x. \tag{4.324}$$

Substitution of the expression for the radial velocity (4.323) into Eq. (4.320) gives an equation that describes the profile of the spreading drop:

$$\frac{\partial h}{\partial t} = \frac{1}{r}\frac{\partial}{\partial r}\left(r\frac{h^3\frac{\partial p}{\partial r}}{3\eta_{ef}\left(h\frac{\partial p}{\partial r}\right)}\right), \tag{4.325}$$

where an effective viscosity, $\eta_{ef}(y)$, is determined as

$$\frac{1}{3}\eta_{ef}^{-1}(y) = y^{-2}G(y) - y^{-3}\int_0^y G(z)dz. \tag{4.326}$$

In the case of the spreading of a cylindrical drop (a plane symmetrical, two-dimensional drop), a similar consideration using a Cartesian coordinate system yields the following equation:

$$\frac{\partial h}{\partial t} = \frac{\partial}{\partial r}\left(\frac{h^3\frac{\partial p}{\partial r}}{3\eta_{ef}\left(h\frac{\partial p}{\partial r}\right)}\right), \tag{4.327}$$

where

$$p = p_a + \rho g(h-z) - \gamma\frac{\partial^2 h}{\partial r^2}. \tag{4.328}$$

In Eqs. (4.327) and (4.328), r is the Cartesian coordinate perpendicular to the z-axis. The effective viscosity, $\eta_{ef}(y)$, is determined by the same relations (4.324) and (4.326) as in the axisymmetric case.

Equations (4.325), (4.327) and (4.321), (4.328) can be rewritten in the following form:

$$\frac{\partial h}{\partial t} = \frac{1}{r^m}\frac{\partial}{\partial r}\left(r^m\frac{h^3\frac{\partial p}{\partial r}}{3\eta_{ef}\left(h\frac{\partial p}{\partial r}\right)}\right), \tag{4.329}$$

$$p = p_a + \rho g(h-z) - \gamma\frac{1}{r^m}\frac{\partial}{\partial r}\left(r^m\frac{\partial h}{\partial r}\right), \tag{4.330}$$

where $m = 0$ and $m = 1$ correspond to the case of the cylindrical and axisymmetric drops, respectively.

Substitution of Eq. (4.330) into Eq. (4.329) results in the following nonlinear differential equation, which describes the evolution of the profile of the spreading drop:

$$\frac{\partial h}{\partial t} = \frac{1}{r^m}\frac{\partial}{\partial r}\left(r^m\frac{h^3\left\{\rho g\frac{\partial h}{\partial r} - \gamma\frac{\partial}{\partial r}\left[\frac{1}{r^m}\frac{\partial}{\partial r}\left(r^m\frac{\partial h}{\partial r}\right)\right]\right\}}{3\eta_{ef}\left(h\left\{\rho g\frac{\partial h}{\partial r} - \gamma\frac{\partial}{\partial r}\left[\frac{1}{r^m}\frac{\partial}{\partial r}\left(r^m\frac{\partial h}{\partial r}\right)\right]\right\}\right)}\right), \tag{4.331}$$

Conservation of the drop volume reads

$$V = 2\pi^m\int_0^{R(t)} h(r,t)r^m dr, \quad h(R(t),t) = 0, \tag{4.332}$$

where V is the drop volume in the axisymmetric case and the cross-section area in the case of cylindrical drops; and $R(t)$ is the location of the three-phase contact line.

In the case of a non-Newtonian liquid [44]:

$$\eta(S) = k\, S^{(n-1)/2}, \; n > 0, \tag{4.333}$$

where $n < 1$ corresponds to pseudo-plastic fluids (can be as low as 0.1 for some natural rubbers [44]) and $n > 1$ corresponds to dilatant fluids.

In this case, $\eta_{ef}(y)$ is

$$\eta_{ef}(y) = \frac{2n+1}{3n} k^{\frac{1}{n}} |y|^{\frac{n-1}{n}}, \tag{4.334}$$

and Eq. (4.331) transforms into

$$\frac{\partial h}{\partial t} = \frac{n}{2n+1} k^{-\frac{1}{n}} \frac{1}{r^m} \frac{\partial}{\partial r} \left\{ r^m h^{\frac{2n+1}{n}} \operatorname{sign}\left(\rho g \frac{\partial h}{\partial r} - \gamma \frac{\partial}{\partial r} \left[\frac{1}{r^m} \frac{\partial}{\partial r} \left(r^m \frac{\partial h}{\partial r} \right) \right] \right) \right.$$
$$\left. \left| \rho g \frac{\partial h}{\partial r} - \gamma \frac{\partial}{\partial r} \left[\frac{1}{r^m} \frac{\partial}{\partial r} \left(r^m \frac{\partial h}{\partial r} \right) \right] \right|^{\frac{1}{n}} \right\}. \tag{4.335}$$

The position of the macroscopic three-phase contact line, $R(t)$, is the time-dependent characteristic horizontal scale. Let us introduce the time-dependent characteristic thickness of the drop, $H(t)$, using the conservation law (4.332) as $V = 2\pi^m H(t) R^{m+1}(t)$, or

$$H(t) = \frac{V}{2\pi^m R^{m+1}(t)}. \tag{4.336}$$

Self-similar solutions of Eq. (4.335) are tried below in the following form:

$$h(r,t) = H(t)\zeta(\hat{r}), \quad r = R(t)\hat{r}, \tag{4.337}$$

According to (4.332) and (4.336), the dimensionless drop profile, $\zeta(\hat{r})$, satisfies the following conditions:

$$\int_0^1 \zeta(\hat{r})\hat{r}^m d\hat{r} = 1, \quad \zeta(1) = 0. \tag{4.338}$$

In the following two sections, two limiting cases of spreading are considered, when either capillary or gravitational forces can be neglected. In the case of gravitational regime of spreading the capillary forces can be neglected if $R(t) \gg a = \sqrt{\frac{\gamma}{\rho g}}$. In the capillary regime of spreading the capillary forces dominate, that is, $R(t) \ll a = \sqrt{\frac{\gamma}{\rho g}}$.

Gravitational Regime of Spreading

In this case, the spreading equation (4.335) transforms into

$$\frac{\partial h}{\partial t} = \frac{n}{2n+1} \left(\frac{\rho g}{k} \right)^{1/n} \frac{1}{r^m} \frac{\partial}{\partial r} \left\{ r^m h^{\frac{2n+1}{n}} \operatorname{sign}\left(\frac{\partial h}{\partial r} \right) \left| \frac{\partial h}{\partial r} \right|^{1/n} \right\}$$

Assuming the solution in the form (4.337), this equation yields

$$-\frac{V}{2\pi^m}\frac{\dot{R}}{R^{m+2}}\left[(m+1)\zeta + \hat{r}\zeta'\right] = \frac{n}{2n+1}\left(\frac{\rho g}{k}\right)^{\frac{1}{n}}\left(\frac{V}{2\pi^m}\right)^{\frac{2n+2}{n}}$$
$$\frac{1}{R^{\frac{(2m+3)(n+1)}{n}}}\frac{1}{\hat{r}^m}\frac{\partial}{\partial \hat{r}}\left(\hat{r}^m \zeta^{\frac{2n+1}{n}}\operatorname{sign}(\zeta')|\zeta'|^{\frac{1}{n}}\right). \tag{4.339}$$

This equation shows that the radius of spreading, $R(t)$, should satisfy the following equation:

$$\dot{R} = \lambda\frac{n}{2n+1}\left(\frac{\rho g}{k}\right)^{1/n}\left(\frac{V}{2\pi^m}\right)^{(n+2)/n}R^{-[(n+2)(m+1)+1]/n}. \tag{4.340}$$

where λ is a dimensionless constant. If $R(t)$ is selected according to Eq. (4.340) then Eq. (4.339), which describes the dimensionless drop profile, $\zeta(\hat{r})$, becomes

$$\lambda\left[(m+1)\zeta + \hat{r}\zeta'\right] = -\frac{1}{\hat{r}^m}\left(\hat{r}^m\zeta^{(2n+1)/n}\operatorname{sign}\zeta'|\zeta'|^{1/n}\right)'. \tag{4.341}$$

Equation (4.341) should be solved with the following boundary conditions:

$$\zeta'(0) = 0, \tag{4.342}$$

which is the symmetry condition in the drop center, and

$$\zeta(1) = 0. \tag{4.343}$$

The conservation law (4.338) gives the third equation for the determination of unknown parameter λ.

The solution of Eq. (4.340) with initial condition $R(0) = R_0$ is

$$R(t) = R_0\left[1 + \frac{n}{2n+1}\frac{\lambda}{\alpha}\left(\frac{\rho g}{k}\right)^{1/n}\left(\frac{V}{2\pi^m}\right)^{(n+2)/n}\frac{1}{R_0^{1/\alpha}}t\right]^\alpha, \tag{4.344}$$

where the spreading exponent α is

$$\alpha = \frac{n}{(m+2)(n+2)-1}. \tag{4.345}$$

In the case of the Newtonian liquid ($n = 1$) this exponent is $1/(3m + 5)$. Thus, at $n < 1$: $\alpha < 1/(3m + 5)$ and the drop spreads slower than the Newtonian liquid; at $n > 1$: $\alpha > 1/(3m + 5)$ and the drop spreads faster than the Newtonian liquid. The dependence of the spreading exponent α on n, according to Eq. (4.345), in the case of axisymmetric spreading ($m = 1$) is shown in Figure 4.28.

Multiplying Eq. (4.341) by $\hat{r}^m$ and after integration considering the symmetry condition (4.342) results in

$$\lambda\hat{r}^{m+1}\zeta = -\hat{r}^m\zeta^{(2n+1)/n}\operatorname{sign}\zeta'|\zeta'|^{1/n},$$

The solution of this equation with boundary condition (4.343) is

$$\zeta = \operatorname{sign}\lambda|\lambda|^{n/(n+2)}\left[\frac{n+2}{n+1}\left(1 - \hat{r}^{n+1}\right)\right]^{1/(n+2)}. \tag{4.346}$$

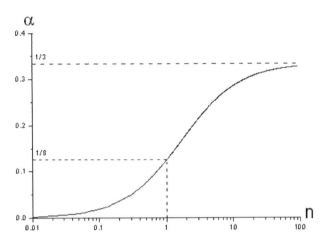

FIGURE 4.28 Axisymmetric ($m = 1$) gravitational regime of spreading. Spreading exponent α (Eq. 4.345) versus n; $n = 1$ Newtonian fluid.

Substitution of solution (4.346) into the conservation law (4.338) gives an equation for the determination of the unknown constant λ:

$$\operatorname{sign}\lambda\,|\lambda|^{n/(n+2)}\frac{(n+2)^{1/(n+2)}}{(n+1)^{(n+3)/(n+2)}}\int_0^1 (1-x)^{1/(n+2)}x^{(m-n)/(n+1)}dx = 1,$$

or, in terms of the gamma function (4.331),

$$\operatorname{sign}\lambda\,|\lambda|^{n/(n+2)}\frac{(n+2)^{1/(n+2)}}{(n+1)^{(n+3)/(n+2)}}\frac{\Gamma\!\left(\dfrac{n+3}{n+2}\right)\Gamma\!\left(\dfrac{m+1}{n+1}\right)}{\Gamma\!\left(\dfrac{(n+3)(n+1)+(m+1)(n+2)}{(n+1)(n+2)}\right)} = 1$$

This equation has the following solution:

$$\lambda = \left(\frac{(n+1)^{n+3}}{n+2}\right)^{\frac{1}{n}}\left\{\frac{\Gamma\!\left(\dfrac{(n+3)(n+1)+(m+1)(n+2)}{(n+1)(n+2)}\right)}{\Gamma\!\left(\dfrac{n+3}{n+2}\right)\Gamma\!\left(\dfrac{m+1}{n+1}\right)}\right\}^{\frac{n+2}{n}}. \tag{4.347}$$

It is possible to check that $\lambda \to m+1,\ if\ n \to \infty$. If $n \to 0$, then at $m = 0$,

$$\lambda \sim 1.0121989\left(\frac{9}{8}\right)^{1/n} \to \infty$$

at $m = 1$,

$$\lambda \sim 1.7635846\cdot\left(\frac{225}{32}\right)^{1/n} \to \infty$$

Dependence of λ on n at $m = 1$ is shown in Figure 4.29.

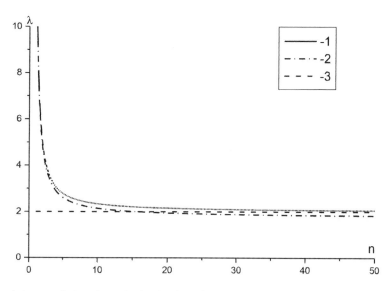

FIGURE 4.29 Axisymmetric ($m = 1$) gravitational regime of spreading. Dimensionless constant λ (Eq. 4.347) versus n; $n = 1$ Newtonian fluid. Solid line: according to Eq. (4.347). Broken line: asymptotic dependence $\lambda \sim 1.7635846 \cdot \left(\frac{225}{32}\right)^{1/n}$ at $n \ll 1$.

Substitution of the expression (4.347) into the solution (4.346) gives the dimensionless drop profile in the case of gravitational regime of spreading

$$\zeta(\hat{r}) = (n+1)\frac{\Gamma\left(\dfrac{(n+3)(n+1)+(m+1)(n+2)}{(n+1)(n+2)}\right)}{\Gamma\left(\dfrac{n+3}{n+2}\right)\Gamma\left(\dfrac{m+1}{n+1}\right)}\left(1 - \hat{r}^{n+1}\right)^{1/(n+2)}. \tag{4.348}$$

In the case of axisymmetric spreading ($m = 1$) of a Newtonian liquid ($n = 1$), this equation gives

$$\zeta = \frac{8}{3}\left(1 - \hat{r}^2\right)^{1/3}, \tag{4.349}$$

which coincides with the solution obtained in Section 4.1. The profiles of axisymmetric spreading drops at different n according to Eq. (4.348) are shown in Figure 4.30.

Capillary Regime of Spreading

In this case, Eq. (4.335) becomes

$$\frac{\partial h}{\partial t} = -\frac{n}{2n+1}\left(\frac{\gamma}{k}\right)^{1/n}\frac{1}{r^m}\frac{\partial}{\partial r}\left\{r^m h^{\frac{2n+1}{n}}\,\text{sign}\left(\frac{\partial}{\partial r}\left[\frac{1}{r^m}\frac{\partial}{\partial r}\left(r^m\frac{\partial h}{\partial r}\right)\right]\right)\left|\frac{\partial}{\partial r}\left[\frac{1}{r^m}\frac{\partial}{\partial r}\left(r^m\frac{\partial h}{\partial r}\right)\right]\right|^{1/n}\right\}. \tag{4.350}$$

Assuming the solution of Eq. (4.350) in the form (4.336) and (4.337) yields

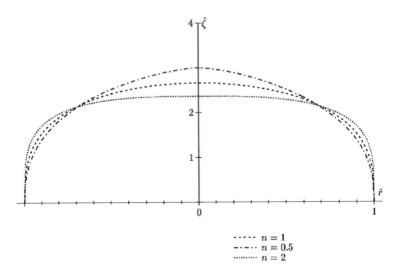

FIGURE 4.30 Axisymmetric ($m = 1$) gravitational regime of spreading. Drop profiles (Eq. 4.348) at different values of n.

$$-\frac{V}{2\pi^m}\frac{\dot{R}}{R^{m+2}}\left[(m+1)\zeta + \hat{r}\zeta'\right] = -\frac{n}{2n+1}\left(\frac{\gamma}{k}\right)^{1/n}\left(\frac{V}{2\pi^m}\right)^{\frac{2n+2}{n}}\frac{1}{R^{(3+2m/n+5/n+2m)}}\frac{1}{\hat{r}^m}$$

$$\times\frac{\partial}{\partial\hat{r}}\left(\hat{r}^m\,\mathrm{sign}\left[\frac{1}{\hat{r}^m}\left(\hat{r}^m\zeta'\right)'\right]\left|\left[\frac{1}{\hat{r}^m}\left(\hat{r}^m\zeta'\right)'\right]\right|^{1/n}\zeta^{(2n+1)/n}\right). \tag{4.351}$$

This equation shows that a self-similar solution exists if the radius of spreading, $R(t)$, satisfies the following equation

$$\dot{R} = \lambda\frac{n}{2n+1}\left(\frac{\gamma}{k}\right)^{1/n}\left(\frac{V}{2\pi^m}\right)^{(n+2)/n}R^{-\frac{5+n+mn+2m}{n}}. \tag{4.352}$$

In this case, the dimensionless drop profile is determined by the following ordinary differential equation

$$\lambda\left[(m+1)\zeta + \hat{r}\zeta'\right] = \frac{1}{\hat{r}^m}\left\{\hat{r}^m\,\mathrm{sign}\left[\frac{1}{\hat{r}^m}\left(\hat{r}^m\zeta'\right)'\right]\left|\left[\frac{1}{\hat{r}^m}\left(\hat{r}^m\zeta'\right)'\right]\right|^{1/n}\zeta^{(2n+1)/n}\right\}', \tag{4.353}$$

where λ is a dimensionless constant, which is different from the previous case (gravitational regime of spreading).

The solution of Eq. (4.352), with the initial condition $R(0) = R_0$, has the following form:

$$R(t) = R_0\left[1 + \frac{n}{2n+1}\frac{\lambda}{\alpha}\left(\frac{\gamma}{k}\right)^{1/n}\frac{V^{(n+2)/n}}{\left(2\pi^m\right)^{(n+2)/n}R_0^{1/\alpha}}t\right]^{\alpha}, \tag{4.354}$$

where the spreading exponent α is

$$\alpha = \frac{n}{mn + 2m + 5 + 2n}, \tag{4.355}$$

which gives 0.1 in the case of the axisymmetric spreading of a Newtonian liquid ($m = n = 1$) and 1/7 in the case the cylindrical drop spreading ($n = 1$, $m = 0$). The dependence (4.355) in the case of axisymmetric spreading, $m = 1$, is shown in Figure 4.31.

Comparison of Eqs. (4.345) and (4.355) shows that capillary and gravitational regimes of spreading give the same dependence: $R(t) \sim t^{1/(m+2)}$, if $n \gg 1$.

Multiplying Eq. (4.353) by $\hat{r}^m$ and integrating yields

$$\lambda \hat{r}^{m+1} \zeta = \hat{r}^m \text{sign} \left[\frac{1}{\hat{r}^m} \left(\hat{r}^m \zeta' \right)' \right]' \left| \left[\frac{1}{\hat{r}^m} \left(\hat{r}^m \zeta' \right)' \right]' \right|^{1/n} \zeta^{(2n+1)/n} + C,$$

where C is an integration constant. Taking into account that the functions $\zeta(\hat{r})$ and $\zeta''(\hat{r})$ should be finite and the symmetry condition in the drop center, $\zeta'(0) = \zeta'''(0) = 0$ this equation becomes

$$\left[\frac{1}{\hat{r}^m} \left(\hat{r}^m \zeta' \right)' \right]' = \text{sign} \lambda |\lambda|^n \hat{r}^n \zeta^{-n-1}. \tag{4.356}$$

According to Section 4.2, an alternative way of solving Eq. (4.350) is as follows. The whole drop is subdivided into two parts: the main spherical part ("outer solution") and the narrow region close to the three-phase contact line ("inner solution"). The volume of the liquid in the narrow "inner zone" can be neglected as compared with the main spherical part. To determine the liquid flow in the inner region, the surface forces action should be introduced in this narrow region. However, the solution in the inner region gives only a pre-exponential factor in the spreading law. Its dependence on the details of the flow has been found insignificant in the case of Newtonian liquids and complete wetting. In the case of non-Newtonian liquids and the complete wetting case, surface forces in the vicinity of the three-phase contact line can be of a very complex nature. We assume that the influence of these complex and unknown surface forces gives only a correction of a pre-exponential factor as in the case of Newtonian liquids. That is why the flow in the inner region is not considered here.

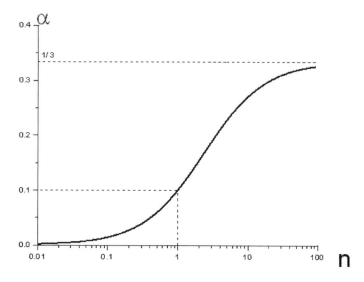

FIGURE 4.31 Axisymmetric ($m = 1$) capillary regime of spreading. Spreading exponent α (Eq. 4.355) versus n.

Accordingly, the right-hand side of Eq. (4.356) is small everywhere except for a narrow vicinity of the three-phase contact line, where ζ approaches zero, and hence, the volume of the liquid in this small region can be neglected. Thus, the central part of the spreading drop is

$$\zeta = \frac{(m+1)(m+3)}{2}\left(1-\hat{r}^2\right),\tag{4.357}$$

that is, the parabolic cap. The dynamic contact angle, θ, is determined by the following relationship $(\tan\theta \approx \theta)$:

$$\theta = \frac{(m+1)(m+3)V}{2\pi^m R^{m+2}}.\tag{4.358}$$

From Eq. (4.352)

$$\left(\frac{U}{\lambda\dfrac{n}{2n+1}\left(\dfrac{\gamma}{k}\right)^{1/n}\left(\dfrac{V}{2\pi^m}\right)^{(n+2)/n}}\right)^{\frac{n(m+2)}{5+n+mn+2m}} = R^{-(m+2)},$$

where $U = \dot{R}(t)$. Substitution of this equation into Eq. (4.358) results in the following dependence of the dynamic contact angle of the spreading drop on the rate of spreading, U:

$$\theta = (m+1)(m+3)\left(\frac{2n+1}{\lambda n}\right)^{\frac{n(m+2)}{5+n+mn+2m}}\left(\frac{V}{2\pi^m}\right)^{\frac{1-n}{5+n+mn+2m}}\left(\frac{k}{\gamma}\right)^{\frac{m+2}{5+n+mn+2m}}U^{\frac{n(m+2)}{5+n+mn+2m}},\tag{4.359}$$

or

$$U = \frac{\lambda n}{2n+1}\left(\frac{\gamma}{k}\right)^{\frac{1}{n}}\left(\frac{2\pi^m}{V}\right)^{\frac{1-n}{n(m+2)}}\left(\frac{\theta}{(m+1)(m+3)}\right)^{\frac{5+n+mn+2m}{n(m+2)}}.\tag{4.360}$$

For the case of an axisymmetric spreading of a drop of Newtonian liquid ($n = m = 1$) Eq. (4.360) gives the Tanner's law (see Sections 4.1 and 4.2). It is interesting to note that the spreading law (4.360) is independent of the drop volume only in the case of Newtonian liquids. If $n \neq 1$, then the right-hand side of Eq. (4.360) depends on the drop volume.

Conclusions

The spreading of drops of non-Newtonian liquids over horizontal solid substrates was theoretically investigated. An equation was deduced describing the liquid profiles of axisymmetric and cylindrical spreading drops. The problem was solved under the following assumptions: (i) complete wetting, (ii) low dynamic contact angle approximation, (iii) low Reynolds number, $Re \ll 1$, and (iv) the rheological properties of the liquid are determined by the viscosity dependency, $\eta(S)$, on the shear deformation rate, S. In the case of complete wetting, the second and the third assumptions are always valid at the final stage of spreading and allows a considerable simplification of the description of the spreading.

Both gravitational and capillary spreading regimes of the spreading were considered. In the gravitational regime case, the spreading law and the profile of the spreading drop have been completely determined and given by Eqs. (4.344) and (4.348), respectively. In the case of the capillary regime, the spreading law and the apparent contact angle of the spreading drop have been calculated and given by Eqs. (4.354) and (4.359), respectively.

If the case of a Newtonian liquid ($n = 1$), the spreading laws for gravitational and capillary regimes, (4.344) and (4.354), coincide with those found earlier in Sections 4.1 and 4.2. If $n < 1$, an axisymmetric drop spreads slower than a drop of a corresponding Newtonian liquid with the same volume. If $n > 1$, the spreading exponents for gravitational and capillary regimes are greater than in the case of Newtonian liquids.

It is interesting to note that both capillary and gravitational axisymmetric regime of spreading give the same power law: $R(t) \sim t^{1/3}$ if $n \gg 1$.

REFERENCES

1. Dussan, E. B. *Ann. Rev. Fluid Mech.*, 11, 371 (1979).
2. Greenspan, H. P. *J. Fluid Mech.*, 84, 125 (1978).
3. Hocking, L. M., and Rivers, A. D. *J. Fluid Mech.*, 121, 425 (1982).
4. de Gennes, P. *Rev. Mod. Phys.*, 57 (3), 827 (1985)
5. Blake, T. D., and Haynes, J. M. *J. Colloid Interface Sci.*, 30 (3), 421 (1969).
6. Kochurova, N. N., and Rusanov, A. I. *J. Colloid Interface Sci.*, 81 (2), 297 (1981).
7. Eggers, J., and Evans, R. *J. Colloid Interface Sci.*, 280, 537 (2004).
8. Blake, T. D., and Shikhmurzaev, Y. D. *J. Colloid Interface Sci.*, 253 196 (2002).
9. Neogi, P., and Miller, C. *J. Colloid Interface Sci.*, 86 (2), 525 (1982).
10. Marmur, A. *Adv. Colloid Interface Sci.*, 19, 75 (1983).
11. Kalinin, V. V., and Starov, V. M. *Colloid J.*, USSR Academy of Sciences, 48 (5), 767 (1986).
12. Summ, B. D., and Goryunov, Y. V. *Physicochemical Principles of Wetting and Spreading* [in Russian]. Khimiya, Moscow, Russia, 1976.
13. Starov, V. M. *Adv. Colloid Interface Sci.*, 39, 147 (1992).
14. Cazabat, A. M., and Cohen-Stuart, M. A. *J. Phys. Chem.*, 90, 5849 (1986).
15. Starov, V. M., Kalinin, V., and Chen, J. D. *Adv. Colloid Interface Sci.*, 50, 187 (1994).
16. Chen, J. D. *J. Colloid Interface Sci.*, 122, 60 (1988).
17. Ausserre, D., Picart, A. M., and Leger, L. *Phys. Rev. Lett.*, 57, 2671 (1986).
18. Leger, L., Erman, M., Guinet-Picart, A. M., Ausserre, D., Strazielle, C., Benattar, J. J., Rieutord, F., Daillant, J., and Bosio, L. *Rev. Phys. Appl.*, 23, 104 (1988).
19. Tanner, L. H. *J. Phys. D.*, 12, 1473 (1979).
20. Kalinin, V. V. and Starov, V. M. *Colloid. J.*, Russian Academy of Sciences, 54 (2), 214 (1992).
21. Nakaja, C. H. *J. Phys. Soc. Jpn. E*, No. 2, 539 (1974).
22. Starov, V. M., Churaev, N. V., and Khvorostyanov, A. G. *Colloid J.*, USSR Academy of Sciences, 39 (1), 176 (1977).
23. Bretherton, F. P. *J. Fluid Mech.*, 10, 166 (1961).
24. Friz, G. *Angew Z. Phys.*, 19, 374 (1965).
25. Ludviksson, V., and Lightfoot, E. N. *AIChE J.*, 14, 674 (1968).
26. Deryaguin, B. V., Churaev, N. V., and Muller, V. M. *Surface Forces*, Consultants Bureau, Plenum Press, New York, 1987.
27. Ivanov, V. I., Kalinin, V. V., and Starov, V. M. *Colloid J.*, USSR Academy of Sciences, 53 (1), 25 (1991).
28. Ivanov, V. I., Kalinin, V. V., and Starov, V. M. *Colloid J.*, USSR Academy of Sceinces, 53 (2), 218 (1991).
29. Starov, V. M., Kalinin, V. V., and Ivanov, V. I. *Colloids Surf. A Physicochem. Eng. Asp.*, 91, 149 (1994).
30. Deryagin, B. V., and Levi, S. M. *Film Coating Theory*, Focal Press, London, UK, 1960.
31. Levich, V. G. *Physicochemical Hydrodynamics*, Prentice-Hall, Englewood Cliffs, NJ, 1962.
32. Quere, D., di Meglio, J.-M., and Brochard-Wyart, F. *Science*, 249, 1256 (1990).
33. Dejaguin, B. V., Karasev, V. V., Starov, V. M., and Khromova, E. N. *J. Colloid Interface Sci.*, 67 (3), 465 (1978).
34. Derjaguin, B. V., Karasev, V. V., Starov, V. M., and Khromova, E. N. *Colloid J.*, USSR Academy of Sciences, 39 (4), 584 (1977).
35. Churaev, N. V. *Colloid J.*, USSR Academy of Sciences, 36, 323 (1974).
36. Derjaguin, B. V., Strakhovskij, G. M., and Malisheva, D. S. *Acta Physicochim. URSS*, 19, 541 (1944).

37. Karasev, V. V., and Derjaguin, B. V. *Zhurn. fiz. khim.*, 33, 100 (1959).

38. Derjaguin, B. V., Zakhavaeva, N. N., Andreev, S. V., and Khomutov, A. M. *Colloid J.*, USSR Academy of Sciences, 24 (3), 289 (1962).

39. Derjaguin, B. V., Zakhavaeva, N. N., Andreev, S. V., Milovidov, A. A., and Khomutov, A. M. *Issledovanie v oblasti poverkhnostnykh sil*, AN SSSR, Moscow, Russia, 1961, p. 139; *Research in Surface Forces*, Consultants Bureau, New York, 1963, p. 110.

40. Derjaguin, B. V., and Zakhavaeva, N. N. *Issledovanie v oblasti vysokomolekulamykh soedinenii*, AN SSSR, Moscow–Leningrad, Russia, 1949, p. 223.

41. Kiseleva, O. A., Starov, V. M., and Churaev, N. V. *Colloid J.*, USSR Academy of Sciences, 39 (6), 1021 (1977).

42. Dzyaloshinskii, I. E., Lifshits, E. M., and Pitaevskii, L. P. *Usp. Fiz. Nauk* (in Russian), 73, 381 (1961).

43. Starov, V. M., Velarde, M. G., Tjatjushkin, A. N., and Zhdanov, S. A. *J. Colloid Interface Sci.*, 257, 284–290 (2003).

44. Pearson, J. R. A. *Mechanics of Polymer Processing*, Elsevier Applied Science Publishers, London, UK, 1985.

45. Carre, A., and Eustache, F. *Langmuir*, 16, 2936 (2000).

5

Spreading over Porous Substrates

Introduction

In this chapter, we consider the kinetics of spreading over porous substrates. The spreading of a liquid over porous substrates is widely used in industry: printing, painting, imbibition into soils, health care and home care products and so on. However, only recently has this process started to develop based on a theoretical, not only on purely empirical, bases.

We show in this chapter that the kinetics of spreading over porous substrates differs substantially from the corresponding kinetics in the case of solid nonporous substrates. First (Section 5.1), we consider the kinetics of spreading over a porous substrate already saturated with the same liquid. This allows extracting an important parameter, which we refer to as an "effective lubrication coefficient." This coefficient turns out to be insensitive to the properties of the porous substrate. This allows us to use its average value for the consideration of the kinetics of spreading over thin porous layers (Section 5.2). We show that the kinetics of spreading over thin porous layers is described by the universal dependency. The unusual finding is that, in the case of complete wetting, the hysteresis of contact angle is present at the spreading on porous substrates. It is not a real hysteresis as seen in the case of partial wetting. We call this "hysteresis" as a hydrodynamic hysteresis, and it is determined by the hydrodynamic flow in the porous substrate.

In the case of spreading over a thick porous substrate (Section 5.3), we are unable to provide a complete theoretical description of the process on the current stage and restrict ourselves to summarizing experimental results only, which surprisingly show some kind of "universal" behavior.

In Section 5.4, spreading from a liquid source is considered.

Spreading of small drops of a non-Newtonian liquid (blood) over a dry porous layer is investigated from both theoretical and experimental points of view (Section 5.5). Both cases of compete and partial wetting are investigated. In the case of *complete wetting*, a system of two differential equations is deduced. The system describes the time evolution of radii of both the drop base and the wetted region inside the porous layer in the course of blood spreading over a dry thin porous substrate. The deduced system of differential equation does not include any fitting parameters and predicts a universal behavior of the dimensionless dependences of the radii of the droplet base and wetted area inside the porous substrate and the contact angle, which are almost completely independent of the rheological properties of blood and which is, however, different from the corresponding dependences for a Newtonian liquid. All the experimental data falls on three universal curves if the appropriate scales are used with plots of the dimensionless radii of the drop base, the wetted region inside the porous layer, and the dynamic contact angle of the drop on dimensionless time. The predicted theoretical relationships are three universal curves accounting quite satisfactorily for the experimental data. The simulated results show a good agreement with experimental data although the bi-porous structure of the filter paper and adsorption of red blood cells inside the porous substrate were not taken into account according to the suggested model.

In the case of *partial wetting* (spreading of blood over nitrocellulose membranes), when pores are sufficiently smaller (less than 0.2 μm) than the size of a red blood cell, red blood cells cannot penetrate the pores and only plasma can penetrate the pores. This observation opens a completely new possibility of investigating red blood cells and plasma separation without damage to the red blood cells.

5.1 Spreading of Liquid Drops over Saturated Porous Layers

In Chapter 4, we considered spreading over smooth homogeneous surfaces. We proved that the singularity at the three-phase contact line is removed by the action of surface forces (see Section 4.2). The vast majority of real solid surfaces are rough to a varying degree and surfaces are frequently either porous or covered with a thin porous skin. These features affect the spreading process. Brinkman's equations are used for the description of the flow in porous media and have a reasonable semi-empirical basis, with physically meaningful coefficients like an *effective* viscosity and a *permeability coefficient*. In this section, we use this approach to study the spreading of liquid drops over thin porous substrates filled with the same liquid, that is, when the thickness of the porous substrate, Δ, is assumed to be much smaller than the drop height, H, $\Delta \ll H_*$ (H_* sets the scale). We follow the evolution of the liquid both in the drop above the porous layer and inside the porous layer.

The spreading of liquids over solid surfaces is one of the fundamental processes with a number of applications such as coating, printing, and painting. In Chapter 4, we considered spreading over smooth homogeneous surfaces. It has been established that singularity at the three-phase contact line is removed by the action of surface forces (see Section 4.2). The vast majority of real solid surfaces are rough to various degrees and in many cases, surfaces are either porous or covered with a thin porous sublayer. The presence of roughness and/or a porous sublayer obviously changes the spreading conditions. The Brinkman's equations [1] are frequently used to describe the flow in porous media and have a reasonable semi-empirical background [2] with physically meaningful coefficients: an *effective viscosity* and a *permeability coefficient*. A new way of calculating these coefficients as functions of the porosity of the porous media has been suggested in reference [3].

An attempt to use the Brinkman's equations to describe the flow inside the porous layer coupled with the drop flow over the layer was undertaken in [4]. Below in this Chapter the same approach is applied to the investigation of the spreading of liquid drops over thin porous substrates filled with the same liquid. The Brinkman's equations are used to describe the liquid flow inside the porous substrate [5].

Theory

The kinetics of the spreading of small liquid drops over thin porous layers saturated with the same liquid is investigated in this section. Theoretical treatment include the kinetics of liquid motion both in the drop above the porous layer and inside the porous layer itself. Consideration of the flow inside the porous layer is based on the Brinkman's equations.

Liquid inside the Drop ($0 < z < h(t, r)$, Figure 5.1)

Let us consider the spreading of an axisymmetric liquid drop over a thin porous layer with thickness Δ saturated with the same liquid. The thickness of the porous layer is assumed to be much smaller than the drop height, that is, $\Delta \ll H_*$, where H_* is the scale of the drop height. The drop profile is assumed to

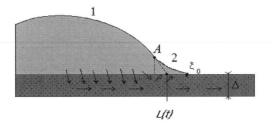

FIGURE 5.1   Spreading of liquid drop over saturated porous layer of thickness Δ. $L(t)$, macroscopic radius of the drop base. (1) Spherical cap (outer region), (2) vicinity of the three-phase contact line (inner region). Inflections point, A, separates inner and outer regions. Inside the outer region, liquid flows from the drop into the porous layer; inside the inner region, the liquid flows from the porous layer into the drop edge.

have a low slope ($H_*/L_* \ll 1$, where L_* is the length scale of the drop base). The influence of gravity is neglected (small drops, Bond number $\ll 1$ or the size of the drop is smaller than the capillary length, a). That is, only capillary forces are taken into account below in this section.

In the case under consideration, the liquid motion inside the drop is described by the following system of equations:

$$\frac{\partial p}{\partial r} = \eta \frac{\partial^2 v}{\partial^2 z^2}, \tag{5.1}$$

$$\frac{\partial p}{\partial z} = 0, \tag{5.2}$$

$$\frac{1}{r}\frac{\partial(rv)}{\partial r} + \frac{\partial u}{\partial z} = 0, \tag{5.3}$$

and boundary conditions:

$$v = v^0, \quad u = u^0, \quad z = +0, \tag{5.4}$$

$$\frac{\partial v}{\partial z} = 0, \quad z = h(t,r), \tag{5.5}$$

$$p = p_a - \frac{\gamma}{r}\frac{\partial}{\partial r}\left(r\frac{\partial h}{\partial r}\right), \tag{5.6}$$

where t is the time; r and z are radial and vertical coordinates, respectively; $z > 0$; and $-\Delta < z < 0$ correspond to the drop and the porous layer, respectively; $z = 0$ is the drop–porous layer interface; p, v, and u are the pressure, radial, and vertical velocity components, respectively; v^0 and u^0 are velocity components at the drop–porous layer interface and are determined by coupling with the flow inside the porous layer; $h(t, r)$ is the drop profile; γ is the liquid–air interfacial tension; and p_a is the pressure in the ambient air.

Equations (5.1) and (5.2) are Stokes equations, in the low slope case; Eq. (5.3) is the incompressibility condition; Eq. (5.5) shows an absence of a tangential stress on the liquid–air interface; Eq. (5.6) presents the pressure jump on the same interface determined by capillary forces only.

The integration of Eqs. (5.1) through (5.3) with boundary conditions (5.4) through (5.6) results in the following equation, which describes evolution of the drop profile:

$$\frac{\partial h}{\partial t} = u^0 - \frac{1}{r}\frac{\partial}{\partial r}\left\{r\left[h^3\frac{\gamma}{3\eta}\frac{\partial}{\partial r}\left(\frac{1}{r}\frac{\partial}{\partial r}\left(r\frac{\partial h}{\partial r}\right)\right) + v^0 h\right]\right\}. \tag{5.7}$$

The liquid velocity components, v^0 and u^0, on the drop–porous layer interface are calculated below in this section.

Inside the Porous Layer beneath the Drop ($-\Delta < z < 0$, $0 < r < L$, Figure 5.1)

If the porous layer is not completely saturated, then the capillary pressure inside the saturated part of the porous layer, p_c, can be estimated as $p_c \approx \gamma/r_*$, where r_* is the scale of capillary radius inside the porous layer. According to Eq. (5.6) the capillary pressure inside the drop, $p - p_a$ can be estimated as $p_a - p \approx \frac{\gamma h_*}{L_*^2} = \frac{h_*}{L_*}\frac{\gamma}{L_*} \ll \frac{\gamma}{L_*} \ll \frac{\gamma}{r_*} \approx p_c$. That is, the capillary pressure inside the spreading drop is substantially smaller (actually several orders of magnitude smaller) than the capillary pressure inside the porous layer in the case of incomplete saturation. This means that the drop pressure cannot disturb in any way the drop–porous layer interface in front of the spreading drop when the porous layer is completely saturated. Hence, this interface always coincides with the surface $z = 0$. It is worth mentioning that everything is going on in time scales much bigger than the initial period considered in [6].

The liquid motion inside the porous layer with thickness Δ is assumed to obey the Brinkman's equations. In this case, the liquid motion inside the porous layer is described by the following system of equations:

$$\frac{\partial p}{\partial r} = \eta_p \frac{\partial^2 v}{\partial z^2} - \frac{v}{K_p}, \tag{5.8}$$

$$\frac{\partial p}{\partial z} = 0, \tag{5.9}$$

$$\frac{1}{r}\frac{\partial(rv)}{\partial r} + \frac{\partial u}{\partial z} = 0, \tag{5.10}$$

and boundary conditions:

$$v = v^0, \quad u = u^0, \quad z = -0 \tag{5.11}$$

$$\frac{\partial v}{\partial z} = u = 0, \quad z = -\Delta, \tag{5.12}$$

$$\eta_p \frac{\partial v}{\partial z}\bigg|_{z=-0} = \eta \frac{\partial v}{\partial z}\bigg|_{z=+0}, \tag{5.13}$$

$$p\big|_{z=-0} = p\big|_{z=+0}, \tag{5.14}$$

where η_p, K_p are viscosity and permeability of the Brinkman's medium, respectively. The boundary condition Eq. (5.12) corresponds to the absence of a tangential stress on the lower boundary of the porous layer, which corresponds to the experimental conditions.

Let us introduce Brinkman's radius, as

$$\delta = \sqrt{\eta_p K_p} \tag{5.15}$$

Solution of Eqs. (5.8) through (5.10) with boundary conditions (5.11) through (5.15) results in

$$v^0 = -\frac{1}{\eta_p}\left(h\delta \coth\frac{\Delta}{\delta} + \delta^2\right)\frac{\partial p}{\partial r}, \quad u^0 = \frac{2}{\eta_p}\frac{1}{r}\frac{\partial}{\partial r}\left[r\left(h\delta^2 + \Delta\delta^2\right)\frac{\partial p}{\partial r}\right]. \tag{5.16}$$

Substitution of these expressions into Eq. (5.7) results in the following equation, which describes the kinetics of the spreading of a liquid drop over a porous substrate:

$$\frac{\partial h}{\partial t} = -\frac{\gamma}{3\eta}\frac{1}{r}\frac{\partial}{\partial r}\left\{r\left(h^3 + 3\alpha \coth\frac{\Delta}{\delta}h^2\delta + 6\alpha h\delta^2 + 3\alpha\delta^2\Delta\right)\frac{\partial}{\partial r}\left(\frac{1}{r}\frac{\partial}{\partial r}\left(r\frac{\partial h}{\partial r}\right)\right)\right\}. \tag{5.17}$$

where $\alpha = \eta/\eta_p$ is the viscosity ratio. According to reference [7], effective viscosity, η_p, is always higher than the liquid viscosity, η, that is, $\alpha < 1$.

If instead of the Brinkman's equation (5.8), a slip condition on the liquid–porous layer interface is used [8], then the vertical velocity component, u^0, should be set to zero in Eq. (5.7) and the following boundary condition should be adopted for the radial velocity component:

$$\frac{\partial v}{\partial z}\bigg|_{z=0} = \frac{\beta}{\delta}\left(v^0 - v_p\right), \tag{5.18}$$

where β is an empirical parameter; $v_p = -K_p/\eta_p\,\partial p/\partial r$ is the velocity inside the porous substrate; and δ/β is a slip length.

If boundary conditions (5.13) and (5.18) are compared, then it is easy to see that this one can be directly obtained from condition (5.18) if we adopt $\eta_p\,\partial v/\partial z\big|_{z=-0} = \eta_p\,v^0 - v_p/\delta$. The combination of this expression and the boundary condition (5.13) gives the following value of the empirical coefficient, β, as

$$\beta = \frac{1}{\alpha}. \tag{5.19}$$

This means that the slip length is equal to $\alpha\delta = \eta\sqrt{K_p/\eta_p}$. K_p and μ_p dependencies on porosity can be calculated according to reference [3].

However, if the slip condition is used, then the omitted contribution of a vertical component, u^0, gives the comparable contribution in the resulting equation.

Equation (5.17) should be solved with the symmetry condition in the drop center,

$$\frac{\partial h}{\partial r} = \frac{\partial^3 h}{\partial r^3} = 0, \quad r = 0, \tag{5.20}$$

and conservation of the drop volume condition,

$$2\pi\int_0^L rhdr = V, \tag{5.21}$$

where $L(t)$ is the macroscopic radius of the drop base (Figure 5.1).

Everywhere at $r < L(t)$ except for a narrow region, ξ, close to three-phase contact line, the following inequality holds: $h \gg \delta$. The size of this narrow region close to the three-phase contact line is calculated below in this section. The same consideration as in Chapter 4 (Section 4.2; see also [9]) shows that the solution of Eq. (5.17) can be presented as "outer" and "inner" solutions (Figure 5.1). The "outer" solution can be deduced in the following way: the left-hand side of Eq. (5.17) should be set to zero and solved with the boundary conditions (5.20) and (5.21) and an additional new boundary condition,

$$h\,(t, L - \xi) \approx 0. \tag{5.22}$$

This procedure gives in the same way as in Chapter 4 (Section 4.2) the "outer" solution in the following form:

$$h(t,r) = \frac{2V}{\pi L^4}(L^2 - r^2), \quad r < L(t) - \xi. \tag{5.23}$$

Equation (5.23) shows that the drop profile retains the spherical shape over the duration of spreading (except for a very short initial stage). Note, the time dependency of the macroscopic position of the apparent three-phase contact line, $L(t)$, is to be determined.

The drop slope at the macroscopic apparent three-phase contact line can be found from Eq. (5.23) as

$$\frac{\partial h}{\partial r}\bigg|_{r=L} = -\frac{4V}{\pi L^3}, \tag{5.24}$$

which is used as a boundary condition for the "inner" solution.

Inside the inner region (Figure 5.1), the solution can be represented in the following form:

$$h(t,r) = \delta \, f(\zeta), \quad \zeta = \frac{r - L(t)}{\chi(t)}, \tag{5.25}$$

where f is a new unknown function; ζ is a similarity variable; and $\chi(t) << L(t)$ is the scale of the "inner" region. This means, that $\xi \approx \chi(t)$. Substitution of the solution in the form Eq. (5.25) into Eq. (5.17) results in the following equation for determining $f(\zeta)$:

$$\frac{df}{d\zeta} \dot{L}(t) = \frac{\gamma}{3\eta} \frac{\delta^3}{\chi^3(t)} \frac{d}{d\zeta} \left[\left(f^3 + 3\alpha \coth \frac{\Delta}{\delta} f^2 + 6\alpha f + 3\alpha \frac{\Delta}{\delta} \right) \frac{d^3 f}{d\zeta^3} \right], \tag{5.26}$$

where the overdot indicates the differentiation with respect to time, t, and small terms are neglected in the same way as in Section 4.2.

Equation (5.26) should not depend on time, this gives two equations:

$$\dot{L}(t) = \frac{\gamma}{3\eta} \frac{\delta^3}{\chi^3(t)} \tag{5.27}$$

and

$$\frac{df}{d\zeta} = \frac{d}{d\zeta} \left[\left(f^3 + 3\alpha \coth \frac{\Delta}{\delta} f^2 + 6\alpha f + 3\alpha \frac{\Delta}{\delta} \right) \frac{d^3 f}{d\zeta^3} \right]. \tag{5.28}$$

The solution of Eq. (5.28) should be matched with the outer solution (5.23). Matching of asymptotic solutions gives the following condition:

$$\frac{\delta}{\chi(t)} \frac{df}{d\zeta} \bigg|_{\zeta \to \infty} = \frac{4V}{\pi L^3(t)}. \tag{5.29}$$

This condition (5.29) should not depend on time, t, which gives

$$\frac{df}{d\zeta} \bigg|_{\zeta \to -\infty} = \lambda \tag{5.30}$$

and

$$\lambda = \frac{4V \chi(t)}{\pi L^3(t) \delta}, \tag{5.31}$$

where λ is a dimensionless constant (see below) in this section. This equation gives

$$\chi(t) = \delta \frac{\pi L^3(t) \lambda}{4V} << L(t). \tag{5.32}$$

This means that the scale of the inner region (Figure 5.1) is proportional to δ and is very small as compared with the size of the drop base.

The combination of Eqs. (5.27) and (5.31) gives the following equation for the radius of spreading, $L(t)$, determination:

$$L^9(t)\dot{L}(t) = \frac{\gamma}{3\eta} \left(\frac{4V}{\pi\lambda} \right)^3 \tag{5.33}$$

The solution of this equation with the initial condition $L(0) = L_0$, where L_0 is the initial drop radius, is

$$L(t) = L_0 \left(1 + \frac{t}{\tau} \right)^{0.1}, \tag{5.34}$$

where $\tau = 3\eta L_0 / 10\gamma \left(\pi\lambda L_0^3 / 4V \right)^3$ is the time scale of the spreading process.

Now, back to the parameter λ determination. Integrating Eq. (5.28) once and setting the integration constant to zero (because of conservation of the drop volume and vanishing of the drop profile in front of the spreading drop) gives

$$\frac{d^3 f}{d\zeta^3} = \frac{f}{f^3 + 3\alpha \coth \frac{\Delta}{\delta} f^2 + 6\alpha f + 3\alpha \frac{\Delta}{\delta}} \tag{5.35}$$

This equation should be solved with the following boundary conditions

$$f(\varsigma_0) = \frac{df(\varsigma_0)}{d\zeta} = 0, \tag{5.36}$$

where ς_0 corresponds in the inner variable to the edge ξ_0 in Figure 5.1, and boundary condition (5.30). We showed in Chapter 4 (see Section 4.2) that this condition cannot be satisfied because Eq. (5.35) does not have a proper asymptotic behavior. An approximate method ("patching" of asymptotic solutions) was suggested in Section 4.2 and allows an approximate determination of parameter λ. Now the unknown constant λ can be calculated in the same approximate way as in Section 4.2. Estimations presented below in this section show, however, that it is not worth doing this.

The second of the two boundary conditions (5.36) indicates a zero micro-contact angle on the microscopic drop boundary.

Equation (5.24) gives the following value of an apparent dynamic contact angle, θ, ($\tan\theta \approx \theta$):

$$\theta = \frac{4V}{\pi L^3}, \tag{5.37}$$

or

$$L = \left(\frac{4V}{\pi\theta}\right)^{1/3}. \tag{5.38}$$

Combination of Eqs. (5.33) and (5.38) results in

$$\frac{dL}{dt} = \omega \frac{\gamma\theta^3}{\mu}, \tag{5.39}$$

where $\omega = 1/3\lambda^3$ is referred as an "effective lubrication coefficient." If the liquid spreads over a solid substrate, an "effective lubrication" is determined by the surface forces action in the vicinity of the three-phase contact line (Section 4.2). In this case, the "effective lubrication coefficient" was calculated in Section 4.2 and its value is 1.36×10^{-2}. The spreading of liquid over a pre-wetted solid substrate was considered in Section 4.3 (see also ref. [10]). Effective lubrication coefficient in this case has been calculated as 1.6×10^{-2}. This shows

1. The effective lubrication coefficient is not very sensitive to experimental conditions. That is, we have chosen not to try to calculate it theoretically although the procedure of its approximate determination is very similar to those presented in Section 4.2.
2. It is reasonable to expect values of effective lubrication coefficient between these two values. Experimentally found values of effective lubrication coefficient agree with our estimations reasonably well (Table 5.1).

TABLE 5.1

Experimental Values Used

Membrane Pore Size (μm)	L_0 (cm)	η (P)	τ (s)	V (cm³)	ω
0.2	0.176	5.58	0.333	0.0039	0.017 ± 0.004
0.2	0.193	0.558	0.0592	0.0040	0.018 ± 0.004
0.2	0.150	1.18	0.00163	0.003	0.014 ± 0.005
0.2	0.150	1.18	0.0026	0.0034	0.014 ± 0.003
0.2	0.165	1.18	0.0086	0.003	0.016 ± 0.005
3	0.119	1.18	0.0546	0.0055	0.015 ± 0.009
3	0.250	1.18	0.461	0.005	0.016 ± 0.009
3	0.253	0.558	0.609	0.005	0.018 ± 0.008
Optical glass	0.226	0.558	0.0378	0.0068	0.012 ± 0.003
Optical glass	0.113	1.18	0.0103	0.002	0.010 ± 0.005

Materials and Methods

In this section we follow ref [5] Silicone oils SO50 (viscosity 0.554 P), SO100 (viscosity 1.18 P) and SO500 (viscosity 5.582 P) purchased from "PROLABO" are used in our spreading experiments. Cellulose nitrate membrane filters purchased from Sartorius (type 113) with average pore sizes of 0.2 and 3 μm, respectively, are used as porous layers. All membranes samples used are a circle plane plate with the radius 25 mm and thickness from 0.0130 to 0.0138 cm. The porosity of membranes ranges between 0.65 and 0.87. Priory to the spreading experiments, the membranes were dried over 3–5 h at 95°C and then stored in a dry atmosphere.

Figure 5.2 shows the sample chamber for monitoring the drop spreading over porous layers and dynamic contact angles. A porous wafer (1) (Figure 5.2) is placed in a thermostatic and hermetically closed chamber (2) with a fixed humidity and temperature. The chamber was made of brass to prevent temperature and humidity fluctuations. In the chamber walls, several channels were drilled to be used for pumping in or out a thermostating liquid. The chamber is equipped with a fan. The temperature is monitored by a thermocouple. Droplets of liquid (3) are placed onto wafer by a dosator (4) (Figure 5.2). The volume of drops is set by the diameter of the separable capillary of the dosator.

The chamber is also equipped with optical glass windows for observation of both the spreading drop shape and size (side view and view from above). Two charged couple device (CCD) cameras and two tape

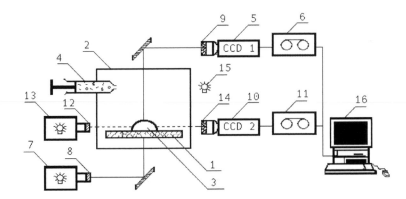

FIGURE 5.2 Experimental setup for monitoring the time evolution of droplets. (1) wafer; (2) sample chamber; (3) tested drop; (4) dosator; (5,10) CCD-cameras; (6,11) VCRs; (7,13) illuminators; (8,9,12,14) interferential light filters [(8,9) with wave length 520 nm, and (12,14) with wave length 640 nm)].

recorders are used for storing of the sequences of spreading. Different colors of monochromatic light are used for the side view and the view from above to eliminate spurious illumination on images. The optical circuit for viewing from above [illuminator (7) as well as the camera (5)] is equipped with interferential light filters (8, 9) with a wave length of 520 nm. The side view circuit [illuminator (13) and camera (10)] are equipped with filters (12, 14) with a wave length of 640 nm. Such an arrangement suppresses the illumination of CCD camera (2) by the diffused light from the membrane and, hence, increases the precision of measurements. The automatic processing of images is carried out using the image-processor "Scion Image." The time discretization in processing ranges from 0.1 to 1 s in different experiments; size of pixel on image corresponds to 0.0125 mm.

The experiments are organized in the following order:

First, a membrane under investigation is placed in the chamber.

Second, a big drop of oil is deposited on the membrane. The volume of the drop, V, is selected as $V = \pi m \Delta R_m^2$, where R_m is the radius of the membrane sample. This choice corresponds to the complete saturation of the membrane by a tested liquid.

Then, after the imbibition process is completed (100–500 s depending on the liquid viscosity), the next tiny drop of the same liquid is deposited on the saturated porous layer and the spreading of this drop is monitored. Drop volumes are measured by direct evaluation of video images. Precision of measurement is around 0.0001–0.0005 cm³.

Results and Discussion. Experimental Determination of the "Effective Lubrication Coefficient" ω

According to the experimental observation [5] in all spreading experiments, the drops retained their spherical shape and no disturbances or instabilities were detected. Experimental data were fitted using the following dependency:

$$L = L_0 (1 + t / \tau)^n \qquad (5.40)$$

It is necessary to comment on the adopted fitting procedure.

If experimentally determined values of the exponent, n, are taken from review [11], then it is easy to see that in most cases this exponent is higher (sometimes considerably higher) than 0.1. Here we present a possible explanation of this phenomenon, which we encountered in our experiments.

In all our experiments (as probably in a number of others' experiments too) drops were placed on the solid substrate using a syringe. Actually, the drops fall from some height. This means that during the very short initial stage of spreading both inertia and a relaxation of the drop shape cannot be ignored. That means during the initial stage of spreading, both the Reynolds number, Re, and the capillary number, Ca, are not small and the capillary regime of spreading is not applicable during this initial stage of spreading. Inertia spreading has been considered in [12], where the following spreading law has been predicted:

$$L(t) = v_\infty t, \qquad (5.41)$$

where $v_\infty = \left(24\gamma / \rho L_0 \right)^{1/2}$; and ρ is the liquid density. This equation shows that the drop spreads much faster during an inertia spreading regime than during a capillary spreading regime. Derivation in [12] is applicable only if the Reynolds number is high enough. Let us estimate the Reynolds number, which is $Re = v_\infty H(t)\rho / \eta$, where $H(t)$ is the maximum drop height. According to [12], $H(t) = V / L^2(t)$ during the inertia period of spreading. Condition $Re \gg 1$ gives

$$\frac{V\rho}{\eta} \left(\frac{\rho L_0}{24\gamma} \right)^{1/2} \frac{1}{t^2} \gg 1$$

or

$$t \ll t_{Re}, \quad t_{Re} = \left(\frac{V\rho}{\eta}\right)^{1/2}\left(\frac{\rho L_0}{24\gamma}\right)^{1/4}. \tag{5.42}$$

The time, t_{Re}, values relevant for our experiments are calculated below in this section. Let us estimate the Reynolds number during the capillary spreading stage. Equation (5.39) gives the velocity of spreading that should be used for calculating the Reynolds number. Simple rearrangement gives $Re = \omega\gamma\theta^3 L/\eta^2 \approx 10^{-2}$. This estimation should be compared with Eq. (5.42), which gives $\left(t_{Re}/t\right)^2 \approx 10^{-2}$ or $t \sim 10 t_{Re} \sim 0.1$ s.

This means that the capillary regime of spreading takes place only at $t > 10\, t_{Re}$.

Using Eq. (5.39) to determine the condition for the fulfillment of the second requirement, $Ca \ll 1$, the latter equation can be rewritten as $Ca = \omega\theta^3$. According to our experimental condition $\omega \approx 10^{-2}, \theta \approx 0.5$. This gives the following estimation of the capillary number: $Ca \sim 10^{-3}$. This means, $U\eta/\gamma \approx L\eta/t\gamma \approx 10^{-3}$ or $t \sim 10^3 t_{Ca} \sim 1$ s, where $t_{Ca} = L\eta/\gamma$. It is necessary to emphasize that t_{Re} is reversibly proportional to $\eta^{1/2}$ (that is, it decreases with the viscosity increase) but t_{Ca} is proportional to the viscosity (that is, it increases with increasing viscosity). This corresponds well to our experimental observations.

In any experimental observation only, a limited number of experimental values of $L(t)$ dependency are measured. If some of these measurements are taken at initial regime of spreading, then it results the fitted exponent having a value higher than 0.1.

For example, let us take the experimental curve in the case of spreading of SO50 over a saturated porous layer (Figure 5.3a and b). In Figure 5.3a, this dependency is presented in a log-log coordinate system. It is easy to see that the whole spreading process consists of two stages corresponding to two different power laws. During the first stage, inertia/shape relaxation cannot be neglected, whereas in the second stage, capillary spreading (exponent close to 0.1) takes place. Figure 5.3b presents the fitting results for this particular spreading experiment. The broken line corresponds to the fitting procedure when all experimental points are taken into account. In this case, the fitted exponent is higher than 0.1 (0.13 ± 0.01). However, if we do not take into account the first three points, which are located within the initial stage of spreading, then the fitted exponent becomes 0.11 ± 0.01, that is, much close to 0.1. Figure 5.3a shows that the initial stage of spreading continues for approximately 0.1 s, which agrees reasonably well with the presented estimations.

The following procedure for the definition of the parameters L_0 and τ was adopted [5]. First, the points, which correspond to the capillary stage of spreading, are selected using the presentation of experimental points in a log-log coordinate system. After this, the fitting procedure using Eq. (5.40) is carried out using only experimental points, corresponding to the capillary stage of spreading. This procedure gives values of L_0 and τ in each run.

After experimental definition of L_0 and τ, the value of ω is calculated in the following way. Equation (5.34) can be rewritten as

$$L = L_0\left(1 + 10\left(\frac{4}{\pi}\right)^3 \frac{V^3\gamma}{L_0^{10}\eta}\omega t\right)^{0.1}. \tag{5.43}$$

Comparison of Eqs. (5.43) and (5.40) gives

$$\omega = \left(\frac{\pi}{4}\right)^3 \frac{1}{\tau}\frac{L_0^{10}}{V^3}\frac{\mu}{10\gamma}. \tag{5.44}$$

The determined values of ω as well as other experimental parameters are presented in the Table 5.1. For comparison in the same table, the results of the spreading over a dry glass (microscope optical glass)

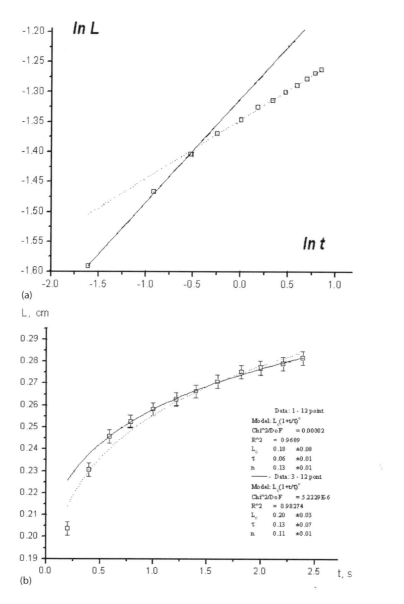

FIGURE 5.3 (a) Radius of the drop base on time in log-log coordinates. SO50 drop, volume 0.0039 cm³, on a porous membrane with average pore size 0.2 μm. (b) Radius of the drop base on time. SO50 drop with volume 0.0039 cm³, on a porous membrane with average pore size 0.2 μm. Broken line, fitted using all experimental points; solid line, fitted using only points which correspond to the capillary stage of spreading. Fitted parameters are given in the insert.

are presented in the last two rows of Table 5.1. The data presented in the Table 5.1 show that (i) the effective lubrication coefficient is higher in the case of spreading over a saturated porous substrate than in the case of "dry spreading" and that (ii) experimentally determined values of "effective lubrication coefficient," ω, agree well with the theoretical estimations. However, the precision of the experimental determination of this parameter does not allow us to extract more information about the effective viscosity of the porous substrate.

5.2 Spreading of Liquid Drops over a Thin Dry Porous Layer: Complete Wetting Case

In this section, we take up the problem treated in the previous section but now we consider the drop spreading over a *dry* porous layer. We consider the case of *complete* wetting. Spreading of small liquid drops over thin dry porous layers is investigated following [5]. The drop motion over a porous layer is caused by an interplay of two processes: (a) the spreading of the drop over already saturated parts of the porous layer, which results in an expanding of the drop base, and (b) the imbibition of the liquid from the drop into the porous substrate, which results in a shrinkage of the drop base and an expanding of the wetted region inside the porous layer. Because of these two competing processes, the radius of the drop goes through a maximum value over time. A system of two differential equations is derived to describe the evolution with time of the radii of both the drop base and the wetted region inside the porous layer. This system includes two parameters: One accounts for the effective lubrication coefficient of the liquid over the wetted porous substrate, and the other is a combination of permeability and effective capillary pressure inside the porous layer. Two additional experiments are used for an independent determination of these two parameters. The system of differential equations does not include any fitting parameter after these two parameters are determined. The experiments were carried out on the spreading of silicone oil drops over various dry microfiltration membranes (permeable in both normal and tangential directions), and the time evolution of the radii of both the drop base and the wetted region inside the porous layer were monitored. All experimental data fell on two universal curves if appropriate scales were used, with a plot of the dimensionless radii of the drop base and of the wetted region inside the porous layer on dimensionless time. The predicted theoretical relationships represent two universal curves accounting quite satisfactory for the experimental data. According to our theoretical prediction, (i) the dynamic contact angle dependence on the same dimensionless time as before should be a universal function, and (ii) the dynamic contact angle should change rapidly over an initial short stage of spreading and should remain a constant value over the duration of the rest of the spreading process. The constancy of the contact angle in this stage has nothing to do with hysteresis of the contact angle: There is no hysteresis in our system because it is a complete wetting case. These conclusions again are in good agreement with our experimental observations [5].

> The spreading of liquids over solid surfaces is a fundamental process with a number of applications in coating, printing, and painting. The spreading over smooth homogeneous surfaces was considered in Chapter 4. It was shown in Section 4.2 that a singularity at the three-phase contact line is removed by the action of surface forces.
>
> However, the vast majority of solid surfaces are rough to some degree and, in many cases, surfaces are either porous or covered with a thin porous sublayer. It was shown in Section 5.1 that the presence of roughness and/or a porous sublayer changes the spreading conditions. In Section 5.1, the spreading of small liquid drops over thin porous layers saturated with the same liquid was investigated. Instead of the "slippage conditions," Brinkman's equations were used in Section 5.1 for the description of the liquid flow inside the porous substrate.

In the present section, we take up the same problem but in the case when a drop spreads over a dry porous layer. The problem is treated under the lubrication theory approximation and in the case of complete wetting. The spreading of "big drops" (but still small enough to neglect the gravity action) over "thin porous layers" is considered.

Theory

The kinetics of liquid motion both in the drop above the porous layer and inside the porous layer itself are considered below in this section. The thickness of the porous layer, Δ, is assumed to be much smaller than the drop height, that is, $\Delta \ll h_*$, where h_* is the scale of the drop height. The drop profile is assumed to have a low slope ($h_*/L_* \ll 1$, where L_* is the scale of the drop base) and the influence of the gravity is neglected (small drops, Bond number $\rho g L_*^2 / \gamma \ll 1$, where ρ, g, and γ are the liquid density, gravity acceleration and the liquid–air interfacial tension, respectively). That is, only capillary forces are considered in this section.

Under these assumptions, a system of two differential equations is obtained to describe the evolution with time of the radius of both the drop base, $L(t)$, and the wetted region inside the porous layer, $l(t)$, (Figure 5.4). Further assumptions are justified in Appendix 1.

As in Section 5.1, the profile of axisymmetric drops spreading over the porous substrate (no difference whether it is dry or saturated with the same liquid) is governed by the following equation:

$$\frac{\partial h}{\partial t} = u^0 - \frac{1}{r}\frac{\partial}{\partial r}\left\{r\left[h^3\frac{\gamma}{3\eta}\frac{\partial}{\partial r}\left(\frac{1}{r}\frac{\partial}{\partial r}\left(r\frac{\partial h}{\partial r}\right)\right) + v^0 h\right]\right\}, \tag{5.45}$$

where $h(t, r)$ is the profile of the spreading drop; t and r are the time and the radial coordinate, respectively; $z > 0$ corresponds to the drop and $-\Delta < z < 0$ corresponds to the porous layer; $z = 0$ is the drop–porous layer interface (Figure 5.4); v, u are the radial and vertical velocity components, respectively; v^0, u^0 are velocity components at the drop–porous layer interface; and η is the liquid viscosity. The liquid velocity components, v^0 and u^0, on the drop–porous layer interface are calculated by matching the flow in the drop with the flow inside the porous layer.

The porous layer is assumed as very thin and the time for saturation in the vertical direction can be neglected relative to other time scales of the process. Let us calculate the time required for a complete saturation of the porous layer in the vertical direction. According to Darcy's equation $u = \frac{K_p}{\eta}\frac{p_c}{z}$, $-\Delta < z < 0$; $u = \frac{dz}{dt}$, where K_p and p_c are the permeability of the porous layer and the effective capillary pressure, respectively; and z is the position of the liquid front inside the porous layer. Solution of this equation results in $\Delta^2 = 2K_p p_c t_\Delta / \eta$, where t_Δ is the time of the complete saturation for the porous layer in the vertical direction and, hence, $t_\Delta = \frac{\Delta^2 \eta}{2K_p p_c}$. Consideration is restricted to $t > t_\Delta$. Estimations show that t_Δ is less than t_0, which is the duration of the initial stage of spreading (see Section 5.1 for details and an estimation of t_0). The capillary spreading regime, which is the only one considered, is not applicable at $t < t_0$. Thus, we must consider times such that $t > max(t_0, t_\Delta)$ when both the initial stage is over and the porous layer is completely saturated in the vertical direction. Accordingly, the porous layer beneath the spreading drop ($0 < r < L(t)$) is always assumed completely saturated.

The capillary pressure inside the porous layer, p_c, can be estimated as $p_c \approx \frac{2\gamma}{r_*}$, where r_* is the scale of capillary radii inside the porous layer. The capillary pressure inside the drop, $p - p_a$, can be estimated as $p - p_a \approx \frac{\gamma h_*}{L_*^2} = \frac{h_*}{L_*}\frac{\gamma}{L_*} << \frac{\gamma}{L_*} << \frac{\gamma}{r_*} \approx p_c$. This means that the capillary pressure inside the pores of the porous layer is several orders of magnitude higher than the capillary pressure in the drop itself, which means that the radius of the wetted region inside the porous layer, $l(t)$, is always bigger than the radius of the droplet base, $L(t)$ (Figure 5.4).

The boundary conditions for Eq. (5.45) are as follows:

Symmetry condition in the drop center:

$$\frac{\partial h}{\partial r} = \frac{\partial^3 h}{\partial r^3} = 0, \quad r = 0, \tag{5.46}$$

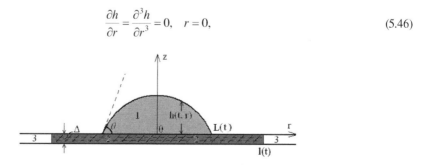

FIGURE 5.4   Cross section of the axisymmetric spreading drop over initially dry thin porous substrate with thickness Δ. (1) Liquid drop; (2) wetted region inside the porous substrate; (3) dry region inside the porous substrate. $L(t)$, radius of the drop base; $l(t)$, radius of the wetted area inside the porous substrate; Δ, thickness of porous substrate; r, z-coordinate system; and $h(t, r)$, profile of the spreading drop.

Conservation of drop volume:

$$2\pi \int_0^L rh\,dr = V(t). \tag{5.47}$$

The drop volume changes over time because of the imbibition of the liquid into the porous layer, which means that

$$V(t) = V_0 - \pi m \Delta l(t)^2, \tag{5.48}$$

where V_0 is the initial volume of the drop; m is the porosity of the porous layer; and $l(t)$ is the radius of the wetted area inside the porous layer. The wetted region is assumed a cylinder with radius $l(t)$ and the height Δ. $l(t)$ is referred to here as the radius of the wetted region inside the porous layer.

Let $t_{\max}$ be the time instant when the drop is completely sucked in by the porous substrate, $V(t_{\max}) = 0 = V_0 - \pi m \Delta l^2_{\max}$, where $l_{\max}$ is the maximum radius of the wetted region in the porous layer. This equation gives

$$l_{\max} = \left(\frac{V_0}{\pi m \Delta} \right)^{1/2}. \tag{5.49}$$

$L_{\max}$ is used then to scale the radius of the wetted region in the porous layer, $l(t)$. It is easy to check that this equation results in $l_{max} > L_*$ in our case.

Combination of Eqs. (5.47) and (5.48) results in

$$2\pi \int_0^L rh\,dr = V_0 - \pi\,m\,\Delta\,l^2(t). \tag{5.50}$$

Everywhere at $r < L(t)$ except for a narrow region, ξ, close to the moving three-phase contact line, we have $h \gg \Delta$, and the liquid motion inside the porous layer under the drop can be neglected both in the vertical and the horizontal directions (see Appendix 1 for details). The size of this narrow region close to the moving three-phase contact line, where the suction of liquid from the drop into the porous substrate takes place, is estimated in Appendix 1.

This means that Eq. (5.45) can be rewritten as

$$\frac{\partial h}{\partial t} = -\frac{\gamma}{3\eta} \frac{1}{r} \frac{\partial}{\partial r} \left\{ rh^3 \frac{\partial}{\partial r} \left(\frac{1}{r} \frac{\partial}{\partial r} \left(r \frac{\partial h}{\partial r} \right) \right) \right\}, \quad r < L(t) - \xi. \tag{5.51}$$

Equation (5.51) should be solved with boundary conditions (5.46) and (5.50) and

$$h(t, L - \xi) \approx 0. \tag{5.52}$$

Following arguments developed in Chapter 4 (Section 4.2), the solution of Eq. (5.51) can be obtained using "outer" and "inner" solutions. The "outer" solution can be deduced in the following way: The left-hand side of Eq. (5.51) should be set to zero. After integration of the resulting equation with boundary conditions (5.46), (5.50) and (5.52) the "outer" solution becomes:

$$h(t,r) = \frac{2V}{\pi L^4} (L^2 - r^2), \quad r < L(t) - \xi. \tag{5.53}$$

Equation (5.53) shows that the drop surface profile remains spherical during the spreading process except for a short initial stage, when the porous layer is not saturated and a final stage, when condition $\Delta \ll h$ is violated everywhere over the whole profile of the drop.

Equation (5.53) gives the following value of the dynamic contact angle, θ, $(\tan\theta \approx \theta)$:

$$\theta = \frac{4V}{\pi L^3}, \tag{5.54}$$

or else

$$L = \left(\frac{4V}{\pi\theta}\right)^{1/3}. \tag{5.55}$$

The drop motion is a superposition of two motions: (a) the spreading of the drop over the already saturated part of the porous layer, which results in an expansion of the drop base, and (b) a shrinkage of the drop base caused by imbibition into the porous layer. Hence, we can write the following equation:

$$\frac{dL}{dt} = v_+ - v_-, \tag{5.56}$$

where v_+, v_- are unknown velocities of the expansion and the shrinkage of the drop base, respectively.

Let us take time derivative of both sides of Eq. (5.55). It gives:

$$\frac{dL}{dt} = -\frac{1}{3}\left(\frac{4V}{\pi\theta^4}\right)^{1/3}\frac{d\theta}{dt} + \frac{1}{3}\left(\frac{4}{\pi V^2\theta}\right)^{1/3}\frac{dV}{dt}. \tag{5.57}$$

Over the whole duration of the spreading over the porous layer, both the contact angle and the drop volume can only decrease with time. Accordingly, the first term in the right-hand side of Eq. (5.57) is positive and the second is negative.

Comparison of these two equations yields

$$v_+ = -\frac{1}{3}\left(\frac{4V}{\pi\theta^4}\right)^{1/3}\frac{d\theta}{dt} > 0$$

$$v_- = -\frac{1}{3}\left(\frac{4}{\pi V^2\theta}\right)^{1/3}\frac{dV}{dt} > 0. \tag{5.58}$$

There are two substantially different characteristic time scales in our problem: $t_{\eta^*} \ll t_{max}$: first, where t_{η^*} and t_{max} are time scales of the viscous spreading and, second, the imbibition into the porous layer, respectively; $\lambda = \frac{t_{\eta^*}}{t_{max}} \ll 1$ is a smallness parameter (around 0.08 under our experimental conditions, see results and discussions in this section). Both time scales are calculated now. Then we have $L = L(T_\eta, T_p)$ [13], where T_η is a fast time of the viscous spreading and T_p is a slow time of the imbibition into the porous substrate. The time derivative of $L(T_\eta, T_p)$ is

$$\frac{dL}{dt} = \frac{\partial L}{\partial T_\eta} + \lambda\frac{\partial L}{\partial T_p}. \tag{5.59}$$

Comparison of Eqs. (5.56), (5.58) and (5.59) shows that

$$v_+ = \frac{\partial L}{\partial T_\eta} = -\frac{1}{3}\left(\frac{4V}{\pi\theta^4}\right)^{1/3}\frac{d\theta}{dt}, \quad v_- = -\lambda\frac{\partial L}{\partial T_p} = -\frac{1}{3}\left(\frac{4}{\pi V^2\theta}\right)^{1/3}\frac{dV}{dt}$$

The smallness of $\lambda = \frac{t_{\eta^*}}{t_{max}} \ll 1$ means that in the case under consideration the two processes act independently: The spreading of the drop over the saturated part of the porous layer and the shrinkage of the drop base caused by the imbibition of the liquid from the drop into the porous layer.

The decrease of the drop volume, V, with time is determined solely by the imbibition into the porous substrate and, hence, the drop volume, V, depends only on the slow time scale.

According to the previous consideration, the whole spreading process can be subdivided into two stages:

1. A first fast stage, when the imbibition into the porous substrate can be neglected and the drop spreads with an approximately constant volume. This stage goes in the same way as the spreading over saturated porous layer, and the arguments developed in Section 5.1 can be used here again.
2. A second slow stage, when the spreading process is already almost over, and the evolution is determined by the imbibition into the porous substrate.

During the first stage Eq. (59) from Section 5.1 can be rewritten in the following form

$$L(t) = \left[\frac{10\gamma\omega}{\eta} \left(\frac{4V}{\pi} \right)^3 \right]^{0.1} (t + t_0)^{0.1}, \tag{5.60}$$

where t_0 is the duration of the initial stage of spreading, when the capillary regime of spreading is not applicable; and ω is an effective lubrication parameter, which has been discussed and estimated in Section 5.1. It is important to emphasize that the effective lubrication parameter, ω, is independent of the drop volume and depends solely on the porous layer properties. According to Eq. (5.60), the characteristic time scale of the first stage of spreading is

$$t_{\eta^*} = \frac{\eta L_0}{10\gamma\omega} \left(\frac{\pi L_0^3}{4V_0} \right)^3, \tag{5.61}$$

where $L_0 = L(t_0)$.

The combination of Eqs. (5.60) and (5.69) gives: $\theta = \left(\frac{4V}{\pi} \right)^{0.1} \left(\frac{\eta}{10\gamma\omega} \right)^{0.3} (t + t_0)^{-0.3}$. The substitution of this expression into the first part of Eq. (5.58) gives the following expression for the velocity of the drop base expansion, v_+:

$$v_+ = 0.1 \left(\frac{4V}{\pi} \right)^{0.3} \left(\frac{10\gamma\omega}{\eta} \right)^{0.1} \frac{1}{(t + t_0)^{0.9}}. \tag{5.62}$$

The substitution of Eq. (5.48) into the second part of Eq. (5.58) gives the following expression for the velocity of the drop base shrinkage, v_-:

$$v_- = \frac{2\pi^{2/3} m\Delta l}{3} \left(\frac{4}{\left(V_0 - \pi m\Delta l^2 \right)^2 \theta} \right)^{1/3} \frac{dl}{dt}. \tag{5.63}$$

The substitution of these two equations into Eq. (5.57) results in

$$\frac{dL}{dt} = 0.1 \left(\frac{4V}{\pi} \right)^{0.3} \left(\frac{10\gamma\omega}{\eta} \right)^{0.1} \frac{1}{(t + t_0)^{0.9}} - \frac{2\pi^{2/3} m\Delta l}{3} \left(\frac{4}{\left(V_0 - \pi m\Delta l^2 \right)^2 \theta} \right)^{1/3} \frac{dl}{dt}. \tag{5.64}$$

The only unknown function now is the radius of the wetted region inside the porous layer, $l(t)$, which is determined below in this section.

Inside the Porous Layer outside the Drop $(-\Delta < z < 0, L < r < l)$

The liquid flow inside the porous layer obeys the Darcy equation:

$$\frac{1}{r}\frac{\partial}{\partial r}\left(r\frac{\partial p}{\partial r}\right) = 0, \quad v = -\frac{K_p}{\eta}\frac{\partial p}{\partial r}.$$

Solving these equations gives

$$p = -(A\eta / K_p)\ln r + B$$

$$v = \frac{A}{r}, \tag{5.65}$$

where A and B are integration constants that should be determined using the boundary conditions for the pressure at the drop edge, $r = L(t)$, and at the circular edge of the wetted region inside the porous layer, $r = l(t)$. This boundary condition is

$$p = p_a - p_c, \quad r = l(t), \tag{5.66}$$

where $p_c \approx \frac{2\gamma}{r_*}$ is the capillary pressure inside the pores of the porous layer, and r_* is a characteristic scale of the pore radii inside the porous layer.

The boundary condition at the drop edge is

$$p = p_a + p_d, \quad r = L(t), \tag{5.67}$$

where p_d is an unknown pressure. It is shown below in this section that $p_d \ll p_c$. However, we keep this small value for a future estimation.

Taking into account these two boundary conditions, both integration constants, A and B, can be determined, giving the following expression for the radial velocity according to Eq. (5.65):

$$v = \frac{K_p\left(p_c + p_d\right)}{\eta r \ln\dfrac{l}{L}}. \tag{5.68}$$

The velocity at the circular edge of the wetted region inside the porous layer is

$$\frac{dl}{dt} = v\Big|_{r=l}.$$

The combination of these two equations gives the evolution equation for $l(t)$:

$$\frac{dl}{dt} = \frac{K_p\left(p_c + p_d\right)}{\eta l \ln\dfrac{l}{L}}. \tag{5.69}$$

An estimation of the time scale t_{max} can be made according to Eq. (5.69) and taking into account Eq. (5.49) as follows: $\frac{l_{max}}{t_{max}} \approx \frac{K_p p_c / \eta}{l_{max}\ln\frac{l_{max}}{L_*}}$, or

$$t_{max} \approx \frac{l_{max}^2 \ln\dfrac{l_{max}}{L_*}}{K_p p_c / \eta} = \frac{V_0 \ln\dfrac{l_{max}}{L_*}}{\pi m \Delta K_p p_c / \eta}. \tag{5.70}$$

The comparison of the estimated values of $t_{\eta*}$ according to Eq. (5.61) and of $t_{\max}$ according to Eq. (5.70) shows that under all our experimental conditions (see below Results and Discussions) the following inequality $t_{\eta*} \ll t_{\max}$ is satisfied.

Omitting the small term, p_d, and substituting Eq. (5.69) into Eq. (5.64) gives the following system of differential equations for the evolution of both the radius of the drop base, $L(t)$, and that of the wetted region inside the porous layer, $l(t)$:

$$\frac{dL}{dt} = 0.1 \left(\frac{4(V_0 - \pi m \Delta l^2)}{\pi} \right)^{0.3} \left(\frac{10\gamma\omega}{\eta} \right)^{0.1} \frac{1}{(t+t_0)^{0.9}} - \frac{2}{3} \frac{\pi m \Delta K_p p_c L / \eta}{(V_0 - \pi m \Delta l^2) \left(\ln \frac{l}{L} \right)}, \quad (5.71)$$

$$\frac{dl}{dt} = \frac{K_p p_c / \eta}{l \ln \frac{l}{L}}. \quad (5.72)$$

Let us make the system of differential equations (5.71) and (5.72) dimensionless using new scales $\bar{L} = L / L_{\max}, \bar{l} = l / l_{\max}, \bar{t} = t / t_{\max}$, where $L_{\max}$ is the maximum value of the drop base, which is reached at the time instant t_m, which is to be determined. The same symbols are used for dimensionless variable as for corresponding dimensional variables (marked with an overbar). The system of Eqs. (5.71) and (5.72) transforms as

$$\frac{d\bar{L}}{d\bar{t}} = \frac{2}{3} \frac{\left(\bar{t}_m + \bar{\tau} \right)^{0.9}}{\left(1 - \bar{l}_m^2 \right)^{1.3} \left(1 + \bar{\chi} \ln \bar{l}_m \right)} \frac{\left(1 - \bar{l}^2 \right)^{0.3}}{\left(\bar{t} + \bar{\tau} \right)^{0.9}} - \frac{2}{3} \frac{\bar{L}}{\left(1 - \bar{l}^2 \right) \left(1 + \bar{\chi} \ln \frac{\bar{l}}{\bar{L}} \right)}, \quad (5.73)$$

$$\frac{d\bar{l}}{d\bar{t}} = \frac{1}{\bar{l} \left(1 + \bar{\chi} \ln \frac{\bar{l}}{\bar{L}} \right)}, \quad (5.74)$$

where $\bar{\tau} = t_0 / t_{\max} \ll 1, \bar{\chi} = 1 / \ln \frac{l_{\max}}{L_{\max}}$. Thus, this system includes only two dimensionless parameters, $\bar{\tau}, \bar{\chi}$: The first one is very small, and the second is insignificantly changed under our experimental conditions because of a weak logarithmic dependence on $l_{\max}/L_{\max}$.

Accordingly, the two dimensionless dependencies $\bar{L}(\bar{t}), \bar{l}(\bar{t})$ should fall on two almost universal curves, which is in the very good agreement with our experimental observations (see the "Results and Discussion" in this section).

According to Eq. (5.73), the dimensionless velocities of the expansion of the drop base, $\bar{v}_+$, and the shrinkage, $\bar{v}_-$, are as follows:

$$\bar{v}_+ = \frac{2}{3} \frac{\left(\bar{t}_m + \bar{\tau} \right)^{0.9}}{\left(1 - \bar{l}_m^2 \right)^{1.3} \left(1 + \bar{\chi} \ln \bar{l}_m \right)} \frac{\left(1 - \bar{l}^2 \right)^{0.3}}{\left(\bar{t} + \bar{\tau} \right)^{0.9}}, \quad \bar{v}_- = \frac{2}{3} \frac{\bar{L}}{\left(1 - \bar{l}^2 \right) \left(1 + \bar{\chi} \ln \frac{\bar{l}}{\bar{L}} \right)}. \quad (5.75)$$

Figure 5.5 shows dimensionless velocity $\bar{v}_+$ and $\bar{v}_-$ calculated according to Eq. (5.75). It appears that

1. The first stage is very short. The capillary spreading prevails on this stage over the drop base shrinkage caused by the liquid imbibition into the porous substrate.
2. The spreading of the drop almost stops after the first stage of spreading, and the shrinkage of the drop base is determined by the suction of the liquid from the drop into the porous substrate.

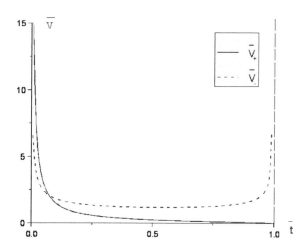

FIGURE 5.5 Dimensionless velocity of spreading ($\bar{v}_+$, solid line) and velocity of drop shrinkage ($\bar{v}_-$, dotted line) on dimensionless time, calculated according to Eq. (5.31). Intersection of these two dependencies determines the value of the dimensionless time $\bar{t}_{max} \approx 0.08$, when the radius of the drop base reaches its maximum value, $\bar{L}_{max} = 1$ (in dimensionless units).

Let us consider the asymptotic behavior of system (5.71) and (5.72) over the second stage of the spreading. According to Figure 5.5, over the second stage of the spreading, velocity of the expansion of the drop, v_+, decreases. To understand the asymptotic behavior, the term corresponding to v_+ in the left-hand site of Eq. (5.71) is omitted. This gives

$$\frac{dL}{dt} = -\frac{2}{3}\frac{\pi m \Delta K_p p_c L / \eta}{\left(V_0 - \pi m \Delta l^2\right)\left(\ln\dfrac{l}{L}\right)},$$ (5.76)

while the second Eq. (5.72) is left unchanged. The system of differential equations (5.72) and (5.76) can be solved analytically. For this purpose, Eq. (5.76) is divided by Eq. (5.72), which gives

$$\frac{dL}{dl} = -\frac{2\pi m \Delta\, L\, l}{3\left(V_0 - \pi m \Delta l^2\right)}.$$

If $V = V_0 - \pi m \Delta l^2$ is used instead of l, this equation takes the following form: $\frac{dL}{dV} = \frac{L}{3V}$, which can be easily integrated, and the solution is

$$V = C\, L^3,$$ (5.77)

where C is an integration constant. Let us rewrite Eq. (5.54) using the same dimensionless variables:

$$\bar{V} = \frac{4L_{max}^3}{\pi V_0}\, \theta\, \bar{L}^3.$$ (5.78)

Comparison of Eqs. (5.77) and (5.78) shows that the dynamic contact angle asymptotically remains constant over the duration of the second stage. This constant value is marked as θ_f. Let us introduce $\theta_m = \frac{\pi V_0}{4L_{max}^3}(1 - \bar{l}_m^2)$, which is the value of the dynamic contact angle at the time instant when the maximum value of the drop base is reached. Then Eq. (5.78) can be rewritten as

$$\frac{\theta}{\theta_m} = \frac{(1 - \bar{l}^2)/(1 - \bar{l}_m^2)}{\bar{L}^3},$$ (5.79)

and this relationship should be a universal function of the dimensionless time, $\bar{t}$. This conclusion agrees well with our experimental observations (see the "Results and Discussion" in this section). It is necessary to emphasize that in the case under consideration, the constancy of the contact angle has nothing to do with the contact angle hysteresis: There is no hysteresis in our system here. θ_f is not a receding contact angle, but forms as a result of a self-regulation of the flow in the drop–porous layer system. Recall that we are considering the complete wetting case, where the hysteresis of contact angle does not take place.

The system of Eqs. (5.71) and (5.72) includes seven parameters, five of which can be measured directly (V_0, γ, η, m, and Δ, which are the initial volume of the drop, the liquid–air interfacial tension, the liquid viscosity, the porosity of the porous layer and the thickness, respectively), and two additional parameters, ω and $K_p p_c$, which should be determined independently. It is noteworthy that the porous layer permeability and the capillary pressure always enter as a product, and this product can thus be considered as a single parameter. A procedure of independent determination of an effective lubrication coefficient, ω, was discussed in Section 5.1.

Experimental Part

Silicone oils SO20 (viscosity 0.218 P), SO50 (viscosity 0.554 P), SO100 (viscosity 1.18 P) and SO500 (viscosity 5.582 P) purchased from PROLABO were used in the spreading experiments. The viscosity of oils was measured using the capillary Engler Viscometer VPG-3 at $20°C \pm 0.5°C$. Cellulose nitrate membrane filters, purchased from Sartorius (type 113), with pore sizes of 0.2 and 3 μm (marked by the supplier) were used as porous layers. These membranes are referred to here as the membrane, 0.2 μm, and the membrane, 3 μm, respectively. Pore size distribution and permeability of membranes were tested using the Coulter Porometer II.

The pore size of the membrane, 0.2 μm, falls in the range of 0.2–0.38 μm, with an average pore size of 0.34 μm. The permeability of the same membrane is equal to 12 L/min*cm² (air flux at the transmembrane pressure 5 bar). The permeability of the membrane, 3 μm is 2.5 L/(min*cm²) at the transmembrane pressure 0.1 bar. All membrane samples used are plane parallel circles of radius 25 mm and thickness in the range of 0.0130–0.0138 cm. The porosity of the membranes ranges between 0.65 and 0.87. The porosity was measured using the difference in the weight of saturated with oil and dry membranes. Membranes were dried over 3–5 h at 95°C and then stored in a dry atmosphere prior to the spreading experiments.

The same experimental device as described in Section 5.1 (Figure 5.2) was used for monitoring the spreading of drops over the initially dry porous layers. The time evolution of the radius of the drop base, $L(t)$, the dynamic contact angle, $\theta(t)$, and the radius of the wetted region inside the porous layer, $l(t)$, were monitored. The porous wafer (1) (Figure 5.2) was placed in a thermostatic and hermetically closed chamber (2), where zero humidity and fixed temperature ($20°C \pm 0.5°C$) were maintained. The distance from the wafer to the tip of the syringe ranged from 0.5 to 1 cm in different experiments. The volume of drops was set by the diameter of the separable replaceable capillary of the syringe in the range of 1–15 μL.

Experiments were carried out in the following order:

- The membrane was placed in the chamber and left in a dry atmosphere for 15–30 min.
- A light pulse produced by a flash gun was used to synchronize the time instant, when the drop started to spread and both videotape recorders (a side view and a view from above) were started.
- A droplet of silicone oil was placed onto the membrane.

Each run was carried out until complete imbibition of the drop into the membrane took place.

Independent Determination of $K_p p_c$

As mentioned above (see Eq. (5.72)), the permeability of the porous layer and the capillary pressure always enter as a product, that is, as a single coefficient. Additional experiments were carried out to determine this coefficient. For this purpose, the horizontal imbibition of the liquid under investigation into the dry porous sheet was undertaken. Rectangular sheets, 1.5×3 cm, were used. Those porous sheets were cut from the same membranes used in the spreading experiments. Each sheet was immersed 0.3–0.5 cm into a liquid container and the position of the imbibition front was monitored over time. In the case under investigation, a unidirectional flow of liquid inside the porous substrate took place. Using Darcy's law, we can conclude that

$$d^2(t) = 2K_p p_c t / \eta, \tag{5.80}$$

where $d(t)$ is the position of the imbibition front inside the porous layer. It was found that in all runs, $d^2(t)/2$ proceeded along a straight line whose slope gives us the $K_p p_c$ value. According to reference [3], $K_p p_c$ should be independent of the tested liquid viscosity.

The measured values of $K_p p_c$ are presented in Table 5.2, which shows that coefficient, $K_p p_c$, for each type of membrane is independent of the tested liquid within the experimental error. Averaged values of $K_p p_c$ for each membrane were used in the calculations.

Results and Discussion

According to our observations, the whole spreading process can be subdivided into two stages (Figure 5.6): first, fast spreading over first several seconds until the maximum radius, L_{max}, of the drop base is reached. Over the duration of the first stage an imbibition front inside the membrane expands slightly ahead of the spreading drop. In the second stage, the drop base starts to shrink slowly and the imbibition front expands until the drop completely disappears. An example of the time evolution of the radius of the drop base and the radius of the wetted region inside the porous layer is provided in Figure 5.7.

In all our spreading experiments, the drops remained spherical over the duration of both the first and the second stages of the spreading process. This was cross-checked by reconstructing the drop profiles at different time instants of spreading and fitting those profiles by a spherical cap: $h = z_{center} + \sqrt{R^2 - (r - r_{center})^2}$, where (r_{center}, z_{center}) is the position of the center of the sphere; and R is the radius of the sphere. r_{center}, z_{center}, and R are used as fitting parameters. The fitting is based on the Levenberg-Marquardt algorithm. In all cases, the reduced Chi-square value was found to be less than 10^{-4}. The fitted parameter, R, gives the radius of curvature of the spreading drops at different times.

The edge of the wetted region inside the porous layers was always circular. Drops remained in the center of this circle over the duration of the spreading process. No deviations from cylindrical symmetry or instabilities were detected.

The spherical form of the spreading drop allowed measuring the evolution of the dynamic contact angle of the drop. In all cases, the dynamic contact angle decreased rapidly over the first stage of spreading until a constant value was reached, which is referred to here as θ_f. The dynamic contact angle remained constant, θ_f, over the main part of the second stage.

An example of the evolution with time of the dynamic contact angle is presented in Figure 5.8 for the same drop as in Figures 5.6 and 5.7. Figure 5.8 shows that the dynamic contact angle decreases rapidly over the duration of the first stage of the spreading when the radius of the drop base expands to its maximum value, L_{max}. The dynamic contact angle remains almost unchanged over the duration of the second stage of spreading. Note again that this behavior has nothing to do with the hysteresis of the contact angle because this does not occur in our system: silicone oils on nitrate cellulose membranes. Solid lines in

TABLE 5.2

Properties of Membranes Used

Membrane Pore Size (μm)	Liquid	$K_p p_c$ (dyn)
3	SO20	$(1.2 \pm 0.4) \times 10^{-4}$
3	SO100	$(1.77 \pm 0.03) \times 10^{-4}$
3	SO500	$(1.6 \pm 0.2) \times 10^{-4}$
0.2	SO5	$(3.4 \pm 0.3) \times 10^{-5}$
0.2	SO100	$(3.1 \pm 0.3) \times 10^{-5}$

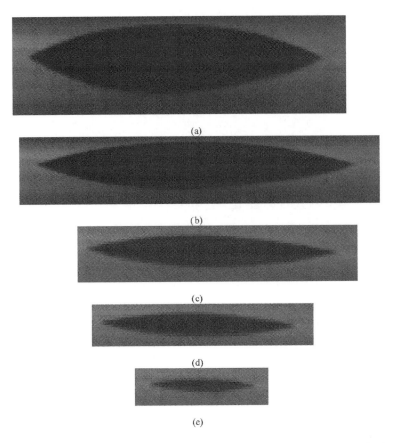

FIGURE 5.6   Time sequence of spreading of SO500, volume 8.7 μL over the membrane with pore size 3 μm (side view): (a) $t = 0.5$ s (after deposition), (b) 3 s, (c) 12 s, (d) 22 s, and (e) 36 s.

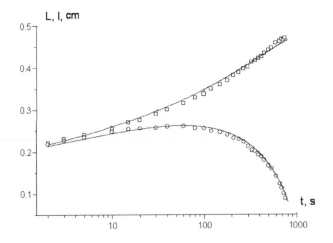

FIGURE 5.7   Development over time of the drop base, L, and the radius of the wetted region inside the porous layer, l. The same drop as in Figure 4.6: SO500 drop, placed onto the membrane with pore size 3 μm. Solid lines are calculated according to Eqs. (5.71) and (5.72).

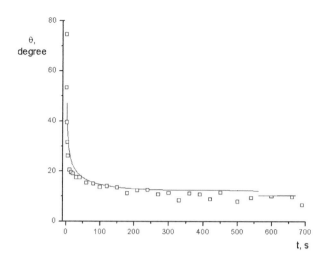

FIGURE 5.8 Development over time of the dynamic contact angle. The same drop as in Figures 5.6 and 5.7: SO500 drop, placed onto the membrane with pore size 3 μm. Solid line is calculated according to Eqs. (5.96) and (5.97).

Figures 5.7 and 5.8 represent the results of numerical integration of the system of Eqs. (5.71) and (5.72). The short final period (just before the drop disappears) is not covered by our calculations. Close to this final stage, the calculation errors increased as a result of a division by a very small quantity in the second term in the right-hand side of Eq. (5.71).

Table 5.3 shows that the final value of the dynamic contact angle, θ_e (last column in Table 5.3), depends on the volume of the drop, as well as on the viscosity of the liquid, and, hence, θ_e is determined solely by hydrodynamics.

Figure 5.9a presents experimentally measured dependencies of radius of the drop base and the wetted region inside the porous layer on time for different silicone oils, porous layers and drop volumes. All relevant values are summarized in Table 5.3. The main result appears in Figure 5.9b, which shows that all experimental data (the same as in Figure 5.9a) fall on two universal curves if dimensionless coordinates are selected as follows: $\bar{L} = L / L_{max}$, $\bar{l} = l / l_{max}$, $\bar{t} = t / t_{max}$, where L_{max} is the maximum value of the drop base, which is reached at the time instant t_m. The same symbols

TABLE 5.3

Experimental Values Measured

Liquid	Notataion on Figures	Membrane pore size, μm	V_0, ml *10^3	Δ, mm	$\bar{m}$ Porosity	t_{max} s	l_{max} cm	l_{max} theory	L_{max} cm	L_{max} theory	θ_m degree	θ_m degree
SO20	▲	0.2	3.1	0.114	0.85	102	0.318	0.317	0.179	0.184	20.0	12
SO20	●	0.2	9.0	0.116	0.72	440	0.585	0.584	0.314	0.31	12.6	11.4
SO20	■	0.2	15.6	0.116	0.73	814	0.77	0.766	0.387	0.39	14.2	12.1
SO20	○	3	3.8	0.136	0.87	10.9	0.345	0.319	0.196	0.198	25.9	22.3
SO20	▷	3	5.5	0.134	0.83	17.1	0.428	0.398	0.223	0.234	20.3	18.6
SO100	✕	3	8.6	0.137	0.82	186	0.493	0.494	0.257	0.274	18.5	11
SO100	☆	3	14.5	0.138	0.77	354	0.659	0.659	0.332	0.34	15.5	17.2
SO500	□	3	3.6	0.138	0.89	296	0.306	0.306	0.174	0.179	22.6	19.4
SO500	△	3	8.7	0.138	0.88	851	0.477	0.478	0.264	0.286	19.1	17.3
SO500	○	3	14.5	0.136	0.78	1660	0.66	0.660	0.339	0.34	15.8	16.5

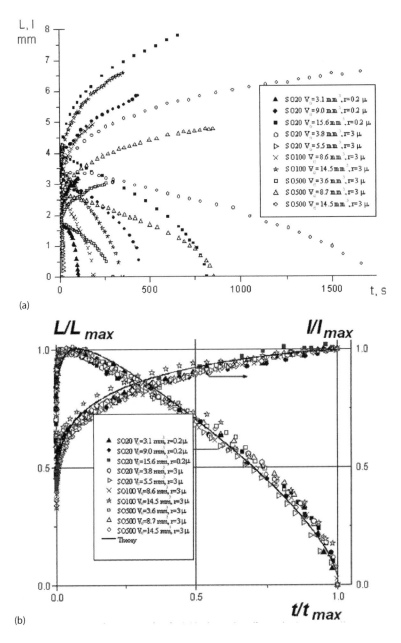

(a)

(b)

FIGURE 5.9 (a) Measured dependencies of radii of the drop base (L, mm) and radii of the wetted region inside the porous layer (l, mm) on time (t, s). All relevant values are summarized in Table 5.3. (b) The same as in Figure 5.13a but using dimensionless coordinates: $\bar{L} = L / L_{max}$, $\bar{l} = l / l_{max}$, $\bar{t} = t / t_{max}$, where L_{max} is the maximum value of the drop base, which is reached at the moment t_m. The same symbols (with overbar) are used for dimensionless values as for dimensional ones. The scale l_{max} is determined by Eq. (5.49), and the time scale t_{max} is given by Eq. (5.70). Solid lines are according to Eqs. (5.73) and (5.74).

(with an overbar) are used for the dimensionless values as for the dimensional. The scale l_{max} is determined by Eq. (5.49) and the time scale t_{max} is given by Eq. (5.70).

The measured values of L_{max}, l_{max}, and t_{max} for all experimental runs are given in Table 5.3. Figure 5.9b shows that the dimensionless time $\bar{t}_m$ is about $0.08 \approx \lambda \ll 1$, as required. The dimensionless time 1 corresponds to the time instant when the drop is completely sucked up by the porous substrate. Solid curves on Figure 5.9b represent the solution of the system of differential Eqs. (5.73) and (5.74). If the parameters $\bar{\tau}$

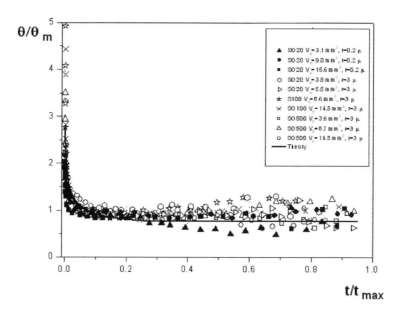

FIGURE 5.10 Dynamic contact angle on the dimensionless time. Solid line is according to Eq. (5.79).

and $\bar{\chi}$ change, then both those theoretical curves remain inside the array of experimental points. In this sense, they represent universal relationships.

The twelfth column in Table 5.3 gives the experimental values of the dynamic contact angle, θ_m, which the drop has when the drop base reaches its maximum value, $L_{\max}$. These values were used for plotting the time evolution of the dynamic contact angle, θ / θ_m. Figure 5.10 shows that all experimental points fall on a single universal curve, as predicted by Eq. (5.79).

The solid line in Figure 5.10 is a result of calculations according to Eq. (5.79), where dimensionless dependencies $\bar{L}(\bar{t})$, $\bar{l}(\bar{t})$ are taken from Figure 5.9b.

APPENDIX 1

The slip boundary condition is used here for simplicity. The slippage coefficient is taken according to Section 5.1. Similar results can be deduced using Brinkman's equations for the description of the liquid flow inside porous layers.

The slippage condition at the drop–porous layer interface is

$$\eta \left. \frac{\partial v}{\partial z} \right|_{z=+0} = \eta_p \frac{v^0 - v_p}{\delta} \tag{A1.1}$$

where $v_p = \frac{K_p}{\eta_p} \frac{\partial p_p}{\partial r}$ is the velocity inside the porous substrate; η_p is an effective viscosity inside the porous layer (see ref. [3]); and $p_a - p_p$ is the pressure inside the porous layer, which may be different from the pressure in the spreading drop. The porous layer thickness, Δ, is assumed to be much bigger than the Brinkman's length, $\delta = \sqrt{K_p}$. Hence, both velocities v^0 and u^0 change stepwise at the drop–porous layer interface: the jump of the first velocity is given by Eq. (A1.1), whereas the jump of the second velocity is

$$u^0 = \frac{K_p}{\eta_p \delta} \left(-\frac{\gamma}{r} \frac{\partial}{\partial r} \left(r \frac{\partial h}{\partial r} \right) + p_p \right). \tag{A1.2}$$

This means that the vertical velocity changes stepwise from the value given by Eq. (A1.2) on the drop–porous layer interface to zero inside the porous layer. Eq. (A1.1) becomes

$$v^0 = \frac{K_p}{\eta_p} \frac{\partial p_p}{\partial r} \frac{\delta h \gamma}{\eta_p} \frac{\partial}{\partial r} \left(-\frac{1}{r} \frac{\partial}{\partial r} \left(r \frac{\partial h}{\partial r} \right) \right) \tag{A1.3}$$

The substitution of Eqs. (A1.2) and (A1.3) into Eq. (5.45) gives

$$\frac{\partial h}{\partial t} = \frac{K_p}{\eta_p \delta} \left(\frac{\gamma}{r} \frac{\partial}{\partial r} \left(r \frac{\partial h}{\partial r} \right) - p_p \right) - \frac{K_p}{\eta_p r \partial r} \left(r h \frac{\partial p_p}{\partial r} \right)$$

$$-\frac{\gamma}{3} \frac{1}{\eta r} \frac{\partial}{\partial r} \left\{ r \left[(h^3 + 3\alpha\delta h^2) \frac{\partial}{\partial r} \left(\frac{1}{r} \frac{\partial}{\partial r} \left(r \frac{\partial h}{\partial r} \right) \right) \right] \right\} \tag{A1.4}$$

where $\alpha = \frac{\eta}{\eta_p} < 1$, according to reference [3]. This equation describes the evolution of the spreading profile of the drop both in space and time. The only unknown dependence left is the pressure inside the porous layer, p_p.

The conservation law inside the porous layer $\frac{1}{r} \frac{\partial(rv)}{\partial r} + \frac{\partial u}{\partial z} = 0$ is used to determine this pressure. This equation is integrated over z from $-\Delta$ to 0 inside the porous layer, using condition Eq. (1.2) and Darcy's law to express the velocity components using the pressure gradient. After some transformations the final equation becomes

$$\frac{1}{r} \frac{\partial}{\partial r} \left(r \frac{\partial p_p}{\partial r} \right) - \frac{p_p}{\Delta\delta} = -\frac{\gamma}{r} \frac{\partial}{\partial r} \left\{ r \left[\frac{h}{\delta} \frac{\partial}{\partial r} \left(\frac{1}{r} \frac{\partial}{\partial r} \left(r \frac{\partial h}{\partial r} \right) \right) \right] \right\} - \frac{1}{\Delta\delta} \frac{\gamma}{r} \frac{\partial}{\partial r} \left(r \frac{\partial h}{\partial r} \right), \tag{A1.5}$$

which describes the radial distribution of the pressure inside the porous layer. Equation (A1.5) shows that

1. Under the spherical part of the spreading drop, the pressure inside the porous layer remains constant and equal to the pressure inside the drop, that is,

$$p_p^\infty = \frac{2\gamma\theta}{L}, \tag{A1.6}$$

   and, hence, the liquid does not flow at all inside the porous layer under the spherical part of the drop.

2. The pressure inside the porous layer changes from the constant value Eq. (A1.6) to the pressure outside the drop close to $r = L$ in a narrow region with a scale

$$\xi = \sqrt{\Delta\delta} \ll L \tag{A1.7}$$

   As stated earlier (see Eqs. 5.51 through 5.53).

Let us introduce a new local dimensionless variable, $\bar{x}$, as follows:

$$\bar{x} = \frac{r - L}{\xi}, \quad -\infty < \bar{x} < 0. \tag{A1.8}$$

After omitting small terms, Eq. (A1.5) becomes

$$p_p'' - p_p = -\frac{\gamma\delta}{\xi^2} \left[\left(\bar{f} \, \bar{f}''' \right)' + \bar{f}'' \right], \tag{A1.9}$$

where $\bar{f} = h/\delta$ is the dimensionless profile of the drop in this narrow transition region. This scale is selected in the same way as in Section 5.1.

Equation (A1.9) can be directly integrated using the following boundary conditions:

$$p_p(-\infty) = p_p^{\infty}, \tag{A1.10}$$

$$p_p(0) = p_d. \tag{A1.11}$$

The boundary condition Eq. (A1.10) follows from Eq. (A1.6), whereas condition (A1.11) still includes the unknown pressure, p_d, which is determined now.

The solution of Eq. (A1.9), which satisfies both boundary conditions (A1.10) and (A1.11) is

$$p_p = p_d e^{\bar{x}} + \frac{\gamma \delta}{\xi^3} \int_x^0 \left[\left(\bar{f}\ \bar{f}''' \right)' + \bar{f}'' \right] \sinh\left(\bar{x} - \bar{y}\right) d\bar{y}, \tag{A1.12}$$

Let us assume that

$$p_d \gg p_p^{\infty}. \tag{A1.13}$$

This assumption means that the pressure inside the porous layer in the narrow region under consideration is much higher than the capillary pressure in the drop. This is reasonable to expect because the pressure inside the porous layer is the sole determinant of the drop suction by the porous layer. In this case Eq. (A1.12) reduces to

$$p_p = p_d e^{\bar{x}}. \tag{A1.14}$$

In order to determine the unknown value, p_d, the solution for the pressure distribution inside the porous layer outside the drop (5.65) and (5.68) is used. The solutions (5.65) and (A1.14) should give the same radial velocity from both sides at $r = L$. Thus we have

$$p_d = \frac{p_c}{1 + \dfrac{L}{\xi} \ln \dfrac{l}{L}} \approx p_c \frac{\xi}{L \ln \dfrac{l}{L}} \ll p_c. \tag{A1.15}$$

This equation justifies the neglect of p_d relative to p_c in this section.

5.3 Spreading of Liquid Drops over Thick Porous Substrates: Complete Wetting Case

Let us extend the study presented in Section 5.2 to the case of spreading of *small* silicone oil drops (capillary spreading regime) over porous solid substrates thicker than the drop size. The spreading of small silicone oil drops (capillary regime of spreading) over various dry thick porous substrates (permeable in both normal and tangential directions) is experimentally investigated in this section. The time evolution of the radii of both the drop base and the wetted region on the surface of the porous substrate were monitored. It was observed that the total duration of the spreading process can be divided into two stages: a first stage, when the drop base expands until its maximum value is reached, and a subsequent second stage, when the drop base shrinks. It was found that the dynamic contact angle remains constant during the second stage of spreading. This fact has nothing to do with contact angle hysteresis because there is no hysteresis in the system. Appropriate scales are used, with a dimensionless time, to plot the dimensionless radii of the drop base and of the wetted circle on the surface of the porous substrate, the relative dynamic contact angle and the effective contact angle inside the porous substrate. All these experimental data fall onto universal curves when the spreading of different silicone oils is done on porous substrates of similar pore size and porosity [14].

However, please note that in this section consideration of flow inside a porous substrate is purely semi-empirical. For some reason this approach results in a very good agreement with experimental data.

The spreading of small liquid drops over thin porous layers saturated with the same liquid (Section 5.1) or a dry porous layer (Section 5.2) was considered by appropriately matching flows in both the spreading drop and the porous substrate.

In Section 5.2, the spreading of silicone oil drops over various dry microfiltration membranes (permeable in both normal and tangential directions) was discussed. Plotting the dimensionless radii of the drop base and of the wetted region inside the porous layer using a dimensionless time, all experimental data falls on two universal curves. According to the theory presented in Section 5.2, (i) the dynamic contact angle dependence on the same dimensionless time should be a universal function, and (ii) the dynamic contact angle should change rapidly over an initial short spreading stage and should remain constant over the remaining duration of the spreading process. This fact has nothing to do with contact angle hysteresis because there was no hysteresis in the system under consideration in Section 5.2. These conclusions were in good agreement with experimental observations (Section 5.2).

In the present section, we extend our study in Section 5.2 to the spreading of small silicone oil drops (capillary spreading regime) over different porous substrates whose thickness is much bigger than the drop size. A number of similarities with the case of the spreading over thin porous substrates (Section 5.2) was found.

Theory

As we mentioned in the introduction to Chapter 3, at small capillary numbers, $Ca = \frac{U\eta}{\gamma} \ll 1$, the drop profile remains in the spherical shape over the main part of the spreading drop. We also concluded in Section 5.1 that at spontaneous spreading, the capillary number is always small, except for a short initial stage. Based on that, we conclude that, during the spreading over a porous substrate, the liquid drops retains their spherical shape (Figure 5.11). Hence, the drop profile over the spherical part (assuming the low slope profile) is

$$h(t,r) = \frac{2V}{\pi L^4}(L^2 - r^2), \quad r < L(t) \tag{5.81}$$

where $V(t)$ is the drop volume; $h(t, r)$ is the drop profile; L is the radius of the drop basis; and r is the radial coordinate.

Equation (5.81) gives the following value of the dynamic contact angle, θ, $(\tan\theta \approx \theta)$:

$$\theta = \frac{4V}{\pi L^3}, \tag{5.82}$$

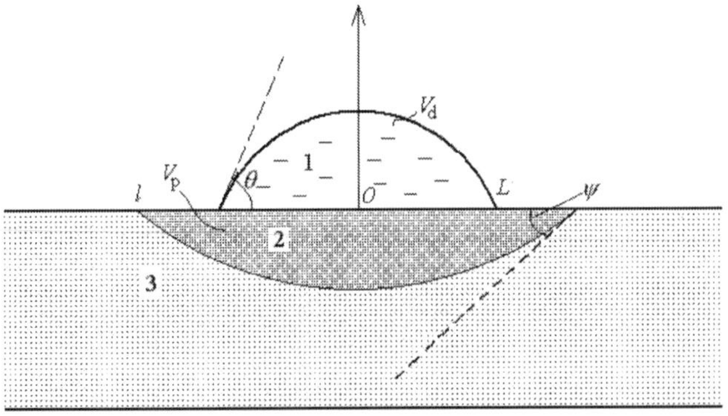

FIGURE 5.11 Spreading of liquid drops over dry porous substrates. (1) Spherical drop, (2) wetted region inside the porous substrate (modeled by a spherical cap), (3) dry part of the porous substrate. $L(t)$, radius of the drop base; $l(t)$, radius of the wetted circle on the surface of the porous substrate; $\theta(t)$, dynamic contact angle of the spreading drop; and $\psi(t)$, effective contact angle inside the porous substrate.

hence,

$$L = \left(\frac{4V}{\pi \theta} \right)^{1/3}.$$ (5.83)

The spreading process is a superposition of two motions: (a) the spreading of the drop over the already saturated part of the surface of the porous substrate, which results in an expansion of the drop base, and (b) the shrinkage of the drop base caused by the imbibition into the porous substrate. Hence, we can write the following balance equation:

$$\frac{dL}{dt} = v_+ - v_-,$$ (5.84)

where v_+ and v_- denote unknown velocities of the expansion and the shrinkage of the drop base, respectively.

Using Eq. (5.83) we get

$$\frac{dL}{dt} = -\frac{1}{3} \left(\frac{4V}{\pi \theta^4} \right)^{1/3} \frac{d\theta}{dt} + \frac{1}{3} \left(\frac{4}{\pi V^2 \theta} \right)^{1/3} \frac{dV}{dt}.$$ (5.85)

Over the whole duration of the spreading over the porous layer, both the contact angle and the drop volume can only decrease with time. Accordingly, the first term in the right-hand side of Eq. (5.85) is positive and the second is negative.

Comparison of the Eqs. (5.84) and (5.85) yields

$$v_+ \equiv -\frac{1}{3} \left(\frac{4V}{\pi \theta^4} \right)^{1/3} \frac{d\theta}{dt} > 0$$

$$v_- \equiv -\frac{1}{3} \left(\frac{4}{\pi V^2 \theta} \right)^{1/3} \frac{dV}{dt} > 0.$$ (5.86)

Let $l(t)$ be the radius of the wetted region on the surface the porous substrate. This unknown quantity cannot be determined without the numerical integration of the Brinkman-Darcy equations inside the porous substrate and coupling with the flow in the drop. However, we can draw some conclusions based on the analysis in Section 5.2. According to this analysis the whole spreading process can be subdivided into two stages: during the first stage v_+ prevails and v_- dominates during the second stage of spreading.

Let us consider the second spreading stage. During this stage, the first term in the right-hand side of Eq. (5.85) can be neglected and this equation reduces to

$$\frac{dL}{dt} = \frac{1}{3} \left(\frac{4}{\pi V^2 \theta} \right)^{1/3} \frac{dV}{dt}.$$ (5.87)

This equation, after inserting the expression of θ, Eq. (5.82), takes the form: $\frac{dL}{dV} = \frac{L}{3V}$, which can be easily integrated. Its solution is

$$V = C L^3,$$ (5.88)

where C is an integration constant. Comparison with Eq. (5.82) yields

$$\theta(t) = \theta_f = \text{const}$$ (5.89)

over the duration of the second stage of the spreading. This conclusion agrees well with experimental observations [14] (see the "Results and Discussion" in this section) as well as with observations in Section 5.2. Note that in the case under consideration, as in Section 5.2, the constancy of the contact angle has nothing to do with the contact angle hysteresis: There is no hysteresis in the system under consideration. θ_f is not a receding contact angle but forms as a result of a self-regulation of the drop–porous layer system.

Inside the Porous Substrate

The analysis of experimental data shows that the radius of the wetted region inside the porous layer is proportional to l^3. If the drop base (Figure 5.11) is assumed to be a point source of liquid, then the shape of the wetted area inside the porous substrate is a hemisphere with an increasing radius and a constant *effective* contact angle $\psi(t)$. Obviously "a point source" assumption is too simple an approximation given that both radii, $l(t)$ and $L(t)$, are of the similar size. Yet it would help understanding the essence of the process. Hence, let us assume that the wetted region inside the porous substrate is a spherical cap with the changing effective contact angle $\psi(t)$ (Figure 5.11).

Let $V_p = (V_0 - V)/m$, where V_p is the volume of the liquid inside the porous substrate at time t; V_0 is the initial volume of the drop; and m is the porosity. Under the above assumption the liquid volume in the porous substrate, V_p, can be expressed as (Figure 5.11):

$$V_p = \frac{\pi}{3} l^3 \frac{\left(1 - \cos\psi\right)^2 (2 + \cos\psi)}{\sin^3 \psi},\tag{5.90}$$

where l is the radius of the wetted circle on the outer surface of the porous substrate. Equation (5.90) enables us to calculate the time evolution of ψ using the experimental data.

Experimental Part

Silicone oils S5 (viscosity 0.05 P), S100 (viscosity 1.0 P) and S500 (viscosity 5.0 P) purchased from Brookfield Engineering Laboratories, Inc., (Middleboro, MA) were used in the spreading experiments. Glass filters (J. Bibby Science Products, Ltd.) and metal filters (Sintered Products, Ltd.), both purchased from Claremont Ltd. (Broseley, UK), were used as porous substrates. The diameters of the glass filters were 5.0, 2.9 and 2.9 cm, and their thicknesses were 2.5, 1.9 and 2.2 mm, respectively. The diameter of metal filters was 5.6 cm, and their thickness was 1.9 mm. The pore size distribution and permeability of membranes were tested using the Coulter Porometer II (Coulter Electronic Ltd., Luton, UK).

Glass filters with three different pore size distributions were used. The average pore sizes were 3.7, 4.7 and 26.8 μm, respectively; their porosities were 0.56, 0.53, and 0.31, respectively. The permeability of the same membranes were 1.8, 1.9 and 11.5 L/min/cm² (air flux, transmembrane pressure 0.1 bar), respectively. The metal filter (Cupro-Nickel) average pore size was 26.1 μm and its porosity 0.32. The porosity of filters was measured using the difference in the weight of the filters saturated with oil and the dry filters. Filters were dried over 3–5 h at 95°C and then stored in a dry atmosphere prior to the spreading experiments. All relevant information and values are summarized in Table 5.4.

Figure 5.2 shows a schematic description of the setup. The time evolution of the radius of the drop base, $L(t)$, the dynamic contact angle, $\theta(t)$, and the radius of the wetted circle on the surface of the porous substrate, $l(t)$, were monitored. Before the experiment, all porous substrates were placed for half an hour in a KOH aqueous solution inside an ultrasonic bath, then they were well rinsed with Milli-Q water (Milli-Q water system, made by MILLIPORE S.A., France) in an ultrasonic bath, after which they were dried for up to 2 h in an 110°C dry atmosphere. Porous substrates were stored in a dry atmosphere prior to starting the experiments.

Experiments were carried out in the following order:

- The dry porous substrate was placed in the dry atmosphere chamber and left there for 15–30 min.
- A light pulse produced by a flash gun was used to synchronize the time instant in both video recorders when the drop started to spread.
- A droplet of silicone oil was placed onto the porous substrate.

Each run was carried out until the complete imbibition of the drop into the porous substrate took place.

TABLE 5.4

Characteristics of Porous Substrates and Drops Used

Material, Figure, Symbol	Porosity	Average Pore Size (μm)	η (P)	V_0 (μL)	t_{max} (s)	L_{max} (mm)	L_{max} (mm)	θ_m (grad)	ψ_{max} (grad)
Glass Figure 5.12 Δs	0.53	4.7	0.05	5.0	0.64	2.42	3.10	25.8	38
Glass Figure 5.12	0.53	4.7	1	5.9	12.0	2.30	3.20	24.4	39
Glass Figure 5.12 O	0.53	4.7	5	8.2	60.0	2.58	3.50	23.6	42
Cupro-Nickel Figure 5.13	0.32	26.1	5	8.2	15.7	2.38	3.20	35.0	52
Glass Figure 5.13 O	0.31	26.8	5	6.8	18.52	2.40	3.20	25.5	44
Glass Figure 5.14	0.56	3.7	5	8.0	36.0	2.53	3.2	21.5	30
Glass Figure 5.14 O	0.31	26.8	5	6.8	18.52	2.40	3.20	25.5	44

Results and Discussion

According to observations, the whole spreading process can be subdivided into two stages (e.g., Figures 5.12 and 5.13): the first stage, when the drop spreads until the maximum radius, L_{max}, of the drop base is reached, and the second stage, when the drop radius decreases. Over the duration of the first stage, the imbibition front inside the membrane grows slightly ahead of the drop spreading front. Subsequently, the drop base starts to shrink until the drop completely disappears and the imbibition front grows until the end of the process. Examples of the time evolution of the radius of the drop base and the radius of the wetted circled on the surface of the porous substrates are presented in Figures 5.12, 5.13, 5.16, and 5.19.

In all experiments, the drops remained spherical over the whole spreading process. This was cross-checked by reconstructing the drop profiles at different time instants of spreading, fitting those drop profiles by a spherical cap: $h = z_{center} + \sqrt{R^2 - (r - r_{center})^2}$, where (r_{center}, z_{center}) is the position of the center of the sphere, and R is the radius of the sphere. r_{center}, z_{center}, and R are used as fitting parameters. The Levenberg-Marquardt algorithm was used for fitting. In all cases, the reduced Chi-squared value was found to be smaller than 10^{-4}. The fitted parameter, R, gives the radius of curvature of the spreading drops at different times.

The wetted area on the surface of the porous substrate was always circular. Drops remained in the center of this circle over the whole duration of the spreading process. No deviations from cylindrical symmetry or instabilities were detected.

The spherical form of the spreading drop allows measuring the evolution of the dynamic contact angle of the spreading drops. In all cases, the dynamic contact angle decreased rapidly during the first spreading stage and remained constant during the second spreading stage. This constant value of the contact angle is denoted here as θ_f.

Examples of the evolution with time of the dynamic contact angle are presented in Figures 5.14, 5.17, and 5.20. These figures show that the dynamic contact angle decreases during the first spreading stage when the radius of the drop base grows to its maximum value, L_{max}. Within experimental error, the dynamic contact angle remained unchanged during the second spreading stage.

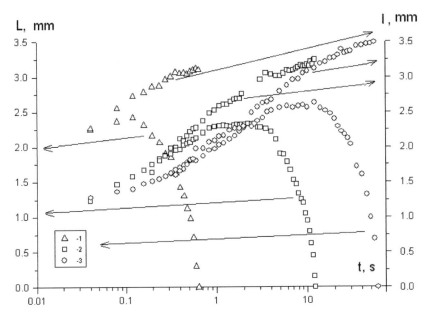

FIGURE 5.12 Spreading of different silicone oils over identical dry porous glass filters. In dimensional form. Glass porous filter: porosity 0.56, average pore size 4.7 μm, permeability 1.9 L/min/cm². (1) silicone oil ($\eta = 0.05$ P); (2) silicone oil ($\eta = $ P); and (3) silicone oil ($\eta = 5$ P) (see insets). The same symbols show the evolution of $l(t)$ and $L(t)$ with time. Upper parts: $L(t)$, radius of the base of the spreading drop; and lower parts: $l(t)$, radius of the wetted circle on the surface of the porous glass filter.

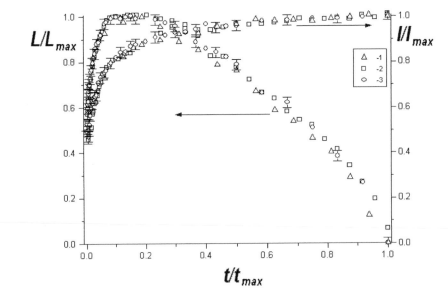

FIGURE 5.13 Spreading of different silicone oils over identical dry porous glass filters. The same data as in Figure 5.12a using appropriate dimensionless coordinates. Glass porous filter: porosity 0.53, average pore size 4.7 μm, permeability 1.9 L/min/cm². (1) Silicone oil ($\eta = 0.05$ P), (2) silicone oil ($\eta = $ P), and (3) silicone oil ($\eta = 5$ P) (see insets). L_{max}, maximum value of the radius of the drop base; l_{max}, maximum value of the radius of the wetted circle on the surface of the glass filter, and t_{max}, total duration of the process (see Table 5.4).

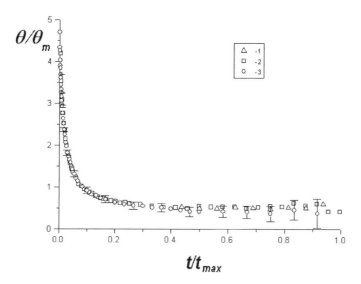

FIGURE 5.14 Spreading of different silicone oils over identical dry porous glass filters. Dynamic contact angle versus time in dimensionless units. Glass porous filter: porosity 0.53, average pore size 4.7 μm, permeability 1.9 L/min/cm². (1) Silicone oil ($\eta = 0.05$ P), (2) silicone oil ($\eta = $ P), and (3) silicone oil ($\eta = 5$ P) (see insets). θ_m, dynamic contact angle value at the time when the maximum value of the radius, L_{max}, of the drop base is reached (see Table 5.4).

Spreading of Silicone Oil Drops of Different Viscosity over Identical Glass Filters

Figures 5.12 through 5.15 present experimental results on the spreading of silicone oil drops of different viscosity over identical glass filter. Silicone oils S5, S100, and S500 were used in these spreading experiments. The diameter of the glass filter was 2.9 cm, its thickness 1.9 mm, its porosity 0.53, and its average pore size 4.7 μm.

In Figures 5.12 through 5.15, the open triangle (Δ) indicates the spreading of the silicone oil S5 (drop volume $V_0 = 5.0$ μL, the maximum radius of the drop base $L_{max} = 2.42$ mm, the maximum (final) radius of the wetted circle on the outer surface of the glass filter $l_{max} = 3.10$ mm, and the total duration of the spreading $t_{max} = 0.64$ s); the open square (□) indicates the spreading of the silicone oil S100 (drop volume $V_0 = 5.9$ μL, the maximum radius of the drop base $L_{max} = 2.30$ mm, the maximum (final) radius of the wetted circle on the outer surface of the glass filter $l_{max} = 3.20$ mm, and the total duration of the spreading $t_{max} = 12.0$ s); and the open circle (○) indicates the spreading of the silicone oil S500 (drop volume $V_0 = 8.2$ μL, the maximum radius of the drop base $L_{max} = 2.58$ mm, the maximum (final) radius of the wetted circle on the outer surface of the glass filter $l_{max} = 3.50$ mm, and the total duration of the spreading $t_{max} = 60.0$ s).

Figure 5.12 presents the time evolution of both the radius of the base of the spreading drops and the radius of the wetted circle on the outer surface of the glass filter for silicone oils of different viscosity using experimental data in dimensional form. Figure 5.12 shows that the kinetics of the spreading and the imbibition varies for drops of different sizes and different viscosities. Consequently, the total duration of the spreading process, the maximum radius of the drop base and the radius of the wetted circle on the outer surface of the glass filter vary considerably. However, if as in Section 5.2 we rescale quantities as L/L_{max}, l/l_{max} and t/t_{max}, then all experimental data falls into two universal curves, as Figure 5.13 shows.

According to Section 5.2, the evolution of reduced dynamic contact angle, θ/θ_m, with the dimensionless time, t/t_{max}, should be universal. Here, θ_m is the value of the dynamic contact angle, which is reached when the radius of the drop base reaches its maximum value (the end of the first stage of spreading). The same procedure in the case of spreading over thick porous substrates is used here. In Figure 5.14, the reduced dynamic contact angle, θ/θ_m, is plotted versus the dimensionless time t/t_{max}.

This plot shows that (i) all three experimental curves fall into a single universal curve, and (ii) the dynamic contact angle remains constant during the second spreading stage as theory predicts.

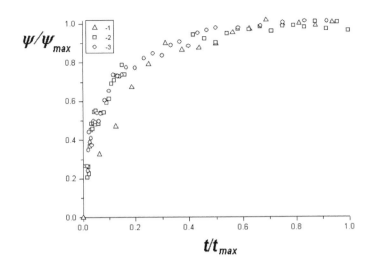

FIGURE 5.15 Spreading of different silicone oils over the same dry porous glass filter. Evolution of the effective contact angle inside the porous glass filters with the dimensionless time. Glass porous filter: porosity 0.53, average pore size 4.7 μm, permeability 1.9 L/min/cm². (1) Silicone oil ($\eta = 0.05$ P); (2) silicone oil ($\eta = $ P); and (3) silicone oil ($\eta = 5$ P) (see insets). ψ_{max}, maximum value of the effective contact angle (see Table 5.4).

In Figure 5.15, the evolution of the relative effective dynamic contact angle, ψ/ψ_{max}, inside the porous glass filter with the dimensionless time is presented. It is noteworthy that in all three cases all data follow a single universal curve.

These three experimental runs show that the spreading behavior of drops of different viscosities and volumes on the same thick porous substrate is identical if appropriate dimensionless coordinates are used.

Spreading of Silicone Oil Drops over Filters with Similar Properties but Made of Different Materials

In this series of experiments, the spreading of drops of the same silicone oil S500, over different substrates was studied. We wanted to check whether the universal behavior found in the previous case remains valid even when different porous substrates made of different material, glass and metal filter, are used.

In Figures 5.16 through 5.18: the open square (□) indicates the spreading of the silicone oil S500 (drop volume $V_0 = 8.2$ μL, the maximum radius of the drop base $L_{max} = 2.38$ mm, the maximum radius of the wetted circle on the outer surface of the porous substrate $l_{max} = 3.20$ mm, the total duration of the spreading $t_{max} = 15.7$ s over the metal filter (Cupro-Nickel) with diameter 5.6 cm, thickness 1.9 mm, average pore size 26.1 μm, porosity 0.32); and the open circle (○) indicates the spreading of the silicone oil S500 (drop volume $V_0 = 6.8$ μL, maximum radius of the drop base $L_{max} = 2.40$ mm, maximum radius of the wetted circle on the outer surface of the porous substrate $l_{max} = 3.20$ mm, the total duration of the spreading $t_{max} = 18.52$ s over the glass filter with diameter 2.9 cm, thickness 2.2 mm, porosity 0.31 and average pore size 26.8 μm).

Figure 5.16 presents the dependence of the dimensionless radius of the drop base (left ordinate) and that of the dimensionless radius of the wetted circle on the outer surface of the porous substrate (right ordinate) on the dimensionless time. Figure 5.17 presents the dependence of the relative dynamic contact angle on the dimensionless time. Figure 5.18 presents the dependence of the effective dynamic contact angle inside the porous substrate on the same dimensionless time.

The curves show that the spreading of drops of different sizes on porous substrates made of different materials with, however, similar porosity and average pore size, fall on universal curves if, as in the previous case, the same dimensionless coordinates are used. Thus, the universal spreading behavior over porous substrates does not depend on the material of the porous substrate.

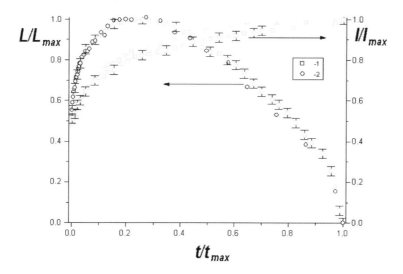

FIGURE 5.16 Spreading of silicone oil ($\eta = 5$ P) over dry porous glass and metal filters. Radii in reduced coordinates. (1) Metal filter: porosity 0.32, average pore size 26.1 μm (see insets), and (2) glass filter: porosity 0.31, average pore size 26.8 μm (see insets). L_{max}, maximum value of the radius of the drop base; l_{max}, maximum value of the radius of the wetted circle on the surface of the glass filter; and t_{max}, total duration of the process (see Table 5.4).

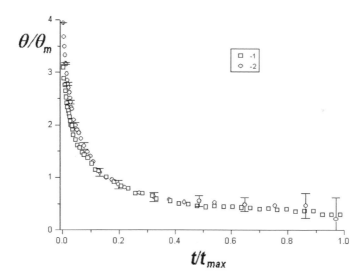

FIGURE 5.17 Spreading of silicone oil ($\eta = 5$ P) over dry porous glass and metal filters. Dynamic contact angle versus time in dimensionless units. (1) Metal filter: porosity 0.32, average pore size 26.1 μm (see insets); and (2) glass filter: porosity 0.31, average pore size 26.8 μm (see insets). θ_m, dynamic contact angle value at the time when the maximum value of the radius of the drop base, L_{max}, is reached (see Table 5.4).

Spreading of Silicone Oil Drops with the Same Viscosity ($\eta = 5$ P) over Glass Filters with Different Porosity and Average Pore Size

In this section, the spreading of silicone oil drops with the same viscosity over glass filters with different porosities and average pore sizes is investigated to check whether the universal behavior found in two previous sections is still applicable.

In Figures 5.19 through 5.21, the open square (□) indicates the spreading of silicone oil S500 (drop volume $V_0 = 8.0$ μL, maximum radius of the drop base $L_{max} = 2.53$ mm, maximum radius of the wetted

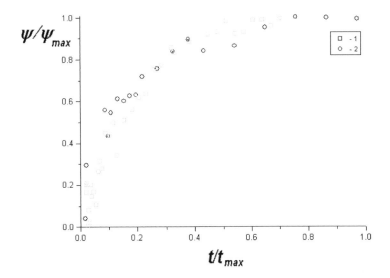

FIGURE 5.18 Spreading of silicone oil ($\eta = 5$ P) over dry porous glass and metal filters. Evolution of the effective contact angle inside the porous filters with the relative time. (1) Metal filter: porosity 0.32, average pore size 26.1 μm (see insets); and (2) glass filter: porosity 0.31, average pore size 26.8 μm (see insets). ψ_{max}, maximum value of the effective contact angle (see Table 5.4).

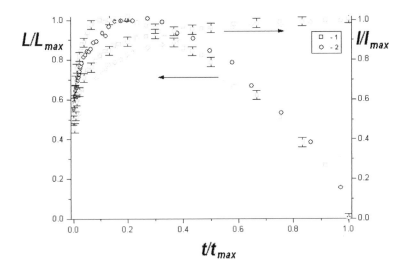

FIGURE 5.19 Spreading of silicone oil ($\eta = 5$ P) over different dry glass filters. Radii of spreading in reduced coordinates. (1) Glass filter: porosity 0.56; pore size 3.7 μm (see insets); and (2) glass filter: porosity 0.31; pore size 26.8 μm (see insets). L_{max}, maximum value of the radius of the drop base; l_{max}, maximum value of the radius of the wetted circle on the surface of the glass filter; and t_{max}, total duration of the process (see Table 5.4).

circle on the surface $l_{max} = 3.20$ mm, the total duration of the spreading $t_{max} = 36.0$ s over the glass filter with diameter 5.0 cm thickness 2.5 mm, porosity 0.56 and average pore sizes 3.7 μm); and the open circle (O) indicates the spreading of the silicone oil S500 (drop volume $V_0 = 6.8$ μL, maximum radius of the drop base $L_{max} = 2.40$ mm, maximum radius of the wetted circle on the outer surface of the glass filter $l_{max} = 3.20$ mm, the total duration of the spreading $t_{max} = 18.52$ s over the glass filter with diameter 2.9 cm, thickness 2.2 mm, porosity 0.31 and average pore sizes 26.8 μm).

Figures 5.20 and 5.21 show that the relationships between the relative dynamic contact angle, θ/θ_m, and the effective dynamic contact angle inside the porous substrate, ψ/ψ_{max}, on the dimensionless time,

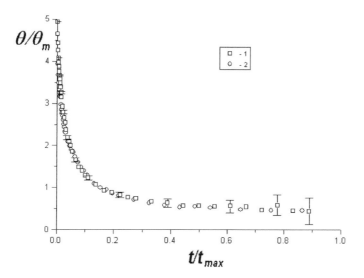

FIGURE 5.20 Spreading of silicone oil ($\eta = 5$ P) over different dry glass filters. Dynamic contact angle versus time in dimensionless units. (1) Glass filter: porosity 0.56; pore size 3.7 μm (see insets); and (2) glass filter: porosity 0.31; pore size 26.8 μm (see insets). θ_m, the dynamic contact angle value at the time when the maximum value of the radius of the drop base, $L_{\max}$, is reached (see Table 5.4).

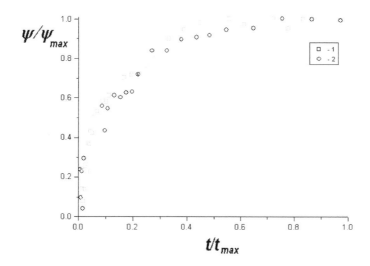

FIGURE 5.21 Spreading of silicone oil ($\eta = 5$ P) over different dry glass filters. Evolution of the effective contact angle inside the porous filters with time in dimensionless units. (1) Glass filter: porosity 0.56; pore size 3.7 μm (see insets); and (2) glass filter: porosity 0.31; pore size 26.8 μm (see insets). $\psi_{\max}$, the maximum value of the effective contact angle (see Table 5.4).

$t/t_{\max}$, again exhibit universal behavior. However, the dependence of the dimensionless radius of the drop base, $L/L_{\max}$, and that of the dimensionless radius of the wetted circle on the surface of the porous substrate, $l/l_{\max}$, on the dimensionless time, $t/t_{\max}$, deviate from universal behavior during the first spreading stage. The duration of the first stage, when the drop base increases with time and reaches its maximum value, is shorter in the case of the spreading of silicone oil drops over the glass filter with a smaller average pore size than in the case of the spreading over the glass filter with a larger average pore size (Figure 5.19).

In conclusion, we can safely say that the spreading behavior over porous substrates is mostly if not all determined by the porosity and the average pore size of the porous substrate and differs if these two characteristics differ.

Conclusions

Experiments were carried out on the spreading of small silicone oil drops (capillary regime of spreading) over various dry thick porous substrates (permeable in both normal and tangential directions). The time evolution of the radii of both the drop base and the wetted region on the surface of the porous substrate were monitored. It was shown that the overall duration of the spreading process can be divided into two stages: a first stage, when the drop base grows until a maximum value is reached, and a second stage, when the drop base shrinks. It was observed that the dynamic contact angle remained constant during the second spreading stage. This fact is supplied by a heuristic argument and has nothing to do with hysteresis of contact angle given that there is no hysteresis in the system. Using appropriate scales, the dimensionless radius of the drop base, the radius of the wetted circle on the surface of the porous substrate, the dynamic contact angle, and the effective contact angle inside the porous substrate have been plotted using a dimensionless time.

Experimental data shows that the spreading of silicone oil drops over dry thick porous substrates show a universal behavior if (i) porous substrates made of different materials with, however, similar porosity and average pore size are used, and if (ii) appropriate dimensionless coordinates are introduced to depict the data. However, if porous substrates with different porosity and average pore size are used, the dynamics of both the radius of the drop base, L/L_{max}, and the radius of the wetted circle on the outer surface of the porous filter, l/l_{max}, behave differently during the spreading process. Yet both the relative dynamic contact angle θ/θ_m, and the effective contact angle inside the porous substrate, ψ/ψ_{max}, show universal behavior.

5.4 Spreading of Liquid Drops from a Liquid Source

In this section, we consider a liquid drop being created and then spreading over a solid substrate with a liquid source. We look at both cases, *complete* and *partial* wetting and for both small and large drops. Then we expect to observe spreading and forced flow caused by the liquid source in the drop center. Both capillary and gravitational regimes of spreading are considered [15].

For conditions of complete wetting, the spreading is an overlapping of two processes: a spontaneous spreading and a forced flow caused by the liquid source in the center. Both capillary and gravitational regimes of spreading are considered, and power laws are deduced. In both cases of small and large droplets, the exponent is a sum of two terms: the first term corresponds to the spontaneous spreading and the second term is determined by the intensity of the liquid source. In the case of a constant flow rate from the source, this gives for the radius of spreading the following law $R(t) \sim t^{0.4}$ in the case of the capillary spreading and $R(t) \sim t^{0.5}$ in the case of gravitational spreading. In the case of partial wetting droplets, spread with a constant advancing contact angle (at small capillary numbers). This yields $R(t) \sim t^{1/3}$. Experimental data are in good agreement with theoretical predictions.

The spreading of liquid drops over solid nonporous substrates was investigated in Section 4.1, Chapter 4: For conditions of complete wetting the spreading of small droplets is governed by the capillary law of spreading Eq. (4.25), which is

$$R(t) = 0.65\lambda_c \left(\frac{\gamma V^3}{\eta} \right)^{0.1} (t + t_{in})^{0.1},$$

where R is the radius of the drop base; t is time; and λ_c is a pre-exponential factor, which is determined by the disjoining pressure isotherm (see Section 4.2). In the same case of complete wetting, the spreading of bigger drops is governed by gravity according to Eq. (5.36) (see Section 4.1):

$$R(t) = 0.78 \left(\frac{\rho g V^3}{\eta} \right)^{1/8} (t + t_c)^{1/8}.$$

For small drops, the capillary law of spreading is in excellent agreement with experimental data as shown in Section 4.2. Over time, small droplets spread out, the radii of the drop bases increase with time and, hence, should be a transition from a capillary regime of spreading to the gravitational regime. This transition was experimentally confirmed [16] in the case of spontaneous spreading. Note, in the case of complete wetting, the dynamic contact angle tends to zero over time and the droplets spread out completely (see Sections 4.1 and 4.2 for details).

The spreading of liquid drops in the case of partial wetting has been less investigated and it is less understood. The main problem in this case is the presence of contact angle hysteresis, which was considered in Section 1.3 and Chapter 3. This phenomenon is usually associated with non-homogeneity/roughness of the solid substrates. However, it was shown in Chapter 3 that an S-shaped disjoining pressure isotherm leads to the presence of the contact angle hysteresis even on smooth homogeneous substrates. In the case of partial wetting, the droplet spreads out until a static advancing contact angle is reached. After that, the droplet does not spread out on a macroscale but still spreads out on microscale (see Chapter 4). If the droplet is "gently" pushed from inside by pumping liquid from the orifice at its center then the droplet spreads out with a constant advancing contact angle, which is equal to the static advancing contact angle. The term "gently" means that the capillary number remains small during the spreading process.

The spreading of liquid drops for both cases, complete and partial wetting, when liquid is injected into the droplets from a small orifice at their center is considered below in this section.

Theory

Let us consider the spreading of a small liquid droplet over a solid substrate in the presence of a liquid source in the drop center (Figure 5.22).

It is assumed that the shape of the drop remains axisymmetrical, and, hence, a cylindrical coordinate system, (r,z), is used, where r is the radial distance from the center, and z is the vertical coordinate. Because of symmetry, the angular component of velocity vanishes, and all other unknowns are independent of the angle. It is assumed that the Reynolds number, $Re \ll 1$, and that the drop profile has a low slope, that is, $\varepsilon = \left(\frac{\partial h}{\partial r}\right)^2 \ll 1$, where $h(t, r)$ is the drop profile. Using this assumption, we arrive to Eq. (110), Section 3.1, which describes the profile of the spreading liquid

$$\frac{\partial h}{\partial t} = \frac{1}{3\eta r} \frac{\partial}{\partial r}\left(r h^3 \frac{\partial p}{\partial r}\right), \tag{5.91}$$

which is referred to as the equation of spreading.

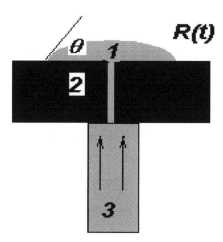

FIGURE 5.22 Schematic presentation of the spreading in the presence of the liquid source in the drop center. $R(t)$, radius of the drop base; and θ, contact angle. (1) Liquid drop, (2) solid substrate with a small orifice in the center, and (3) liquid source (syringe).

The liquid is assumed nonvolatile and injected from the center according to a prescribed time rate $V'(t)$; hence, the drop volume obeys the following conservation law:

$$2\pi \int\limits_{0}^{R(t)} r\,h\,dr = V(t), \tag{5.92}$$

where now $V(t)$ is fixed by the pumping rate.

Two unknown functions are to be determined: the liquid profile, $h(t, r)$, and the radius of the spreading drop base, $R(t)$, which is referred to here as "the radius of spreading." The pressure inside the spreading drop, $p(t, r)$, can be determined via the liquid profile, $h(t, r)$. This expression includes several components (see Sections 3.1 and 3.2), where we keep now only two components: capillary and gravitational parts:

$$p = p_a - \gamma K + \rho g h, \tag{5.93}$$

where p_a is the pressure in the ambient air; γ is the liquid–air interfacial tension; ρ and g are the liquid density and gravity acceleration, respectively; and K is the curvature of the liquid–air interface. In the low slope approximation ($\varepsilon \ll 1$) the curvature is $K = \frac{1}{r}\frac{\partial}{\partial r}\left(r\frac{\partial h}{\partial r}\right)$, hence, Eq. (5.93) becomes

$$p = p_a - \gamma \frac{1}{r}\frac{\partial}{\partial r}\left(r\frac{\partial h}{\partial r}\right) + \rho g h. \tag{5.94}$$

Let us introduce scales in radial and vertical directions, r_*, h_*, respectively. Then the following dimensionless quantities are introduced:

$$\bar{r} = \frac{r}{r_*}, \quad \bar{h} = \frac{h}{h_*} \quad or \quad r = \bar{r}r_*, \quad h = \bar{h}h_*,$$

where dimensionless values are marked by an overbar. In the dimensionless form, Eq. (5.94) can be rewritten as

$$p = p_a - \frac{\gamma\, h_*}{r_*^2}\frac{1}{\bar{r}}\frac{\partial}{\partial \bar{r}}\left(\bar{r}\frac{\partial \bar{h}}{\partial \bar{r}}\right) + \rho g h_* \bar{h}$$

In this expression, we have two parameters: $\frac{\gamma h_*}{r_*^2}$ and $\rho g h_*$. The first dimensionless parameter estimates the intensity of capillary forces and the second that of the gravity force.

If capillary forces prevail, then we have the capillary regime of spreading, that is,

$$if \quad \frac{\gamma\, h_*}{r_*^2} \gg \rho g h_* \quad or \quad r_* \ll \sqrt{\frac{\gamma}{\rho g}}.$$

If the gravity force dominates, then the gravitational regime of spreading takes place, that is,

$$if \quad \frac{\gamma\, h_*}{r_*^2} \ll \rho g h_* \quad or \quad r_* \gg \sqrt{\frac{\gamma}{\rho g}}.$$

The length $a = \sqrt{\frac{\gamma}{\rho g}}$, which was introduced in Section 3.1, is the capillary length. Hence, the capillary regime ($R(t) < a$) is the initial stage of spreading of small drops, whereas the gravitational regime is the final stage of spreading of small drops ($R(t) > a$) or the overall regime of spreading of big drops. We consider below in this section the spreading of small drops with a transition from the capillary to the gravitational regime of spreading over time.

Note, that in Eqs. (5.93) and (5.94) we ignored the role played by the Derjaguin's pressure and, hence, we cannot consider the drop profile in a vicinity of the three-phase contact line. However, now we are not going to calculate the drop profile. Instead we try similarity solutions of the equation of spreading (5.91) in the case of both capillary and gravitational regime of spreading (that is, initial and final stages of spreading of small drops). This method allows us to calculate the time evolution of the radius of spreading, but the pre-exponential factor includes an unknown dimensionless integration constant. This constant can be calculated base on the consideration presented in Section 4.2; however, in this section, it is extracted from experimental data.

If the initial drop size is small enough, then the effect of gravity can be ignored. Accordingly, the drop radius, $R(t)$, has to be smaller than the capillary length, $R(t) \le \sqrt{\frac{\gamma}{\rho g}}$. The liquid is injected through the orifice in the drop center with the flow rate $V'(t)$, where $V(t)$ is an imposed function of time. In the case of spontaneous spreading, $V(t) = V_0 = $ const, where V_0 is the constant volume of the spreading drop. In the case of constant liquid flow rate from the liquid source, $V(t) = I\,t$, where I is the intensity of the liquid source.

Mass conservation of liquid Eq. (5.92) demands that $V(0) = 0$ if there was not a drop at the initial time.

In Appendix 2, we deduce a condition when the drop spreads according to the power law, and the general possible form of $V(t)$ compatible with the power law of spreading is also deduced. In Appendix 2, we, deduce dependences of the radius of spreading with time in the case of complete wetting (both capillary and gravitational regime of spreading) and in the case of partial wetting (small capillary numbers, $Ca = \frac{U\eta}{\gamma}$).

In the case of the constant source of liquid I in the drop center, the following dependencies for the radius of the drop base are deduced in Appendix 2.

During the initial (capillary) stage of spreading of small drops, the radius of the drop base should follow

$$R(t) = \alpha_c \left(\frac{\gamma I^3}{\eta} \right)^{0.1} t^{0.4}, \tag{5.95}$$

whereas at the final (gravitational) stage of spreading of small drops, the radius of the drop base follows the law:

$$R(t) = \alpha_g \left(\frac{\rho g I^3}{\eta} \right)^{1/8} t^{0.5}, \tag{5.96}$$

where α_c, α_g are the dimensionless constants to be determined from experimental data.

In the case of partial wetting, the static hysteresis of the contact angle determines the spreading behavior at low capillary numbers, $Ca \ll 1$, where $U = \dot{R}(t)$ is the rate of spreading. This condition was always satisfied under our experimental conditions.

Because of the contact angle hysteresis, the drop does not move if the contact angle, θ, is in the range $\theta_r < \theta < \theta_a$, where θ_a and θ_r are the static advancing and static receding contact angles, respectively (see Chapter 3). In our experimental procedure (Figure 5.22), we are interested in only the static advancing contact angle.

If the capillary number, Ca, is very small, which is the case in our experiments, then the advancing contact angle does not vary significantly. It is assumed that the contact angle, θ, does not vary over the duration of the spreading experiment and remains equal to its static value θ_a.

In this case, the radius of the drop base should follow

$$R(t) = \left(\frac{I}{\Phi(\theta_a)} \right)^{1/3} t^{1/3}. \tag{5.97}$$

Despite the similarity between expressions for the radius of spreading [Eqs. (5.95) and (5.96) in the case of complete wetting and Eq. (5.97) in the case of partial wetting], there is one significant difference between these two spreading processes: If the liquid source is closed, then in the case of complete wetting, the drop will continue to spread out according to the law $R(t) \sim t^{0.1}$ (in the case of capillary regime) or $R(t) \sim t^{1/8}$ (in the case of gravitational regime). However, in the case of partial wetting, the drop will stop spreading as soon as the liquid source is closed.

Experimental Setup and Results

Materials and Methods

The spreading of silicone oil and aqueous droplets over glass substrates were investigated. Silicone oil was purchased from BROOKFIELD. Its viscosity was measured using the rheometer AR1000 (TA Instruments) at 25°C. Density was measured by the weight method and for measuring surface tension the Tensiometer (White, Elec. Inst., Co. Ltd.) was used. The following values were found: dynamic viscosity $\eta = 91.0$ cP, density $\rho = 0.96$ g/cm^3, and surface tension $\gamma = 22.5$ dyn/cm.

Microscope glass slides (76 × 26 mm, Menzel-Glaser, Germany) were used for spreading experiments. Circular orifices of 0.5-mm diameter were drilled in their centers. A liquid was injected through those orifices in the glass substrate with a constant flow rate using a Harvard Apparatus syringe pump. This produced a liquid drop over the solid substrate. Constant flow rate resulted in a linear increase of the drop volume with time. The time evolution of the radius of the base of the spreading drops was monitored.

The glass slides were cleaned by immersing them in a chromic acid solution for 2 h followed by rinsing 10 times with distilled water and 2 times with ultra-pure water, then they were dried in an oven at 70°C for 30 min. Each cleaned and dried slide was used only once. At least three runs were conducted for each experimental condition and average values are reported here.

The diagram of the experimental setup is shown in Figure 5.23. All experiments were carried out at 25°C ± 0.5°C. The solid substrate (1) (Figure 5.23) was fixed in the ring; a syringe (4,) was positioned in the center of the substrate, and connected to the Harvard Apparatus syringe pump (18). The droplets of silicone oil or water (3) were formed due to the injection. The following flow rates were used 0.005, 0.01 and 0.02 mL/min.

The spreading process was recorded using CCD cameras (5, 10) (Figure 5.23) and a VHS video recorder (6, 11). The camera was equipped with filters with a wavelength of 640 nm. Such an arrangement

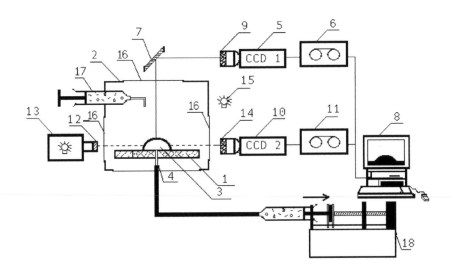

FIGURE 5.23 Schematic presentation of the experimental setup for monitoring the advancing and receding contact angles on a smooth substrate. (1) Porous substrate, (2) hermetically closed, thermostated chamber, (3) liquid drop, (4) glued-in syringe needle, positioned in the center of the solid substrate 1 connected to the Harvard Apparatus syringe pump (18,5), (10) CCD cameras, (6,11) VCRs, (7) mirror, (8) PC (9,14) telephoto objectives, (12) collimating lenses, (13) light source, (15) flash gun, (16) optical windows, (17) upper syringe, and (18) Harvard apparatus syringe pump.

suppresses illumination of the CCD camera by the scattered light from the substrate and, hence, results in a higher precision of the measurements. A source of light (13) was used during experiments. The camera and video recorder were connected to a computer (8). Images were analyzed using Drop Shape Analysis "FTA 32."

Results and Discussion

The kinetics of spreading of silicone oil (complete wetting) and aqueous droplets (partial wetting) over glass substrates was investigated. The spreading process was caused by both the spontaneous spreading and the injection of liquid through the liquid source in the center of drops.

Complete Wetting

Two stages of spreading of small silicone oil drops have been observed: the first initial stage, the capillary regime, and the second final stage, the gravitational regime. The experimental time dependences of radius of spreading of silicone oil drops over a glass surface are presented in Figure 5.24. Three experiments with different injection velocities 0.005, 0.01, 0.02 mL/min were conducted. In each experiment, the two aforementioned stages of spreading are presented in Figure 5.24.

The data obtained from the initial stage of spreading, correspond to the capillary stage $(R(t) < a)$. Equation (5.95) can be rewritten as follows:

$$\lg R(t) = \lg\left[\alpha_c \left(\frac{\gamma I^3}{\eta}\right)^{0.1}\right] + 0.4 \cdot \lg t, \tag{5.98}$$

which is a linear function of time in *log-log* coordinates.

According to Eq. (5.98), the experimental data were fitted according to

$$\lg R(t) = X + n \lg t, \tag{5.99}$$

where intercept, X, and slope, n, are the fitting parameters. The fitted exponents, n (Table 5.6), show good agreement with the theoretically predicted exponent, 0.4.

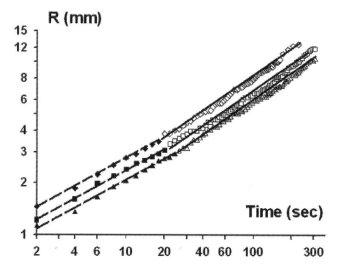

FIGURE 5.24 Radius of spreading versus time for the spreading of silicone oil drops on glass surface (log-log plot), diameter of the orifice 0.5 mm: ▲ capillary stage; △ gravitational stage, $I = 0.005$ mL/min, experiment 1; ■ capillary stage; □ gravitational stage, $I = 0.01$ mL/min, experiment 2; ◆ capillary stage; ◊ gravitational stage, $I = 0.02$ mL/min, experiment 3; dashed line fitted according to Eq. (5.103), solid line fitted according to Eq. (5.104).

TABLE 5.5

Capillary Radius Profile

x (cm)	R(x) [(cm) × 10⁴, measured]	R(x) [(cm) × 10⁴, fitted]
0		4.26
0.50	4.31	4.31
2.07	4.47	4.47
3.07		4.59
4.05	4.72	4.71
6.97	5.11	5.12
10.07	5.64	5.65
13.52		6.34

TABLE 5.6

Spreading of Silicone Oil Drops over Glass Surface with the Orifice in the Center

Injection Flow Rate	Capillary Stage		Gravitational Stage	
I mm³/s	α_c	n	α_g	n
0.0833	2.7589	0.4007	2.6402	0.4893
0.1667	2.5426	0.4005	2.1811	0.5000
0.3333	2.4040	0.4005	1.8833	0.4991

The fitted value of X was compared with Eq. (5.98), and the unknown dimensionless constant, ω_c, was determined as follows:

$$\alpha_c = \left(\frac{\eta}{\gamma I^3} \right)^{0.1} \exp(X) \tag{5.100}$$

Using this equation, the constant α_c was calculated for each set of data (Table 5.6). The average value $\pm$ standard deviation (S.D.) obtained was $\alpha_c = 2.57 \pm 0.18$. The model predictions are shown in Figure 5.24 by dashed lines.

In the case of the gravitational regime of spreading ($R(t) > a$), the experimental data were compared with the theoretical predictions according to Eq. (5.96). This equation can be rewritten as

$$\lg R(t) = \lg \left[\alpha_g \left(\frac{\rho g I^3}{\eta} \right)^{1/8} \right] + 0.5 \cdot \lg t, \tag{5.101}$$

Following to Eq. (5.101), experimental date were fitted according to

$$\lg R(t) = Y + n \lg t, \tag{5.102}$$

where Y and n are the fitting parameters. The theoretical predictions are shown in Figure 5.24 by solid lines. The fitting of the data resulted in an average value of the exponent, n, in Eq. (5.102) equal to 0.4961, which shows good agreement with the theoretically predicted exponent 0.5. Using the fitted value Y and Eq. (5.101), the unknown dimensionless constant, α_g, was calculated as:

$$\alpha_g = \left(\frac{\eta}{\rho g I^3} \right)^{1/8} \exp(Y). \tag{5.103}$$

The average value of the constant $\pm$ S.D. was $\alpha_g = 2.64 \pm 0.38$. The average drop height was 3.6 ± 0.3 mm, which is in good agreement with the calculated value, 3.9 mm.

Partial Wetting

The partial wetting case was investigated using the spreading of water droplets over the same glass surfaces. Three experiments with different injection velocities 0.005, 0.01, 0.02 mL/min are presented in Figure 5.25. The spreading behavior was compared with the theoretical prediction according to Eq. (5.97), which can be written as

$$\lg R(t) = \lg\left(\frac{I}{f(\theta_a)}\right)^{1/3} + \frac{1}{3}\lg t. \tag{5.104}$$

During the spreading of water droplets over glass surfaces, the advancing contact angle does not vary significantly. The average value of the contact angle was calculated for each experimental run, and the average advancing contact angle was determined from three experimental runs. The average advancing contact angle, $\theta_a \pm S.D.$, is $54 \pm 2°$. This value was used in Eq. (2.34) for the calculation of the function $f(\theta_a)$, which was substituted in Eq. (5.104). The dashed line in Figure 5.25 was plotted according to Eq. (5.104). Note, that Eq. (5.104) does not include any fitting parameters. Figure 5.25 shows that our experimental data are in good agreement with the theoretically predicted law Eq. (5.104).

Conclusions

The spreading of liquid over solid substrates when there is liquid injection through an orifice was investigated from both theoretical and experimental points of view. Two cases of spreading over a glass substrate with a diameter of the orifice 0.5 mm were studied: the spreading of silicone oil droplets (complete wetting) and the spreading of water droplets (partial wetting). In the case of silicone oil spreading, two regimes of spreading were observed: the capillary regime and the gravitational regime. A theory has been developed for the cases of complete wetting and partial wetting at low capillary numbers. In all, three case power laws of spreading were deduced: the capillary regime of spreading according to Eq. (5.95), the gravitational regime of spreading according to Eq. (5.96), and the partial regime of spreading according to Eq. (5.97). Experimental data validated our theoretical dependences of the radius of spreading on both time and injection velocity in both cases of complete and partial wetting.

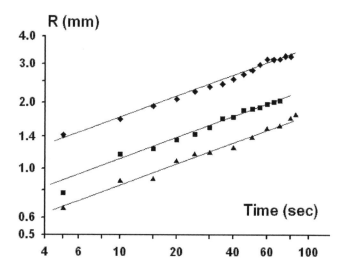

FIGURE 5.25 Radius of spreading versus time for the spreading of water droplets on glass surface (log-log plot), diameter of the orifice 0.5 mm: ▲ $I = 0.005$ mL/min, experiment 1; ■ $I = 0.01$ mL/min, experiment 2; ♦ $I = 0.02$ mL/min, experiment 3; solid line-drawn according to Eq. (5.105).

APPENDIX 2

Let us introduce the following similarity coordinate and function:

$$\xi = \frac{r}{R(t)}, \quad \phi(\xi) = \frac{h(t,r)}{H(t)}, \tag{A2.1}$$

where $\phi(\xi)$, $H(t)$ are two new unknown functions. Note, we are interested only in the time evolution of the radius of spreading, $R(t)$.

The substitution of Eq. (A2.1) into the conservation law, Eq. (5.92), results in

$$2\pi R^2(t) H(t) \int_0^1 \xi \phi(\xi) d\xi = F(t). \tag{A2.2}$$

Let us select the unknown function $H(t)$ as

$$H(t) = \frac{V(t)}{2\pi R^2(t)}. \tag{A2.3}$$

Then Eq. (A2.2) can be reduced to

$$\int_0^1 \xi \phi(\xi) d\xi = 1. \tag{A2.4}$$

To move further, it is necessary to specify the equation of spreading (5.91), which determines the drop profile, $h(t, r)$, or $\varphi(\xi)$. Two different cases are under consideration: complete and partial wetting cases.

Capillary Regime, Complete Wetting

In this case according to Eq. (5.94), the pressure inside the spreading drop can be written as

$$p = p_a - \gamma \frac{1}{r} \frac{\partial}{\partial r} \left(r \frac{\partial h}{\partial r} \right).$$

This expression should be substituted into Eq. (5.91), which yields the time evolution of the drop profile, $h(t, r)$:

$$\frac{\partial h}{\partial t} = -\frac{1}{3\eta} \frac{1}{r} \frac{\partial}{\partial r} \left[r h^3 \frac{\partial}{\partial r} \left(\frac{\gamma}{r} \frac{\partial}{\partial r} \left(r \frac{\partial h}{\partial r} \right) \right) \right]. \tag{A2.5}$$

Note that the omission of the surface forces action results in the well-known singularity on the moving three-phase contact line (see Section 3.1).

Substitution of the similarity coordinate and function using Eqs. (A2.1) and (A2.3) into Eq. (A2.5) results in

$$\dot{H}(t)\phi(\xi) - \frac{H(t)\dot{R}(t)}{R(t)} \xi \phi'(\xi) = -\frac{\gamma}{3\eta} \frac{H^4(t)}{R^4(t)} \frac{1}{\xi} \left[\xi \phi^3(\xi) \left(\frac{1}{\xi} \left(\xi \phi'(\xi) \right)' \right)' \right]',$$

where an overdot denotes differentiation with time; and the symbol ' indicates differentiation with respect to the similarity variable, ξ. This equation can be rewritten as

$$\frac{3\eta\dot{H}(t)R^4(t)}{\gamma H^4(t)}\phi(\xi) - \frac{3\eta R^3(t)\dot{R}(t)}{\gamma H^3(t)}\xi\phi'(\xi) = -\frac{1}{\xi}\left[\xi\phi^3(\xi)\left(\frac{1}{\xi}\left(\xi\phi'(\xi)\right)'\right)'\right]' \qquad (A2.6)$$

Equation (A2.6) should depend on the similarity coordinate only and it should not include any time dependence. This is possible only if the following two relations are satisfied simultaneously:

$$\frac{3\eta\dot{H}(t)R^4(t)}{\gamma H^4(t)} = B_1, \quad \frac{3\eta R^3(t)\dot{R}(t)}{\gamma H^3(t)} = B_2, \qquad (A2.7)$$

where B_1 and B_2 are unknown constants. Both constants should be positive because $H(t)$ and $R(t)$ are both increasing functions of time.

Let $\alpha = B_1/B_2$, and divide the first equation in (A2.7) by the second equation. This results in $\frac{\dot{H}}{H} = \alpha\frac{\dot{R}}{R}$, which upon integration yields

$$H(t) = CR^{\alpha}(t), \qquad (A2.8)$$

where C is an integration constant and α is still an unknown exponent.

Substitution of Eq. (A2.8) into both parts of Eq. (A2.7) results in the following time evolution of the radius of spreading, $R(t)$,

$$R(t) = \left(\frac{(4-3\alpha)\gamma B_2 C^3}{3\eta}\right)^{1/(4-3\alpha)} t^{1/(4-3\alpha)}, \qquad (A2.9)$$

which shows that $4 - 3\alpha$ should be positive, that is, $\alpha < 4/3$.

Equations (A2.9) and (A2.3) allow determination of the unknown function $H(t)$:

$$H(t) = C\left(\frac{(4-3\alpha)\gamma B_2 C^3}{3\eta}\right)^{\alpha/(4-3\alpha)} t^{\alpha/(4-3\alpha)}. \qquad (A2.10)$$

Using Eqs. (A2.10), (A2.9) and (A2.3), we can conclude that the following relation should be satisfied:

$$V(t) = 2\pi C\left[\frac{(4-3\alpha)\gamma B_2 C^3}{3\eta}\right]^{\frac{2+\alpha}{4-3\alpha}} t^{\frac{2+\alpha}{4-3\alpha}}. \qquad (A2.11)$$

This relation shows that the similarity mechanism considered is possible only if the dependency $V(t)$ is defined by the power law (A2.11).

Let us assume now that

$$V(t) = a' t^b, \qquad (A2.12)$$

where a' and b are constants imposed by the source. The exponents in Eqs. (A2.11) and (A2.12) should be equal, and hence $\alpha = \frac{4b-2}{1+3b}$. Substitution of this expression into Eq. (A2.9) gives the following spreading law:

$$R(t) = C \cdot t^{0.1+0.3b}, \qquad (A2.13)$$

Accordingly, the exponent is the sum of two terms: the first term, 0.1, stems from spontaneous spreading (see Chapter 3), and the second term, 0.3b, is determined by the liquid source.

In the case of a constant flow rate of the liquid,

$$V(t) = I t,$$ (A2.14)

and the comparison of exponents in Eqs. (A2.14) and (A2.9), yields $1 = \frac{2+\alpha}{4-3\alpha}$, or $\alpha = 1/2$. On the other hand, $\alpha = B_1/B_2$, or $B_2 = 2B_1$ though B_1 is the only unknown constant. The comparison of the pre-exponential factors in Eqs. (A2.14) and (A2.9) gives

$$C = \left(\frac{3\eta I}{10\pi\gamma B_1} \right)^{1/4}.$$ (A2.15)

The spreading law according to Eq. (A2.13) now takes the following form:

$$R(t) = \left(\frac{5\gamma B_1 I^3}{24\eta\pi^3} \right)^{0.1} t^{0.4}.$$ (A2.16)

The only unknown constant in Eq. (A2.16) is the dimensionless constant B_1.

The equation of spreading Eq. (A2.6) can be rewritten now as

$$B_1\phi(\xi) - 2B_1\xi\phi'(\xi) = -\frac{1}{\xi}\left[\xi\phi^3(\xi)\left(\frac{1}{\xi}\left(\xi\phi'(\xi)\right)'\right)' \right]'.$$ (A2.17)

Recall that this equation describes the behavior of the drop profile, $\varphi(\xi)$, not too close to the moving three-phase contact line (valid only away from the range of the surface forces action). Let us try to include the disjoining pressure action into Eq. (A2.17). In the case of complete wetting, the disjoining pressure isotherm is $\Pi(h) = \frac{A}{h^3}$, where A is the Hamaker constant. Now Eq. (5.8) should be rewritten as

$$p = p_a - \gamma\frac{1}{r}\frac{\partial}{\partial r}\left(r\frac{\partial h}{\partial r} \right) + \rho g h - \frac{A}{h^3}$$

and this expression should be substituted into the equation of spreading (A2.5). Using the same similarity coordinate, ξ, and the drop profile, φ, according to Eq. (A2.1), we arrive after similar to the previous consideration to the following equations for the determination of the unknown function, φ:

$$B_1\phi(\xi) - 2B_1\xi\phi'(\xi) = -\frac{1}{\xi}\left[\xi\phi^3(\xi)\left(\frac{1}{\xi}\left(\xi\phi'(\xi)\right)' + \frac{\chi B_1}{\phi^3(\xi)} \right)' \right]',$$ (A2.18)

where $\chi = \frac{10\pi A}{3\eta I}$ is a dimensionless constant, which characterizes the intensity of the surface forces action. Eq. (A2.18) shows that the same similarity property is valid for the full Eq. (A2.18) when the surface forces action is taken into account as for Eq. (A2.17).

Equation (A2.18) must satisfy the following boundary conditions:

$$\phi'(0) = \phi'''(0) = 0,$$ (A2.19)

which are the symmetry conditions in the drop center. It also should satisfy

$$\phi(\xi) \to 0, \quad \xi \to \infty,$$ (A2.20)

where "infinity" means the drop edge. It was shown in Section 3.2 that the drop profile tends asymptotically to zero inside "the inner region," which is the meaning of the boundary condition (A2.20). The conservation law (A2.10) should also be taken into account.

The unknown parameter, B_1, should be small according to the previous consideration in Chapter 3. In this case, the whole drop can be subdivided into two regions: the "outer region," which is of spherical shape, and the "inner region," where the inner coordinate should be introduced. Matching these two regions allows determination of the unknown parameter, B_1, via the dimensionless Hamaker constant, χ. It was shown in Section 3.2 that the dependence of B_1 on the parameter χ is a weak one. It was also shown in Section 3.2 that the lack of the proper asymptotic behavior of Eq. (A2.18) does not allow determining precisely the unknown constant B_1.

That is why we use Eq. (A2.16) in the following form:

$$R(t) = \alpha_c \left(\frac{\gamma I^3}{\eta} \right)^{0.1} t^{0.4}, \quad \alpha_c = \left(\frac{5B_1}{24\pi^3} \right)^{0.1}, \tag{A2.21}$$

where the dimensionless parameter α_c is determined below in experimental section using experimental data.

Gravitational Regime, Complete Wetting

In this case, Eq. (5.94) becomes $p = p_a + \rho g h$, and substitution of this equation into the equation of spreading Eq. (5.91) results in

$$\frac{\partial h}{\partial t} = \frac{\rho g}{3\eta r} \frac{\partial}{\partial r} \left(r h^3 \frac{\partial h}{\partial r} \right). \tag{A2.22}$$

Using the same similarity coordinate and function Eq. (A2.1) we conclude that relations Eqs. (A2.2) through (A2.4) are still valid. Using the same procedure as above in this Appendix, we can transform Eq. (A2.22) to

$$\dot{H}(t)\phi(\xi) - \frac{H(t)\dot{R}(t)}{R(t)} \xi\phi'(\xi) = \frac{\rho g}{3\eta} \frac{H^4(t)}{R^2(t)} \frac{1}{\xi} \left[\xi\phi^3(\xi)\phi'(\xi) \right]'.$$

This equation can be rewritten as

$$\frac{3\eta \dot{H}(t)R^2(t)}{\rho g H^4(t)} \phi(\xi) - \frac{3\eta R(t)\dot{R}(t)}{\rho g H^3(t)} \xi\phi'(\xi) = \frac{1}{\xi} \left[\xi\phi^3(\xi)\phi'(\xi) \right]'. \tag{A2.23}$$

Equation (A2.23) should depend on the similarity coordinate only, that is, it should not include any time dependence. This is possible only if the following relations are satisfied simultaneously:

$$\frac{3\eta \dot{H}(t)R^2(t)}{\rho g H^4(t)} = D_1, \quad \frac{3\eta R(t)\dot{R}(t)}{\rho g H^3(t)} = D_2, \tag{A2.24}$$

where D_1 and D_2 are unknown dimensionless constants. Both constants should be positive (or zero) because $H(t)$ and $R(t)$ are both increasing functions of time.

Let $\beta = D_1/D_2$ and divide the first equation in (A2.24) by the second equation. That results in $\frac{\dot{H}}{H} = \beta \frac{\dot{R}}{R}$, which upon integration yields

$$H(t) = GR^\beta(t), \tag{A2.25}$$

where G is an integration constant and β is still an unknown exponent.

Substitution of Eq. (A2.25) into both parts of Eq. (A2.24) results in the following time evolution of the radius of spreading, R:

$$R(t) = \left(\frac{(2-3\beta)\rho g D_2 G^3}{3\eta} \right)^{1/(2-3\beta)} t^{1/(2-3\beta)}, \qquad (A2.26)$$

which shows that $2 - 3\beta$ should be positive, that is, $\beta < 2/3$.

Equations (A2.26) and (A2.25) allow for determination of the unknown function $H(t)$:

$$H(t) = G \left(\frac{(2-3\beta)\rho g D_2 G^3}{3\eta} \right)^{\beta/(2-3\beta)} t^{\beta/(2-3\beta)}. \qquad (A2.27)$$

Using Eqs. (A2.26), (A2.27) and (A2.3), we can conclude

$$V(t) = 2\pi G \left[\frac{(2-3\beta)\rho g D_2 G^3}{3\eta} \right]^{\frac{2+\beta}{2-3\beta}} t^{\frac{2+\beta}{2-3\beta}}. \qquad (A2.28)$$

Thus, we see that the similarity mechanism considered above in this Appendix is possible only if the dependency $V(t)$ is defined by the power law Eq. (A2.28).

Let us assume now that the liquid source produces the liquid in the same way as in the case of the capillary regime, that is, according to Eq. (A2.12).

The exponents in Eqs. (A2.28) and (A2.12) should be equal, and hence in $\beta = \frac{2b-2}{1+3b}$. The substitution of this expression into Eq. (A2.26) gives the following spreading law:

$$R(t) = \text{const} \cdot t^{1/8+3b/8}, \qquad (A2.29)$$

that is, the exponent is the sum of two terms: the first term, 1/8, stems from the spontaneous gravitational spreading (see Section 3.1), and the second term, 3b/8, is determined by the liquid source.

In the case of constant flow rate of the liquid,

$$V(t) = I t \qquad (A2.30)$$

and the comparison of exponents in Eqs. (A2.30) and (A2.12) results in $1 = \frac{2+\beta}{2-3\beta}$, or $\beta = 0$. On the other hand, $\beta = D_1/D_2$; hence, $D_1 = 0$. Then D_2 is the only unknown constant. According to the first part of Eqs. (A2.24) and (A2.25), $D_1 = \beta = 0$ means that, in the case of gravitational spreading, the maximum height of the spreading drop remains constant when the source of liquid follows Eq. (A2.30). The comparison of the pre-exponential factors in Eqs. (A2.30) and (A2.28) gives

$$G = \left(\frac{3\eta I}{4\pi\rho g D_2} \right)^{1/4}$$

The spreading law according to Eq. (A2.26) takes now the following form:

$$R(t) = 2^{1/6} \left(\frac{\rho g D_2 I^3}{3\eta\pi^3} \right)^{1/8} t^{0.5}. \qquad (A2.31)$$

The only unknown in Eq. (A2.31) is the dimensionless constant D_2, which is left unknown and was determined experimentally. Note, the constant D_2 can be, in general, determined in the following way: (i) a narrow zone close to the drop edge, where capillary forces become important should be considered; and (ii) matching of asymptotic solutions (capillary zone as an "inner zone" and the gravitational zone as an "outer zone") should be made, which allows for determining the numerical constant D_2.

To determine the unknown constant, we rewrite Eq. (A2.31) as

$$R(t) = \alpha_g \left(\frac{\rho g I^3}{\eta} \right)^{1/8} t^{0.5}, \quad \alpha_g = 2^{1/6} \left(\frac{D_2}{3\pi^3} \right)^{1/8},$$

(A2.32)

and the constant α_g is obtained using experimental data.

This expression allows for determining the constant thickness of the spreading drop during the gravitational regime of spreading. Combining Eqs. (A2.3) and (A2.32) results in

$$H = \frac{1}{\alpha_g^2} \left(\frac{I\eta}{\rho g} \right)^{1/4},$$

(A2.33)

which is independent of time as predicted. It means that during the gravitational regime of spreading from a liquid source with constant flow rate intensity, the drop spreads like a "pancake."

Partial Wetting

The drop volume can be expressed in terms of the spreading radius and the contact angle as follows:

$$V(t) = R^3(t) f(\theta_a), \quad f(\theta_a) = \frac{\pi}{6} \tan \frac{\theta_a}{2} \left(3 + \tan^2 \frac{\theta_a}{2} \right),$$

(A2.34)

where θ_a is the static advancing contact angle. We assume that the capillary number is very small and, hence, the contact angle does not change during spreading. Hence, $f(\theta_a)$ also remains constant. Combination of Eqs. (A2.10) and (A2.34) results in $R(t) = \left(\frac{V(t)}{f(\theta_a)} \right)^{1/3}$.

In the case of the constant flow rate from the liquid source (Figure 5.22), according to Eq. (A2.30) this equation gives

$$R(t) = \left(\frac{I}{f(\theta_a)} \right)^{1/3} t^{1/3},$$

(A2.35)

which is compared with our experimental observations in the case of partial wetting.

5.5 Spreading of Non-Newtonian Liquids over Dry Porous Layer: Complete and Partial Wetting Cases

A droplet placed on a porous substrate, as shown in Section 5.2, involves two competing processes: (i) spreading over already wetted part of the porous substrate and (ii) simultaneous imbibition into the porous substrate. The process (i) results in the increase of radius of the droplet base, the second process (ii) results in a shrinkage of the droplet base. In Section 5.2, we investigated the case of complete wetting. It is not the universal case: some liquids wet the porous material only partially. However, even in the case of partial wetting, simultaneous spreading and imbibition take place. How is it possible to distinguish complete and partial wetting in the case of spreading over porous substrates? In the case of a solid non-deformable, nonporous substrate, the situation is relatively simple: in the case of complete wetting, a droplet deposited on a substrate spreads out until complete disappearance. In the case of partial wetting, the droplet will stop when the contact angle reaches the static advancing contact angle. We start with partial wetting case. As in Section 5.2, we consider a spreading of "big" drops over a thin porous substrate but still the droplet size is small enough to neglect the gravity action.

Partial and Complete Wetting Cases

The spreading and imbibition behavior of a droplet over a porous substrate in the case of partial wetting can be subdivided into three subsequent stages, as shown schematically in Figure 5.26. Stage 1: During this stage, the droplet spreads relatively fast over thin porous substrate until the radius of the droplet base reaches the maximum value, L_{ad}, and the contact angle decreases to the value of the static advancing contact angle, θ_{ad}. Stage 2: During this stage, the three-phase contact line remains fixed at the maximum value while the contact angle decreases from the static advancing contact angle, θ_{ad}, to static receding contact angle, θ_r, due to the loss of droplet volume caused by the imbibition into the porous substrate. Stage 3: During the third stage of spreading/imbibition, the drop base shrinks at approximately constant static receding, θ_r, contact angle until the moment when the droplet is completely sucked up by the porous substrate.

The characteristic feature of partial wetting case is the presence of contact angle hysteresis: this results in the existence of Stage 2, during which the edge of the droplet is pinned. The presence of Stage 2 allows the conclusion that we are dealing with partial wetting case.

There is no contact angle hysteresis in the case of complete wetting; therefore, Stage 2, present in the partial wetting case, is absent in the complete wetting case and there are only two stages of spreading (Figure 5.27). Stage 1: During this stage, the droplet spreads quickly over the porous substrate and the radius of the droplet base reaches its maximum value, L_m. Stage 3: During this stage, the imbibition prevails over the spreading and the radius of the droplet base shrinks until complete disappearance. Note, over most of Stage 3 in the case of complete wetting, the contact angle retains the constant value, which has nothing to do with contact angle hysteresis (there is no contact angle hysteresis in the case of complete wetting, see Section 5.2) but is determined by a pure hydrodynamic reason. An interesting observation is as follows: The duration of Stage 1 in the case of partial wetting is much shorter than the duration of the same stage in the case of complete wetting. The reason, as explained in Chapters 3 and 4, is that in the case of complete wetting the spreading proceeds with a "precursor" film in front, where a lot of energy is burned. However, in the case of partial wetting the spreading proceeds after deposition until the static advancing contact angle is reached through caterpillar motion, which requires much lower energy and, hence, results in faster spreading. Below in this section we follow publications [17,18].

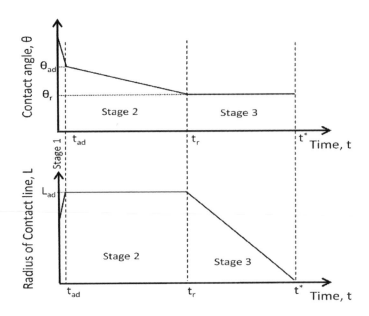

FIGURE 5.26 Three stages of spreading/imbibition of droplet over porous substrate in the case of partial wetting: L_{ad}, maximum radius of droplet base; θ_{ad}, advancing contact angle; t_{ad}, time when θ_{ad} is reached; θ_r, receding contact angle; t_r, time when θ_r is reached; and t^*, time when imbibition is finished completely.

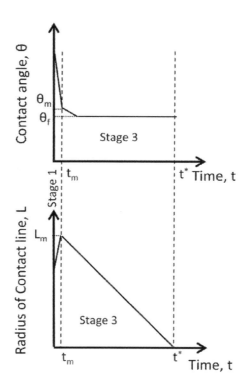

FIGURE 5.27 Two stages of spreading/imbibition of droplet over porous substrate in the case of complete wetting. L_m, maximum radius of droplet base; t_m, time when L_m is reached; θ_m, contact angle at t_m; t^*, time when complete imbibition is finished; and θ_f, final contact angle at t^*. Note, in the case of complete wetting, Stage 2 is absent (Figure 5.19).

5.5.1 Complete Wetting Case

Introduction

Dried blood spot (DBS) sampling is a method of blood collection, transportation and storage which has been investigated and used over the recent decades [19–23]. However, most studies of DBS have been focused on the clinical aspects, such as, developing new analytical method and improving their sensitivity of measurements, and the detection of new analytes, and so on. The mechanisms of blood spreading behavior during DBS sampling on DBS filter paper has not yet been investigated. The basic procedure of DBS sampling is to deposit a small but known amount of blood droplet on a filter paper where the droplet will spread, penetrate into and slowly dry out as a spotted sample. Hence, the sampling process can be described in terms of the spreading of blood, which is a non-Newtonian fluid, over a thin porous substrate. In order to investigate the influence of the spreading/imbibition behavior on the DBS analysis, the development of a theoretical model is presented below in this section.

The spreading of Newtonian liquids over smooth homogeneous surfaces has been investigated in [24–27]. It was established in Section 4.2 that a singularity at the three-phase contact line is removed by the action of surface forces. The presence of roughness and/or a porous sublayer changes substantially the wettability of the substrate [28] and, hence, the spreading behavior [29–31]. The theoretical description of spreading over real surfaces is usually based on an ad hoc empirical "slippage condition" [8,32–35]. In [36] the spreading of small drops of Newtonian liquids over thin porous layers saturated with the same liquid has been investigated. Instead of the "slippage conditions" Brinkman's equations were used in Section 5.1 for the description of the liquid flow inside the porous substrate. In Sections 5.2 and 5.3 the spreading of Newtonian liquid over dry porous substrates was investigated in the case of complete wetting. In Chapter 7, simultaneous spreading and evaporation of droplets of Newtonian liquids are considered.

The spreading of droplets of non-Newtonian liquids over smooth solid surfaces was considered in [37] in the case of complete wetting (see Section 4.10). Considerable deviations from the spreading of Newtonian liquids are found in Section 4.10.

The process under investigation here is the spreading/imbibition of a blood drop, which is a non-Newtonian power-law liquid, over a filter paper. Hence, the problem under investigation is similar to that considered in Section 5.2 when a drop of Newtonian liquid spreads over a dry porous layer; however, the difference is that now the liquid is a non-Newtonian liquid, blood. The experimental result on blood drops spreading over Whatman 903 filter paper (GE healthcare, UK) has been presented earlier in [38] and is presented below in this section. A brief summary of experimental results on blood rheology from [38] is also presented.

The experimental results presented show that the blood spreading deviates from the corresponding Newtonian liquid in two ways: (i) the droplets spreading is governed by a different law as compared with Newtonian liquid and (ii) non-Newtonian liquid imbibition into a porous substrate differs from that of Newtonian liquids.

The problem is treated in this part of the section under the lubrication theory approximation and in the case of complete wetting. Spreading of "big drops," that is, bigger as compared with thickness of the porous substrate but still small enough to neglect the gravity action over "thin porous layers," is considered.

Theory

The kinetics of blood flow both inside the drop above the porous layer and within the porous layer itself are taken into account here. It is assumed that the thickness of the porous layer, Δ, is much smaller than the drop height, that is, $\Delta \ll h^*$, where h^* is the scale of the drop height. The drop profile is assumed to have a low slope, $h^*/L^* \ll 1$, where L^* is the scale of the drop base, and the influence of the gravity is neglected (small drops, Bond number $\rho g L^{*2}/\gamma \ll 1$, where ρ, g, and γ are the liquid density, gravity acceleration, and the liquid–air interfacial tension, respectively). That is, only capillary forces are taken into account.

Under such assumptions, a system of two differential equations is deduced below in this section, which describes the time evolution of the radii of both the drop base, $L(t)$, and the wetted region inside the porous layer, $\ell(t)$, (Figure 5.28).

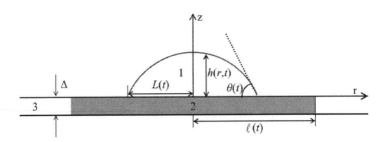

FIGURE 5.28 Cross section of the axisymmetric spreading drop over initially dry filter paper with thickness Δ. (1) Liquid drop; (2) wetted region inside the porous substrate; and (3) dry region inside the porous substrate. $L(t)$, radius of the drop base; $\ell(t)$, radius of the wetted area inside the porous substrate; Δ, thickness of porous substrate; $\theta(t)$, contact angle; and r, z coordinate system.

Droplet Profile

According to Sections 5.1 through 5.3 in the case of the capillary spreading, the droplet profile in the central part of the droplet, except for a small vicinity of the three-phase contact line, can be presented as a spherical cap even in the case of non-Newtonian liquids, which is similar to the case of Newtonian liquid (Section 5.1 through 5.3):

$$h(t,r) = \frac{2V}{\pi L^4}(L^2 - r^2), \quad r < L(t). \tag{5.105}$$

This expression is independent of the power law for the bulk viscosity. The kinetics of the spreading of power law non-Newtonian liquid is considered in more detail in Section 4.10. We use results from this section now.

The porous layer is assumed to be thin enough and the time for saturation in the vertical direction can be neglected relative to other time scales of the process (see Section 5.2).

Darcy's law cannot be used for description of the flow of a non-Newtonian liquid in the thin porous layer; hence, a modified approach is used (see Appendix 3 for further details).

In the same way as in Section 5.2, the time required for a complete saturation of the porous layer in the vertical direction, t_Δ, can be estimated. Accordingly, the porous layer beneath the spreading drop $(0 < r < L(t))$ is always assumed completely saturated.

The capillary pressure inside the porous layer, p_c, can be estimated as $p_c \approx \frac{2\gamma}{a^*}$, where a^* is the scale of capillary radii inside the porous layer and γ is the liquid–vapor surface tension. The capillary pressure inside the drop, $p - p_g$, can be estimated as $p - p_g \approx \frac{\gamma h^*}{L^{*2}} = \frac{h^*}{L^*}\frac{\gamma}{L^*} \ll \frac{\gamma}{L^*} \ll \frac{\gamma}{a^*} \approx p_c$, where p_g is the pressure in the ambient air. That is, the capillary pressure inside the drop can be neglected as compared with the capillary pressure inside the filter paper.

The drop volume, $V(t)$, changes over time, t, because of the imbibition of the liquid into the porous layer, this means

$$V(t) = V_0 - \pi m \Delta \ell^2(t), \tag{5.106}$$

where V_0 is the initial volume of the drop; m is the porosity of the porous layer; and $\ell(t)$ is the radius of the wetted circle on the surface of the porous layer. The wetted region is a cylinder with radius $\ell(t)$ and the height Δ. $\ell(t)$ is referred to here as the radius of the wetted region inside the porous layer.

Let t^* be the time instant when the drop is completely sucked up by the porous substrate, $V(t^*) = 0 = V_0 - \pi m \Delta \ell^{*2}$, where ℓ^* is the maximum radius of the wetted region in the porous layer. This equation gives

$$\ell^* = \left(\frac{V_0}{\pi m \Delta}\right)^{1/2}. \tag{5.107}$$

ℓ^* is used here to scale the radius of the wetted region in the porous layer, $\ell(t)$.

Equation (5.105) gives the following value of the dynamic contact angle, θ, $(\tan\theta \approx \theta)$:

$$\theta = \frac{4V}{\pi L^3}, \tag{5.108}$$

or

$$L = \left(\frac{4V}{\pi\theta}\right)^{1/3}. \tag{5.109}$$

Spreading above Porous Substrate

The drop motion is a superposition of two motions: (a) the spreading of the drop over an already saturated part of the porous layer, which results in an expansion of the drop base, and (b) a shrinkage of the drop base caused by the imbibition into the porous layer. Hence, the following equation can be written:

$$\frac{dL}{dt} = v_+ - v_-,$$ (5.110)

where v_+, v_- are unknown velocities of the expansion and the shrinkage of the drop base, respectively.

Let us take the time derivative of both sides of Eq. (5.109). This results in

$$\frac{dL}{dt} = -\frac{1}{3}\left(\frac{4V}{\pi\theta^4}\right)^{1/3}\frac{d\theta}{dt} + \frac{1}{3}\left(\frac{4}{\pi V^2\theta}\right)^{1/3}\frac{dV}{dt}.$$ (5.111)

Over the whole duration of the spreading over the porous layer, both the contact angle and the drop volume can only decrease with time. Accordingly, the first term in the right-hand side of Eq. (5.111) is positive and the second is negative.

Comparison of the Eqs. (5.110) and (5.111) yields

$$v_+ = -\frac{1}{3}\left(\frac{4V}{\pi\theta^4}\right)^{1/3}\frac{d\theta}{dt} > 0, \quad v_- = -\frac{1}{3}\left(\frac{4}{\pi V^2\theta}\right)^{1/3}\frac{dV}{dt} > 0$$ (5.112)

The derivation presented can be justified as follows. There are two substantially different characteristic time scales in the problem under consideration: $T_\eta = t_\eta^* \ll t_p^* = T_p$, where t_η^* and t_p^* are time scales of the viscous spreading and the imbibition into the porous layer, respectively; $\delta = \frac{t_\eta^*}{t_p^*} \ll 1$ is a smallness parameter (around 0.08 for Newtonian fluid [5]). It is possible to write down $L = \tilde{L}(T_\eta, T_p)$, where T_η is a time of the fast viscous spreading and T_p is a time of the slower imbibition into the porous substrate. The time derivative of $L(T_\eta, T_p)$ is

$$\frac{dL}{dt} = \frac{\partial L}{\partial T_\eta} + \frac{\partial L}{\partial T_p}.$$ (5.113)

Comparison of Eqs. (5.110), (5.112) and (5.113) shows that

$$v_+ = \frac{\partial L}{\partial T_\eta} = -\frac{1}{3}\left(\frac{4V}{\pi\theta^4}\right)^{1/3}\frac{d\theta}{dt}, \quad v_- = \frac{\partial L}{\partial T_p} = -\frac{1}{3}\left(\frac{4}{\pi V^2\theta}\right)^{1/3}\frac{dV}{dt}$$ (5.114)

The decrease of the drop volume, V, with time is determined solely by the imbibition into the porous substrate and, hence, the drop volume, V, only depends on the slow time scale.

According to the previous consideration the whole spreading process can be subdivided into two stages:

1. First fast stage, when the imbibition into the porous substrate is small, and the drop spreads with approximately constant volume. This stage goes in the same way as the spreading over a saturated porous layer and the arguments developed in [36] and Section 5.2 can be used here.
2. Second slower stage, when the spreading process is almost over, and the evolution is determined mostly by the imbibition into the porous substrate.

During the first stage, the dependency of the droplet base radius can be rewritten in the following form [37] (see Section 4.10 for details):

$$L(t) = \left[\frac{n}{2n+1} \frac{\lambda}{\alpha} \left(\frac{\sigma}{k} \right)^{1/n} \frac{V_0^{(n+2)/n}}{(2\pi)^{(n+2)/n}} (t+t_0) \right]^{\alpha}, \tag{5.115}$$

where n is the flow behavior index from the known Otswald–de Waele relationship; $\alpha = \frac{n}{3n+7}$ and t_0 is the duration of the initial stage of spreading, when the capillary regime of spreading is not applicable; λ is a constant, which has been already determined from experimental data [36] for Newtonian liquids; and $\sigma \cong 50 \pm 5$ dyn/cm is the blood/air interfacial tension. The same value as in [36] was adopted for λ in the case of non-Newtonian liquids because the actual viscosity of blood investigated was not substantially different from those liquids used in [36]. Note, the parameter λ is independent of the droplet volume [36]. The parameter α increases from 0 at $n = 0$ to 1/3 at $n\to\infty$ and equals to 0.1 in the case of Newtonian liquid at $n = 1$; that is, the limits are $0 < \alpha < 1/3$.

In the case of Newtonian liquids, Eq. (5.115) gives the well-known results of spreading in the case of complete wetting [25] (Section 5.2): $L(t) \sim (t+t_0)^{0.1}$. Equation (5.115) shows that pseudoplastic fluids, $n < 1$, spread slower than Newtonian liquids because in this case $0 < \alpha < 0.1$, however, dilatant fluids, $n > 1$, spread faster than Newtonian liquid because in this case $\alpha > 0.1$.

According to Eq. (5.115) the characteristic time scale of the first stage of spreading is

$$t_\eta^* = \frac{(2\pi)^{(n+2)/n} L_0^{1/\alpha}}{\frac{n}{2n+1} \frac{\lambda}{\alpha} \left(\frac{\sigma}{k} \right)^{1/n} V_0^{(n+2)/n}}, \tag{5.116}$$

where $L_0 = L(0)$ is the radius of the droplet base in the end of the fast initial stage of spreading; $\sigma \cong 50 \pm 5$ dyn/cm is the blood/air interfacial tension. Eq. (5.115) can be rewritten now as

$$L(t) = L_0 \left[\frac{t+t_0}{t_\eta^*} \right]^{\alpha}. \tag{5.117}$$

Combination of Eqs. (5.117), (5.112) and (5.108) gives

$$v_+ = \left(\frac{V(t)}{V_0} \right)^{1/3} \frac{\alpha L_0}{\left(\frac{t+t_0}{t_\eta^*} \right)^{1-\alpha} t_\eta^*}. \tag{5.118}$$

Substitution of Eq. (5.106) into the Eq. (5.112) gives the following expression for velocity of the drop base shrinkage, v_-:

$$v_- = \frac{2\pi^{2/3} m\Delta\ell}{3} \left(\frac{4}{\left(V_0 - \pi m\Delta\ell^2 \right)^2 \theta} \right)^{1/3} \frac{d\ell}{dt}. \tag{5.119}$$

Substitution of these two equations into Eq. (5.110) results in:

$$\frac{dL}{dt} = \frac{\alpha \left(\frac{V}{V_0} \right)^{1/3} \left(\frac{n}{2n+1} \frac{\lambda}{\alpha} \left(\frac{\sigma}{k} \right)^{1/n} \frac{V_0^{(n+2)/n}}{(2\pi)^{(n+2)/n}} \right)^{\alpha}}{(t+t_0)^{1-\alpha}} - \frac{2\pi^{2/3} m\Delta\ell}{3} \left(\frac{4}{\left(V_0 - \pi m\Delta\ell^2 \right)^2 \theta} \right)^{1/3} \frac{d\ell}{dt}, \tag{5.120}$$

where $\sigma \cong 50 \pm 5$ dyn/cm. The only unknown function left now is the radius of the wetted region inside the porous layer, $\ell(t)$, which is determined below in this section.

Inside the Porous Layer outside the Drop ($-\Delta < z < 0, L < r < l$)

The liquid flow inside the porous layer obeys the modified Darcy's law (Eq. A3.11 in Appendix 3):

$$\frac{1}{r}\frac{\partial}{\partial r}\left(r\frac{\partial v}{\partial r}\right)=0, \quad v=K_n\left(\frac{1}{k}\left|\frac{\partial p}{\partial r}\right|\right)^{1/n}. \tag{5.121}$$

Solution of these equations is

$$p=-\left(\frac{c}{K_n}\right)^n k\frac{r^{1-n}}{1-n}+b, \quad v=\frac{c}{r}, \tag{5.122}$$

where b and c are integration constants, which should be determined using the boundary conditions for the pressure at the drop edge, $r = L(t)$, and at the circular edge of the wetted region inside the porous layer, $r = \ell(t)$.

Note that in the case of the imbibition of a Newtonian liquid [5], the solution is different from the case of power law non-Newtonian liquids:

$$p=-(c\mu/K_p)\ln r+b; \quad v=\frac{c}{r} \tag{5.123}$$

because the solution, Eq. (5.122), becomes singular at $n = 1$, that is, in the case of a Newtonian liquid.

The boundary condition at the drop edge is

$$p=p_g-p_c, \quad r=\ell(t) \tag{5.124}$$

where $p_c \approx \frac{2\gamma}{a^*}$ is the capillary pressure inside the pores of the porous layer, and a^* is a characteristic scale of the pore radii inside the porous layer.

The other boundary condition is

$$p=p_g+p_d, \quad r=L(t) \tag{5.125}$$

where p_d is an unknown pressure [5] inside the drop. It was estimated and in [5] that $p_d << p_c$. That is, p_d is omitted in expression (5.125).

Taking into account these two boundary conditions, both integration constants, b and c, can be determined, giving the following expression for the radial velocity according to Eq. (5.122):

$$v=\frac{K_n}{r}\left[\frac{(1-n)(p_d+p_c)}{k(\ell^{1-n}-L^{1-n})}\right]^{1/n} \tag{5.126}$$

The velocity at the circular edge of the wetted region inside the porous layer is

$$\frac{d\ell}{dt}=v\Big|_{r=\ell} \tag{5.127}$$

Combination of these two equations gives the evolution equation for $\ell(t)$:

$$\frac{d\ell}{dt}=\frac{K_n}{\ell}\left[\frac{(1-n)p_c}{k(\ell^{1-n}-L^{1-n})}\right]^{1/n} \tag{5.128}$$

An estimation of the time scale t_p^* can be made according to Eq. (5.128) in the same way as in [5] and in Section 5.2:

$$t_p^* = \frac{\ell^{*(1+1/n)} k^{1/n}}{K_n p_c^{1/n}}. \tag{5.129}$$

Comparison of the estimated values of t_η^* according to Eq. (5.116) and of t_p^* according to Eq. (5.129) shows that under all our experimental conditions the following inequality $t_\eta^* \ll t_p^*$ is really satisfied.

Substitution of Eq. (5.128) into Eq. (5.120) gives the following system of differential equations for the evolution of both the radius of the drop base, $L(t)$, and that of the wetted region inside the porous layer, $\ell(t)$:

$$\frac{dL}{dt} = \frac{\alpha \left(\frac{V}{V_0}\right)^{1/3} \left[\frac{n}{2n+1}\frac{\lambda}{\alpha}\left(\frac{\sigma}{k}\right)^{1/n}\frac{V_0^{(n+2)/n}}{(2\pi)^{(n+2)/n}}\right]^{\alpha}}{(t+t_0)^{1-\alpha}} - \frac{2L\pi m\Delta}{3}\frac{K_n}{V_0 - \pi m\Delta\ell^2}\left[\frac{(1-n)p_c}{k(\ell^{1-n}-L^{1-n})}\right]^{1/n} \tag{5.130}$$

$$\frac{d\ell}{dt} = \frac{K_n}{\ell}\left[\frac{(1-n)p_c}{k(\ell^{1-n}-L^{1-n})}\right]^{1/n} \tag{5.131}$$

where $\sigma \cong 50 \pm 5$ dyn/cm is the blood/air interfacial tension. Let us make the system of differential Eqs. (5.130) and (5.131) dimensionless using new scales $\bar{L} = L/L_m$, $\bar{\ell} = \ell/\ell^*$, $\bar{t} = t/t^*$, where L_m is the maximum value of the drop base, which is reached at the unknown time instant t_m, that is, at the moment when the right-hand side of Eq. (5.130) is equal to zero. Note that the time instant t^* is unknown but according to the previous estimations is not very much different from the characteristic time t_p^* according to Eq. (5.129). That is, we adopt that $t^* = v\, t_p^*$, where v is an unknown parameter, which can depend on n.

The same symbols are used here for dimensionless variables (marked with an overbar) as for corresponding dimensional variables. The system of Eqs. (5.130) and (5.131) transforms as

$$\frac{d\bar{L}}{d\bar{t}} = \frac{v\, A\, (1-\bar{\ell}^2)^{\frac{1}{3}}}{(\bar{t}+\bar{t}_0)^{1-\alpha}} - \frac{2\bar{L}}{3}\frac{v}{1-\bar{\ell}^2}\left[\frac{(1-n)}{(\bar{\ell}^{1-n}-\chi\bar{L}^{1-n})}\right]^{1/n}, \tag{5.132}$$

$$\frac{d\bar{\ell}}{d\bar{t}} = \frac{v}{\bar{\ell}}\left[\frac{(1-n)}{(\bar{\ell}^{1-n}-\chi\bar{L}^{1-n})}\right]^{1/n}, \tag{5.133}$$

where

$$\bar{t}_0 = t_0/t^* \ll 1, \quad \chi = \left(\frac{L_m}{\ell^*}\right)^{1-n}, \quad A = \frac{2}{3}\frac{(\bar{t}_m+\bar{t}_0)^{1-\alpha}}{\left(1-\bar{\ell}_m^2\right)^{4/3}\left(\frac{\bar{\ell}_m^{1-n}-\chi}{1-n}\right)^{1/n}}, \quad \bar{\ell}_m = \frac{\ell_m}{\ell^*}, \tag{5.134}$$

where ℓ_m is the radius of wetting region at the time instant t_m.

Equations (5.132) and (5.133) are a system of two first-order differential equations, and, hence, two initial conditions should be imposed for solving them. These initial conditions should be initial conditions for both $\bar{L}(\bar{t}_0)$ and $\bar{\ell}(\bar{t}_0)$. The system of differential Eqs. (5.132) and (5.133) is singular at zero values of both $\bar{L}(\bar{t}_0)$ and $\bar{\ell}(\bar{t}_0)$. Hence, small but nonzero values were selected instead:

$$\bar{L}(\bar{t}_0) = 0.001, \tag{5.135}$$

$$\overline{\ell}(\overline{t}_0) = 0.001, \tag{5.136}$$

that is, very small values. The small initial time, $\overline{t}_0$, was selected as $\overline{t}_0 = 0.00001$. The calculation presented below in this section shows that the result almost independent of the small initial values selected.

The solution of Eqs. (5.132) and (5.133) should satisfy the following four extra conditions:

$$\overline{L}(1) = 0, \tag{5.137}$$

$$\overline{L}(\overline{t}_m) = 1, \tag{5.138}$$

$$\frac{d\overline{L}(\overline{t}_m)}{d\overline{t}} = 0, \tag{5.139}$$

$$\overline{\ell}(1) = 1, \tag{5.140}$$

Conditions (5.138) and (5.139) are used to select an expression for A and conditions (5.137) and (5.140) to be satisfied. However, the system of differential Eqs. (5.132) and (5.133) includes two unknown dimensionless parameters, χ, ν. Two extra conditions (5.137) and (5.140) are used to determine these unknown parameters.

According to Eqs. (5.114) and (5.132), the dimensionless velocities of the expansion of the drop base, $\overline{v}_+$, and the shrinkage, $\overline{v}_-$, are

$$\overline{v}_+ = \frac{\nu \, A \, (1 - \overline{\ell}^2)^{\frac{1}{3}}}{(\overline{t} + \overline{t}_0)^{1-\alpha}}, \overline{v}_- = \frac{2\overline{L}}{3} \frac{\nu}{1 - \overline{\ell}^2} \left[\frac{(1-n)}{(\overline{\ell}^{1-n} - \chi \overline{L}^{1-n})} \right]^{1/n}, \tag{5.141}$$

Let us rewrite Eq. (5.108) using the same dimensionless variables and $\overline{V} = V/V_0 = 1 - \overline{\ell}^2$:

$$\theta = \frac{4V_0}{\pi L^3} (1 - \overline{\ell}^2). \tag{5.142}$$

Let us introduce $\theta_m = \frac{4V_0}{\pi L_m^3}(1 - \overline{\ell}_m^2)$, which is the value of the dynamic contact angle at the time instant when the maximum value of the drop base is reached. Then Eq. (5.142) can be rewritten as

$$\frac{\theta}{\theta_m} = \frac{(1 - \overline{l}^2)/(1 - \overline{l}_m^2)}{\overline{L}^3} \tag{5.143}$$

and this relationship should be a universal function of the dimensionless time, $\overline{t}$.

Let us consider the asymptotic behavior of system (5.130) and (5.131) during the second stage of the spreading. According to the previous consideration and as confirmed in Figure 5.30, over the second stage of the spreading velocity of the expansion of the drop base, v_+, becomes small and, hence, to understand the asymptotic behavior the term corresponding to v_+ in the left-hand side of Eq. (5.130) is omitted. This results in

$$\frac{dL}{dt} = -\frac{2L\pi m\Delta}{3} \frac{K_n}{V_0 - \pi m\Delta \ell^2} \left[\frac{(1-n)p_c}{k(\ell^{1-n} - L^{1-n})} \right]^{1/n}, \tag{5.144}$$

while Eq. (5.131) is left unchanged. The system of differential Eqs. (5.131) and (5.144) can be solved analytically. For this purpose, Eq. (5.144) is divided by Eq. (5.131), which gives $\frac{dL}{d\ell} = -\frac{2\pi m\Delta \, L \, \ell}{3(V_0 - \pi m\Delta \ell^2)}$. If $V = V_0 - \pi m\Delta \ell^2$ is used instead of ℓ, this equation takes the following form: $\frac{dL}{dV} = \frac{L}{3V}$, which can be easily integrated and the solution is

$$V = C \, L^3 \tag{5.145}$$

where C is an integration constant. Let us rewrite Eq. (5.145) using the same dimensionless variables and $\bar{V} = V/V_0$:

$$\bar{V} = \frac{\pi L_m^3}{4V_0}\,\theta\,\bar{L}^3 \tag{5.146}$$

Comparison of Eqs. (5.145) and (5.146) shows that the dynamic contact angle asymptotically remains constant during the second stage. This constant value is marked as θ_f. It is necessary to emphasize that in the case under consideration the constancy of the contact angle during the second stage has nothing to do with the contact angle hysteresis: There is no hysteresis in our system here: θ_f is not a receding contact angle but forms as a result of a self-regulation of the flow in the drop–porous layer system [5].

Experimental Data

Spreading Behavior

All experimental data for our theoretical calculations were taken from [38], where a series of experiments on blood spreading were reported. Pig's blood was collected in EDTA anti-coagulated tubes (International Scientific Supplies Ltd., Bradford, UK) from a local butcher and used to prepare blood samples of different hematocrit levels [38]: 30%, 50%, and 70%. Figure 5.28 shows a cross section of the axisymmetric spreading drop over the porous medium with thickness Δ. The time evolution of the spreading radius, $L_{exp}(t)$, the droplet height, $h_{exp}(t)$, and the wetting region radius, $\ell_{exp}(t)$, were monitored [38]. According to [38], the pore structure of DBS sampling cards (filter paper) proves that the porous medium (DBS card) has a bi-porous structure, that is, there are macro sized pores between the fibers and micro sized pores inside the fibers. The fast penetration in large pores, between the fibers, currently is the only process taken into account according to the model in this section. However, there are two additional processes that might influence the penetration of blood inside the porous medium: (i) the penetration inside the fibers, and (ii) adsorption of red blood cells on/inside fibers.

Blood Rheology

Blood is a non-Newtonian liquid, which shows a shear-thinning behavior [38–40]. Different rheological models have been used based on curve fitting of experimental data for a description of the blood rheology [41–43]. The power law equation known as the Ostwald–de Waele relationship has been used as the rheological model of blood (see Appendix 4):

$$\eta = k\dot{\gamma}^{n-1}. \tag{5.147}$$

Independent Determination of $K_n p_c^{1/n}/k^{1/n}$

As mentioned above (see Eq. (5.72)), the permeability of the porous layer and the effective capillary pressure are considered as a single coefficient in our theoretical model. In order to determine this coefficient, additional experiments were carried out to obtain this coefficient based on the modified Darcy's law. The horizontal imbibition of blood samples with different hematocrit levels into the filter paper was used similar to [5]. The filter paper used in the spreading experiment was cut into rectangular sheets of 1.5×3.0 cm. The depth of the immersed part of each sheet was around 0.3–0.5 cm into the blood container, and the position of imbibition front was recorded by a high-speed camera over time. According to Eq. (3.8) the position of the imbibition front can be expressed in the following way:

$$d^{1+1/n}(t) = \frac{K_n p_c^{1/n}}{k^{1/n}}(1+1/n)\,t, \tag{5.148}$$

where $d(t)$ is the position of the imbibition front inside the filter paper. It was found that in all runs, $d^{1+1/n}(t)$ proceeds along a straight line, whose slope gives the value $\frac{K_n p_c^{1/n}}{k^{1/n}}$ as shown in Figure 5.29.

The measured values of $\frac{K_n p_c^{1/n}}{k^{1/n}}$ are presented in Table 5.7.

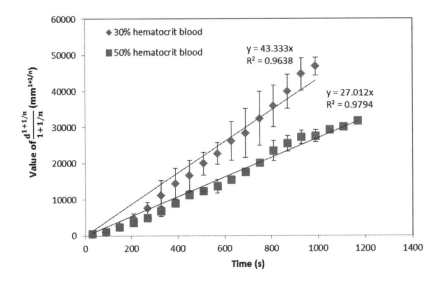

FIGURE 5.29 The time evolution of $d^{1+1/n}(t)/1 + 1/n$ for different hematocrit level pig's blood.

TABLE 5.7

The Value of $K_n p_c^{1/n}/k^{1/n}$ for Different Hematocrit Level of Blood

Hematocrit Level	$K_n p_c^{1/n}/k^{1/n}(\mathrm{mm}^{(1+1/n)} \cdot \mathrm{s}^{-1})$
30%	43.333 ± 6.504
50%	27.012 ± 1.407
70%	N/A

Numerical Solution of Eqs. (5.132) and (5.133)

The system of differential Eqs. (5.132) and (5.133) was solved using the initial conditions Eqs. (5.135) and (5.136). The two unknown parameters χ, v were determined using additional conditions Eqs. (5.137) and (5.140). Note that, after that, the system does not include any fitting parameters. According to our calculations, both parameters, χ, v, vary insignificantly for all three n values used (Table 5.8).

Hence, the three dimensionless dependencies, $\bar{L}(\bar{t})$, $\bar{\ell}(\bar{t})$ and θ/θ_m, should fall on three almost universal curves, see the solid lines in Figures 5.31 through 5.33. This conclusion is in a good agreement with our experimental observations (see below): Results and discussions experimental dependencies also show a universal behavior.

TABLE 5.8

Calculated Values of Two Unknown Dimensionless Parameters
χ, v. The Values of n Were Determined

Hematocrit Level	n	v	χ
30%	0.394	0.281	0.376047
50%	0.368	0.2439	0.3747172
70%	0.325	0.214	0.3454006

Source: Chao, T.C. et al., *Colloids Surf. A: Physicochem. Eng. Asp.*, 451, 38–47, 2014.

Results and Discussion

The time evolution of the radius of a blood droplet at constant volume, 10 ± 0.5 μL, spreading on Whatman 903 filter paper and the time evolution of the radius of wetting region have been presented. The droplet radius of all the blood samples reached zero at the end of the spreading/imbibition process. Although there are some differences of spreading behavior between the blood samples with different hematocrit levels spreading on DBS filter paper, in general, the spreading results of all samples showed similar features during the whole spreading process. The whole process can be subdivided into two stages as in the case of Newtonian liquids (see Section 5.2): Over the duration of the first stage, the expansion of droplet radius, due to the capillary regime of spreading, is faster than the shrinkage, due to the imbibition into the filter paper, until they counterbalance themselves as the maximum radius is reached. After that, the second stage begins where the droplet base starts to shrink and the wetted region expands until the droplet disappears.

In all spreading experiments the drops shape remained a spherical cap over the entire duration of spreading, showing good agreement with our theoretical assumption. This spherical cap shape allowed calculating the dynamic contact angle of each spreading droplet. The dynamic contact angle of all samples decreased rapidly during first stage of spreading as well as the radius of droplet base increased. After the spreading radius reached its maximum value, $L_{m,exp}$, the dynamic contact angle leveled off and remained a constant value over the duration of the second stage.

Table 5.9 provides the experimental data obtained from the spreading experiment [38], which include the maximum spreading radius, $L_{m,exp}$, the time of complete imbibition, t^*_{exp}, contact angle, $\theta_{m,\,exp}$, and ℓ^*_{exp}.

Experimental dependences of the radius of the drop spreading, L_{exp}, the radius of the wetted area, ℓ_{exp}, and the contact angle, θ_{exp}, were made dimensionless using the values presented in Table 5.9.

Figure 5.30 shows the dimensionless velocities, $\bar{v}_+$ and $\bar{v}_-$, of different hematocrit level blood spreading over Whatman 903 filter paper calculated according to Eq. (5.141). Figure 5.30 confirms that

1. The first stage is very short as compared with the second stage of spreading/imbibition. The capillary spreading prevails on this stage over the drop base shrinkage caused by the liquid imbibition into the porous substrate.
2. The spreading of the drop almost stops after the first stage of spreading, and the shrinkage of the drop base is determined by the suction of the liquid from the drop into the porous substrate.

The main results of comparison of experimental and calculated dependences according to Eqs. (5.132), (5.133), and (5.143) are presented in Figures 5.31 through 5.33. These figures show that all the data fall into corresponding universal curves when the dimensionless values are used: $L / L_m, \ell / \ell^*, t / t^*$, and θ / θ_m. The measured values of the corresponding experimental values $L_{m,exp}$, ℓ^*_{exp} and t^*_{exp} for all experimental runs are given in the Table 5.8. It shows that the dimensionless time $\bar{t}_{m,exp}$ is about $0.14 \approx \delta$. Solid curves in Figures 5.31 through 5.33 represent the solution of the system of differential Eqs. (5.132) and (5.133) and Eq. (5.143) for different values of n. Figures 5.31 through 5.33 shows that the developed theory results in a reasonable prediction of experimental dependences. However, the main conclusion

TABLE 5.9

The Experimental Values Used to Make Experimental Dependences Dimensionless for Comparison with Predicted Time Dependences According to Eqs. (5.132) and (5.143)

	Time of Complete Imbibition [t^*_{exp} (s)]	Maximum Spreading Radius of Droplet Base [$L_{m,\,exp}$ (mm)]	Contact Angle at Maximum Spreading [$\theta_{m,exp}$ (degree)]	Maximum Radius of Wetting Region [ℓ^*_{exp} (mm)]
30% Hematocrit level	0.351 ± 0.014	2.18 ± 0.07	32.80 ± 1.88	3.36 ± 0.09
50% Hematocrit level	0.508 ± 0.022	2.18 ± 0.08	30.94 ± 2.50	3.32 ± 0.06
70% Hematocrit level	0.663 ± 0.027	2.09 ± 0.09	33.48 ± 2.37	3.20 ± 0.04

Source: Chao, T.C. et al., *Colloids Surf. A: Physicochem. Eng. Asp.*, 451, 38–47, 2014.

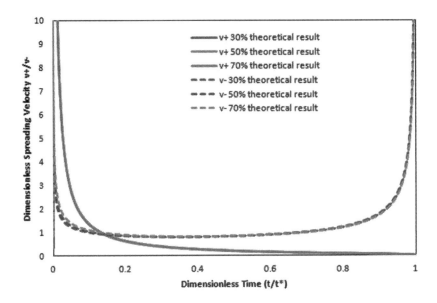

FIGURE 5.30 Complete wetting. The time evolution of the dimensionless velocity $\bar{v}_+$ and $\bar{v}_-$ according to Eq. (5.141) for blood samples with 30%, 50%, and 70% hematocrit level.

Dimensionless Spreading radius

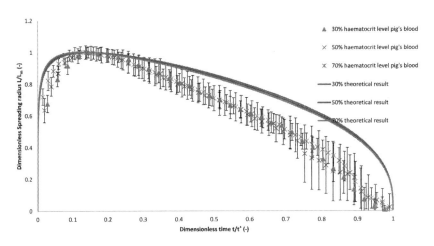

FIGURE 5.31 Complete wetting. The simulation result of the dimensionless time evolution of radius of blood drop-let. Theoretical results according to Eqs. (5.132) and (5.133). (Experimental data from Chao, T.C. et al., *Colloids Surf. A Physicochem. Eng. Asp.*, 451, 38–47, 2014.)

is that both experimental dependences and the calculated theoretical dependences show a remarkable universal behavior independent of n.

It is necessary to take into account that we used a simple model of blood rheology and its penetration into the porous substrate: We did not take into account absorption of red blood cells inside the porous substrate, the bi-porous structure of the substrate, and evaporation.

Calculated dependences presented in Figures 5.31 through 5.33 (solid curves) show that the shape of these dependences is independent of the small values selected as initial conditions according to Eqs. (5.135) and (5.136).

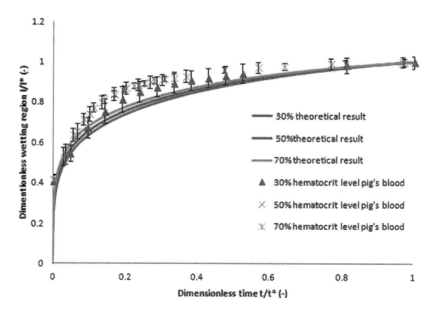

FIGURE 5.32 Complete wetting. Time evolution of radius of the wetted area inside porous substrate. Theoretical results according to Eqs. (5.132) and (5.133). (Experimental data from Chao, T.C. et al., *Colloids Surf. A Physicochem. Eng. Asp.*, 451, 38–47, 2014.)

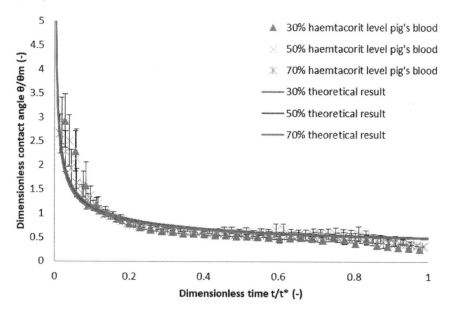

FIGURE 5.33 Complete wetting. The time evolution of contact angle of blood droplet with different hematocrit levels on porous substrate. The solid curve calculated according to Eq. (5.143). (Experimental data from Chao, T.C. et al., *Colloids Surf. A Physicochem. Eng. Asp.*, 451, 38–47, 2014.)

In order to completely predict the blood spot behavior, the four experimental data—time of complete imbibition t^*, maximum radius of droplet base, L_m, maximum radius of wetting region, ℓ^* and contact angle at maximum spreading, θ_m, have to be determined from our simulated results and physical parameters, which are obtained independently from different experiments. These physical parameters are the viscosity of blood, $\eta = k\dot{\gamma}^{n-1}$, namely, n and k; the interfacial tension, σ; and the physical parameters of porous substrate: the permeability coefficient, $K_n p_c^{1/n}$, the thickness, Δ, and the porosity, m.

TABLE 5.10

The Predicted Values of Blood Spot Spreading Behavior Calculated from Simulated Result and Physical Properties of Blood and Porous Substrate

	Time of Complete Imbibition $[t^*_{\text{sim}}$ (s)]	Maximum Spreading Radius of Droplet Base $[L_{m,\text{sim}}$ (mm)]	Contact Angle at Maximum Spreading $[\theta_{m,\text{sim}}$ (degree)]	Maximum Radius of Wetting Region $[\ell^*_{\text{sim}}$ (mm)]
30% Hematocrit level	0.464 ± 0.220	2.050 ± 0.096	44.30 ± 8.37	3.344 ± 0.255
50% Hematocrit level	0.803 ± 0.297	1.957 ± 0.082	49.61 ± 8.65	3.344 ± 0.255
70% Hematocrit level	N/A	N/A	N/A	3.344 ± 0.255

The universal dimensionless constants, namely, v, χ, $\bar{t}_m$, $\bar{\ell}_m$, could be determined from the result of our dimensionless model.

The value ℓ^* can be determined directly according to Eq. (5.107), where the values of V_0, m and Δ are equal to 10 ± 0.5, 0.57 ± 0.03 and 500 ± 50 μm, respectively [38]. The imbibition time, t^*, can be obtained by $t^* = v\, t^*_p$, where t^*_p is determined by Eq. (5.129) and the measured values of $\frac{K_n p_c^{1/n}}{k^{1/n}}$ in Table 5.7.

Due to the fast expansion of blood drop spreading, the imbibition of blood could be neglected during the first stage of spreading; hence, the maximum spreading radius can be approximately approached by Eq. (5.115) and accordingly, the contact angle, θ_m, can be determined by Eq. (5.142) as the following equations:

$$L_m = \left[\frac{n}{2n+1} \frac{\lambda}{\alpha} \left(\frac{\sigma}{k} \right)^{1/n} \frac{V_0^{(n+2)/n}}{(2\pi)^{(n+2)/n}} t^* (\bar{t}_m + \bar{t}_0) \right]^\alpha , \tag{5.149}$$

$$\theta_m = \frac{4V_0}{\pi L_m^3} (1 - \bar{\ell}_m^2), \tag{5.150}$$

where $\sigma \cong 50 \pm 5$ dyn/cm and the values of the k and n for different hematocrit levels are presented in [38].

All the calculated data for the maximum spreading radius, $L_{m,\text{sim}}$, the time of complete imbibition, t^*_{sim}, contact angle, $\theta_{m,\text{sim}}$, and ℓ^*_{sim} based on the simulation result and independent measurements of physical parameters are shown in Table 5.10. The comparison of simulated and experiment result demonstrates that the simulated prediction is in a good agreement with our experimental observation.

Conclusions on Section "Complete Wetting Case"

The process of dried blood spot sampling involves simultaneous spreading and penetration of blood into a porous filter paper with subsequent evaporation and drying in the case of complete wetting is investigated. Spreading of small drops of blood, which is a non-Newtonian liquid but completely wets the porous substrate, over a dry porous layer is investigated from both theoretical and experimental points of view. A system of two differential equations is derived from the combination of the model of spherical cap spreading over porous layer and a modified Darcy's law for power law fluids. It describes the time evolution of radii of both the drop base and the wetted region inside the porous layer in the course of blood spreading over a dry porous substrate. The deduced system of the differential equation does not include any fitting parameters and predicts a universal behavior of the dimensionless dependences of the radii of the droplet base, wetted area inside the porous substrate, and contact angle, which are almost completely independent of the rheological properties of the blood. Experimentally, the time evolution of the radii of the drop base and the wetted region inside the porous layer and dynamic contact angle were monitored for power law dependency of the viscosity of blood. All experimental data fell on three universal curves if appropriate scales were used with a plot of the dimensionless radii of the drop base and the wetted region inside the porous layer, and the dynamic contact angle of the drop on dimensionless time. The predicted theoretical relationships were three universal curves accounting quite satisfactorily for the experimental data. The simulated results show good agreement with experiment data although the bi-porous structure of the filter paper and adsorption of red blood cells inside the porous substrate were not taken into account according to the suggested model.

5.5.2 Partial Wetting Case

In this section, we investigate spreading of a blood droplet over a porous substrate in the case when blood wets only partially the porous substrate. The spreading of a droplet over a porous substrate in this case can be subdivided into three stages (Figure 5.26): (1) relatively fast spreading of the droplet until the radius of the droplet base reached the maximum value and the contact angle reaches the value of static advancing contact angle; (2) the contact line remains fixed at the maximum value, but the volume decreases due to loss caused by the imbibition into porous substrate. Contact angle on this stage decreases from the static advancing contact angle to static receding contact angle; and (3) the shrinkage of the drop base at constant static receding contact angle until complete disappearance of the droplet.

Note that in the case of completely wetting, the second stage does not take place according to the previous section (Figure 5.27) This means that the appearance of pinning of the droplet radius during stage two is a manifestation of contact angle hysteresis and partial wetting and can be used to determine whether it is a partial wetting or a completely wetting case.

The models of Newtonian liquid droplet spreading over a dry porous substrate in the case of complete wetting were investigated in Section 5.2. A model of blood spreading (non-Newtonian liquid) over a porous substrate in the case of complete wetting was presented in Section "Complete Wetting Case".

Below, in this section a modified model of non-Newtonian liquid spreading over porous substrate is presented in this section for the case of partial wetting.

According to Figure 5.38, the viscosity of blood shows a shear-thinning behavior as a function of shear rate applied.

The spreading behavior of blood over nitrocellulose membranes with different average pore sizes in the case of partial wetting is investigated below in this section. The spreading of blood with different hematocrit levels is investigated. The spreading behavior of blood observed from experiment data shown that it shows partial wetting spreading/imbibition behavior on all nitrocellulose membranes investigated. A model of partial wetting spreading/imbibition is developed here to describe the spreading behavior of blood over nitrocellulose membrane. The comparison of simulated result with experimental demonstrate the reliability of this model in the prediction of the time evolution of spreading profile, wetted area inside the porous substrate and contact angle.

Theory

In the case of partial wetting, contact angle hysteresis is the most important feature: In the presence of contact angle hysteresis, the spreading/imbibition of a sessile droplet goes through three consecutive stages (Figure 5.26). *Stage 1:* During this short stage immediately after a deposition, both the contact angle and radius change simultaneously, reaching in the end values θ_{ad}, which is a static advancing contact angle, and L_{ad}, which is the maximum radius of the droplet base, respectively.

These values are used as initial values for the following stage of spreading/imbibition process. It is possible to neglect imbibition during the short spreading stage and, hence, it can be described using a conventional hydrodynamic approach (see Chapter 3). There is a remarkable similarity between the two processes: (i) the spreading/evaporation of droplets, and (ii) the spreading/imbibition over a porous substrate (see Chapter 7). Below in this section we neglect the fast first stage of spreading (Figure 5.26). *Stage 2*: The contact angle decreases from a static advancing contact angle, θ_{ad}, down to a static receding contact angle, θ_r at a constant radius of the droplet base, $L = L_{ad}$. *Stage 3*: The contact angle remains constant and equal to the static receding value, θ_r, while the radius of the droplet base, L, decreases until zero value.

Two of the longest stages of spreading/imbibition, 2 and 3 (Figure 5.26), are considered here.

Droplet Profile

According to Section 5.2, in the case of capillary spreading, the droplet profile in the central part of the droplet, except for a small vicinity of the three-phase contact line, can be presented as a spherical cap even in the case of non-Newtonian liquids, which is similar to the case of Newtonian liquid (see Section 7.5).

This expression is independent of the power law for the bulk viscosity. Hence, during both stages, the droplet retains the spherical shape. That is, the volume of the droplet, V, can be presented as follows:

$$V = L^3 f(\theta), \qquad f(\theta) = \frac{\pi}{3} \frac{(1-\cos\theta)^2 (2+\cos\theta)}{\sin^3\theta}. \tag{5.151}$$

The porous layer is assumed to be thin enough and the time for saturation in the vertical direction can be neglected relative to other time scales of the process (Figure 5.28 and Section 5.2).

In the same way as in Section 5.2, the time required for a complete saturation of the porous layer in the vertical direction, t_Δ, can be estimated. Accordingly, the porous layer beneath the spreading drop $(0 < r < L(t))$ is always assumed completely saturated.

The capillary pressure inside the porous layer, p_c, can be estimated as $p_c \approx \frac{2\gamma}{a^*}$, where a^* is the scale of capillary radii inside the porous layer and γ is the surface tension. The capillary pressure inside the drop, $p - p_g$, can be estimated as $p - p_g \approx \frac{\gamma h^*}{L^{*2}} = \frac{h^*}{L^*} \frac{\gamma}{L^*} \ll \frac{\gamma}{L^*} \ll \frac{\gamma}{a^*} \approx p_c$, where p_g is the pressure in the ambient air. That is, the capillary pressure inside the drop can be neglected as compared with the capillary pressure inside the filter paper.

The drop volume, $V(t)$, changes over time, t, because of the imbibition of the liquid into the porous layer, this means (Figure 5.28)

$$V(t) = V_0 - \pi m \Delta \ell^2(t), \tag{5.152}$$

where V_0 is the initial volume of the drop; m is the porosity of the porous layer; and $\ell(t)$ is the radius of the wetted region on the surface of the porous layer. The wetted region is a cylinder with radius $\ell(t)$ and the height Δ. $\ell(t)$ is referred to here as the radius of the wetted region inside the porous layer.

Blood is a non-Newtonian liquid and shows a shear-thinning behavior (see Section 5.51). The power law equation known as the Otswald–de Waele relationship has been used as the rheological model of blood according to Eq. (5.147).

The developed theory includes two unknown dependences: static advancing contact angle, θ_{ad}, and static receding contact angle, θ_r. Both of them could be functions of concentrations of red blood cells and can be extracted from experimental data only on this stage. Note, the theory developed in Chapter 3, which allowed calculating both static advancing and static receding contact angle can be applied in the case under consideration if the Derjaguin's pressure isotherm of blood at various hematocrit levels is known. Unfortunately, such isotherm has not been measured yet. This is why both hysteresis contact angles static advancing contact angle, θ_{ad}, and static receding contact angle, θ_r on this stage can be extracted from only experimental data.

Below in this section we consider two possible scenarios of blood spreading: (i) large pore size, when the red blood cells can penetrate into the porous substrate with the plasma flow and (ii) small pore size, when red blood cells cannot penetrate into the porous substrate and accumulate on the surface of the porous substrate.

According to [17,44], the blood droplets spreading/imbibition on nitrocellulose membranes shows a partial wetting behavior, that is, the spreading/imbibition behavior is of the type shown in Figure 5.26, that is, the presence of contact angle hysteresis, which is a characteristic feature of partial wetting.

Pores of the Porous Substrate Are Large Enough and Red Blood Cells Penetrate into the Substrate with Plasma Flow

Let t^* be the time instant when the drop is completely sucked up by the porous substrate,

$$V(t^*)=0=V_0-\pi m\Delta\ell*^2, \tag{5.153}$$

where $\ell*$ is the maximum radius of the wetted region in the porous layer. This equation gives

$$\ell* =\left(\frac{V_0}{\pi m\Delta}\right)^{1/2}. \tag{5.154}$$

$\ell*$ is used to scale the radius of the wetted region inside the porous layer, $\ell(t)$.

The combination of Eqs. (5.151) and (5.152) results in

$$L^3 f\left(\theta\right)+\pi\ell^2(t)\,\Delta m=V_0 \tag{5.155}$$

Equation (5.155) should be combined with an equation that describes the flow inside the porous substrate. It is necessary to take into account that the liquid (blood) has a non-Newtonian behavior. This was taken into account and the following equation has been deduced in Section 5.51:

$$\frac{d\ell}{dt}=\frac{K_n}{\ell}\left[\frac{(1-n)p_c}{k(\ell^{1-n}-L^{1-n})}\right]^{1/n}. \tag{5.156}$$

The Second Stage of the Process

During this stage, the radius of the contact line remains constant and equal to L_{ad} (Figure 5.26). Hence, Eqs. (5.155) and (5.156) results in

$$f\left(\theta\right)=\left(V_0-\pi\ell^2(t)\,\Delta m\right)/L_{ad}^3, \tag{5.157}$$

$$\frac{d\ell}{dt}=\frac{K_n}{\ell}\left[\frac{(1-n)p_c}{k(\ell^{1-n}-L_{ad}^{1-n})}\right]^{1/n} \tag{5.158}$$

with the initial conditions

$$\theta\big|_{t=t_{ad}}=\theta_{ad}, \tag{5.159}$$

and

$$\ell(t_{ad}) = \sqrt{\frac{V_0 - L_{ad}^3 f(\theta_{ad})}{\pi \Delta m}}.$$ (5.160)

The end of the second stage is the moment t_r, which is determined as the moment when

$$\theta(t_r) = \theta_r.$$ (5.161)

At this moment,

$$\ell(t_r) = \sqrt{\frac{V_0 - L_{ad}^3 f(\theta_r)}{\pi \Delta m}}.$$ (5.162)

The procedure of solution during the second stage is as follows:

1. Solve Eq. (5.168) with initial condition (5.160).
2. Calculate $\theta(t)$ according to Eq. (5.157).
3. Carry on until condition Eq. (5.161) is satisfied. This determines the moment t_r.

Stage 3 (Figure 5.26).
  During this stage $\theta(t_r) = \theta_r$, hence, from Eq. (5.155) we conclude

$$L(t) = \left(\frac{V_0 - \pi \ell^2(t) \Delta m}{f(\theta_r)} \right)^{1/3}.$$ (5.163)

Substituting Eq. (5.163) into Eq. (5.156) results in

$$\frac{d\ell}{dt} = \frac{K_n}{\ell} \left[\frac{(1-n)p_c}{k \left(\ell^{1-n} - \left(\frac{V_0 - \pi \ell^2(t) \Delta m}{f(\theta_r)} \right)^{(1-n)/3} \right)} \right]^{1/n}$$ (5.164)

The procedure of solution during the third stage is as follows:

1. Solve Eq. (5.164) with initial condition (5.162).
2. Calculate $L(t)$ according to Eq. (5.163).
3. The end of the process, t^*, is the moment when $L(t^*) = 0$.

Pores inside the Porous Substrate Are Small and Red Blood Cells Do Not Penetrate into the Porous Substrate

In this case, only plasma penetrates into the porous substrate, that is, $V_0 = V_{0p} + V_{bc}$, where V_{0p} is the initial volume of plasma and V_{bc} is the volume of blood cell, which does not change over time. Hence, Eq. (5.152) can be rewritten as

$$V_p(t) + V_{bc} + \pi \ell^2(t) \Delta m = V_0$$ (5.165)

In the end of the process, t^{**}, $V_p(t^{**}) = 0$, hence, $V_{bc} + \pi \ell *^2 \Delta m = V_0$, and

$$\ell^* = \left(\frac{V_0 - V_{bc}}{\pi \Delta m} \right)^{1/2}.$$ (5.166)

Note, the moment t^{**} is different from the moment t^* (Figures 5.31 through 5.33) because in the case under consideration the blood droplet cannot penetrate into the porous substrate completely, and the part, V_{bc}, will remain on the surface of the porous substrate.

The theory in this case is rather straightforward and presented in [17,44]. That is, below in this section this theory is omitted and only conclusions are presented.

In order to compare the spreading behavior of blood on different nitrocellulose membranes, the following dimensionless parameters were used: $\bar{L} = L(t)/L_{ad}$, $\bar{\ell} = \ell(t)/\ell^*$, $\bar{\theta} = \theta(t)/\theta_{ad}$ and $\bar{t} = t/t^*$, where L_{ad} is the maximum radius of droplet base, ℓ^* is the radius of wetted region at the end of the process, θ_{ad} is the advancing contact angle, t^* is the time when imbibition is finished. In Figures 5.34 through 5.36, the time evolution of the dimensionless radius of the droplet base, dynamic contact angle, and radius of the wetted area are presented in the case of partial wetting. A corresponding universal behavior independent of hematocrit level of blood and n values is clearly demonstrated by Figures 5.34 through 5.36 for different nitrocellulose membranes. The whole spreading process on nitrocellulose membranes can be subdivided into three stages, as shown in Figure 5.26, which is a characteristic feature of the partial wetting case. However, note that in the case of spreading over 0.2-μm pore size nitrocellulose membranes, the red blood cells did not penetrate inside the membrane pores and only plasma could penetrate inside. That is, the final radius of the droplet base did not vanish (Figure 5.34) and only Stages 1 and 2 are present in this case. This observation opens a completely new possibility to (i) investigate red blood cells and plasma separately and (ii) use this method for nondestructive separation of living cells from aqueous solutions.

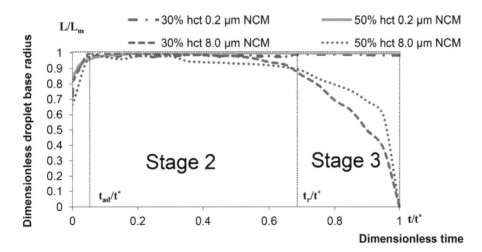

FIGURE 5.34 Partial wetting. Dimensionless radius of the droplet base in the case of spreading over nitrocellulose membranes. In the case of nitrocellulose membrane (NCM) with 8.0-μm pore size, the red blood cells are smaller than the pore size and penetrate the membrane with the plasma flow. In this case, all three stages are detected. However, in the case of membrane with smaller pores, 0.2-μm, the red blood cells are unable to penetrate the membrane and, hence, only Stages 1 and 2 are present. (Reprinted from *Colloids Surf. A: Physicochem. Eng. Asp.*, 505, Chao, T.C. et al., Simultaneous spreading and imbibition of blood droplets over poroussubstrates in the case of partial wetting, 9–17, Copyright 2016, with permission from Elsevier; *Curr. Opin. Colloid Interface Sci.*, 36, Trybala, A. and Starov, V., Kinetics of spreading wetting of blood over porous substrates, 84–89, Copyright 2018, with permission from Elsevier.)

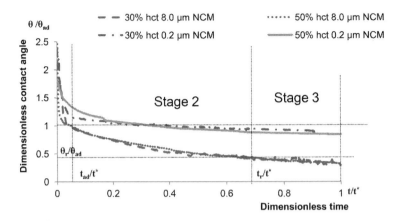

FIGURE 5.35 Partial wetting. Dimensionless dynamic contact angle in the case of spreading over nitrocellulose membrane. Note, in the case of 0.2-μm nitrocellulose membrane (NCM), there is no Stage 3, it is a continuation of Stage 2. The contact angle remained almost constant after Stage 1. (Reprinted from *Colloids Surf. A: Physicochem. Eng. Asp.*, 505, Chao, T.C. et al., Simultaneous spreading and imbibition of blood droplets over porous substrates in the case of partial wetting, 9–17, Copyright 2016, with permission from Elsevier; *Curr. Opin. Colloid Interface Sci.*, 36, Trybala, A. and Starov, V., Kinetics of spreading wetting of blood over porous substrates, 84–89, Copyright 2018, with permission from Elsevier.)

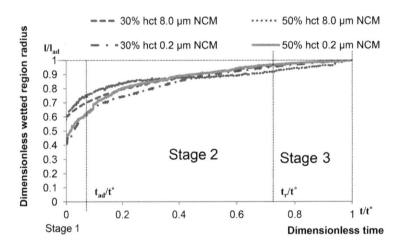

FIGURE 5.36 Partial wetting. Dimensionless radius of the wetted area inside the membrane in the case of spreading over nitrocellulose membrane. Note, in the case of 0.2-μm nitrocellulose membrane (NCM), Stage 3 is only a continuation of Stage 2. (Reproduced from *Colloids Surf. A: Physicochem. Eng. Asp.*, 505, Chao, T.C. et al., Simultaneous spreading and imbibition of blood droplets over poroussubstrates in the case of partial wetting, 9–17, Copyright 2016, with permission from Elsevier; *Curr. Opin. Colloid Interface Sci.*, 36, Trybala, A. and Starov, V., Kinetics of spreading wetting of blood over porous substrates, 84–89, Copyright 2018, with permission from Elsevier.)

Conclusions on Section "Partial Wetting Case"

Theory and experimental investigations of spreading of blood droplets over porous substrates in the case of partial wetting is presented. According to our measurement, both blood (in the range of the hematocrit levels investigated) and plasma showed non-Newtonian shear thinning power law behavior. Both blood and plasma wet partially nitrocellulose membranes.

In the case of partial wetting (spreading of blood over nitrocellulose membranes) when pores are smaller enough (less than 0.2 μm) than the size of red blood cell, the red blood cells could not penetrate the pores and only plasma could penetrate the pores. This observation opens a completely new possibility of investigating red blood cells and plasma separation without damaging the red blood cells.

APPENDIX 3

A radial velocity profile for flow in a pipe of a power law fluid is given by the following expression [17]:

$$u(r) = \frac{n}{n+1} \left(\left| \frac{dp}{dx} \right| \frac{1}{2k} \right)^{1/n} \left(R^{1+1/n} - r^{1+1/n} \right).$$

(A3.1)

Hereafter, we assume that the flow is directed along x.
  Hence, the flow rate, Q, will be

$$Q = \pi \frac{n}{3n+1} \left(\left| \frac{dp}{dx} \right| \frac{1}{2k} \right)^{1/n} R^{3+1/n}.$$

(A3.2)

This equation allows determining the average velocity of flow, v,

$$v = \frac{Q}{\pi R^2} Q = \frac{n}{3n+1} \left(\left| \frac{dp}{dx} \right| \frac{1}{2k} \right)^{1/n} R^{1+1/n}.$$

(A3.3)

Capillary Imbibition of Non-Newtonian Liquid into a Thin Capillary

Figure 5.37 shows the schematic of capillary imbibition of a non-Newtonian liquid into a thin capillary. The rate of the imbibition is given by Eq. (A3.3). Hence,

$$\frac{d\ell}{dt} = v = \frac{n}{3n+1} \left(\frac{dp}{dz} \frac{1}{2k} \right)^{1/n} R^{1+1/n}.$$

(A3.4)

The pressure gradient is equal to capillary pressures/ℓ:

$$\frac{dp}{dz} = \frac{\sigma \cos\theta / R}{\ell},$$

(A3.5)

where σ is the liquid–air interfacial tension; and θ is the contact angle. Substitution of Eq. (A3.4) into Eq. (A3.5) results in

$$\frac{d\ell}{dt} = \frac{n}{3n+1} \left(\frac{\sigma \cos\theta}{\ell} \frac{1}{2k} \right)^{1/n} R,$$

(A3.6)

or

$$\ell^{1/n} \frac{d\ell}{dt} = \frac{n}{3n+1} \left(\frac{\sigma \cos\theta}{2k} \right)^{1/n} R, \quad \ell(0) = 0.$$

(A3.7)

FIGURE 5.37 Capillary imbibition of a non-Newtonian liquid into a thin capillary.

The solution of this equation with the initial condition is as follows:

$$\ell(t) = \left(\frac{nR}{3n+1}\right)^{\frac{n}{n+1}} \left(\frac{\sigma\cos\theta}{2k}\right)^{\frac{1}{n+1}} t^{\frac{n}{n+1}}.$$ (A3.8)

In the case of a Newtonian liquid, that is $n = 1$, Eq. (A3.8) results in a well-known Washburn solution,

$$\ell(t) = \left(\frac{R}{4}\right)^{\frac{1}{2}} \left(\frac{\sigma\cos\theta}{2\eta}\right)^{\frac{1}{2}} t^{\frac{1}{2}}.$$ (A3.9)

Comparison of these two solutions shows that pseudoplastic fluids ($n < 1$) penetrate into thin capillaries slower than Newtonian liquids, while dilatant fluids ($n > 1$) penetrate faster as compared with Newtonian fluids.

It is easy to check that the same conclusion remains valid for the imbibition of non-Newtonian fluids into a porous medium.

One-Dimensional Penetration of a Non-Newtonian Liquid into a Porous Medium

In the case of one-dimensional penetration of a non-Newtonian liquid into a porous medium, we can rewrite Eq. (A3.3) as follows:

$$v = K_n \left(\left|\frac{dp}{dx}\right|\frac{1}{k}\right)^{1/n},$$ (A3.10)

where K_n is the permeability of the porous medium towards the power law non-Newtonian liquid of power n. In the case of Newtonian liquid, we recover from (A3.10) the well-known Darcy's law

$$v = \frac{K}{\eta}\frac{dp}{dx}.$$ (A3.11)

APPENDIX 4

The blood rheology measurements were made in [38] using the rheometer with plane geometry (4-cm diameter, stainless steel) and a 250-μm gap. The viscosity measurements were made in the shear range 0.2–100 s^{-1} in [38]. The power law equation known as the Ostwald–de Waele relationship was used as the rheological model of blood:

$$\eta = k\dot{\gamma}^{n-1}.$$ (5.167)

where k is the "flow consistency index," n is the "flow behavior index" and $\dot{\gamma}$ is the shear rate. Eq. (5.167) showed a good agreement with experimental measurement of blood viscosity [38], where both k and n were determined for all hematocrit levels under investigation. Eq. (5.167) can be rewritten as $\ln\mu = \ln k - n\ln\dot{\gamma}$ the parameters k and n were fitted as shown in Figure 5.38. The good fitting proves that the blood viscosity is a power law shear thinning liquid.

Fitted values of k and n are presented in Table 5.11 [38].

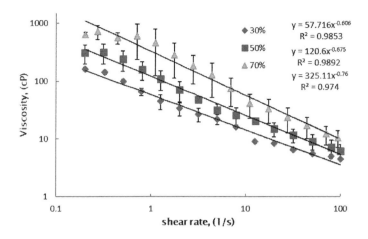

FIGURE 5.38 The dependency of blood viscosity with 30%, 50% and 70% hematocrit levels on shear rate from 0.5 to 100.0 s^{-1}.

TABLE 5.11

Fitted Values of k and n According to Eq. (4.1) in Appendix 4

Hematocrit Levels in Blood (%)	k	n
0	23.99	0.38
30	57.72	0.394
50	120.60	0.325
70	325.11	0.240

REFERENCES

1. Brinkman, H. The viscosity of concentrated suspensions and solutions. *J. Chem. Phys.*, 20, 571 (1952); Brinkman, H. A calculation of the viscous force exerted by a flowing fluid on a dense swarm of particles. *Appl. Sci. Res.*, A1, 27 (1947).
2. Whitaker, S. *The Method of Volume Averaging*, Kluwer Academic Publishers, London, UK, 1999, p. 221.
3. Starov, V. M., and Zhdanov, V. G. Effective viscosity and permeability of porous media. Colloids *Surf. A: Physicochem. Eng. Asp.*, 192, 363 (2001).
4. Kalinin, V. V., and Starov, V. M. Spreading of viscous drops over porous substrata. *Colloid J.* (USSR Academy of Sciences, English Translation), 51 (5), 860 (1989).
5. Starov, V. M., Kosvintsev, S. R., Sobolev, V. D., Velarde, M. G., and Zhdanov, S. A. Spreading of liquid drops over dry porous layers: Complete wetting case. *J. Colloid Interface Sci.*, 252, 397–408 (2002).
6. Kornev, K. G., and Neimark, A. V. Spontaneous penetration of liquids into capillaries and porous membranes revisited. *J. Colloid Interface Sci.*, 235, 101 (2001).
7. Snijdewind, I. J. M., van Kampen, J. J. A., Fraaij, P. L. A., van der Ende, M. E., Osterhaus, A. D. M. E., and Gruters, R. A. Current and future applications of dried blood spots in viral disease management. *Antiviral Res.*, 93 (3), 309–321 (2012).
8. Neogi, P., and Miller, C. A. Spreading kinetics of a drop on a rough solid surface. *J. Colloid Interface Sci.*, 92 (2), 338–349 (1983).
9. Starov, V. M., Kalinin, V. V., and Chen, J.-D. Spreading of liquid drops over dry surfaces. *Adv. Colloid Int. Sci.*, 50, 187 (1994).

10. Kalinin, V., and Starov, V. The viscous spread of drops over a wetted surface. *Colloid J.* (USSR Academy of Sciences, English Translation), 48 (5), 767 (1986).

11. Marmur, A. Equilibrium and spreading of liquids on solid surfaces. *Adv. Colloid Interface Sci.*, 19, 75 (1983).

12. Joanny, J.-F. Spreading of superfluid drops. *J. Physique*, 46, 807 (1985).

13. Nayfeh, A. H. *Perturbation Methods*,Wiley-Interscience, New York, 1973.

14. Starov, V. M., Zhdanov, S. A., and Velarde, M. G. Spreading of liquid drops over thick porous layers: Complete wetting case. *Langmuir*, 18, 9744 (2002).

15. Holdich, R., Starov, V., Prokopovich, P., Njobuenwu, D., Rubio, R., Zhdanov, S., and Velarde, M. Spreading of liquid drops from a liquid source, *Colloids Surf. A: Physicochem. Eng. Asp.*, 282–283, 247–255 (2006).

16. Cazabat, A. M., and Cohen Stuart, M. A. Dynamics of wettlng: effects of surface roughness. *J. Phys. Chem.*, 90, 5849 (1986).

17. Chao, T. C., Arjmandi-Tash, O., Das, D. B., and Starov, V. M. Simultaneous spreading and imbition of blood droplets over porous substrates in the case of partial wetting. *Colloids Surf. A: Physicochem. Eng. Asp.*, 505, 9–17 (2016).

18. Chao, T. C., Arjmandi-Tash, O., Das, D. B., and Starov, V. M. Spreading of blood drops over dry porous substrate: Complete wetting case. *J. Colloid Interface Sci.*, 446, 218–225 (2015).

19. Meesters, R., and Hooff, G. State-of-the-art dried blood spot analysis: An overview of recent advances and future trends. *Bioanalysis*, 5, 2187–2208 (2013).

20. Edelbroek, P. M., Van der Heijden, J., and Stolk, L. M. L. Dried blood spot methods in therapeutic drug monitoring: Methods, assays, and pitfalls. *Ther. Drug Monit.*, 31 (3), 327–336 (2009).

21. Demirev, P. Dried blood spots: Analysis and applications. *Anal. Chem.*, 85 (2), 779–789 (2013).

22. Tanna, S., and Lawson, G. Analytical methods used in conjunction with dried blood spots. *Anal. Methods*, 3 (8), 1709 (2011).

23. Lehmann, S., Delaby, C., Vialaret, J., Ducos, J., and Hirtz, C. Current and future use of "dried blood spot" analyses in clinical chemistry. *Clin. Chem. Lab. Med.*, 51 (10), 1897–909 (2013).

24. De Gennes, P. Wetting: Statics and dynamics. *Rev. Mod. Phys.*, 57, 827–863 (1985).

25. Starov, V., Kalinin, V., and Chen, J. Spreading of liquid drops over dry surfaces, *Adv. Colloid Interface Sci.*, 50, 187–221 (1994).

26. Blake, T. and Haynes, J. Kinetics of liquidliquid displacement, *J. Colloid Interface Sci.*, 80 (3), 421–423 (1969).

27. Teletzke, G. F., Ted Davis, H. andScriven, L. E. How liquids spread on solids, *Chem. Eng. Commun.*, 55 (1–6), 41–82 (1987).

28. Taniguchi, M., Pieracci, J., and Belfort, G. Effect of undulations on surface energy: A quantitative assessment. *Langmuir*, 10, 4312–4315 (2001).

29. Aradian, A., Raphael, E., and De Gennes, P. Dewetting on porous media with aspiration. *Eur. Phys. J. E*, 376, 367–376 (2000).

30. Zadražil, A., Stepanek, F., and Matar, O. K. Droplet spreading, imbition and solidification on porous media. *J. Fluid Mech.*, 562, 1 (2006).

31. Bacri, F., and Brochard-Wyart, L. Droplet suction on porous media. *Eur. Phys. J. E*, 3, 87–97 (2000).

32. Beavers, G., and Joseph, D. Boundary conditions at a naturally permeable wall. *J. Fluid Mech*, 30 (1), 197–207 (1967).

33. Greenspan, H. On the motion of a small viscous droplet that wets a surface. *J. Fluid Mech.*, 84, 125–143 (1978).

34. Davis, S. H., and Hocking, L. M. Spreading and imbition of viscous liquid on a porous base. *Phys. Fluids*, 11 (1), 48 (1999).

35. Davis, S. H., and Hocking, L. M. Spreading and imbition of viscous liquid on a porous base. II. *Phys. Fluids*, 12 (7), 1646 (2000).

36. Starov, V. M., Kosvintsev, S. R., Sobolev, V. D., Velarde, M. G., and Zhdanov, S. A. Spreading of liquid drops over saturated porous layers. *J. Colloid Interface Sci.*, 246 (2), 372–379 (2002).

37. Starov, V. M., Tyatyushkin, A. N., Velarde, M. G., and Zhdanov, S. A. Spreading of non-Newtonian liquids over solid substrates. *J. Colloid Interface Sci.*, 257 (2), 284–290 (2003).

38. Chao, T. C., Trybala, A., Starov, V., and Das, D. B. Influence of haematocrit level on the kinetics of blood spreading on thin porous medium during dried blood spot sampling. *Colloids Surf. A: Physicochem. Eng. Asp.*, 451, 38–47 (2014).

39. Baskurt, O. K., and Meiselman, H. J. Blood rheology and hemodynamics. *Semin. Thromb. Hemost.*, 29 (5), 435–450 (2003).

40. Merrill, E. Rheology of blood. *Physiol. Rev.*, 49 (4), 863–888 (1969).

41. Yilmaz, F. A critical review on blood flow in large arteries; Relevance to blood rheology, viscosity models, and physiologic conditions. *Korea-Australia Rheol. J.*, 20 (4), 197–211 (2008).

42. Marcinkowska-Gapińska, A., Gapinski, J., Elikowski, W., Jaroszyk, F., and Kubisz, L. Comparison of three rheological models of shear flow behavior studied on blood samples from post-infarction patients. *Med. Biol. Eng. Comput.*, 45 (9), 837–844 (2007).

43. Sequeira, A., and Janela, J. An overview of some mathematical models of blood rheology. In *A Portrait of State-of-the-Art Research at the Technical University of Lisbon*, pp. 65–87 (2007).

44. Trybala, A., and Starov, V. Kinetics of spreading wetting of blood over porous substrates. *Curr. Opin. Colloid Interface Sci.*, 36, 84–89 (2018).

6

Wetting of Wetting/Spreading in the Presence of Surfactants

Introduction

In this chapter, we consider the kinetics of spreading of surfactant solutions over hydrophobic, porous substrates and over thin aqueous layers.

Despite the wide use of spreading of surfactant solutions over hydrophobic substrates, we are not in the position to answer even basic questions in this area such as how surfactant molecules are transferred in the vicinity of an apparent three-phase contact line. More than that, we are currently unable to answer the question, "What does the apparent three-phase contact line mean in the case of surfactant solutions?" Our knowledge of the behavior of the transition zone from the droplet/meniscus to thin films in front is very limited in the case of surfactant solutions and hydrophobic substrates: disjoining pressure isotherms in the presence of surfactant solutions are investigated only in the case of free liquid films [1], and much less is known in the case of liquid films on solid substrates. Beyond that, on the current stage, we are unable present a clear physical picture of the equilibrium of droplets/menisci in the presence of surfactants and on surfactant transfer in the vicinity of an apparent three-phase contact line.

In Chapters 1 and 2 we showed that the Newman-Young equation does not have any theoretical basis and should be replaced by the Deriaguin-Frumkin equation for the equilibrium contact angle. In spite of this, in view of our limited knowledge of the surfactant behavior in a vicinity of the moving three-phase contact line, we use the Newman-Young equation for the quasi-equilibrium contact angle for the description of spreading processes over hydrophobic surfaces in this chapter. We realize that any conclusion based on this semi-empirical relation should be accordingly understood as semi-empirical. However, our hypothesis on the adsorption of surfactants on a bare hydrophobic substrate in front of the moving meniscus and/or droplet allows us to develop some theoretical predictions that are in reasonable agreement with known experimental data (Sections 6.2, 6.4, and 6.6 and Appendix 1). However, the main question of how this transfer goes on is left unanswered.

The situation is even less investigated in the case of simultaneous spreading and imbibition into a porous substrate (Section 6.1). We present some theoretical and experimental investigations of the process (Section 6.1), which should be considered as a first step in this direction.

In Section 6.5, we consider the much more theoretically understood process of flow caused by the surface tension gradient (Maramgoni flow). We show that the flow caused by the point source of surfactant on the surface of thin aqueous film is driven by the surface tension gradient only and all other forces can be neglected.

6.1 Spreading of Aqueous Surfactant Solutions over Porous Layers

In this section, we follow the track of Chapter 5 for the case of *surfactant aqueous solutions* such as an *anionic* surfactant sodium dodecyl sulfate (SDS). We start with the spreading problem of "big" drops, albeit small enough to allow neglecting gravity, over *porous* solid substrates [2]. This should be "thin" in the sense used in Chapter 5. We consider various SDS concentrations: zero (pure water) and concentrations below, near and above the *critical* micelle concentration (CMC). In all cases under consideration, we deal with partial wetting cases. However, the spreading of surfactant solutions is a very special case of partial wetting: the static receding contact angle is zero, which is the same as in the case complete wetting. According to Chapter 5, the overall spreading process can be divided into three stages (Figure 5.19): In the first stage, the drop base expands until its maximum value is attained; the contact angle decreases fast. In the second stage, the radius of the drop base remains constant while the contact angle decreases linearly with time but not to zero. In this case the constancy of the receding contact angle during the third stage has nothing to do with the static receding contact angle (which is zero) but is determined by pure hydrodynamic conditions as in the case of complete wetting. Finally, in the third stage, the drop base shrinks, but the contact angle remains constant. The wetted area inside the porous solid substrate expands all the time [2].

Appropriate scales were used with a plot of the dimensionless radii of the drop base, of the wetted area inside the porous substrate and the dynamic contact angle on the dimensionless time.

The experimental data show that the overall time of the spreading of drops of SDS solution over dry thin porous substrates decreases with increasing surfactant concentrations; the difference between advancing and hydrodynamic receding contact angles decreases as the surfactant concentration increases; the constancy of the contact angle during the third stage of spreading has nothing to do with the hysteresis of contact angle but is determined by hydrodynamic reasons. It is shown using independent spreading experiments of the same drops on nonporous nitrocellulose substrate that the static receding contact angle is equal to zero, supporting our conclusion on the hydrodynamic nature of the receding contact angle on porous substrates.

In Chapter 5, the spreading of liquid drops over thin porous layers saturated with the same liquid (Section 5.1) or dry (Section 5.2) was investigated in the case of complete wetting. Brinkman's equations were used for the description of the liquid flow inside the porous substrate.

In this section, we take up the same problem as in the case when a drop spreads over a dry porous layer in the case of partial wetting. Spreading of "big drops" (but still small enough to neglect the gravity action) of aqueous SDS solutions over "thin porous layers" (nitrocellulose membrane) is considered below in this section.

Experimental Methods and Materials

Spreading on Porous Substrates (Figure 5.4)

Aqueous solutions of SDS (an anionic surfactant) were used in the spreading experiments. The following SDS concentrations were used: zero concentration (pure water), concentrations below CMC, near CMC, and above CMC [1].

SDS was purchased from Fisher Scientific and used as obtained, without further purification. Nitrocellulose membranes, purchased from Millipore, with average pore size (a) of 0.22, 0.45, and 3.0 μm (marked by the supplier) were used as a model of thin porous layers.

The same experimental chamber as in Chapter 5 (Figure 5.2) was used for monitoring the spreading of liquid drops over the initially dry porous substrates. The time evolution of the radius of the drop base, $L(t)$, the dynamic contact angle, $\theta(t)$, and the radius of the wetted area inside the porous substrate, $l(t)$, were monitored (Figure 5.4). The porous substrate (1 in Figure 5.2) was placed in a thermostated and hermetically closed chamber (2 in Figure 5.2), where 100% relative humidity and fixed temperature were maintained. The initial volume of the droplets ranged between 1.47 and 10.45 μL.

Care was taken so that all interfaces inside the syringe and the attached replaceable capillary (4 in Figure 5.2) had been saturated with surfactants before the actual solutions were used. The drop was applied on the surface by manually pumping the piston of the syringe. The distance between the forming drop and the surface was kept minimal to avoid collateral inertia effects.

Experiments were carried out in the following order:

- The dry membrane (initially stored in 0% humidity atmosphere) was placed in the chamber with a 100% humidity atmosphere and left for 15–30 min.
- A light pulse produced by a flash gun was used to synchronize both videotape recorders (a side view and a view from above).
- A droplet of liquid was placed onto the membrane.

Each run was carried out until the complete imbibition of the drop into the porous substrate took place.

Measurement of Static Advancing and Receding Contact Angles on Nonporous Substrates

A modified experimental setup (as compared with the previous case) was used for measuring static advancing and receding contact angles on a nonporous nitrocellulose substrate. Figure 5.16 shows the schematic presentation of the sample chamber for monitoring of the advancing and receding contact angles during droplet spreading over a smooth nonporous nitrocellulose substrate. The time evolution of the radius of the drop base, $L(t)$ and the dynamic contact angle, $\theta(t)$, were monitored.

The nonporous substrate under investigation (1 in Figure 5.16) was attached by double-sided tape to a solid substrate and was placed into a thermostated and hermetically closed chamber (2 in Figure 5.16), where controlled fixed humidity and fixed temperature were maintained. A glued-in syringe needle (4 in Figure 5.16) was positioned in the center of the solid substrate and connected to the Harvard Apparatus syringe pump (18 in Figure 5.16).

The nonporous nitrocellulose substrate was prepared as follows:

- The nitrocellulose substrate was attached to the solid substrate (1 in Figure 5.16) using double-sided sticky tape, ensuring that no air bubbles could be trapped.
- The nitrocellulose substrate attached to the solid substrate was placed into an acetone atmosphere. The acetone vapor was used to seal any pores that may have been open, and the nitrocellulose substrate was left in the acetone atmosphere for approximately 20 min, until the surface became transparent.
- After that, the nitrocellulose substrate attached to the solid substrate (1 in Figure 5.16) was placed into a sealed container filled with air, allowing the surface to come to equilibrium after the acetone treatment.
- A hole was made in the center of the nitrocellulose substrate, allowing a penetration of the liquid through a glued-in syringe needle (4 in Figure 5.16) connected to the Harvard Apparatus syringe pump (18 in Figure 5.16). This allowed for drawing the fluid in or out of the surface and monitoring the droplet volume according to a prescribed rate.

The experimental runs were carried out in the following order:

- The speed at which the fluid would be drawn out of the drop was fixed in the Harvard Apparatus syringe pump (18 in Figure 5.16). The refill rate was varied between 0.01 and 0.1 µL/min.
- The nitrocellulose substrate (1 in Figure 5.16) was placed into the experimental chamber (2), and the needle was connected to the pump (18).
- The experimental chamber was closed, and the fan was switched on to equilibrate atmosphere inside the chamber.
- After approximately 20 min, the fan was switched off, and the video recorders were switched on.
- The droplet of the liquid was deposited onto the surface under investigation using the upper syringe (17 in Figure 5.16) into the center of syringe needle (4).
- After 1 min of recording, the pump was started, to draw off the liquid from the drop.

The volume of the spreading drops should remain constant in the case of spreading over a nonporous nitrocellulose surface. The constancy of the volume of the spreading drops was monitored during these experiments and which confirmed that the substrate used was nonporous.

Results and Discussion

In all our spreading experiments, the drops remained the spherical shape over the whole spreading process. This was cross-checked by reconstructing of the drop profiles at different time instants of spreading, fitting those drop profiles by a spherical cap (see Sections 5.1 and 5.2). In all cases, the reduced chi-squared value was found to be less than 10^{-4}.

In all cases of spreading over porous substrates, saturation of the membrane in the vertical direction was much faster than the whole duration of the spreading-imbibition process, that is, the membrane was assumed always saturated in the vertical direction (see Section 5.2). A schematic presentation of the process is presented in Figure 5.4.

According to our observations, the whole spreading process can be subdivided into three stages (see Figure 6.1). During the *first stage*, the contact angle, θ, rapidly decreases while the radius of the drop base, L, increases until its maximum value, L_{max}, is reached. The first stage is followed by the *second stage*, when the radius of the drop base, $L = L_{max}$, is constant, and the contact angle, θ, decreases linearly with time. During the *third stage*, the radius of the drop base decreases, and the contact angles remains constant. At the final, third stage, the drop base shrank until the drop completely disappears and the imbibition front expands until the end of the process. The spherical form of the spreading drop allows

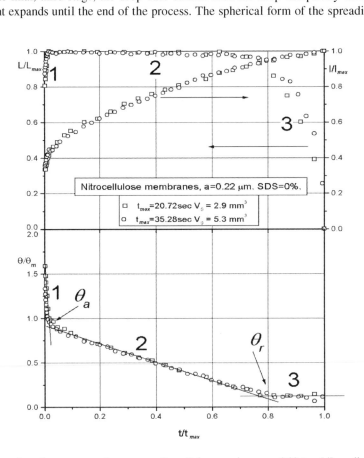

FIGURE 6.1 Spreading of pure aqueous drops over a nitrocellulose membrane $a = 0.22\ \mu$m. L/L_{max}, dimensionless radius of the drop base; l/l_{max} radius of the wetted area; θ/θ_m, dynamic contact angle; and t/t_{max} dimensionless time. (1) First stage: θ rapidly decreases while the L increases until the maximum value L_{max}. (2) Second stage: L is constant; $L = L_{max}$; and θ decreases linearly with time. (3) Third stage: L decreases and θ remains constant.

for measuring of the evolution of the contact angle of the spreading drops. During the first stage, the contact angle, θ, decreases very rapidly; during the second stage, it decreases much slower and linearly; and during the third stage, it remains a constant value. This constant value of the contact angle is referred to as θ_m.

All relevant experimental data are summarized in the Table 6.1.

Here we present a brief theoretical explanation why the dependency of the contact angle during the second stage of the spreading is a linear function of time.

During the second stage:

$$L = L_{max}, \quad t > t_m, \tag{6.1}$$

where t_m corresponds to the beginning of the second stage. This means that according to Eq. (5.72) from Section 5.2,

TABLE 6.1

Experimental Values Used

Pore Size, Figure, Symbols	Average Pore Size, a (mm)	SDS Concentration (%)	Initial Volume of the Drop (V_0 mL)	t_{max}(s)	L_{max}(mm)	L_{max}(mm	θ_m(grad)	t_m(s)
Figure 6.1 squares	0.22	0	2.9	20.72	1.49	3.38	51	0.36
Figure 6.1 circles	0.22	0	5.3	35.28	1.85	4.13	50	0.36
Figure 6.2 squares	3.0	0.1	2.1	2.96	1.47	2.30	31	0.36
Figure 6.2 triangles	3.0	0.1	3.7	4.92	2.01	3.40	24	0.64
Figure 6.2 circles	3.0	0.1	9.7	13.12	2.59	5.28	35	0.48
Figure 6.3 diamonds	0.22	0	2.9	20.72	1.49	3.38	51	0.36
Figure 6.3 circles	0.22	0.1	10.5	75.2	2.78	5.79	32	0.4
Figure 6.3 squares	0.22	0.2	7.4	33.84	2.45	4.67	27	0.44
Figure 6.3 triangles	0.22	0.5	2.6	8.2	2.03	2.90	20	0.12
Figure 6.4 diamonds	0.45	0	6.1	8.44	2.83	4.35	44	0.52
Figure 6.4 circles	0.45	0.1	9.5	10.28	2.46	4.92	38	0.35
Figure 6.4 squares	0.45	0.2	7.2	8.44	2.83	4.35	21	0.36
Figure 6.4 triangles	0.45	0.5	7.1	6.84	2.75	4.80	23	0.08
Figure 6.5 diamonds	3.0	0	5.9	108.04	1.89	4.21	46	21
Figure 6.5 circles	3.0	0.1	9.7	13.12	2.59	5.28	35	0.48
Figure 6.5 squares	3.0	0.2	6.8	3.96	2.65	4.43	21	0.28
Figure 6.5 circles	3.0	0.5	5.1	2.12	2.39	3.97	22	0.12

$$\frac{dl}{dt} = \frac{K_p p_c / \eta}{l \ln \dfrac{l}{L_{\max}}}, \quad l(t_m) = l_m, \tag{6.2}$$

where l is the radius of the circular edge of the wetted region inside the porous layer; K_p is the permeability of the porous layer in the tangential direction; p_c is the capillary pressure on the wetting front inside the porous layer; l_m is the radius of the circular edge of the wetted region inside the porous layer at the moment t_m; and η is the liquid dynamic viscosity. This equation can be easily integrated, giving

$$l^2 \ln \frac{l}{\sqrt{e} L_{\max}} - l_m^2 \ln \frac{l_m}{\sqrt{e} L_{\max}} = \frac{2 K_p p_c}{\eta}(t - t_m),$$

or

$$l^2 = \frac{2 K_p p_c}{\eta \ln \dfrac{l}{\sqrt{e} L_{\max}}}(t - t_m) + \frac{l_m^2 \ln \dfrac{l_m}{\sqrt{e} L_{\max}}}{\ln \dfrac{l}{\sqrt{e} L_{\max}}}. \tag{6.3}$$

The term $\ln \frac{l}{\sqrt{e} L_{\max}}$ in the denominator of this equation is a slow changing function of l, that is, the right-hand side of this equation is almost indistinguishable from the linear function of time. Equation (6.3) can be rewritten as

$$l^2 = At + B, \quad A = \frac{2 K_p p_c}{\eta \ln \dfrac{l}{\sqrt{e} L_{\max}}}, \quad B = \frac{l_m^2 \ln \dfrac{l_m}{\sqrt{e} L_{\max}}}{\ln \dfrac{l}{\sqrt{e} L_{\max}}} - t_m \frac{2 K_p p_c}{\eta \ln \dfrac{l}{\sqrt{e} L_{\max}}}. \tag{6.4}$$

According to Eqs. (5.48) and (5.54) from Section 5.2,

$$V(t) = V_0 - \pi m \Delta l^2(t)$$

and

$$\theta = \frac{4V}{\pi L_m^3} = \frac{V_0 - \pi m \Delta l^2(t)}{\pi L_m^3} = \frac{V_0}{\pi L_m^3} - \frac{\pi m \Delta}{\pi L_m^3} l^2(t), \tag{6.5}$$

where Δ is the thickness of the porous layer. Combining Eqs. (6.4) and (6.5) results in

$$\theta = \frac{V_0}{\pi L_{\max}^3} - \frac{\pi m \Delta}{\pi L_{\max}^3}(At + B) = -\frac{\pi m \Delta A}{\pi L_{\max}^3} t + \left[\frac{V_0}{\pi L_{\max}^3} - B \frac{\pi m \Delta}{\pi L_{\max}^3} \right]. \tag{6.6}$$

This equation shows that the contact angle decreases linearly with time during the second stage of spreading.

Hereafter the time evolution of both the radius of the base of the spreading drops and the radius of the wetted area is presented using dimensionless coordinates. The drops used were of different volumes and different SDS concentrations. The total duration of the spreading process, $t_{\max}$, the maximum radius of the drop base, $L_{\max}$, and the final radius of the wetted area, $l_{\max}$, vary considerably depending on the drop volume, SDS concentration, the average pore size, and the porosity of nitrocellulose membranes. It was suggested in Section 5.2 to use the following dimensionless values: $L/L_{\max}$, $l/l_{\max}$ and $t/t_{\max}$. The same dimensionless values are used here. According to Section 5.2, the reduced contact angle dependency, θ/θ_m, against dimensionless time $t/t_{\max}$ is used, where θ_m is the value of the dynamic contact angle, which is now reached, $t = t_m$ (at the end of the first stage of the spreading process).

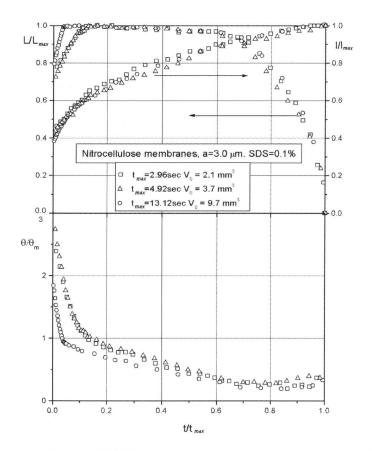

FIGURE 6.2 Spreading of droplets of 0.1% SDS solution over nitrocellulose membrane, $a = 3.0$ μm. L/L_{max}, dimensionless radius of the drop base; l/l_{max} radius of the wetted area; θ/θ_m, dynamic contact angle; and t/t_{max} dimensionless time.

Figure 6.1 shows that the spreading behavior of drops of different volumes over the same porous substrate has a universal character in dimensionless coordinates. However, not all the experimental data have shown such universal behavior. Some experimental runs differ during the first stage of spreading in dimensionless coordinates (Figure 6.2). This difference becomes bigger if the overall time of spreading is shorter (i.e., due to droplets of different volumes).

In Figures 6.3 through 6.5 (0.22-, 0.45-, and 3.0-μm nitrocellulose membranes, respectively) the time evolution is presented of the radius of the drop base, the radius of the wetted area inside the porous substrates, and the contact angle at different SDS concentrations.

Figures 6.3 through 6.5 show that the second stage of spreading becomes shorter in dimensionless coordinates with the increase in SDS concentration. However, contact angles show the universal constant behavior during the third stage of spreading for each of SDS concentrations.

Advancing and Hydrodynamic Receding Contact Angles on Porous Nitrocellulose Membranes

Using presented experimental data on the spreading of a drop of an aqueous SDS solution over dry porous substrates, the values of the advancing, θ_a, and hydrodynamic receding, θ_{rh}, contact angles were extracted as functions of SDS concentration. We are using the term "hydrodynamic receding contact angle" and the symbol θ_{rh} to distinguish it from the static receding contact angle, which is found equal to zero (see section "Static hysteresis of the contact angle of SDS solution drops on smooth nonporous nitrocellulose substrate"). The advancing contact angle, θ_a, was defined at the end of the first stage when

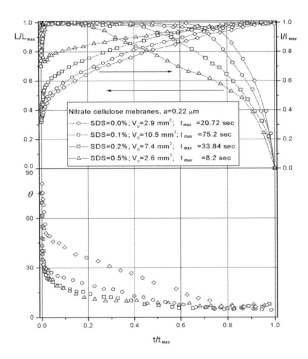

FIGURE 6.3 Spreading of droplets of SDS solutions over nitrocellulose membrane, $a = 0.22$ µm. L/L_{max}, dimensionless radius of the drop base; l/l_{max} radius of the wetted area; θ/θ_m *dynamic* contact angle; and t/t_{max} dimensionless time.

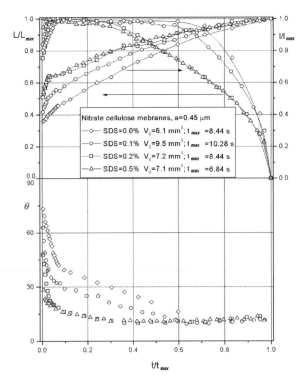

FIGURE 6.4 Spreading of droplets of SDS solutions over nitrocellulose membrane, $a = 0.45$ µm. L/L_{max}, dimensionless radius of the drop base; l/l_{max} radius of the wetted area; θ/θ_m, dynamic contact angle; and t/t_{max} dimensionless time.

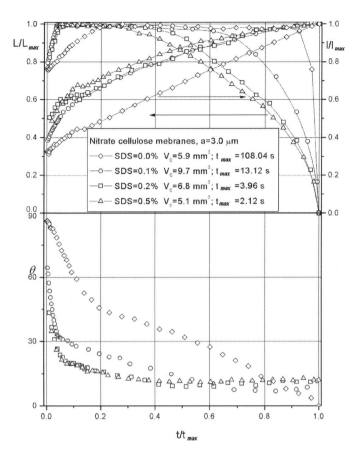

FIGURE 6.5 Spreading of droplets of SDS solutions over nitrocellulose membrane, $a = 3.0$ μm. L/L_{max}, dimensionless radius of the drop base; l/l_{max} radius of the wetted area; θ/θ_m, dynamic contact angle; and t/t_{max} dimensionless time.

the drop stopped spreading (when the radius of the drop base reached its maximum value). The hydrodynamic receding contact angle, θ_{rh}, was defined at the moment when the drop base started to shrink.

Figure 6.6 summarizes the experimental data on the apparent contact angle hysteresis. This figure shows that the advancing contact angle, θ_a, decreases with a increasing SDS concentration, whereas, on the contrary, the hydrodynamic receding contact angle, θ_{rh}, slightly increases with increasing SDS concentration.

These experimental runs show that the difference between advancing and receding contact angles becomes smaller with the increase in the SDS concentration (Figure 6.6); the dimensionless time interval when the drop base does not move also decreases with the increase in the SDS concentration.

Static Hysteresis of the Contact Angle of SDS Solution Drops on Smooth Nonporous Nitrocellulose Substrate

In the previous parts of this section, the spreading of drops at different SDS concentrations on nitrocellulose membranes of various pore sizes was considered. In all cases during the third stage of spreading, the radius of the drop base, L, shrank and the hydrodynamic receding contact angle, θ_{rh}, remained constant. The duration of the third stage of spreading increases with the SDS concentration increase. It is necessary to note that the behavior of drops of aqueous SDS solutions during the third stage of spreading (partial wetting) is remarkably similar to the behavior during the third stage of spreading in the case of complete wetting (see Section 5.2).

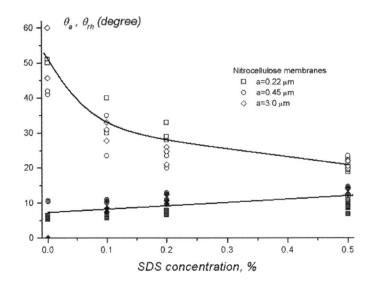

FIGURE 6.6 Porous nitrocellulose substrates. Apparent contact angle hysteresis variation with SDS concentration. Nitrocellulose membranes of different average pore sizes. Open symbols correspond to the advancing contact angle, θ_a. The same filled symbols correspond to the hydrodynamic receding contact angle, θ_{rh}.

It was found above in this section, that the advancing and hydrodynamic receding contact angles are strongly dependent on the SDS concentration on porous nitrocellulose membranes.

In this part, results of measurements of static advancing and static receding contact angles on smooth nonporous nitrocellulose substrate for different SDS concentrations are presented. The idea is to compare the hysteresis of contact angles on the smooth nonporous nitrocellulose substrate with the hysteresis contact angles obtained earlier on porous nitrocellulose substrates at corresponding SDS concentrations.

Static advancing/receding contact angle values were obtained using the experimental procedure described in Section 5.4 (Figure 5.16). The droplet was pumped using the syringe, and the dynamics of the droplet spreading were monitored. The final value of the contact angle under this experimental condition was equal to the static value of the advancing contact angle.

The static advancing contact angle of pure water on nonporous nitrocellulose substrate was found approximately equal to 70°. The static advancing contact angle decreases with the increase of SDS concentration (Figure 6.7). This trend continues until the CMC is reached. At concentrations above the CMC, the advancing contact angle remains constant and approximately equal to 35°.

The receding contact angle values were obtained using the same experimental setup (Figure 5.16). In this case, the contact angle dynamics was investigated using a linearly decreasing droplet volume.

The nonzero value of the static receding contact angle was found only in the case of pure water droplets. In all other cases (even at the smallest SDS concentrations used, 0.025%) the static receding contact angle was found equal to zero in the all concentration range used: from 0.025% (one-tenth the CMC) to 1% (five times the CMC).

Figure 6.7 presents both static receding and static advancing contact angles on smooth nonporous nitrocellulose substrate against SDS concentrations.

The results highlight a linear decline of the static advancing contact angle from 70° at 0% SDS (pure water) to approximately 35° at the value of the CMC (2.4%); after that, the static advancing contact angle reaches a steady value that remains constant irrespective of further increase in the SDS concentration. In contrast to that, the static receding contact angle is approximately equal to 45° for the pure water and is equal to zero in the presence of SDS, even at concentrations as low as 0.025%.

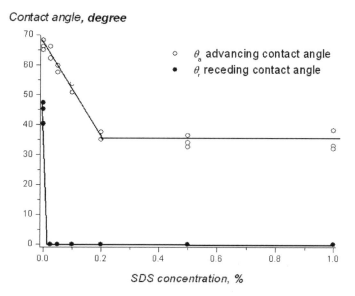

FIGURE 6.7 Nonporous nitrocellulose substrate. Advancing and receding contact angles variation with SDS concentration. Open symbols correspond to the static advancing contact angle, θ_a. Filled symbols correspond to the static receding contact angle, θ_r.

A comparison of Figures 6.6 and 6.7 shows

- The advancing contact angle dependence on SDS concentration on porous nitrocellulose substrates is significantly different from the static advancing contact angle dependence on nonporous nitrocellulose substrates. This means that, in the case of porous substrates, the influence of both the hydrodynamic flow caused by the imbibition into the porous substrate and the substrate roughness significantly change the advancing contact angle.
- The hydrodynamic receding contact angle in the case of the porous substrates has nothing to do with the hysteresis of the contact angle and determined complete by the hydrodynamic interactions in the way similar to the complete wetting case (see Section 5.3).

Conclusions

Experimental investigations were carried out on the spreading of small drops of aqueous SDS solutions (capillary regime of spreading) over dry nitrocellulose membranes (permeable in both normal and tangential directions) in the case of partial wetting. The nitrocellulose membranes were chosen because of their partial hydrophilicity. The time evolution was monitored of the radii of both the drop base and the wetted area inside the porous substrate.

The total duration of the spreading process was subdivided into three stages:

- *The first stage*, when the drop base expands until the maximum value of the drop base is reached, the contact angle rapidly decreases during this stage.
- *The second stage*, when the radius of the drop base remains constant and the contact angle decreases linearly with time until a hydrodynamic receding contact angle is reached.
- *The third stage*, when the drop base shrinks and the contact angle remains constant and equal to the hydrodynamic receding contact angle.

The wetted area inside the porous substrate expends during the whole spreading process.

Appropriate scales of the wetted area inside the porous substrate and the contact angle on the dimensionless time were used with a plot of the dimensionless radii of the drop base.

Experimental data presented in this section show

- The overall time of the spreading of drops of SDS solution over dry thin porous substrates decreases with the increase of surfactant concentration.

- The difference between advancing and hydrodynamic receding contact angles decreases with the surfactant concentration increase.

- The constancy of the contact angle during the third stage of spreading has nothing to do with the hysteresis of contact angle but is determined by the hydrodynamic reasons and the spreading behavior becomes similar to the case of the complete wetting.

- Contact angle hysteresis on porous and corresponding nonporous substrates are substantially different.

6.2 Spontaneous Capillary Imbibition of Surfactant Solutions into Hydrophobic Capillaries

In this section, a theory is developed to describe a spontaneous imbibition of surfactant solutions into hydrophobic capillaries, considering the micelle disintegration and solution concentration reduction close to the moving meniscus because of adsorption as well as spontaneous adsorption of surfactant molecules on the base hydrophobic interface in front of the moving apparent three-phase contact line (see Appendix 1). The theory predictions are in good agreement with the experimental investigations on the spontaneous imbibition of the nonionic aqueous surfactant solution, Syntamide-5, into hydrophobized quartz capillaries [3,4]. A theory of the spontaneous capillary rise of surfactant solutions in hydrophobic capillaries is presented, which connects the experimental observations with an adsorption of surfactant molecules in front of the moving meniscus on a bare hydrophobic interface [5]. In Appendix 1, we prove that the adsorption of water molecules in front of the apparent three-phase contact line is a spontaneous process on hydrophobic substrates.

Pure water does not penetrate spontaneously into a hydrophobized quartz capillary; however, surfactant solutions do penetrate spontaneously, and the penetration rate depends on the concentration of surfactant. Both the air–liquid interfacial tension, γ, and the contact angle of the moving meniscus, θ_a, are concentration dependent, where a subscript a indicates the advancing contact angle.

It is obvious that adsorption of surfactant molecules behind the moving meniscus results in a decrease of the bulk surfactant concentration from the capillary inlet in the direction of the moving meniscus. However, as we show in this section, the major process, which determines penetration of surfactant solutions into hydrophobic capillaries or spreading of surfactant solutions over hydrophobic substrates is the adsorption of surfactant molecules onto a bare hydrophobic substrate in front of the moving apparent three-phase contact line. This process results in a partial hydrophilization of the hydrophobic surface in front of the meniscus/drop, which, in turn, determines the spontaneous imbibition/spreading (see Appendix 1).

It is easy to understand why the adsorption in front of the moving meniscus on a hydrophobic substrate is so vital in the case of a hydrophobic substrate. Let us consider the very beginning of the imbibition process when a meniscus of a surfactant solution touches for the first time an inlet of a hydrophobic capillary. The contact angle, θ_a, at this moment, is bigger than $\pi/2$ and the liquid cannot penetrate the hydrophobic capillary. Solid–liquid and liquid–air interfacial tensions, γ_{sl} and γ_{lv}, do not vary with time on the initial stage because the adsorption of surfactant molecules onto these surfaces is a relatively fast process as compared with the rate of the imbibition, which is very slow. The only interfacial tension, which can vary, is the solid–air interfacial tension, γ_{sv}. If the adsorption on the solid–air interface does not occur, then the spontaneous imbibition into a hydrophobic capillary cannot take place spontaneously because the advancing contact angle remains above $\pi/2$. However, if the adsorption of surfactant molecules on the bare hydrophobic surface in a vicinity of the three-phase contact line takes place, then the solid–air interfacial tension, γ_{sv}, grows with time. After some critical surface adsorption, Γ_{svcr}, is reached, then the advancing contact angle becomes below $\pi/2$. Only after that can the spontaneous imbibition process start. This consideration shows that there is a critical bulk concentration, C_*, below which Γ_{sv} remains lower than its critical value, Γ_{svcr}, and the spontaneous imbibition process does not take place.

Let us consider expression (6.2) from Section 1.1 for the excess free energy, Φ, of the droplet on a solid substrate:

$$\Phi = \gamma_{lv}S + P_eV + \pi a^2 \left(\gamma_{sl} - \gamma_{sv}\right),$$

where S is the area of the liquid–air interface; $P_e = P_a - P_l$ is the excess pressure inside the liquid; P_a is the pressure in the ambient air; P_l is the pressure inside the liquid; and the last term in the right-hand side gives the difference between the energy of the part of the bare surface covered by the liquid drop as compared with the energy of the same solid surface without the droplet.

This expression shows that the excess free energy decreases if (i) the liquid–air interfacial tension, γ_{lv}, decreases; (ii) the liquid–solid interfacial tension, γ_{sl}, decreases; and (iii) the solid-vapor interfacial tension, γ_{sv}, increases (see Appendix 1).

Let us assume that in the absence of surfactant, the drop forms an equilibrium contact angle above $\pi/2$. If the water contains surfactants, then three transfer processes take place from the liquid onto all three interfaces: surfactant adsorption at both (i) the inner liquid–solid interface, which results in a decrease of the solid–liquid interfacial tension, γ_{sl}; (ii) the liquid–vapor interface, which results in a decrease of the liquid air interfacial tension, γ_{lv}; and (iii) transfer from the drop onto the solid–vapor interface just in front of the drop. As noted, all three processes result in a decrease of the excess free energy of the system (Appendix 1). Adsorption processes (i) and (ii) result in a decrease of corresponding interfacial tensions, γ_{sv} and γ_{lv}; however, the transfer of surfactant molecules onto the solid–vapor interface in front of the drop results in an increase of a local free energy, but the total free energy of the system decreases (Appendix 1). That is, surfactant molecule transfer (iii) goes via a potential barrier and, hence, goes slower than adsorption processes (i) and (ii). Hence, processes (i) and (ii) are "fast" processes as compared with the third process (iii).

In the case of partial wetting, the capillary imbibition in the horizontal direction proceed according to the following dependency:

$$\ell = \sqrt{\frac{r\gamma \cos \theta_a}{2\eta}} t,\tag{6.7}$$

where l is the length of the part of the capillary filled with the liquid; r is the radius of the capillary; η is the liquid viscosity; θ_a, the advancing contact angle, is below $\pi/2$: $0 < \theta_a < \pi/2$; and t is time.

However, pure water does not penetrate spontaneously into hydrophobic capillaries and shows the advancing contact angle, $\theta_a > \pi/2$. This means that the liquid can be only forced into the capillary. Note, the advancing contact angle is a decreasing function of the surfactant concentration and at some critical concentration, C_*, is equal to $\pi/2$. This means, that above C_*, the surfactant solution penetrates spontaneously into the hydrophobic capillaries.

In the case of the imbibition of surfactant solutions into hydrophobic capillaries, the penetration is controlled by both the surfactant molecules transfer and the liquid viscosity according to Eq. (6.7). In this case, the aim is to reveal the mechanism of the penetration and to determine the concentration of surfactant molecules near the meniscus $C_m < C_0$, where C_0 is the surfactant concentration at the capillary inlet. Figure 6.8 shows that in the case of Syntamid-5, the advancing contact angle, θ_a, exceeds 90° at concentration C_*, which is slightly under 0.05%.

We show that both the bulk diffusion and the surface diffusion of surfactant molecules (including that on the unwetted portion of the capillary in front of the moving meniscus) play an important role and a theory is presented for this case.

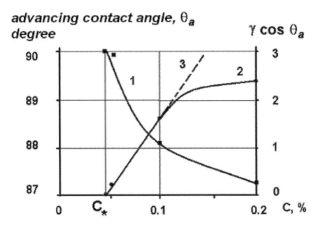

FIGURE 6.8 The influence of concentration of aqueous surfactant solutions (Syntamide-5, molecular weight 420) on the advancing contact angle, θ_a (curve 1) and $\gamma \cos \theta_a$ (curve 2) measured on a flat hydrophobized quartz surface. C_* marks the critical surfactant concentration, below which surfactant solution does not spread. Broken line (3) according to Eq. (6.12).

Theory

As mentioned in the introduction to Chapter 6, we know surprisingly little about the behavior of surfactant solutions in a vicinity of the apparent three-phase contact line. This is one reason why we are using the Newman-Young equation for the theoretical treatment in this chapter. We showed in Chapter 2 that this equation does not have a firm theoretical basis. Hence, all conclusions made based on the Newman-Young equation are qualitative and semi-empirical at best.

Let us consider a dependency of $\Psi(C_m) = \gamma(C_m)\cos\theta_a(C_m)$ on the concentration of surfactant, C_m, on the moving meniscus. According to the triangle rule, it can be calculated as

$$\Psi(C_m) = \gamma_{sv}(C_m) - \gamma_{sl}(C_m), \tag{6.8}$$

where γ_{sv} and γ_{sl} are solid–vapor and solid–liquid interfacial tensions. According to Antonov's rule, the dependency of the two latter interfacial tensions on the concentration can be presented as

$$\gamma_{sv}(C_m) = \gamma_{sv}^0\left(1 - \frac{\Gamma_+}{\Gamma^\infty}\right) + \gamma_{sv}^\infty \frac{\Gamma_+}{\Gamma^\infty}; \ \gamma_{sl}(C_m) = \gamma_{sl}^0\left(1 - \frac{\Gamma_-}{\Gamma^\infty}\right) + \gamma_{sl}^\infty \frac{\Gamma_-}{\Gamma^\infty}, \tag{6.9}$$

where superscripts 0 and ∞ mark zero and complete coverage of hydrophobic adsorption sites, respectively; subscripts − and + mark adsorption just behind and just in front of the moving meniscus, respectively; and Γ is the surface adsorption of surfactant molecules.

Note that adsorption of surfactant molecules results in a decreasing of solid–liquid interfacial tension, that is, $\gamma_{sl}^0 - \gamma_{sl}^\infty > 0$. However, adsorption of surfactant molecules on the bare hydrophobic interface in front of the moving meniscus results in a local increase of the solid–vapor interfacial tension, that is, $\gamma_{sv}^0 - \gamma_{sv}^\infty < 0$. The initial contact angle on the bare hydrophobic interface is assumed bigger than $\pi/2$, that is, $\gamma_{sv}^0 - \gamma_{sl}^0 < 0$.

It is assumed that both adsorption isotherms are linear functions of the surfactant concentration below the CMC (which is the only case considered on this stage) and remain constant above the CMC. This means that

$$\Gamma_- = G_{sl}C_m, \tag{6.10}$$

at concentrations below the CMC.

Both a spontaneous imbibition and a spontaneous capillary rise into hydrophobic capillaries are sufficiently slow processes, that is, we assume here a condition of local equilibrium on the moving apparent three-phase contact line. According to this assumption, the equality of chemical potentials of adsorbed surfactant molecules should be satisfied across the apparent three-phase contact line, that is, $\ln\Gamma_- + \Phi_{sl} = \ln\Gamma_+ + \Phi_{sv}$, where Γ_- and Γ_+ are jumps of adsorption across the meniscus surface (Figure 6.9), and Φ_{SL}, Φ_{SV} (in kT units) are corresponding values of the energy of surfactant molecules at solid–water and solid–air interfaces, respectively. From this equation, we conclude: $\Gamma_+ = \frac{\Gamma_-}{\exp(\Phi_{sv} - \Phi_{sl})}$. It is obvious that Φ_{sv} is higher than Φ_{sl}, and, hence, $\Gamma_+ < \Gamma_-$. This relation can be rewritten using Eq. (6.10) as

$$\Gamma_+ = G_{sv}C_m, \quad G_{sv} = \frac{G_{sl}}{\exp(\Phi_{sv} - \Phi_{sl})}, \tag{6.11}$$

Substituting Eqs. (6.9) and (6.11) into Eq. (6.8) and bearing in mind these inequalities, we can conclude after rearrangements, that

$$\Psi(C_m) = \alpha(C_m - C_*), \tag{6.12}$$

where

$$\alpha = \left[\frac{G_{sv}}{\Gamma^\infty}\left(\gamma_{sv}^\infty - \gamma_{sv}^0\right) + \frac{G_{sl}}{\Gamma^\infty}\left(\gamma_{sl}^0 - \gamma_{sl}^\infty\right)\right] > 0, \quad C_* = \frac{\alpha^0}{\alpha}, \quad \alpha^0 = \gamma_{sl}^0 - \gamma_{sv}^0 > 0.$$

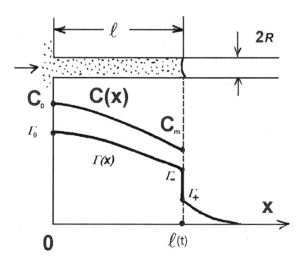

FIGURE 6.9 Distribution of surfactant concentrations along the capillary length during a spontaneous imbibition process. *C*, volume concentration; and *a*, surface concentration.

According to Eq. (6.12), $\Psi(C_m)$ dependency should be a linear function of concentration at $C_m > C_*$, which is in good agreement with experimental observations (line 3 in Figure 6.8 at concentrations $C > C_*$) in a range of surfactant concentrations under consideration in this section.

We now try to solve theoretically the problem of a spontaneous imbibition of surfactant solutions into hydrophobic cylindrical capillaries considering the transfer and the surface diffusion of surfactant molecules as well as adsorption on the bare hydrophobic surface in front of the moving meniscus. The location of the moving meniscus in the capillary is $l(t)$ (Figure 6.9).

The transfer of surfactant molecules in the filled portion of the capillary is described by the convective diffusion equation:

$$\frac{\partial C(t,x,r)}{\partial t} = D\frac{\partial^2 C(t,x,r)}{\partial x^2} + D\frac{1}{r}\frac{\partial}{\partial r}\left(r\frac{\partial C(t,x,r)}{\partial r}\right) - \frac{\partial}{\partial x}\left(v(r)C(t,x,r)\right),$$

where $C(t, x, r)$ is the local concentration of surfactant; D is the diffusion coefficient; t, x, r are time, axial and radial coordinates, respectively; and $v(r)$ is the axial velocity distribution.

Integration of this equation over the radius from 0 to R, where R is the capillary radius results in

$$\frac{\partial}{\partial t}\left(\int_0^R rC(t,x,r)dr\right) = D\frac{\partial^2}{\partial x^2}\left(\int_0^R rC(t,x,r)dr\right) + R\left(D\frac{\partial C(t,x,r)}{\partial r}\right)_{r=R} - \frac{\partial}{\partial x}\left(\int_0^R rv(r)C(t,x,r)dr\right).$$

The second term in the right-hand side of this equation is equal to

$$-D\frac{\partial C(t,x,r)}{\partial r}\bigg|_{r=R} = \frac{\partial \Gamma}{\partial t} - D_{sl}\frac{\partial^2 \Gamma}{\partial x^2},$$

where D_{sl} is the surface diffusion coefficient over the filled portion of the capillary. Hence, these two equations result in

$$\frac{\partial}{\partial t}\left(\int_0^R rC(t,x,r)dr\right) = D\frac{\partial^2}{\partial x^2}\left(\int_0^R rC(t,x,r)dr\right) - R\left(\frac{\partial \Gamma}{\partial t} - D_{sl}\frac{\partial^2 \Gamma}{\partial x^2}\right) - \frac{\partial}{\partial x}\left(\int_0^R rv(r)C(t,x,r)dr\right).$$

A characteristic time scale of the equilibration of the surfactant concentration in a cross section of the capillary, $\tau \sim R^2/D \approx 0.1\,\mathrm{s}$, if we use $R \sim 10\,\mu\mathrm{m}$ and $D \sim 10^{-5}\,\mathrm{cm^2/s}$ for estimations. A characteristic time scale of the spontaneous capillary imbibition or rise into hydrophobic capillaries is much bigger than 0.1 s (Figures 6.10 and 6.11) by at least three orders of magnitude. This means that the surfactant concentration is equilibrated across the cross section of the capillary and remains constant in any cross section. Hence, the concentration depends only on the position, x (Figure 6.9), that is, $C = C(t, x)$. We also assume also that the adsorption equilibrium in any cross section is also reached. Taking this into account, this equation can be rewritten after both sides are divided by $R^2/2$ as

$$\frac{\partial(a+C)}{\partial t} = D\frac{\partial^2 C}{\partial x^2} + D_{sl}\frac{\partial^2 a}{\partial x^2} - v\frac{\partial C}{\partial x}, \qquad 0<x<l(t), \tag{6.13}$$

where D and D_{sl} are diffusion coefficients of surfactant molecules in the volume and over the wetted capillary surface; $v = dl/dt$ is the meniscus velocity; and

$$a(x,t) = \frac{2}{R}G_{sl}C(x,t) = F_{sl}C(x,t), \tag{6.14}$$

where $F_{sl} = (2/R)\,G_{sl}$.

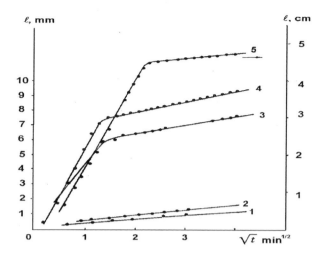

FIGURE 6.10 The time evolution of the imbibition length *l* (mm) with time, *t* (min) for aqueous solutions of Syntamide-5 in a horizontal hydrophobized quartz capillary, $R = 16$ μm. (1) $C_0 = 0.05\%$; (2) $C_0 = 0.1\%$; (3) $C_0 = 0.4\%$; (4) $C_0 = 0.5\%$; (5) $C_0 = 1\%$.

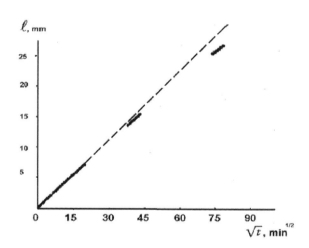

FIGURE 6.11 Spontaneous capillary rise in a vertical hydrophobised quartz capillary ($R = 11$ μm), Syntamid-5 surfactant solution ($C_0 = 0.1\%$). Time evolution of the imbibition length, *l* (mm), on time, *t* (min). (From Churaev, N.V. et al., *Colloid Polymer Sci.*, 259, 747, 1981.)

Concentration below the CMC

Let $C(x, t)$ and $a(x, t)$ be the local surfactant concentrations in the bulk solution and in the adsorbed state on the capillary surface. A constant surfactant concentration $C(0, t) = C_0 < $ CMC is kept at the capillary inlet. In this case, the surfactant transport in the filled portion of the capillary obeys Eq. (6.13).

Substitution of expression (6.14) into Eq. (6.13) results in

$$(1+F_{sl})\frac{\partial C}{\partial t} = D_{ef}\frac{\partial^2 C}{\partial x^2} - \frac{dl}{dt}\frac{\partial C}{\partial x}, \qquad 0 < x < l(t), \tag{6.15}$$

where $D_{ef} = (D + F_{sl} D_{sl})$ is the effective diffusion coefficient. On the non-wetted portion of the capillary surface, $l(t) < x$, only surface diffusion takes place, that is

$$\frac{\partial a}{\partial t} = D_{sv} \frac{\partial^2 a}{\partial x^2}, \tag{6.16}$$

where D_{sv} is the diffusion coefficient of surfactant molecules on the non-wetted hydrophobic capillary surface.

Equations (6.15) and (6.16) are then solved using the following boundary and initial conditions:

$$C(0, t) = C_0; \, a(\infty, t) = 0, \tag{6.17}$$

$$l(0) = 0; \, a(x, 0) = 0. \tag{6.18}$$

The following condition of the mass balance on the moving meniscus surface should be satisfied:

$$\left(D_{sv} \frac{\partial a}{\partial x} \right)_{l+} - \left(D \frac{\partial C}{\partial x} + D_{sv} \frac{\partial a}{\partial x} \right)_{l-} = (a_- - a_+) \frac{dl}{dt}, \tag{6.19}$$

where l_- and l_+ represent two points located on the opposite sides of the meniscus: on the liquid phase side, l_-, and on the unwetted side in front of the moving meniscus, l_+ (Figure 6.9). Condition (6.19) expresses the conservation of mass at the moving meniscus and the moving apparent three-phase contact line.

As mentioned, the imbibition process is the slow one, the duration is shown in Figures 6.10 and 6.11. It allows us to assume a condition of local equilibrium on the moving apparent three-phase contact line. From this condition, we conclude in the same way as in Eq. (6.11) that $a_+ = \frac{a_-}{\exp(\Phi_+ - \Phi_-)}$. The energy of a surfactant molecule on a bare hydrophobic substrate in front of the moving meniscus, Φ_+, is higher than the corresponding energy in the aqueous solution, Φ_-, and, hence, $a_+ < a_-$.

Equations (6.15) and (6.16) with boundary (6.19) and initial conditions (6.17) and (6.18) have a solution only if the concentration on the moving meniscus remains constant, that is, $C_m = $ const. This corresponds to the experimentally observed law of spontaneous imbibition, $l = K\sqrt{t}$, where the constant K is to be determined.

The solution of Eqs. (6.15) and (6.16) is tried in the following form $C = C(\xi)$, $a = a(\xi)$, where $\xi = \frac{x}{\sqrt{t}}$. Transformations of Eqs. (6.15) and (6.16) yield

$$-\frac{\xi}{2} \frac{dC}{d\xi} (1 + F_{sl}) = D_{ef} \frac{d^2 C}{d\xi^2} - \frac{K}{2} \frac{dC}{d\xi}, \tag{6.20}$$

$$-\frac{\xi}{2} \frac{da}{d\xi} = D_{sv} \frac{d^2 a}{d\xi^2}. \tag{6.21}$$

The solution of these equations have the following forms:

$$C(\xi) = A_1 \int_0^\xi \exp\left[-\frac{(1 + F_{sl})\xi^2}{4D_{ef}} + \frac{K\xi}{2D_{ef}} \right] d\xi + B_1. \tag{6.22}$$

$$a(\xi) = A_2 \int_K^\xi \exp\left[-\frac{\xi^2}{4D_{sv}} \right] d\xi + B_2. \tag{6.23}$$

Integration constants A_1, A_2, B_1 and B_2 should be determined using initial (6.17) and (6.18) and boundary conditions (6.19). This results in

$$A_1 = -\frac{C_0 - C_m}{K \int_0^K \exp\left[-\frac{(1+F_{sl})\xi^2}{4D_{ef}} + \frac{K\xi}{2D_{ef}}\right]d\xi}; \quad B_1 = C_0, \tag{6.24}$$

$$A_2 = \frac{-F_{sv}C_m}{\int_K^\infty \exp\left[-\frac{\xi^2}{4D_{sv}}\right]d\xi}; \quad B_2 = F_{sv}C_m. \tag{6.25}$$

Substitution expressions from Eqs. (6.24) and (6.25) into boundary condition Eq. (6.19) yields the following transcendental equation:

$$\frac{K}{2}(F_{sl} - F_{sv}) = \frac{\left(\dfrac{C_0}{C_m} - 1\right)D_{ef}\exp\left[\dfrac{(1-F_{sl})K^2}{4D_{ef}}\right]}{K \int_0^K \exp\left[-\dfrac{(1+F_{sl})\xi^2}{4D_{ef}} + \dfrac{K\xi}{2D_{ef}}\right]d\xi}.$$

$$- \frac{F_{sv}D_{sv}\exp\left[-\dfrac{K^2}{4D_{sv}}\right]}{\int_K^\infty \exp\left[-\dfrac{\xi^2}{4D_{sv}}\right]d\xi}. \tag{6.26}$$

The K value governs the imbibition rate according to Eq. (6.7) and can be rewritten using Eq. (6.12) as

$$K = \sqrt{\frac{r\alpha(C_m - C_*)}{2\eta}}. \tag{6.27}$$

Substitution of this equation into Eq. (6.26) gives the required equation for determining the unknown concentration on the moving meniscus, C_m.

Equations (6.26) and (6.27) give the solution of the problem under consideration. To solve Eq. (6.26), the following coefficients should be known: diffusion coefficients, D, D_{sl}, and D_{sv}; adsorption constants, G_{sl} and G_{sv}; capillary radius, R; solution viscosity, η; concentration at the capillary inlet, C_0; and the coefficient α in Eq. (6.12) or (6.27).

Numerical analysis of the final equation for K shows that its values increases as R, C_0, D and D_{sl} increase or D_{sv} decreases. This is because the high surface mobility of the surfactant on the unwetted portion of the capillary reduces the concentration on the meniscus, C_m, thereby inhibiting the imbibition.

Concentration above the CMC

In the case of the surfactant concentration at the capillary inlet, C_0, above the CMC, diffusion and adsorption are accompanied by the destruction of the micelles. If the surfactant concentration at the capillary inlet is higher than the CMC, than the total surfactant concentration, C, can be presented as $C = C_{mol} + C_M$, where C_{mol} and C_M are concentrations of free surfactant molecules and molecules inside micelles, respectively. Concentration of free molecules, C_{mol}, remains approximately constant above the CMC and equal to the CMC (which is marked as C_c below). This means, that (i) decrease if the surfactant concentration goes through disintegration of micelles, whereas the concentration of the free surfactant molecules remain constant, and (ii) the wetting depends on the concentration of the free molecules and, hence, remains independent of the concentration unless the concentration is below the CMC: $\gamma(C_c)\cos\theta(C_c) = \varphi_c$ = const. This means that the K value, which determines the imbibition rate, is constant and is equal to

$$K_c = \sqrt{r\alpha(C_c - C_*)/2\eta} = \text{const}, \qquad (6.28)$$

until the concentration at the meniscus, C_m, is higher than the CMC.

The values of K_c for $C_0 > CMC \sim 1\%$ obtained in experiments with Syntamide-5 [3,4] were of the order of 10^{-1}, which corresponds to the contact angle $\theta_a = 89°$, that is, only slightly different from 90°.

Unlike the case when the concentration at the capillary inlet is below the CMC, the meniscus moves with velocity $l = K_c\sqrt{t}$, where K_c is given by Eq. (6.28) and does not vary with time. In this case, the surfactant adsorption on the capillary surface is accompanied by a continuous decrease of concentration, C_m, near the meniscus from $C_m = C_0 > CMC$ at the beginning of the imbibition process at $t = 0$ to $C_m = CMC$ or $C_m = C_c$ at the end of this first fast stage.

After C_m reaches the CMC, the condition K_c = const is no longer valid. The K value decreases below K_c, and the imbibition rate slows down. After $C_m = CMC$ is reached on the meniscus, a further reduction of concentration, C_m, causes separation of the micelle front (where $C = CMC$) from the meniscus surface. The micelle front movement is governed, as shown below, by the same law, $l_M = K_M\sqrt{t}$, where $K_M < K$. The further stage of imbibition occurs when $C_m < C_c$ is described in the similar way to the previous section: the concentration no longer varies, but remains *const* and smaller than the CMC.

The variations of the imbibition process according to the micelles dissociation is illustrated in Figure 6.10. A sharp change in the rate of imbibition at some distance $l = l_c$ was indeed observed for Syntamide-5 solutions at $C_0 > CMC$.

Values of l_c, which correspond to the first fast stage of the imbibition process are estimated below. As mentioned above in this section, the first fast stage of the imbibition is determined by the dissociation of micelles in the close vicinity of a moving meniscus. Let us assume an adsorption of micelles on the meniscus according to [6], which is adopted according to the linear law $\Gamma = G_M(C_m - C_c)$, where $C_m - C_c$ is the micelles concentration at $C_0 > C_c$ and G_M is the corresponding adsorption constant. Diffusion of micelles is neglected because of the short duration of the first stage and the much smaller diffusion coefficient of micelles. The mass balance on the moving meniscus during the first fast stage of the imbibition is

$$(F_{sl} - F_{sv})C_c\frac{dl}{dt} = -G_M\frac{d(C_m - C_c)}{dt}, \quad C_m(0) - C_c = C_0 - C_c, \qquad (6.29)$$

where $(F_{sl} - F_{sv})C_c$ = const is the limiting adsorption on solid surface (micelles do not adsorb); the difference $C_m - C_c$ is equal to the micelle concentration. The left-hand side of Eq. (6.29) characterizes the surfactant adsorption rate on the newly wetted surface, and the right-hand side the micelle disintegration rate from the adsorbed layer of micelles.

The solution of Eq. (6.29) under the initial condition $t = 0$, $C_m - C_c = C_0 - C_c$ has the following form:

$$C_m(t) - C_c = (C_0 - C_c) - \frac{2C_c(G_{sl} - G_{sv})K_c\sqrt{t}}{RG_M}. \qquad (6.30)$$

The left-hand side of this equation vanishes as all the micelles near the meniscus disintegrate, that is, the concentration C_m reduces to the CMC. Equation (6.30) allows determining the instant $t = t_c$ when $C_m = C_c$, which is the end of the first fast stage of the imbibition. Using this condition and Eq. (6.30), the length of the fast imbibition can be determined as

$$l_c = K_c \sqrt{t_c} = \left(\frac{C_0}{C_c} - 1 \right) \frac{rG_M}{2\left(G_{sl} - G_{sv}\right)}. \tag{6.31}$$

For $R \sim 2 \times 10^{-3}$cm, $C_0 = 2\%$, $C_c = 0.1\%$, $l_c = 0.5$ cm (in agreement with experimental observations), we get $\frac{G_M}{(G_{sl}-G_{sv})} \sim 25$ in an agreement with [6].

The first fast stage of the imbibition is followed by a second slower stage, and the concentration on the moving meniscus, C_m, is below the CMC. Two regions can be now identified inside the capillary: the first region, from the capillary inlet to some position, which we mark as $l_M(t)$, where the concentration is above the CMC and the solution includes both micelles and individual surfactant molecules; and the second region, from $l_M(t)$ to $l(t)$, where the concentration is below the CMC and only individual molecules are transferred. The concentration is equal to the CMC at $x = l_M(t)$. Consideration in the second region, $l_M(t) < x < l(t)$, is similar to that at a concentration below the CMC, which is why only the transport in the first region is considered here.

Inside the first region, $0 < x < l_M(t)$, concentration of free surfactant molecules is constant and equal to the CMC [7]. Hence, the transfer is determined by the diffusion of micelles and convection of all molecules. The total concentration, $C = C_{mol} + C_M$, and C_{mol} remains constant and equal to the CMC, hence,

$$\frac{\partial C}{\partial t} = D_M \frac{\partial^2 C}{\partial x^2} - \frac{dl}{dt}\frac{\partial C}{\partial x}, \tag{6.32}$$

where D_M is the diffusion coefficient of micelles; and $C = C_c + C_M$ is the total concentration. Adsorption on membrane pores is determined by the concentration of free molecules, which is constant in the first region and, therefore, that is where adsorption occurs. This is why the diffusion of adsorbed molecules in the first region is omitted in Eq. (6.32). The transfer of surfactant molecules in the second region (a micelle-free region) is described by Eq. (6.15).

Boundary conditions on the moving boundary between the first and second regions, $l_M(t)$, are as follows:

$$D_M \left(\frac{\partial C}{\partial x} \right)_{x = l_M -} = D_{ef} \left(\frac{\partial C}{\partial x} \right)_{x = l_M +}, \quad C(l_M.t) = C_c. \tag{6.33}$$

As before, we assume that

$$l(t) = K\sqrt{t}, \quad l_M(t) = K_M \sqrt{t}, \tag{6.34}$$

where K is given by Eq. (6.27), that is, K is expressed via an unknown concentration on the moving meniscus, C_m; and K_M is a new unknown constant.

Let a similarity variable be introduced now in the same way as in the case of concentration below the CMC, that is, $\xi = x/\sqrt{t}$ in Eqs. (6.32), (6.15), and (6.16). Using boundary conditions (6.33) and (6.19) the following system of two nonlinear algebraic equations can be deduced:

$$\frac{K}{2}\left(F_{sl} - F_{sv}\right) = \frac{\left(\frac{C_c}{C_m} - 1\right)D_{ef}\exp\left[\frac{(1-F_{sl})K^2}{4D_{ef}}\right]}{\int_{K_M}^{K}\exp\left[-\frac{(1+F_{sl})\xi^2}{4D_{ef}} + \frac{K\xi}{2D_{ef}}\right]d\xi} - \frac{F_{sv}D_{sv}\exp\left[-\frac{K^2}{4D_{sv}}\right]}{\int_{K}^{\infty}\exp\left[-\frac{\xi^2}{4D_{sv}}\right]d\xi}. \tag{6.35}$$

$$D_M \frac{(C_0 - C_c)\exp\left(\dfrac{K_M^2}{4D_M}\right)}{\displaystyle\int_0^{K_M} \exp\left[-\frac{1}{2D_M}\left(\frac{\xi^2}{2} - K_M\xi\right)\right]d\xi} = D_{ef} \frac{(C_c - C_m)\exp\left(\dfrac{(1 - F_{sl})K^2}{4D_{ef}}\right)}{\displaystyle\int_{K_M}^{K} \exp\left[-\frac{(1 + F_{sl})\xi^2}{4D_{ef}} + \frac{K\xi}{2D_{ef}}\right]d\xi}.$$

Two unknown values in this system of equations are the concentrations on the moving meniscus, C_m, and the unknown constant K_M. It is necessary to recall that K is expressed via C_m according to Eq. (6.27).

Using Eqs. (6.35), it is possible to show that $K_M < K$, that is, the border between the first and the second regions moves slower than the meniscus. The diffusion coefficient of micelles is much smaller than that of individual molecules. This means that the system of equations (Eq. 6.35) includes a small parameter, $D_M / D_{ef} \ll 1$. This means that the second equation can be omitted in the zero approximation and the constant K is only slightly different from the same value in the case of concentration below the CMC. That is, during the second stage, the rate of the spontaneous imbibition is only slightly higher than in the case when the concentration is below the CMC. Experimental data in Figure 6.10 confirms this conclusion.

Now we can conclude that the two-stage process of imbibition (as presented in Figure 6.10) takes place only if the concentration on the capillary inlet is above the CMC and if adsorption on the liquid–gas interface (that is, on the moving meniscus) is much faster than on solid–liquid interface (~25 faster in our case). It is necessary to emphasize that the duration of the fast stage is the shorter the thinner the capillary is. K values during the second stage of the imbibition are only slightly higher than in the case when concentration is below the CMC.

Spontaneous Capillary Rise in Hydrophobic Capillaries

A further confirmation of the adsorption of surfactant molecules in front of the moving meniscus on the bare hydrophobic substrate is a phenomenon of the spontaneous capillary rise of surfactant solutions in hydrophobic capillaries, which is considered in this section.

Figure 6.11 shows the results of one of the capillary rise experiments with a Syntamide-5 solution in a vertical hydrophobized quartz capillary [5]. The observed time evolution of the imbibition length, $l(t)$, follows $l(t) = K\sqrt{t}$ dependency at the initial stage of the process. The value of K determined from the slopes of the $l / \sqrt{t}$ dependencies correspond to the advancing contact angle value, θ_a, being only a few seconds less than 90°. At such a θ_a value, the capillary rise would be expected to stop as soon as the liquid reached a height of $l_{max} = 10^{-3}$ cm. However, it does not stop at this height; instead it goes up to a height of 3–4 cm with an almost constant K. The only explanation of this phenomenon is that the meniscus rises in the capillary following the surface diffusion front of surfactant molecules that hydrophilize the bare hydrophobic capillary surface in front of the moving meniscus. At each position, $l(t)$, the meniscus curvature must satisfy the following equilibrium condition:

$$\frac{2\Psi(C_m)}{R} = \rho g l(t), \tag{6.36}$$

where $\Psi(C_m) = \gamma(C_m)\cos\theta_a(C_m)$; ρ is the density of the solution; and g is the gravity acceleration. Because θ_a is very close to $\pi/2$ according to Eq. (6.12), we can use a linear dependency (6.12): $\Psi(C_m) = \alpha(C_m - C_*)$.

Using Eq. (6.12), we can rewrite Eq. (6.36) as

$$l(t) = \frac{2\alpha}{\rho g R}(C_m - C_*). \tag{6.37}$$

During the initial stage of the capillary rise, $l(t) = K\sqrt{t}$ (Figure 6.11). This is possible only if

$$C_m = C_* + B\sqrt{t}, \text{ and } B = K\frac{\rho g R}{2\alpha}, \tag{6.38}$$

where the constant K is to be determined.

This equation shows that the case under consideration is governed by a completely different mechanism as compared with the case of the horizontal imbibition (where C_m remains constant over time). In the case of a spontaneous capillary rise in hydrophobic capillaries, $C_m(t)$ does not remain constant but must, instead, increase as the capillary rise progresses. The comparison of Figures 6.10 and 6.11 shows that the time scale of the spontaneous capillary rise is around 100 times larger than that of the corresponding time scale in the case of the capillary imbibition into horizontal capillaries.

The maximum height of the capillary rise, $l_{\max}$, is reached after the concentration on the meniscus, C_m, becomes equal to the concentration at the capillary entrance, C_0. After that, the capillary rise stops. Using Eq. (6.37), $l_{\max}$ is determined as

$$l_{\max} = \frac{2\alpha}{\rho g R}(C_0 - C_*). \tag{6.39}$$

This means that the experimental observation presented in Figure 6.11 corresponds to $l(t) \ll l_{\max}$, which is the initial stage of the capillary rise.

Now the problem of the spontaneous capillary rise of surfactant solutions in hydrophobic capillaries is considered in the case when concentration at the capillary inlet is below the CMC. In this case, the transport of surfactant molecules is described by Eqs. (6.15) and (6.16) and boundary conditions (6.17) and (6.19). The substantial difference from the spontaneous capillary imbibition is that now the relation between $l(t)$ and the concentration on the moving meniscus, C_m, is given by relation (6.37), which shows that C_m is an unknown function of time. Using these equations and boundary conditions, we show in this section that $l(t)$ dependency on time can be calculated and that it is proportional to the square root of time at the initial stage of the capillary rise (see Appendix 1 for details).

The solution in Appendix 1 shows that, at the initial stage of the capillary rise, $l(t)$ develops as

$$l(t) = K\sqrt{t}, \quad K = l_{\max}\sqrt{\frac{2\kappa}{\omega}}, \tag{6.40}$$

where κ and ω are defined in Appendix 1. This dependency agrees with experimental observations in [4] (Figure 6.11). At the final stage of the capillary rise, $l(t)$ levels off as

$$l(t) = l_{\max}\left(1 - \frac{\omega+1}{\kappa t}\right).$$

According to Eq. (6.39) and Figure 6.11, $l_{\max}^{\exp} \sim 35\,\text{mm}$ in the experiment presented in Figure 6.11. The experimental value of $K^{\exp}$, in Eq. (6.40), calculated according to Figure 6.11, is $K^{\exp} \sim 5 \cdot 10^{-3}\,\text{cm/s}^{1/2}$. In experiments presented in Figure 6.11, $\omega \sim 1$. The comparison of these estimations gives $\kappa \sim 10^{-5}\,\text{sec}^{-1}$, which coincides with the value calculated in Appendix 2 according to Eq. (2.11); if we assume that $F_{sv} \ll F_{sl}$, $D \sim 10^{-5}\,\text{cm}^2/\text{s}$, use estimation $G_{SL} \sim 10^{-5}\,\text{cm}$ [6], and value of α is taken directly from Figure 6.8.

APPENDIX 1
Excess Free Energy of an Aqueous Surfactant Droplet on a Hydrophobic Substrate

In this appendix, we consider the possibility of a spontaneous adsorption of surfactant molecules on bare hydrophobic substrate in front of the moving apparent three-phase contact line. We consider the case of an aqueous surfactant droplet on a hydrophobic substrate because the case of a meniscus can be conserved similarly.

It is proven in this Appendix that the adsorption of surfactants onto a bare hydrophobic substrate in front of the apparent three-phase contact line is a favorable process from the thermodynamic point of view and, hence, it proceeds spontaneously.

Let us consider a free energy of an aqueous surfactant droplet on a solid hydrophobic substrate as shown in Figure 6.12. As mentioned in the introduction to Chapter 6, we know close to nothing about surface forces in the presence of surfactants.

Hence, we neglect the surface forces action in the vicinity of a three-phase contact line and the free energy of the system, Φ, can be written as

$$\Phi = \gamma_{lv}S + PV + \pi a^2\left(\gamma_{sl} - \gamma_{sv}\right) \tag{A1.1}$$

where γ_{lv}, γ_{sl}, and γ_{sv} are interfacial tensions of the liquid–vapor, the solid–liquid, and the solid–vapor interfaces, respectively; a is the radius of the droplet base; S is the area of the liquid–vapor interface; P is the excess pressure inside the droplet; and V is the liquid volume, θ is the contact angle.

Note, the derivation is identical for both cases $0<\theta<\pi/2$ and $\pi/2<\theta< \pi$.

It is well known that the requirement of the equilibrium results in an equation for the droplet profile (see Chapter 1). Let $\mathfrak{R}$ be the radii of the curvature of the droplet (Figure 6.12). The only equilibrium solution of Eq. (A1.1) is a spherical segment (see Chapter 1 for details). Let us express P_e, S, a and V via radius, $\mathfrak{R}$, and the contact angle, θ. The following expressions can be verified

$$P = -2\gamma_{lv}/\mathfrak{R}$$

$$S = 2\pi\mathfrak{R}^2\left(1-\cos\theta\right)$$

$$a^2 = \mathfrak{R}^2\left(1-\cos^2\theta\right)$$

$$V = \frac{1}{3}\pi\mathfrak{R}^2\left(2-3\cos\theta+\cos^3\theta\right)$$

The substitution of all the latter expressions into Eq. (A1.1) results in

$$\Phi = \frac{\pi\gamma_{lv}\mathfrak{R}^2}{3}f(\theta),\ f(\theta)\left(2-3\cos\theta+\cos^3\theta\right). \tag{A1.2}$$

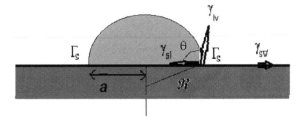

FIGURE 6.12 Equilibrium or quasi-equilibrium droplet on hydrophobic solid substrate.

Considering the assumption of constancy of the drop volume at equilibrium, Eq. (A1.2) can be rewritten as follows:

$$\Phi = \frac{\gamma_{lv} \, \pi^{1/3} \, V^{2/3}}{3^{1/3}} f^{1/3} (\cos\theta).$$ (A1.3)

The transversality condition at equilibrium (see Chapter 1 for details) results in the well-known Young equation $\cos\theta = \frac{\gamma_{sv} - \gamma_{sl}}{\gamma_{lv}}$, which in a combination with Eq. (A1.3) yields the following equation for the free energy

$$\Phi = \gamma_{lv} \left(\frac{\pi V^2}{3} \right)^{1/3} f^{1/3} \left(\frac{\gamma_{sv} - \gamma_{sl}}{\gamma_{lv}} \right).$$ (A1.4)

Equation (A1.4) gives the dependency of the excess free energy of the droplet via three interfacial tensions, γ_{lv}, γ_{sl} and γ_{sv}.

Let us calculate derivatives of the excess free energy of the droplet with all three interfacial tensions γ_{lv}, γ_{sl} and γ_{sv} according to Eq. (A1.4). Simple calculations show that

$$\frac{\partial \Phi}{\partial \gamma_{sl}} = A \frac{\sin^2\theta}{f^{2/3}(\cos\theta)} > 0,$$ (A1.5)

$$\frac{\partial \Phi}{\partial \gamma_{lv}} = A \frac{2(1 - \cos\theta)}{f^{2/3}(\cos\theta)} > 0,$$ (A1.6)

$$\frac{\partial \Phi}{\partial \gamma_{sv}} = -A \frac{\sin^2\theta}{f^{2/3}(\cos\theta)} > 0,$$ (A1.7)

where $A = \left(\frac{\pi V^2}{3} \right)^{1/3}$

Equations (A1.5) and (A1.6) show that the excess free energy of the droplet decreases if a solid–liquid interfacial tension decreases and/or a liquid–vapor interfacial tension decreases. These two conclusions are well known and well confirmed experimentally. As a result, two spontaneous processes take place in the presence of surfactants: adsorption of surfactants on the liquid–hydrophobic solid interface and adsorption on the liquid–vapor interface.

Equation (A1.7) shows that the excess free energy of the droplet decreases if the solid–vapor interfacial tension *increases*. Note, the adsorption of surfactant molecules on a bare hydrophobic surface in front of the three-phase contact line (autophilic phenomenon [8]) results in a local increase of the solid–vapor interfacial tension, that is, in hydrophilization of the originally hydrophobic bare solid substrate. However, according to Eq. (A1.7), the total excess free energy *decreases* because of this process. This means that the adsorption of surfactant molecules on a bare hydrophobic substrate in front of the three-phase contact line is *a spontaneous process*. However, because locally that process results in an increase of the local solid–vapor interfacial tension, this process goes via a potential barrier, that is, it is slower than the previous two spontaneous processes.

It is interesting to note, that the simple theoretical conclusion presented in this section has never been noticed before in the literature.

The presence of adsorption of surfactant molecules on a bare hydrophobic surface in front of the advancing apparent three-phase contact line has been proven experimentally in [8] and referred to as autophilic phenomenon.

As the autophilic phenomenon results in an increase of the surface energy locally, then the adsorption goes via a potential barrier, which is an energy required for a jump of a single surfactant molecule from a liquid–vapor interface on a bare hydrophobic surface:

$$\Delta E = A_{sv}\left[\gamma_{sv}\left(\Gamma_{sv}\right)-\gamma_{sv}\left(0\right)\right]-A_{lv}\left[\gamma_{iv}\left(\Gamma_{iv}\right)-\gamma_{iv}\left(0\right)\right],\tag{A1.8}$$

where A_{sv} and A_{lv} are the surface areas of per surfactant molecule at the solid–vapor and the liquid–vapor interfaces, respectively; $\gamma_{sv}(0)$ is the interfacial surface tension of the bare hydrophobic solid surface; $\gamma_{lv}(0)$ is the surface tension of pure water on the liquid–vapor interface; $\gamma_{sv}\left(\Gamma_{sv}\right)$ and $\gamma_{lv}\left(\Gamma_{lv}\right)$ are the interfacial tensions at the solid–vapor and the liquid–vapor interfaces with adsorbed surfactants molecules, respectively.

Let us estimate the energy barrier for an aqueous solution of $C_{12}E_8$ at a hydrophobized silica surface using data presented by Kumar et al. [8]. Using the available data for $C_{12}E_8$ from [8,9] we conclude: $\gamma_{sv}\left(\Gamma_{sv}\right)-\gamma_{sv}(0)\approx 24.5$ mN/m is the surface pressure, $\gamma_{lv}\left(\Gamma_{lv}\right)=35.2$ mN/m is the liquid–vapor surface tension, and $A_{lv}=62\text{Å}^2$ is the area per molecule. Assuming that $A_{sv}=A_{lv}$ [8], Eq. (A1.8) gives the following value of the local energy barrier: $\Delta E\approx 10^{kT}$. This estimation shows that the energy barrier for the particular case considered is not unreasonably high. However, the height of the barrier determines the slow rate of the surfactant molecule transfer and the slow rate of spreading determined by that mechanism.

According to our experience nonionic surfactants (at least some of them, trisiloxanes for example) demonstrate the autophilic phenomenon, but anionic surfactants (e.g., SDS) do not. In the case of anionic surfactants, the potential barrier is probably too high for transfer because the charged surfactant molecules should be transferred with counter-ions to maintain electroneutrality.

APPENDIX 2

To simplify the mathematical treatment of the problem, we neglect in this section the diffusion of surfactant molecules in front of the moving meniscus. That means, only the adsorption in front of the moving meniscus is considered. Under this simplification, the process of the capillary rise in a hydrophobic capillary is governed by Eq. (6.15) with the boundary condition (6.19).

The concentration of surfactant changes considerably only in close proximity to the moving meniscus. That means the surfactant concentration differs from the concentration at the capillary inlet, C_0, in a narrow region between $l(t)-\delta(t)$ and $l(t)$, where $\delta(t)$ is to be determined. At the boundary $l(t)-\delta(t)$, the concentration of surfactant is equal to the concentration at the capillary inlet, C_0, and the derivative of the concentration at this point is zero (a smooth transition to the constant concentration).

Let us introduce, for convenience, a new unknown function, $Z=C_0-C$, which satisfies the following equation and boundary conditions:

$$\left(1+F_{sl}\right)\frac{\partial Z}{\partial t}=D_{ef}\frac{\partial^2 Z}{\partial x^2}-\frac{dl}{dt}\frac{\partial Z}{\partial x},\quad 0<x<l(t).\tag{A2.1}$$

From Eq. (6.37): $C_m=C_*+Al(t)$, where $A=\dfrac{\rho g R}{2\alpha}$. The boundary condition (6.19) takes the following form using these equations:

$$D_{ef}\left(\frac{\partial Z}{\partial x}\right)_{x=l}=\left(F_{sl}-F_{sv}\right)\left(C_*+Al\right)\frac{dl}{dt}.\tag{A2.2}$$

Other boundary conditions are

$$Z(l-\delta)=0,\quad Z(l)=C_0-C_*-Al,\tag{A2.3}$$

$$\left(\frac{\partial Z}{\partial x}\right)_{x=l-\delta}=0.\tag{A2.4}$$

The integration of Eq. (A2.1) after simple manipulations using the boundary conditions (A2.3) and (A2.4) results in

$$\frac{d}{dt}\int_{l-\delta}^{l} Z dx = \frac{D_{ef}}{1+F_{sl}}\left(\frac{\partial Z}{\partial x}\right)_{x=l} + \frac{F_{sl}}{1+F_{sl}}(C_0 - C_* - Al)\frac{dl}{dt}. \tag{A2.5}$$

Now, we find a solution that satisfies the integral balance Eq. (A2.5) and the boundary conditions (A2.2 and A2.4). The simplest solution, which satisfied boundary conditions (A2.3) and (A2.4) is as follows:

$$Z = (C_0 - C_* - Al)\left(1 - \frac{l-x}{\delta}\right), \tag{A2.6}$$

where both $l(t)$ and $\delta(t)$ dependencies are to be determined. Substitution of this expression into boundary condition (A2.2) and Eq. (A2.5) gives, after some rearrangements, a system of two differential equations for the determination of two unknown dependencies, $l(t)$ and $\delta(t)$:

$$\begin{cases} \frac{d}{dt}\left[(C_0 - C_* - Al)\delta\right] = \frac{6D_{ef}}{1+F_{sl}}\frac{(C_0 - C_* - Al)}{\delta} + \frac{3F_{sl}}{1+F_{sl}}(C_0 - C_* - Al)\frac{dl}{dt}, \\ \frac{dl}{dt} = \frac{2D_{ef}}{\Delta F}\frac{(C_0 - C_* - Al)}{\delta(C_* + Al)} \end{cases} \tag{A2.7}$$

where $\Delta F = F_{sl} - F_{sv} > 0$. The following initial conditions should be satisfied:

$$l(0) = \delta(0) = 0. \tag{A2.8}$$

If the first equation in (A2.7) is divided by the second equation, then an equation for $\delta(l)$ dependence can be obtained. This equation can be solved, and the solution substituted into the second equation in system (A2.7), which then gives the following equation for $l(t)$ determination:

$$\frac{du}{dt} = \lambda \frac{(1-u)^2}{u(\omega+u)(2+\chi-u)}; \quad u(0) = 0, \tag{A2.9}$$

where

$$u = l/l_{max}, \quad l_{max} = \frac{2\alpha}{\rho g r}(C_0 - C_*), \quad \omega = \frac{C_*}{C_0 - C_*}, \quad \lambda = \frac{4D_{ef}(1+F_{sl})}{3\Delta F F_{sv}l_{max}^2}, \quad \chi = \frac{2\Delta F}{F_{sv}}\frac{C_0}{(C_0 - C_*)}. \tag{A2.10}$$

These definitions show that $\chi \gg 1$, hence, Eq. (A2.9) can be rewritten as

$$\frac{du}{dt} = \kappa \frac{(1-u)^2}{u(\omega+u)}; \quad \kappa = \frac{\lambda}{2+\chi}, \quad u(0) = 0. \tag{A2.11}$$

This equation can be easily solved, and the solution is as follows:

$$u + \frac{(\omega+1)u}{1-u} + (\omega+2)\ln(1-u) = \kappa t. \tag{A2.12}$$

If $u \ll 1$ (initial stage of the process), then from Eq. (A2.12), we conclude

$$u = \sqrt{\frac{2\kappa}{\omega}t}. \tag{A2.13}$$

At the final stage of the process, $1-u \ll 1$, Eq. (A2.12) gives

$$u = 1 - \frac{\omega+1}{\kappa t}.$$

(A2.14)

6.3 Capillary Imbibition of Surfactant Solutions in Porous Media and Thin Capillaries: The Partial Wetting Case

Let us now consider the imbibition of surfactant solutions into porous substrates, which pores are *partially* wetted by water. We shall see that this case is considerably different from that of *hydrophobic* porous media. We continue studying the imbibition in a *cylindrical* capillary, whose walls are partially wetted by water, recalling that such a capillary can be used to model a corresponding porous medium. At variance with the case of hydrophobic capillaries here water can penetrate the capillary even in the absence of surfactants on the moving meniscus. The presence of surfactant molecules on the moving meniscus lowers the contact angle and, hence, a higher capillary pressure builds behind the meniscus. Consequently, the imbibition rate increases with the increase of concentration of surfactant molecules on the moving meniscus.

In this section, as in the whole of Chapter 6, the influence of the Derjaguin's pressure is not considered because we know close to nothing about this phenomenon in the presence of surfactants. This is why, again, the Neuman-Young equation is used.

The moving meniscus covers fresh parts of the capillary walls, where surfactant molecules have not adsorbed yet. Thus, with the imbibition process, there is the simultaneous adsorption of surfactant molecules onto fresh parts of the capillary walls in a vicinity of the moving meniscus. The amount adsorbed is inversely proportional to the radius of the capillary, that is, the thinner the capillary, the higher the adsorption. On the other hand, the imbibition rate is lower in thinner capillaries (due to higher friction). This gives more time for diffusion to bring new surfactant molecules to cover the fresh part of the capillary walls. Thus, we have two competing, opposite trends. Indeed, if the capillary radius is smaller than some critical value then adsorption goes faster than the imbibition process and all surfactant molecules are adsorbed onto the capillary walls, leaving nothing for the meniscus where the concentration vanishes. In such circumstances, the imbibition rate of a surfactant solution becomes independent of the surfactant concentration in the feed solution and has a value equal to that of pure water [10]. These theoretical conclusions are in agreement with experimental observations [10].

The kinetics of the capillary imbibition of aqueous surfactant solutions into hydrophobic capillaries was investigated in Section 6.2. It was shown that the rate of imbibtion is controlled by the adsorption of the surfactant molecules in front of the moving meniscus on the bare hydrophobic surface of the capillary. This process results in a partial hydrophilization of the surface of the capillary in front of the moving meniscus and gives a possibility to aqueous surfactant solution to penetrate the initially hydrophobic capillary. That is, no surfactant molecules on the meniscus—no imbibition. In this section the imbibition of surfactant solutions into the porous substrates, which are partially wetted by water is considered. It is shown that situation in this case is considerably different from the case of hydrophobic porous media.

Theory

The imbibition of aqueous surfactant solutions into a single cylindrical capillary, whose walls are partially wetted by water, is considered in this section. A single capillary is used as a model of a porous medium.

The situation in this case is different from the case of hydrophobic capillaries (Section 6.2): Water can penetrate the capillary even in the absence of surfactant molecules on the moving meniscus. However, the presence of surfactant molecules on the moving meniscus results in a lower contact angle (as compared with pure water) and, hence, in a higher capillary pressure behind the meniscus. As a result, the imbibition rate increases with the concentration of surfactant molecules on the moving meniscus.

The moving meniscus covers fresh parts of the capillary walls, where surfactant molecules have not yet been adsorbed. This means that the imbibition process is accompanied by the simultaneous adsorption of surfactant molecules onto fresh parts of the capillary walls in the vicinity of a moving meniscus. The adsorption is reversely proportional to the radius of the capillary, that is, the thinner capillary, the higher adsorption. From the other hand, the rate of the imbibition is lower in thinner capillaries (higher friction). This gives more time to diffusion to bring new surfactant molecules and cover the fresh part of the capillary walls. That is, there are two competing, opposite trends. This means that if the capillary radius is smaller than some critical value, then adsorption goes faster than the imbibition process, all surfactant molecules are adsorbed on the capillary walls, that is, nothing is left for the meniscus and the concentration on the moving meniscus remains zero. This qualitative consideration shows that if the capillary radius is smaller than some critical value, then the rate of the imbibition of surfactant solutions remains independent of the surfactant concentration in the feed solution and is equal to that of pure water.

This qualitative conclusion is justified using theoretical consideration of the capillary imbibition of aqueous surfactant solutions into cylindrical capillaries, those walls are partially wet by water.

Let us consider that an imbibition of surfactant solution forms a reservoir with a fixed surfactant concentration, C_0, (feed solution) into a thin capillary with the radius $R \ll L$ (Figure 6.12), where L is the capillary length. The capillary walls are partially wet by pure water (at zero concentration of surfactant), that is,

$$\gamma(0)\cos\theta_a(0) = \gamma_{sv}(0) - \gamma_{sl}(0) > 0, \tag{6.41}$$

where $\gamma, \gamma_{sv}, \gamma_{sl}, \theta_a$ are the liquid–air, the solid substrate-vapor, the solid substrate–liquid interfacial tensions, and the advancing contact angle, respectively. All these values are concentration dependent.

Let C_m be the bulk concentration of surfactant behind the moving meniscus, then

$$\gamma(C_m)\cos\theta_a(C_m) = \gamma_{sv}(0) - \gamma_{sl}(C_m) > \gamma_{sv}(0) - \gamma_{sl}(0) \tag{6.42}$$
$$= \gamma(0)\cos\theta_a(0).$$

It is assumed that the imbibition process goes sufficiently fast and that the transfer of the surfactant molecules on the bare surface in front of the moving meniscus can be neglected because this process goes much slower (see Section 6.2). Hence, the solid–vapor interfacial tension, γ_{sv}, does not depend on the surfactant concentration and remains equal to its value at zero concentration. It is also taken into account in Eq. (6.42) that $\gamma_{sl}(C_m)$ is a decreasing function of the surfactant concentration. Equation (6.42) shows that $\Psi(C_m) = \gamma(C_m)\cos\theta_a(C_m)$ is an increasing function of the concentration, with the maximal value, $\Psi_{max} = \Psi(C_{CMC})$, reached at the CMC and the minimal value, $\Psi_{min} = \gamma_{sv}(0) - \gamma_{sl}(0)$, reached at the zero surfactant concentration.

Concentration below the CMC

Let the surfactant concentration at the capillary entrance be below the CMC, $C_0 < C_{CMC}$.

The transfer of surfactant molecules in the filled portion of the capillary is described by the convective diffusion equation as in Section 6.2

$$\frac{\partial C(t,x,r)}{\partial t} = D\frac{\partial^2 C(t,x,r)}{\partial x^2} + D\frac{1}{r}\frac{\partial}{\partial r}\left(r\frac{\partial C(t,x,r)}{\partial r}\right) - \frac{\partial}{\partial x}\left(v(r)C(t,x,r)\right),$$

where $C(t, x, r)$ is the local concentration of surfactant; D is the diffusion coefficient; t, x, r are time, axial and radial coordinates, respectively; $v(r)$ is the axial velocity distribution.

Integration of this equation over the radius from 0 to R, where R is the capillary radius results in

$$\frac{\partial}{\partial t}\left(\int_0^R rC(t,x,r)dr\right) = D\frac{\partial^2}{\partial x^2}\left(\int_0^R rC(t,x,r)dr\right) + R\left(D\frac{\partial C(t,x,r)}{\partial r}\right)_{r=R} - \frac{\partial}{\partial x}\left(\int_0^R rv(r)C(t,x,r)dr\right).$$

The second term in the right-hand side of this equation is equal to

$$-D\frac{\partial C(t,x,r)}{\partial r}\bigg|_{r=R} = \frac{\partial \Gamma}{\partial t} - D_{sl}\frac{\partial^2 \Gamma}{\partial x^2},$$

where D_{sl} is the surface diffusion coefficient over the filled portion of the capillary and Γ is the surface concentration. Hence, these two equations result in:

$$\frac{\partial}{\partial t}\left(\int_0^R rC(t,x,r)dr\right) = D\frac{\partial^2}{\partial x^2}\left(\int_0^R rC(t,x,r)dr\right) - R\left(\frac{\partial \Gamma}{\partial t} - D_{SL}\frac{\partial^2 \Gamma}{\partial x^2}\right) - \frac{\partial}{\partial x}\left(\int_0^R rv(r)C(t,x,r)dr\right).$$

A characteristic time scale of the equilibration of the surfactant concentration in a cross section of the capillary, $\tau \sim R^2/D \approx 10^{-3}$ s, if we use for estimations $R \sim 1$ μm and $D \sim 10^{-5}$ cm²/s. A characteristic time scale of the spontaneous capillary rise into partially hydrophilic capillaries is around 10 s (Figure 6.16), which is much bigger than 10^{-3}s. Hence, the surfactant concentration is constant in any cross section of the capillary and depends only on the position, x (Figure 6.13), that is, $C = C(t, x)$.

Taking this into account, this equation can be rewritten after both sides are divided by $R^2/2$ as

$$\frac{2}{R}\frac{\partial \Gamma}{\partial t} + \frac{\partial C}{\partial t} = D\frac{\partial^2 C}{\partial x^2} + D_{SL}\frac{2}{R}\frac{\partial^2 \Gamma}{\partial x^2} - v\frac{\partial C}{\partial x}, \quad 0 < x < l(t), \qquad (6.43)$$

where $v = \frac{2}{R^2}\int_0^R rv(r)dr$ is the average velocity (the same symbol is used for the average velocity as for the local one). The average velocity, v, is equal to the meniscus velocity, that is:

$$v = \frac{d\ell}{dt}, \qquad (6.44)$$

where $\ell(t)$ is the position of the moving meniscus. Substitution of Eq. (6.44) into Eq. (6.43) and neglecting the surface diffusion results in the following equation, which describes the concentration profile inside the filled portion of the capillary:

$$\frac{2}{R}\frac{\partial \Gamma}{\partial t} + \frac{\partial C}{\partial t} = D\frac{\partial^2 C}{\partial x^2} - \frac{dl}{dt}\frac{\partial C}{\partial x}, \quad 0 < x < l(t). \qquad (6.45)$$

Because of adsorption on the capillary wall, the concentration of the surfactant molecules on the moving meniscus, C_m, is lower than at the capillary entrance and should be determined in a self-consistent way.

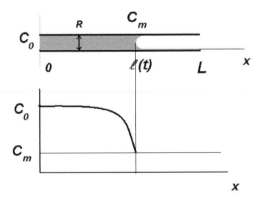

FIGURE 6.13 Imbibition of a surfactant solution in thin capillary. C_0, concentration at the capillary entrance; C_m, concentration on the moving meniscus; $\ell(t)$, position of the moving meniscus.

Solution of Eq. (6.45) should be subjected the following boundary conditions:

$$C(t,0) = C_0,$$ (6.46)

$$C(t,l(t)) = C_m,$$ (6.47)

and the last boundary condition on the moving front is

$$\frac{2\Gamma(C_m)}{R}\frac{dl}{dt} = -D\frac{\partial C}{\partial x}\bigg|_{x=l(t)}.$$ (6.48)

This condition expresses the conservation of mass at the moving meniscus and the three-phase contact line. To deduce this boundary condition, the following procedure should be undertaken: (i) The local coordinate should be introduced in a narrow region close to the moving meniscus; (ii) local transport equations should be integrated over this narrow region; and (iii) limits from the filled portion of the capillary should be calculated, which gives condition (6.48).

Because $R \ll L$, the liquid flow inside the capillary is the simple Poiseuille flow, that means

$$\frac{dl}{dt} = \frac{R^2}{8\eta}\frac{2\gamma(C_m)\cos\theta_a(C_m)}{R}\frac{1}{l} = \frac{R\gamma(C_m)\cos\theta_a(C_m)}{4\eta l},$$

where η is the dynamic viscosity, which is assumed to be independent of surfactant concentration. It is assumed that the surfactant concentration remains constant at the moving meniscus (similar to the case of hydrophobic capillaries in Section 6.2). Solution of this equation using initial condition $l(0) = 0$ gives

$$l(t) = K\sqrt{t}, \quad K = \sqrt{\frac{R\gamma(C_m)\cos\theta_a(C_m)}{2\eta}}.$$ (6.49)

Substitution of Eq. (6.49) into Eq. (6.45) results in

$$\frac{2}{r}\frac{\partial\Gamma}{\partial t} + \frac{\partial C}{\partial t} = D\frac{\partial^2 C}{\partial x^2} - \frac{K}{2\sqrt{t}}\frac{\partial C}{\partial x},$$ (6.50)

with boundary conditions (6.46) and (6.47) and

$$\frac{\Gamma(C_m)K}{K\sqrt{t}} = -D\frac{\partial C}{\partial x}\bigg|_{x=K\sqrt{t}},$$ (6.51)

which is the consequence of Eq. (6.48).

Let us introduce the following similarity coordinate, $\xi = \frac{x}{K\sqrt{t}}$, and the solution of Eq. (6.50) is assumed to depend on the similarity coordinate only. In this case, Eq. (6.50) and the boundary conditions (6.46), (6.47), and (6.51) take the following form:

$$\lambda^2 C'' = C'\left[1 - \left(\frac{2}{R}\Gamma'(C) + 1\right)\xi\right],$$ (6.52)

$$C(0) = C_0,$$ (6.53)

$$C(1) = C_m,$$ (6.54)

$$C'(1) = -\frac{\Gamma(C_m)\Psi(C_m)}{2D\eta},$$ (6.55)

where $\lambda^2 = \frac{2D}{K^2} = \frac{4D\eta}{R\Psi(C_m)}$ is a dimensionless parameter. In the case of aqueous surfactant solutions $D \sim 10^{-5} \text{cm}^2\text{s}$, $\eta \sim 10^{-2} P$, $\gamma \sim 10^2 \text{dyn/cm}$, that is, this parameter can be estimated as $\lambda^2 \sim \frac{10^{-9} \text{cm}}{R}$. We consider only capillaries with $R > 0.1 \ \mu\text{m} = 10^{-5}$ cm. In this case, $\lambda^2 < 10^{-4} \ll 1$.

The first consequence of Eqs. (6.52) through (6.55) is that the problem under consideration is really a similarity one.

To further simplify the mathematical treatment of the problem under consideration, the simplest adsorption isotherm is adopted as shown:

$$\Gamma(C) = \begin{cases} \Gamma_\infty, & C > 0 \\ 0, & C = 0 \end{cases}.$$

(6.56)

In this case, Eqs. (6.52) and (6.55) can be rewritten as follows if concentration on the moving meniscus is above zero (the case of the zero concentration is considered separately):

$$\lambda^2 C'' = C'(1 - \xi), \quad 0 < \xi < 1,$$

(6.57)

$$C'(1) = -\frac{\Gamma_\infty \Psi_{max}}{2D\eta},$$

(6.58)

with boundary conditions (6.53) and (6.54). Note, according to Eq. (6.42), Ψ in the case under consideration (according to Eq. 6.56) is independent of the concentration and equal to its maximal value, Ψ_{max}.

The smallness of the parameter λ^2 is utilized. Now this small parameter is a multiplier at the highest derivative in Eq. (6.57). This means that matching of asymptotic solutions can be used. Let us introduce the following local variable, z:

$$z = \frac{1 - \xi}{\lambda},$$

(6.59)

Using the new variable, the inner solution of Eq. (6.57) satisfies the following system:

$$C'' = -zC', \quad 0 < z < \infty$$

$$C'(0) = \frac{\Gamma_\infty \Psi_{max}\lambda}{2D\eta},$$

(6.60)

$$C(0) = C_m$$

and the boundary condition

$$C(\infty) = C_0.$$

(6.61)

Solution of the problem (6.60) is

$$C(z) = C_m + \frac{\Gamma_\infty \Psi_{max}\lambda}{2D\eta} \int_0^z \exp(-z^2/2) dz.$$

(6.62)

The following equation for the determination of the unknown concentration on the moving meniscus, C_m, yields using boundary condition (6.21) and solution (6.62):

$$C_0 = C_m + \frac{\Gamma_\infty \Psi_{max}\lambda\sqrt{\pi}}{2^{3/2} D\eta},$$

or

$$C_m = C_0 - \Gamma_\infty \sqrt{\frac{\pi \Psi_{max}}{2D\eta R}}. \tag{6.63}$$

The concentration on the moving meniscus should be positive, $C_m > 0$, that is, the following requirement should be satisfied:

$$C_0 > \Gamma_\infty \sqrt{\frac{\pi \Psi_{max}}{2D\eta R}}, \tag{6.64}$$

or

$$R > \frac{\Gamma_\infty^2 \pi \Psi_{max}}{2D\eta C_0^2}. \tag{6.65}$$

Let us introduce the following notation

$$R_{cr} = \frac{\Gamma_\infty^2 \pi \Psi_{max}}{2D\eta C_{CMC}^2}. \tag{6.66}$$

Two cases are considered: (i) $R < R_{cr}$ and (ii) $R > R_{cr}$.

In the first case, (i), condition (6.64) is violated at any concentrations in the feed solution between zero and the CMC. This means that the concentration of surfactant molecules on the moving meniscus is equal to zero at any concentration from this range. Hence, there are two regions behind the moving meniscus: The first region, close to the capillary entrance where the concentration is changing from C_0 in the feed solution to zero on the moving border between two regions; the first region is followed by the second region where concentration remains zero over the duration of the whole process. Let the moving border between these two regions be

$$\ell_1(t) = K\beta\sqrt{t}, \tag{6.67}$$

where $\beta < 1$ is a value to be determined.

In this case, the concentration on the meniscus remains zero and the meniscus moves "slowly" according to Eq. (6.49)

$$l(t) = \sqrt{\frac{R\Psi_{min}}{2\eta}} \sqrt{t}. \tag{6.68}$$

The concentration profile in the first region is a solution of the following problem (using the same as before similarity variable):

$$\lambda^2 C'' = C'(1-\xi), \quad 0 < \xi < \beta, \tag{6.69}$$

$$C'(\beta) = -\frac{\Gamma_\infty \Psi_{min}\beta}{2D\eta}, \tag{6.70}$$

$$C(0) = C_0, \tag{6.71}$$

$$C(\beta) = 0. \tag{6.72}$$

Condition $\lambda \ll 1$ is assumed to be satisfied, this means

$$\beta = 1 - \lambda\chi, \tag{6.73}$$

where χ is a new unknown value. Solution of the problem presented by Eqs. (6.69) and (6.73) gives the following equation for the determination of an unknown value χ

$$C_0 = \Gamma_\infty \sqrt{\frac{\pi \Psi_{\min}}{2D\eta R}} \exp(-\chi^2),$$

or

$$\chi = \left[\ln\left(\frac{\Gamma_\infty}{C_0} \sqrt{\frac{2\pi \Psi_{\min}}{D\eta R}} \right) \right]^{1/2}. \tag{6.74}$$

The main conclusion from this consideration is that the adsorption process in sufficiently thin capillaries consumes all surfactant and the imbibition is not influenced by the presence of surfactants in the feed solution at any concentration.

The second case, (ii), is when the capillary radius is bigger than the critical value determined by Eq. (6.66), that is, $R > R_{cr}$.

If the concentration in the feed solution is low enough, that is, condition (6.65) is violated, then the concentration on the moving meniscus, C_m, is equal to zero and the meniscus moves "slowly," according to Eq. (6.68). It is worth noting that the capillary radius is assumed to be bigger than in the previous case.

If, however, the concentration in the feed solution, C_0, is high enough, that is, condition (6.65) is satisfied, then the concentration of the surfactant molecules is different from zero on the moving meniscus and the imbibition process goes "faster" according to

$$l(t) = \sqrt{\frac{R\Psi_{\max}}{2\eta}} \sqrt{t}. \tag{6.75}$$

Hence, if the capillary radius is bigger than the critical value, then the whole concentration range in the feed solution can be subdivided into two parts: (i) the low concentration range, $C_0 < C_{cr} = \Gamma_\infty \sqrt{\frac{\pi \Psi_{\max}}{2D\eta R}}$, when the adsorption consumes all surfactant molecules and the concentration on the moving meniscus is equal to zero, and the meniscus moves "slowly" according to (6.68); and (ii) the high concentration range, $C_0 > C_{cr} = \Gamma_\infty \sqrt{\frac{\pi \Psi_{\max}}{2D\eta R}}$, when the adsorption does not consume all surfactant molecules and the concentration on the moving meniscus is different from zero, and the meniscus moves "faster" according to (6.75).

In Figure 6.14, ℓ^2/t against the concentration in the feed solution, C_0, is schematically plotted according to the presented simplified theoretical model.

If a more realistic case of adsorption isotherm (approximated by a Langmuir type isotherm),

$$\Gamma(c) = \begin{cases} \omega c, & c < c* \\ \\ \omega c* = \Gamma_\infty, \end{cases}$$

is adopted, then the dependency presented in Figure 6.14 changes continuously from lowest to the highest value instead of the stepwise change.

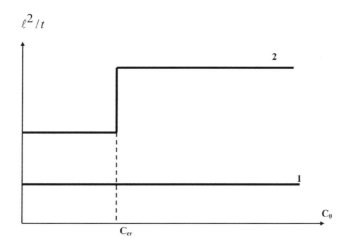

FIGURE 6.14 Dependency of permeability on surfactant concentration: theoretical predictions. (1) Thin capillary, the radius is below critical value; (2) Thick capillary, the radius is above critical.

Concentration above the CMC

If the concentration in the feed solution, C_0, is above the CMC, then after some short initial period inside the capillary, two zones form (similar to Section 6.2): in the first region, close to the capillary entrance, the concentration inside the capillary is higher than the CMC, this region is followed by the second region where concentration is below the CMC. The concentration is equal to the CMC at the border between these two regions. Consideration similar to that in Section 6.2 and presented in this section shows that the main conclusion remains unchanged in this case: There is a critical radius of the capillary below which the concentration on the moving meniscus remains zero at any concentration in the feed solution.

Concentration on the moving meniscus, C_m, is below the CMC. Two regions can be identified inside the capillary: the first region, from the capillary inlet to some position, $l_M(t)$, where concentration is above the CMC and the surfactant solution includes both micelles and individual surfactant molecules, the second region, from $l_M(t)$ to $l(t)$, where concentration is below the CMC and only individual surfactant molecules are transferred. Concentration is equal to the CMC at $x = l_M(t)$. Consideration in the second region, $l_M(t) < x < l(t)$, is like that at a concentration below the CMC. That is why only the transport in the first region is considered here.

Inside the first region, $0 < x < l_M(t)$, the concentration of free surfactant molecules is constant and equal to the CMC (see Section 6.2). Hence, the transfer is determined by the diffusion of micelles and convection of all molecules. The total concentration, $C = C_{mol} + C_M$, and C_{mol} remains constant and equal to the CMC, hence:

$$\frac{\partial C}{\partial t} = D_M \frac{\partial^2 C}{\partial x^2} - \frac{dl}{dt}\frac{\partial C}{\partial x}, \tag{6.76}$$

where D_M is the diffusion coefficient of micelles and $C = C_c + C_M$ is the total concentration of surfactant molecules. Adsorption onto membrane pores is determined by the concentration of free molecules, which is constant in the first region and so this is where adsorption occurs. This is why the diffusion of adsorbed molecules in the first region is omitted in Eq. (6.76). Transfer of surfactant molecules in the second region (micelles free region) is described by Eq. (6.45).

Boundary conditions on the moving boundary between the first and second regions, $l_M(t)$, are as follows:

$$D_M \left(\frac{\partial C}{\partial x} \right)_{x = l_M -} = D \left(\frac{\partial C}{\partial x} \right)_{x = l_M +} , \quad C(l_M, t) = C_c. \tag{6.77}$$

As before, we assume that

$$l(t) = K\sqrt{t}, \quad l_M(t) = K_M \sqrt{t}, \tag{6.78}$$

where K is given by Eq. (6.49), that is, K is expressed via an unknown concentration on the moving meniscus, C_m; and K_M is an unknown value to be determined.

Let a similarity variable be introduced in the same way as in the case of concentration below the CMC, that is, $\xi = x / K\sqrt{t}$ in Eqs. (6.45) and (6.76). Using boundary conditions (6.77), expressions (6.78) and boundary conditions (6.54) and (6.55) the system of two nonlinear algebraic equations can be deduced for the determination of two unknown values: concentration on the moving meniscus, C_m, and the position of the boundary, K_M. This system includes a small parameter $\varepsilon = \frac{D_M}{D} \ll 1$, which is the ration of diffusion coefficients of micelles and free surfactant molecules. Using this new small parameter, it is possible to show that solution of the mentioned system only slightly deviates from the solution in the previous case when concentration is below the CMC.

This means that the constant K and the expression for the critical radius (6.66) are only slightly different from the same values in the case of concentration below the CMC. Hence, the previous conclusion concerning the existence of the critical radius remains valid even at concentrations above the CMC, which is confirmed by our experimental data.

Experimental Part

Figure 6.15 shows the sample chamber for monitoring the permeability of the initially dry porous layers. The time evolution of the permeability front was monitored. The membrane (1) was fastened on a lifting up/down device (2) and placed in a thermostated and hermetically closed chamber (3), where 100% humidity (to prevent evaporation from the wetted part of the membrane) and fixed temperature (20°C ± 0.5°C) were maintained. To prevent temperature fluctuations, the chamber was made from brass and in the chamber walls several channels were drilled which were used for pumping of a thermostating liquid. The chamber was equipped with a fan. The temperature was monitored by a thermocouple. A box with water was used to maintain absolute humidity inside the chamber. On the bottom of the chamber, a small Petri dish (4) with different water solutions of SDS was placed.

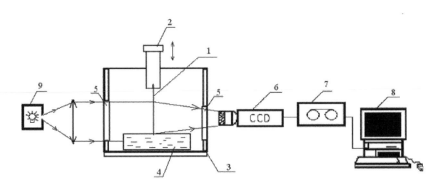

FIGURE 6.15 Schematic presentation of the experimental setup. (1) Membrane; (2) Lifting up/down device; (3) Thermostated chamber; (4) Petri dish with SDS solution; (5) Optical glass windows 5; (6) CCD camera; (7) Video tape-recorder; (8) PC; (9) Light source.

The chamber was equipped with optical glass windows (5) for observation of the imbibition front of the surfactant solution. A CCD camera (6) and VCR video recorder (7) were used for storing the sequences of the imbibition. Automatic processing of images was carried out on PC (8) using Scion Image image-processing software. The duration of each experimental run was in the range of 2.5–30 s. A discretization of time in the processing ranged from 0.04 to 2 s in different experimental runs; the size of a pixel in an image was 0.01 mm.

Experiments were carried out in the following order:

- The membrane was placed in the chamber and left in an atmosphere of 100% humidity for several minutes.
- The membrane was immersed vertically (0.1–0.2 cm) into a container with the SDS solution. After that, the position of the imbibition front was monitored over time.
- Several runs for each membrane type and each concentration of SDS solution were carried out.

A 1.5 × 3 cm rectangular membrane sample was used. Those porous samples were cut from Cellulose Nitrate membranes purchased from Millipore. Three different membranes with average pore size were used: 0.22, 0.45, and 3.0 μm were used. Each membrane sample was immersed 0.1–0.2 cm into a liquid container, and the position of the imbibition front was monitored over time.

Results and Discussions

It was checked that the gravity action could be neglected in all our experimental runs. A unidirectional flow of liquid inside the porous substrate took place. Using Darcy's law, we can conclude that

$$\ell^2(t) = K_p p_c t / \eta, \tag{6.79}$$

where $\ell(t)$ now is the position of the imbibition front inside the porous layer; K_p is the permeability of the porous membrane; and p_c is an effective capillary pressure inside the porous sample. The permeability of the porous layer and the capillary pressure enter as a product in Eq. (6.79), that is, as a single coefficient. Experiments were carried out to determine this coefficient and its dependency on the surfactant concentration if any. It was found that in all runs $d^2(t)/2$ proceeded along a straight line, whose slope gives us the $K_p p_c$ value.

According to our previous notations, $K_p p_c = K^2 \eta$.

Figure 6.16 is an example of the imbibition of a 0.1% SDS solution into a 0.22-μm nitrocellulose membrane. This dependency is in a good agreement with Eq. (6.79).

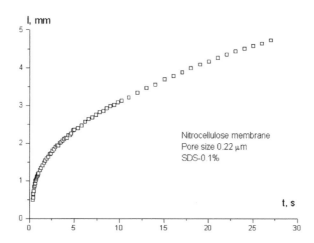

FIGURE 6.16 Example of the time evolution of the imbibition front. SDS concentration 0.1%, nitrocellulose membrane with average pore size 0.22 μm.

386   *Wetting and Spreading Dynamics*

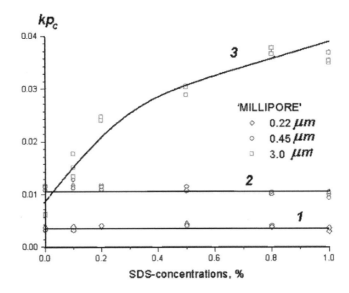

FIGURE 6.17 kp_c dependency on concentration of SDS solutions for nitrocellulose membranes with different average pore sizes. Remains constant in the case of membranes with average pore size both 0.22 μm (line 1) and 0.45 μm (line 2). Increases with surfactant concentration in the case of membrane with average pore size 3 μm (curve 3 is drawn simply as a guide).

Figure 6.16 presents the $K_p p_c$ dependency on the concentration of the SDS in the feed solution for three different membranes. The $K_p p_c$ in the case of membranes with 0.22 and 0.45-μm average pore size is independent of concentration. However, in the case of membranes with 3.0-μm average pore size, the $K_p p_c$ increases with the SDS concentration. This means that the critical radius, R_{cr}, is somewhere between 0.45 and 3.0 μm. Figure 6.17 confirms our conclusion concerning the existence of the critical pore radius below which permeability is independent of surfactant concentration.

6.4 Spreading of Surfactant Solutions over Hydrophobic Substrates

In this section, we study the spreading of aqueous surfactant solutions over *hydrophobic* surfaces. The spreading of surfactant solutions over hydrophobic surfaces is considered from both theoretical and experimental points of view. Water droplets do not wet a virgin solid hydrophobic substrate. It is shown in this section that the transfer of surfactant molecules from the water droplet onto the hydrophobic surface changes the wetting characteristics in front of the drop on the three-phase contact line (see Appendix 1). The surfactant molecules increase the solid–vapor interfacial tension and hydrophilize the initially hydrophobic solid substrate just in front of the spreading drop. It is proven in Appendix 1 that the transfer of surfactant molecules from the droplet on a bare hydrophobic substrate in front of the moving apparent three-phase contact line results in a decrease of the total free energy of the system. That is, this process proceeds spontaneously. This process causes water drops to spread over time. The time of evolution of the spreading of the aqueous surfactant solution droplets is predicted and compared with experimental observations. The assumption that surfactant transfer from the drop surface onto the solid hydrophobic substrate controls the rate of spreading is confirmed by our experimental observations.

Surfactant adsorption onto solid–liquid and liquid–vapor interfaces changes the corresponding interfacial tensions. Liquid motion caused by surface tension gradients on a liquid–vapor interface (the Marangoni effect) is the most investigated process (see Section 6.5). The phenomena produced by the presence of surfactant molecules on a solid–vapor interface have been less studied. In Section 6.2, the imbibition of surfactant solutions into thin quartz hydrophobic capillaries was investigated.

The spreading and imbibition of surfactant solutions into both hydrophobic and hydrophilic surfaces (Sections 6.2 and 6.3, respectively) revealed various new and intriguing phenomena. In the present section, we address the problem of aqueous surfactant solutions spreading over hydrophobic surfaces from both the theoretical and experimental points of view [11].

Theory

Let a small water drop be placed on a hydrophobic surface. If the drop is small enough, then the effect of gravity can be ignored. Accordingly, the radius of the drop base, $R(t)$, has to be smaller than the capillary length, a, and hence, $R(t) \leq a = \sqrt{\frac{\gamma}{\rho g}}$, where ρ, and γ are the liquid density and liquid–vapor interfacial tension, respectively; and g is the gravity acceleration.

First, let us consider expression (6.2) from Section 1.1 for the excess free energy, Φ, of the droplet on a solid substrate:

$$\Phi = \gamma_{lv} S + PV + \pi a^2 \left(\gamma_{sl} - \gamma_{sv} \right),$$

where S is the area of the liquid–air interface; $P = P_a - P_l$ is the excess pressure inside the liquid, P_a is the pressure in the ambient air, P_l is the pressure inside the liquid; and the last term in the right-hand side gives the difference between the energy of the part of the bare surface covered by the liquid drop as compared with the energy of the same solid surface without the droplet (Figure 6.17).

Detailed examination of this expression for excess free energy (Appendix 1) shows that the excess free energy decreases if (i) the liquid air interfacial tension, γ_{lv}, decreases; (ii) the liquid–solid interfacial tension, γ_{sl}, decreases; and (iii) the solid-vapor interfacial tension, γ_{sv}, increases.

Let us assume that in the absence of surfactant the aqueous drop does not spread over the hydrophobic substrate. However, if the water contains surfactants, then three transfer processes take place from the liquid onto all three interfaces: surfactant adsorption at (i) the inner liquid–solid interface, which results in a decrease of the liquid air interfacial tension, γ; (ii) the liquid–vapor interface, which results in a decrease of the solid–liquid interfacial tension, γ_{sl}; and (iii) transfer from the drop onto the solid–vapor interface just in front of the drop. As noted, all three processes result in a decrease of the excess free energy of the system. Adsorption processes (i) and (ii) result in a decrease of corresponding interfacial tensions, γ_{sv} and γ; however, the transfer of surfactant molecules onto the solid–vapor interface in front of the drop results in an increase of a local free energy, but the total free energy of the system decreases. That is, the surfactant molecule transfer (iii) goes via a relatively high potential barrier (Appendix 1) and, hence, goes considerably slower than adsorption processes (i) and (ii). Thus, they are "fast" processes as compared with the third process (iii).

The transfer of surfactant molecules onto the unwetted (hydrophobic) solid–vapor interface in front of the liquid was shown in Section 6.2 to play an important role in the wetting of hydrophobic surfaces.

All three surfactant transfer processes are favorable to spreading because they result in both an increase of the spreading power, $\gamma_{sv} - \gamma - \gamma_{sl}$, and, hence, a decrease of the contact angle. The transfer of surfactant molecules from the drop onto the solid–vapor interface in front of the drop results in an increase of local surface tension, γ_{sv}. Hence, it is the slowest process that will be the rate determining step. Let us define the initial contact angle by

$$\cos \theta^0 = \frac{\gamma_{sv}^0 - \gamma_{sl}^0}{\gamma^0} \geq \frac{\pi}{2}, \tag{6.80}$$

with $\gamma_{sv}^0, \gamma_{sl}^0, \gamma^0$ the initial values of solid–vapor, solid–liquid, and liquid–vapor interfaces. The term "initial" means that although the adsorption process on the liquid–vapor and solid–liquid interfaces has been completed (they are fast processes), the solid–vapor interface still has its initial condition as a bare hydrophobic interface without any surfactant adsorption. At this "initial" instant of time a process of slow transfer of surfactant molecules starts from the drop onto the solid–vapor interface. Let $\Gamma_s(t)$ be the instantaneous value of surfactant adsorption onto the solid surface in front of the liquid drop on the three-phase contact line (Figure 6.12) and let Γ_e be the equilibrium surface density of adsorbed surfactant molecules that will eventually be reached. The driving force of the process is proportional to the difference $\Gamma_s(t) - \Gamma_e$. Hence, the surfactant adsorption behavior with time is described by

$$\frac{d\Gamma_s(t)}{dt} = \alpha\left[\Gamma_e - \Gamma_s(t)\right], \tag{6.81}$$

with the initial condition that

$$\Gamma_s(0) = 0 \text{ at } t = 0 \tag{6.82}$$

and $\tau_s = 1/\alpha$ is the time scale of surfactant transfer from the drop onto the solid–liquid interface at three-phase contact line. Let us assume that

$$\alpha = \alpha_T \Xi \exp\left(\frac{-\Delta E}{kT}\right), \tag{6.83}$$

where the pre-factor α_T is determined by thermal fluctuations only; ΔE is an energy barrier for surfactant transfer from the liquid drop onto the solid–liquid interface; k and T are Boltzmann's constant and absolute temperature, respectively; and Ξ is a fraction of the drop liquid–vapor interface covered with surfactant molecules. Note, we assume that the position of surfactant molecules on a bare hydrophobic interface is "hydrophobic tails down."

We have assumed that the transfer of surfactant molecules onto the hydrophobic solid interface takes place only from the liquid–vapor interface. It is difficult to assess the contribution of surfactant molecule transfer along the solid surface from beneath the liquid. However, experimental data presented below in this section support this assumption (although they do not prove it decisively). The drop surface coverage, Ξ, is an increasing function of the bulk surfactant concentration inside the drop, whose maximum is reached close to the CMC. It follows from Eq. (6.83) that at low surfactant concentrations inside the drop, the characteristic time scale of the surfactant molecules transfer, τ_s, decreases with increased concentration, whereas above the CMC, τ_s levels off and reaches its lowest value. Both effects are observed in experimental results below (compare Figure 6.21).

As the drop adopts a position according to the triangle rule, the contact angle, $\theta(t)$, is determined by the relationship,

$$\cos\theta(t) = \frac{\gamma_{sv}(t) - \gamma_{sl}^0}{\gamma^0}, \tag{6.84}$$

where $\gamma_{sv}(t)$ is the instantaneous solid–vapor interfacial tension at the three-phase contact line. This dependency is determined by $\Gamma_s(t)$. According to Antonov's rule

$$\gamma_{sv}(t) = \gamma_{sv}^\infty \frac{\Gamma_s(t)}{\Gamma^\infty} + \gamma_{sv}^0\left(1 - \frac{\Gamma_s(t)}{\Gamma^\infty}\right), \tag{6.85}$$

where γ_{sv}^∞ is the solid–vapor interfacial tension of the surface completely covered by surfactants, and Γ^∞ is the total number of sites available for adsorption. Hence, the final value of the contact angle can be determined from Eq. (6.84) as

$$\cos\theta^\infty = \frac{\gamma_{sv}^\infty - \gamma_{sl}^0}{\gamma^0}. \tag{6.86}$$

According to Eq. (6.85), the solid–vapor interface in front of the spreading drop changes its wettability with time: from highly hydrophobic at the initial stage to partially hydrophilic at the final stage.

Using Eq. (6.85) in Eq. (6.84) yields the instantaneous contact angle,

$$\cos\theta(t) = \cos\theta^0 + \lambda\frac{\Gamma_s(t)}{\Gamma^\infty},\tag{6.87}$$

where $\cos\theta^0$ is given by Eq. (6.80), and the positive value of λ is $\lambda = \frac{\gamma_{sv}^\infty - \gamma_{sv}^0}{\gamma^0}$.

Equation (6.81) with initial condition (6.82) yields the solution

$$\Gamma_s(t) = \Gamma_e(1 - \exp(-\alpha t))\tag{6.88}$$

Using (6.88) in Eq. (6.87) gives the final expression for the instantaneous contact angle

$$\cos\theta(t) = \cos\theta^0 + \lambda\frac{\Gamma_e}{\Gamma^\infty}(1 - \exp(-\alpha t)).\tag{6.89}$$

A simple geometrical consideration shows that the radius of the wetted spot, $R(t)$, occupied by the drop can be expressed as

$$R(t) = \left(\frac{6V}{\pi}\right)^{1/3}\frac{1}{\left[\tan\dfrac{\theta}{2}\left(3 + \tan^2\dfrac{\theta}{2}\right)\right]^{1/3}},\tag{6.90}$$

where V is the drop volume, which is supposed to remain constant; and the contact angle, θ, is given by Eq. (6.89).

Equations (6.89), and (6.90) include two parameters: (i) the dimensionless parameter $\beta = \lambda\Gamma_e/\Gamma^\infty$ and the parameter α with the dimension of the inverse of time. It follows from Eq. (6.89) that $\beta = \cos\theta^\infty - \cos\theta^0 > 0$, where θ^∞ is the contact angle after the spreading process is completed. If both values of the contact angle, θ^0 and θ^∞, have been measured, then β can be determined. Hence, only α is used to fit the experimental data.

Let us introduce a dimensionless wetted area, $S(t)$, occupied by the spreading droplet as

$$S(t)\bigg/\left(\frac{6V}{\pi}\right)^{2/3} = \frac{R^2(t)}{\left(\dfrac{6V}{\pi}\right)^{2/3}} = \frac{1}{\tan^{2/3}\dfrac{\theta}{2}\left(3 + \tan^2\dfrac{\theta}{2}\right)^{2/3}} = \frac{1+X}{(1-X)^{1/3}(4+2X)^{2/3}},$$

where $X = \cos\theta$, ($\cos\theta^0 \le X \le \cos\theta^\infty$), which using Eq. (6.89) becomes $X = \cos\theta^\infty - \beta e^{-\alpha t}$.

It follows that both $dS(t)/dt$ and $dS(X)/dX$ are always positive, and the second time derivative is

$$\frac{d^2S(t)}{dt^2} = \frac{\alpha^2}{(1-X)(2+X)}(\cos\theta^\infty - X)\frac{dS(X)}{dX}\left[-2X^2 + 3X\cos\theta^\infty - (2 - \cos\theta^\infty)\right].\tag{6.91}$$

Two different situations are possible: (A) if the second derivative (6.91) changes sign, then the spreading rate can go via a maximum/minimum value, whereas (B) if the second derivative (6.91) is always negative, the spreading rate $dS(t)/dt$ decreases with time. Case A corresponds to a "high surfactant activity,"

$$\cos\theta^\infty \ge \frac{4}{9}\left(\sqrt{10} - 1\right) \approx 0.961,$$

whereas case B corresponds to "a low surfactant activity,"

$$\cos \theta^{\infty} < \frac{4}{9}\left(\sqrt{10}-1\right) \approx 0.961$$

Using Eq. (6.86) these two conditions can be rewritten as

$$\gamma_{sv}^{\infty} > 0.961\,\gamma^0 + \gamma_{sl}^0,$$

at "a high surfactant activity," and

$$\gamma_{sv}^{\infty} < 0.961\,\gamma^0 + \gamma_{sl}^0,$$

at a "low surfactant activity."

Under experimental conditions only the case B was observed and, hence, "low-surface activity" surfactants were used, whereas in [12] "high-surface activity" surfactants (superspreaders were apparently used.

Experiment: Materials

Two types of substrate were used, a PTFE film and a polyethylene (PE) wafer. This substrate was prepared by crushing granules of the PE composition (softening point is 100°C) between two clean glass plates under an applied pressure 1 kg/cm² at 110°C. Transparent wafers of circular section with a radius of 1.5 cm and thickness of 0.01 cm were used [11].

The cleaning procedure of PTFE and PE wafers was as follows: the surfaces were rinsed with alcohol and water, then the substrates were soaked in a sulfochromic acid from 30 to 60 min at 50°C. The surfaces then were washed with distilled water and dried with a strong jet of nitrogen. The equilibrium macroscopic contact angles obtained were 105° and 90° for PTFE and PE substrates, respectively (for pure water droplets).

Aqueous solutions of SDS from Merck with weight concentration from 0.005% to 1% (the CMC of the SDS solution is 0.2%) were used in spreading experiments.

Monitoring Method

The time evolution of the contact line was monitored by following VCR images of drops. The images were stored using a CCD camera and a recorder at 25 frames per second. The automatic processing of images was carried out using the image-processor Optimas. In the case of spreading over PE, the initial contact angle of the drop was less than 90° and the drop was observed from above. The observed wetting area of the drop was monitored, and the wetting radius was calculated. For the PTFE substrate, we used a side view of the drop and, hence, the wetting radius was determined directly.

Simple mass balance estimations show that time variation of surfactant concentration inside the spreading drops can be neglected in our experiments (although it may become important in experiments of longer duration).

Water adsorption in front of the spreading drops was neglected because of the hydrophobic nature of the substrates used (in contrast to spreading of evaporating drops; see Chapter 7). The Peclet number in all experiments was so small that surfactant diffusion in front of the drop was neglected.

Results and Discussion

According to observations in [11], all spreading drops were of spherical shape; no disturbances or instabilities were detected. Immediately after deposition of the drops on the hydrophobic substrate, the drops had a contact angle that differed only slightly from the contact angle of pure water on the same substrate. After a very short initial time, the drops reached a position which was referred to as "the initial" position.

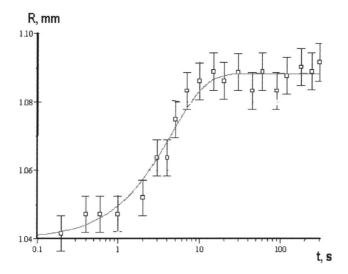

FIGURE 6.18 Time evolution of the spreading of a water drop (aqueous solution $C = 0.05\%$ SDS; 2.5 ± 0.2 μL drop volume) over PTFE wafer. Error bars correspond to the error limits of video evaluation of images (pixel size).

After that, for 1–15 s, depending on the SDS concentration, drops remained at the initial position. Then drops started to spread until a final value of the contact angle was reached and the spreading process was completed.

In Figure 6.18, the evolution of the spreading radius of a drop over PTFE film at 0.05% SDS concentration is plotted. In Figure 6.19 a similar plot is given for 0.1% SDS concentration. In both figures, the solid lines correspond to the fitting of the experimental data by Eqs. (6.89) and (6.90), with $\tau_s = 1/\alpha$ used as a fitting parameter.

Figure 6.20 shows that qualitatively the τ_s dependency agrees with the theoretical prediction and tends to support our assumption concerning the mechanism of surfactant molecule transfer onto the hydrophobic surface in front of the drop.

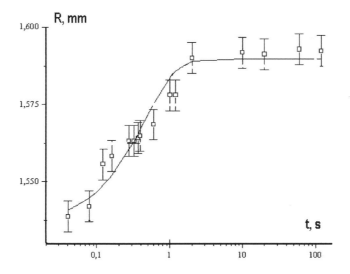

FIGURE 6.19 Time evolution of the spreading of a water drop (aqueous solution $C = 0.1\%$ SDS; 2.5 ± 0.2 μL volume) over PTFE wafer. Error bars correspond to the error limits of video evaluation of images (pixel size).

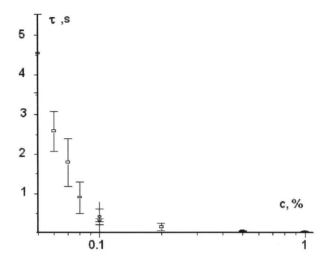

FIGURE 6.20 Fitted dependency of τ_s on surfactant concentration inside the drop (spreading over PTFE wafer). Error bars correspond to the experimental points scattering in different runs; squares are average values.

Similar results were obtained for the spreading over the PE substrate for concentrations below the CMC. However, in this case the spreading behavior of drops at concentrations above the CMC is drastically different with increasing SDS concentration (Figure 6.21). The rate of spreading was increased so much that at 1% concentration, the power law with the exponent 0.1 (solid line) fit the experimental data reasonably well. This clearly shows a transition to a different mechanism of spreading, which can be understood in the following way. In our previous considerations the influence of the viscous flow inside the drop was completely ignored. This means that it was assumed $\tau_s \gg \tau_{vis}$, where τ_{vis} is a time scale of viscous relaxation. In this case τ_s decreases so considerably that the mentioned inequality becomes invalid and instead, now $\tau_s \sim \tau_{vis}$ becomes valid.

Unfortunately, we know close to nothing about the Derjaguin's isotherms in the case of aqueous surfactant solutions on any interfaces, include hydrophobic ones. However, this observation (Figure 6.21) shows that, in this particular case at high surfactant concentration, the Derjaguin's isotherm approaches the one for complete wetting.

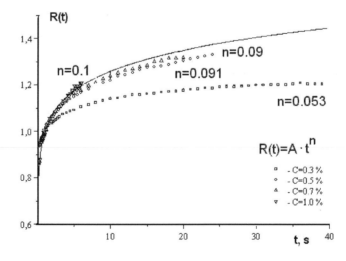

FIGURE 6.21 Spreading of SDS solution over polyethylene substrate, concentration above the CMC. Dependency of spreading radius on time $R(t) = A \cdot t^n$, where A and n are fitted parameters. The case $n = 0.1$ is shown by a solid line.

6.5 Spreading of Insoluble Surfactants over Thin Viscous Liquid Layers

In this section, we present results of the theoretical and experimental study of the spreading of an insoluble surfactant over a thin liquid layer. Initial concentrations of surfactant above and below the CMC have been considered. If the concentration is above the CMC, two distinct stages of spreading are found (i) a fast one that relates to micelle dissolution; and (ii) a slower one that relates to when the surfactant concentration is below the CMC. In the second stage, the formation of a dry spot in the center of the film is observed. A similarity solution of the corresponding equations for spreading gives good agreement with the experimental observations [13].

When a drop of a surfactant solution is deposited on a clean liquid–air interface, tangential stresses on the liquid surface develop. They are caused by a nonuniform distribution of surfactant concentration, Γ, over a part of the liquid surface covered by the surfactant molecules, thus, leading to surface stresses and flow (the Marangoni effect) [14]:

$$\eta \frac{\partial u(r,h)}{\partial z} = \frac{d\gamma}{d\Gamma} \frac{\partial \Gamma}{\partial r}, \qquad (6.92)$$

where η and u are the liquid dynamic shear viscosity and tangential velocity on the liquid surface located at height h, respectively; (r, z) are radial and vertical coordinates; and $\gamma(\Gamma)$ is the liquid–air interfacial tension whose linear dependency on surfactant surface concentration we assume below in this section. The surface tension gradient-driven flow induced by the Marangoni effect moves surfactant along the surface, and a dramatic spreading process takes place. Then the liquid–air interface deviates from an initially flat position to accommodate with the normal stress also occurring during motion.

We restrict consideration in this section to insoluble surfactants. Note, that although a surfactant may be soluble, there are cases such that nonsolubility conditions can be used during a certain initial period in the spreading process. Let us consider two characteristic time scales associated with surfactant transfer: (i) τ_d, which accounts for the transfer from the liquid–air interface to the bulk, and (ii) τ_a, which accounts for the transfer from the bulk back to the interface. In both cases, these characteristic time scales depend on an energy difference between corresponding states. Let us consider an aqueous surfactant solution with both a hydrophilic head and a hydrophobic tail. Let E_{hl}, E_{ta}, and E_{tl} be the energies (in kT units) of head-water, tail-air, and tail-water interactions. Using these notations, the energy of a molecule in an adsorbed state at the interface is $E_{ad} = E_{ta} + E_{hl}$, whereas for the same molecule in the bulk it is $E_b = E_{tl} + E_{hl}$. Then $\tau_d = \tau \cdot \tau \exp\left(E_b - E_{ad}\right) = \tau \cdot \tau \exp\left\{E_{tl} - E_{ta}\right\}$ and $\tau_a = \tau \cdot \tau \exp\left(E_{ad} - E_b\right) = \tau \cdot \tau \exp\left\{-(E_{tl} - E_{ta})\right\}$, where τ is determined by thermal fluctuations. Generally, the tail–water interaction energy is considerably higher than that of the tail–air interaction ($E_{tl} \approx nE_{tl}^1$, where E_{tl}^1 is an interaction energy per hydrophobic unit and n is a number of those units in each tail). Consequently, $\frac{\tau_d}{\tau_a} = \exp\left\{2(E_{tl} - E_{ta})\right\} \gg 1$ and transfer from the interface to bulk is a much slower process than the reverse one. If the duration of a spreading experiment is shorter than τ_d, then during that experiment, the surfactant can be considered as insoluble. Otherwise, if $t > \tau_d$, the solubility of the surfactant in the liquid must be taken into account. In this case, surfactant transfer to the bulk liquid tends to make concentration uniform both in the bulk and at the interface, and the result is a substantial decrease of the surfactant influence of that type (Marangoni flow).

Usually surface diffusion can be neglected as compared with convective transfer. Indeed, from Eq. (6.92), we have as a characteristic scale of surface velocity: $u_* \approx \frac{\gamma_* h_*}{\eta L_*}$, where γ_*, h_*, L_* are characteristic scales of interfacial tension, initial film thickness, and length in a tangential direction, respectively. The diffusion process over the liquid surface scales as: $D_s \frac{\Gamma_*}{L_*^2}$, where D_s, Γ_* are the surface diffusion coefficient and a characteristic scale of surfactant concentration on the surface, respectively. Then, the ratio of diffusion to convective flux can be estimated as $1/P_e = D_s \eta / \gamma_* H_* \sim 10^{-8} \ll 1$, for $D_s \sim 10^{-5}$ cm^2/s (6.114) and $\gamma_* \approx 10^2$ dyn/cm. Here P_e is the mean Peclet number. This estimation shows that surface diffusion can be neglected everywhere except for a small diffusion layer, which is disregarded.

We consider two different cases: (i) when concentration in a droplet of surfactant solution, which is placed in the center of a liquid film is above the CMC, and (ii) when such concentration is below the CMC. In the first case, the spreading process involves two stages: (i) The faster one is when the surfactant concentration is determined by the dissolution of micelles. This stage yields the maximum attainable surfactant concentration in the film center and it is independent of time. The duration of that stage is fixed by the initial number of micelles in the drop; (ii) the second slower stage takes place when the surfactant concentration changes in the film center, but the total mass of surfactant remains constant. In both cases a similarity solution provides a power law predicting the position of the moving front $r_f(t)$ as time proceeds.

Theory and Relation to Experiment

The motion of a thin liquid layer with initial thickness H_* is considered under the action of an insoluble surfactant on its open surface. For simplicity, we assume that surface tension varies linearly with surface concentration of surfactant,

$$\gamma(\Gamma) = \gamma_* - \alpha\Gamma, \qquad \text{at} \qquad 0 < \Gamma < \Gamma_m, \tag{6.93}$$

where γ_* is the interfacial tension of the pure water–air interface, and Γ_m corresponds to the maximum attainable surface concentration (in equilibrium with a micelle solution).

We use here dimensionless parameters and variables and the same symbols as for dimensional quantities. The subscript $*$ is used to mark initial or characteristic values.

We further assume that $\varepsilon = H_*/L_* \ll 1$ and, hence, neglect the nonlinear part of the interface curvature. A dimensionless Bond number, $\rho g H_*^2/\alpha\Gamma_*$, accounts for the ratio of the gravitational force to the Marangoni forcing, where ρ is the liquid density, and g is the gravity acceleration. In our experiments, a water film with $H_* \approx 0.1$ mm is used, the Bond number is about $5 \times 10^{-3} \ll 1$ and, hence, the gravity action can be safely neglected.

Under these conditions, the evolution equations for mass balance for water and surfactant concentration on the surface are as follows:

$$\frac{\partial h}{\partial t} + \frac{1}{r}\frac{\partial}{\partial r}\int_0^h r u\, dz = 0, \tag{6.94}$$

$$\frac{\partial \Gamma}{\partial t} + \frac{1}{r}\frac{\partial}{\partial r}\left(r u(t,h)\Gamma\right) = 0, \tag{6.95}$$

where $h(t, r)$ is the film thickness at time t; r is the radial coordinate and $\Gamma(t, r)$ the surfactant concentration on the surface. Equations (6.94) and (6.95), after performing the integration, become (see Appendix 2 for details)

$$\frac{\partial h}{\partial t} = -\frac{1}{r}\frac{\partial}{\partial r}\left\{r\left[\frac{\gamma h^3}{3\eta}\frac{\partial}{\partial r}\left(\frac{\gamma}{r}\frac{\partial}{\partial r}\left(r\frac{\partial h}{\partial r}\right)\right) - \frac{\alpha h^2}{2\eta}\frac{\partial \Gamma}{\partial r}\right]\right\}, \tag{6.96}$$

$$\frac{\partial \Gamma}{\partial t} = -\frac{1}{r}\frac{\partial}{\partial r}\left\{r\Gamma\left[\frac{\gamma h^2}{2\eta}\frac{\partial}{\partial r}\left(\frac{\gamma}{r}\frac{\partial}{\partial r}\left(r\frac{\partial h}{\partial r}\right)\right) - \frac{\alpha h}{\eta}\frac{\partial \Gamma}{\partial r}\right]\right\}, \tag{6.97}$$

that are to be solved subject to the following boundary conditions:

$$\frac{\partial h}{\partial r} = \frac{\partial^3 h}{\partial r^3} = 0, \quad \text{at} \qquad r = 0, \tag{6.98}$$

$$h \to 1, \qquad \text{at} \quad r \to \infty, \tag{6.99}$$

$$\frac{\partial \Gamma}{\partial r} = 0, \qquad \text{at} \qquad r = 0, \tag{6.100}$$

$$\Gamma \to 0, \qquad \text{at} \qquad r \to \infty. \tag{6.101}$$

Let us introduce the following dimensionless variables and values:

$$h \to \frac{h}{H_*}, \qquad r \to \frac{r}{L_*}, \qquad \Gamma \to \frac{\Gamma}{\Gamma_*}, \qquad t \to \frac{t}{t_*},$$

$$t_* = \frac{\eta L_*^2}{H_* \alpha \Gamma_*}, \qquad \beta = \varepsilon^2 \frac{\gamma_*}{\alpha \Gamma_*} \ll 1$$

The time scale t_* deserves a comment. The capillary number for the spreading process under consideration is very small: $Ca = \frac{\eta U}{\gamma_*} \sim 10^{-2}\, 10^{-1}/10^2 = 10^{-5} \ll 1$. On the other hand, $Ca = \frac{\eta U_*}{\gamma_*} = \frac{\eta}{\gamma_*} \frac{L_*}{\tau_*}$, hence a new time scale can be introduced as $\tau_* = \frac{\eta L_*}{\gamma_* Ca}$, where τ_*, U_* are, respectively, the time scale which actually governs the spreading process and the characteristic velocity scale. Using these estimations, we conclude: $\frac{t_*}{\tau_*} = \frac{\gamma_* Ca}{\alpha \Gamma_* \varepsilon} = \delta \approx 10^{-3} \ll 1$ for $\varepsilon \approx 10^{-2}$, as is the case in the experiments presented. If we now introduce the dimensionless time, $\tau = \frac{t}{\tau_*}$, then in Eqs. (6.94) and (6.95), we have $\delta \frac{\partial}{\partial \tau} = \frac{\partial}{\partial t}$. We show that the time evolution of the position of the moving film front is $r_f(t) \sim t^{0.5 \div 0.25} = (\frac{\tau}{\delta})^{0.5 \div 0.25}$. If we take into account the initial value, ℓ, of $r_f(t)$, this dependence takes the form $r_f(\tau) \approx (\frac{\tau}{\delta} + \ell)^{0.5 \div 0.25}$, where $\ell \sim 1$ represents the contribution of the initial condition. According to our choice $\tau \sim 1$, then, $\frac{\tau}{\delta} \gg \ell$, and thus $r_f(t) \sim (\frac{\tau}{\delta})^{0.5 \div 0.25} = t^{0.5 \sqrt{0.25}}$. Multiplying Eq. (6.96) by r, we conclude after integration that

$$\int_0^\infty r(h-1)dr = 0, \tag{6.102}$$

which reflects conservation law for the liquid.

In our experiments, a droplet of surfactant solution with concentration above the CMC was placed in the center of a water film. Experimental observations show that two distinct stages of spreading take place: (i) a first faster stage, and (ii) a second slower stage. During the first stage, there is dissociation of micelles, and hence, surface concentration of single molecules is kept constant during that stage and $\Gamma(t,0) = \Gamma_m$. Choosing $\Gamma_* = \Gamma_m$ as a characteristic scale for surfactant concentration, we have in dimensionless form the following boundary condition during the first stage:

$$\Gamma(t,0) = 1. \tag{6.103}$$

The first stage lasts until all micelles are dissolved. The duration t_{1*} of that stage is considered now. Past t_{1*} a second stage starts. During the second stage the total mass of surfactant remains constant, hence the following boundary conditions apply

$$\frac{\partial \Gamma}{\partial r} = 0, \qquad \text{at} \qquad r = 0, \tag{6.104}$$

and

$$\int_0^\infty r\Gamma dr = 1, \tag{6.105}$$

where the characteristic scale for Γ is now selected as $\frac{Q_*}{2\pi L_*^2}$, with Q_* being the total amount of surfactant molecules in the droplet.

The first spreading stage: The spreading process in this case is described by Eqs. (6.46) through (6.97) with boundary conditions (6.98) through (6.101) and (6.103). According to Appendix 3, the influence of capillary forces can be neglected for $t \gg \beta$, and Eqs. (6.96) and (6.97) become

$$\frac{\partial h}{\partial t} = \frac{1}{r} \frac{\partial}{\partial r} \left\{ r \left[\frac{h^2}{2} \frac{\partial \Gamma}{\partial r} \right] \right\},$$ (6.106)

$$\frac{\partial \Gamma}{\partial t} = \frac{1}{r} \frac{\partial}{\partial r} \left\{ r \Gamma \left[h \frac{\partial \Gamma}{\partial r} \right] \right\}.$$ (6.107)

Equations (6.106) and (6.107) cannot satisfy the boundary conditions at $r \to \infty$ and a shock-like spreading front forms (for the derivation of these conditions see Appendix 3).

In our case, $h_+ = 1$, $\Gamma_+ = 0$, hence, $\frac{\partial \Gamma_+}{\partial r} = 0$. Then using conditions (A4.3) and (A4.4) we get

$$\dot{r}_f (h_- - 1) = -\frac{1}{2} h_-^2 \frac{\partial \Gamma_-}{\partial r},$$ (6.108)

$$\Gamma_- (\dot{r}_f + h_- \frac{\partial \Gamma_-}{\partial r}) = 0.$$ (6.109)

Equation (6.109) implies two conditions:

$$\Gamma_- = 0,$$ (6.110)

$$\dot{r}_f = -h_- \frac{\partial \Gamma_-}{\partial r}.$$ (6.111)

The matching of asymptotic expansions at the moving shock front (see Appendix 5) shows that both conditions (6.110) and (6.111) must be satisfied.

Let us now introduce a new variable, $\xi = \frac{r}{t^{1/2}}$. Then we have

$$h(t,r) = f(\xi), \qquad \Gamma(t,r) = \varphi(\xi),$$ (6.112)

where $0 < r < r_f(t) = v\sqrt{t}$; and the constant v is determined now. In condition (6.102), the upper limit of integration must be replaced by v, hence

$$\int_0^v \xi f(\xi) d\xi = \frac{v^2}{2},$$ (6.113)

which is compatible with definitions (Eq. 6.112). Equations (6.106) and (6.107), after using (6.112), become

$$-\xi f'(\xi) = \frac{1}{\xi} \left(\xi f^2(\xi) \varphi'(\xi) \right)',$$ (6.114)

$$-\frac{1}{2} \xi \varphi'(\xi) = \frac{1}{\xi} \left(\xi f(\xi) \varphi(\xi) \varphi'(\xi) \right)',$$ (6.115)

with the corresponding boundary conditions which follow from (6.108), (6.110), and (6.111). We have

$$v \left[f(v) - 1 \right] = -f^2(v) \varphi'(v),$$
$$\varphi(v) = 0,$$
$$\frac{v}{2} = -f(v) \varphi'(v).$$

These boundary conditions after simple transformations become:

$$\varphi(v) = 0,$$
$$f(v) = 2,$$
$$\varphi'(v) = -\frac{v}{4}.$$

(6.116)

If we again change variables using

$$\mu = \frac{\xi}{v}, \quad f = \Psi(\mu), \quad \varphi = v^2 G(\mu),$$

(6.117)

then the new functions, Ψ and G, satisfy the same conditions (6.114) and (6.115) with the variable μ, where $0 < \mu < 1$, and the following boundary conditions at $\mu = 1$

$$\Psi(1) = 2, \ G(1) = 0 \text{ and } G'(1) = -\frac{1}{4}.$$

(6.118)

Then the problem does not depend on the unknown parameter v.

Calculated dependencies of the dimensionless film thickness $\Psi(\mu)$ and surface concentration $G(\mu)$ are presented in Figure 6.22, which shows that a substantial depression is formed in the film center. During the second stage, the depression becomes a dry spot right in the middle of the film.

To determine the unknown constant, ν, let us consider the total mass of surfactant, $Q(t)$, during the first spreading stage. We have

$$Q(t) = 2\pi \int\limits_0^\infty r\Gamma(t,r)dr = 2\pi L_*^2 \Gamma_* t \int\limits_0^v \xi\varphi(\xi)d\xi,$$

or

$$\int\limits_0^v \xi\varphi(\xi)d\xi = \frac{Q(t)}{2\pi L_*^2 \Gamma_* t}.$$

(6.119)

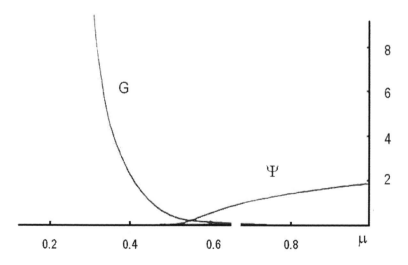

FIGURE 6.22 Theoretical predictions for dimensionless profile of the film, $\Psi(\mu)$, and surfactant concentration on the surface of the film, $G(\mu)$, during the first spreading stage [calculated according to equations and boundary conditions (114), (115), and (118)].

The left-hand side of Eq. (6.119) does not depend on time t, hence, the same is true for the right-hand side. Let us denote by q a constant value to be experimentally determined from the duration of the first stage. Then, Eq. (6.119) takes the form

$$\int_0^v \xi \varphi(\xi) d\xi = q. \tag{6.120}$$

On the other hand, according to definition (6.117), $v^4 \int_0^1 \eta G(\eta) d\eta = q$, that may fix the v value. Unfortunately, the integral $\int_0^1 \eta G(\eta) d\eta$ diverges due to the singularity of the dependence $G(\mu)$ at $\mu = 0$. Hence, we can only write that $v^4 \sim q$, or

$$v \approx q^{1/4}. \tag{6.121}$$

Let t_{1*} be a dimensional time scale of the duration of the first spreading stage; t_{1*}/t_* is the corresponding dimensionless time. If we choose $t_* = t_{1*}$ then using the definition of the time scale, we get the corresponding value of the tangential length scale $L_* = \left(\dfrac{t_{1*} H_* \alpha \Gamma_*}{\eta} \right)^{1/2}$.

From Eq. (6.119) we conclude

$$q = \frac{Q_*}{2\pi L_*^2 \Gamma_*}, \tag{6.122}$$

where Q_* is the total amount of surfactant initially placed on the film surface, which is supposed to be known, $Q_* = V_* C_*$, where V_* and C_* are the droplet volume and surfactant concentration in the droplet respectively.

Unfortunately, the derived similarity solution cannot satisfy boundary condition (6.103) at the origin as the concentration dependence on radial coordinate diverges in the vicinity of the origin. Thus, it is more convenient to redefine a characteristic scale of surfactant surface concentration from Eq. (6.122) using the condition $q = 1$ in Eq. (6.122) (this choice gives the same characteristic scale during both stages of the spreading process). The v value is still undetermined, but we shall show later how to determine it. At time t_{1*} the second stage of the spreading process starts.

The second spreading stage. In this stage the film profile and the surfactant concentration obey the same system of Eqs. (6.96) and (6.97) with boundary conditions (6.98) through (6.101), (6.104) and (6.105).

Let us introduce $\xi = \dfrac{r}{t^{1/4}}$, then the solution of Eqs. (6.96) and (6.97) is $h(t,r) = f(\xi)$, $\Gamma(t,r) = \dfrac{\varphi(\xi)}{t^{1/2}}$, where two unknown functions $f(\xi)$ and $\varphi(\xi)$ obey the following system of equations

$$\xi^2 f'(\xi) = \left\{ \xi \left[\frac{4\beta f^3(\xi)}{3} \left(\frac{1}{\xi} (\xi f'(\xi))' \right)' - 2f^2(\xi)\varphi'(\xi) \right] \right\}', \tag{6.123}$$

$$(\xi^2 \varphi(\xi))' = \left\{ \xi \varphi(\xi) \left[2\beta f^2(\xi) \left(\frac{1}{\xi} (\xi f'(\xi))' \right)' - 4f(\xi)\varphi'(\xi) \right] \right\}'. \tag{6.124}$$

Equation (6.124) can be integrated using condition (6.101), which gives $\varphi(\xi) \to 0$, at $\xi \to \infty$, that is:

$$\xi \varphi(\xi) \left\{ \xi - \left[2\beta f^2(\xi) \left(\frac{1}{\xi} (\xi f'(\xi))' \right)' - 4f(\xi)\varphi'(\xi) \right] \right\} = 0.$$

Thus, either

$$\varphi(\xi) = 0, \tag{6.125}$$

or

$$\varphi'(\xi) = -\frac{\xi}{4f(\xi)} + \frac{\beta}{2}f(\xi)\left(\frac{1}{\xi}\left(\xi f'(\xi)\right)'\right)'. \tag{6.126}$$

In the first case, from Eq. (6.123) we conclude that

$$\xi^2 f'(\xi) = \frac{4\beta}{3}\left\{\xi f^3(\xi)\left(\frac{1}{\xi}\left(\xi f'(\xi)\right)'\right)'\right\}', \tag{6.127}$$

which is valid at the periphery of the spreading part of the film.

Equation (6.127) describes decaying capillary waves on the film surface. Indeed, if we introduce a new local variable, $\zeta = (\xi - \lambda)/\chi$, with $\chi = (\frac{4\beta}{3\lambda})^{1/3}$ near the moving edge λ (to be defined) we get from Eq. (6.127)

$$f'''(\zeta) = \frac{f(\zeta) - 1}{f^3(\zeta)}$$

The asymptotic behavior of this equation, according to Chapter 3 (Section 3.5), yields $f(\zeta) \approx 1 + e^{-\frac{\zeta}{2}}(A_1 \cos\frac{\sqrt{3}\zeta}{2} + A_2 \sin\frac{\sqrt{3}\zeta}{2})$, at $\zeta \to \infty$, which describes decaying capillary waves ahead the advancing front.

In the opposite case, when Eq. (6.126) is valid, we obtain from Eq. (6.123) using Eq. (6.126) that

$$\xi^2 f'(\xi) = \left\{\xi\left[\frac{\beta f^3(\xi)}{3}\left(\frac{1}{\xi}\left(\xi f'(\xi)\right)'\right)' + \frac{f(\xi)\xi}{2}\right]\right\}'. \tag{6.128}$$

The value of λ is determined as a point where $\varphi(\lambda) = 0$, or from Eq. (6.126) and condition (6.105):

$$2 = \int_0^\lambda \left(\frac{\xi}{4f(\xi)} - \frac{\beta}{2}f(\xi)\left(\frac{1}{\xi}\left(\xi f'(\xi)\right)'\right)'\right)d\xi, \tag{6.129}$$

The solution of Eq. (6.126) is

$$\varphi(\xi) = \int_\xi^\lambda \left[\frac{\xi}{4f(\xi)} - \frac{\beta}{2}f(\xi)\left(\frac{1}{\xi}\left(\xi f'(\xi)\right)'\right)'\right]d\xi,$$

where the integration constant is vanishing in accordance with Eq. (6.129).

It is easy to find a solution of the problem under consideration in the zeroth approximation by setting $\beta = 0$ in Eqs. (6.126), (6.127), (6.129) (see Appendix 5). It is also possible to find an exact expression for λ using a zeroth-order solution from Appendix 5 and Eqs. (6.126), (6.127), and (6.129), but it is not our aim.

In conclusion of the theoretical consideration, let us summarize the results obtained. For the dimensional radius of the moving axisymmetric front, we have for the first stage of spreading

$$r_f = \nu L_* \left(\frac{t}{t_1 *} \right)^{1/2} \text{ (cm)}, \ t \leq t_1 *, \tag{6.130}$$

with ν still undetermined. For the second stage, we find

$$r_f = \lambda L_* \left(\frac{t}{t_1 *} \right)^{1/4} \text{ (cm)}, \quad \lambda \approx 2^{5/4}, \quad t \geq t_1 *, \tag{6.131}$$

where $t_{1*}(s)$ is the duration of the first stage of spreading,

$$L_* = \left(\frac{H_* \alpha \Gamma_* t_1 *}{\mu} \right)^{1/2}, \qquad \Gamma_* = \frac{Q_*}{2\pi L_*^2}$$

Because the front position must be the same at time t_{1*} according to both Eqs. (6.130) and (6.131), then $\nu = \lambda$ and Eq. (6.130) becomes

$$r_f = \lambda L_* \left(\frac{t}{t_1 *} \right)^{1/2} \text{ (cm)}, \quad t \leq t_1 *. \tag{6.132}$$

Our theory predicts that the layer thickness decreases in the center with a vanishing value at the origin, which is the dry spot. When comparing with experiments we must take into account both the finite precision in the measurements of the film thickness and the possibility of evaporation during the experiment that may lead to discrepancy between our theory predictions and the experimental results. We can measure the film thickness only for values higher than a certain thickness h_*, hence we can consider that thicknesses below h_* constitute the dry spot. Using a zeroth-order solution, Eq. (6.3), we obtain that

$$2H_* \left(\frac{r_d}{\lambda t^{1/4}} \right)^2 = h_*, \quad \text{or} \quad r_d(t) = \lambda \left(\frac{h_*}{2H_*} \right)^{1/2} t^{1/4}, \tag{6.133}$$

where $r_d(t)$ is the dry spot radius, whose motion according to this equation obeys the same power law as the front of the film.

Moreover, evaporation has a more pronounced influence at smaller thicknesses ($h < h_*$) and that influence progressively increases with time. Hence, evaporation in the liquid film near the center of the layer helps further film thinning during the second stage.

Experimental Results

Observations of the spreading of a surfactant on a thin layer of liquid have been performed using aqueous solutions of SDS at a concentration $c = 20$ g/L above CMS (the critical micellar concentration of the SDS solution is 4 g/L). The thin layer of liquid is prepared by coating the bottom of a 20-cm diameter borosilicate glass Petri dish with 10 mL of distilled water. The resulting thickness is then $H = 0.32 \pm 0.01$ mm. If carefully washed, this layer does not de-wet during the time of the experiment. With a syringe, a drop of the surfactant solution, volume 3 µL, is put on the surface of this water layer. When touching the water surface, the surfactant spreads on it and this motion is followed using a small amount of talcum powder as a marker, and a 25 Hz-video camera to record it.

The spreading of surfactants makes the water flow away from the initial location of the drop thus creating a depression where only a thin film of liquid subsists. The periphery of this depression, that is, the liquid front has a sharp increase in thickness. If the layer is horizontal and the drop is carefully placed, there is no preferred direction, and the edge is circular, but some modulation may appear in the experiment after a few seconds. Note that the surfactant occupies more surface area than the depressed zone because the talcum powder is pushed ahead of it. The dependence on time of the radius of the surfactant patch is given in Figure 6.23a using a log-log plot. The two successive stages earlier described in our theory are clearly seen in Figure 6.23a. First, the short period when the surfactant spreads following the power-law, $r_f(t) \sim t^{0.60 \pm 0.15}$. The exponent is not determined with a high precision because this first stage is too fast (about 0.1 s), which is only three times time resolution (0.04 s). The value 0.60 ± 0.15 agrees well with the earlier given theoretical prediction 0.5, Eq. (6.132).

At the end of the first stage the motion abruptly slows and the moving front follows a different power law, $r_f(t) \sim t^{0.17 \pm 0.02}$. The new exponent is smaller than the theoretical prediction of 0.25 given by Eq. (6.131). In Figure 6.23b, the radius of the shallow region, that is, the dry spot radius, is plotted for the same six experimental runs. We observe the power law dependence, $r_d(t) \sim t^{0.25 \pm 0.05}$, which is the value predicted by the theory Eq. (6.133). According to our measurements, the ratio $\frac{r_d(t)}{r_f(t)}$ is about 1/3 although it slowly changes with time during the second stage. Thus, from Eq. (6.133) we conclude that $h_* \approx 0.07$ cm and it is the lowest thickness, which can be detected by our experimental method.

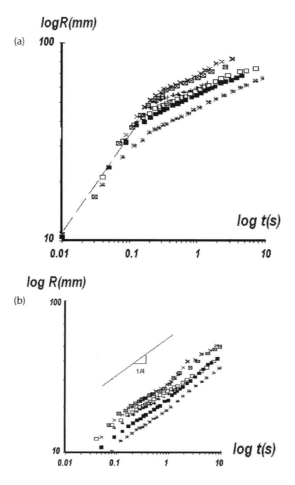

FIGURE 6.23 (a) Radius of the spreading front versus time. Points correspond to six different experimental runs. (b) Radius of the dry spot versus time. Points correspond to six different experimental runs.

In conclusion, the radius of the dry spot moves with the speed predicted by the theory while the front edge of the moving part of the layer proceeds slower than theoretically predicted during the second spreading stage. The discrepancy may be due to one or both of the following reasons:

1. Gravity action, which is much more pronounced at the front higher edge of the layer than at the lower edge (dry spot): Although the Bond number is very low in our experiments, flow reversal onset cannot be ruled out during the second stage;

2. Our assumption that the surfactant used is insoluble during the whole duration of the experiment may not be fully correct. Unfortunately, we have been unable to estimate the desorption time, τ_d. If that time is reached during our experiments, dissolution of surfactant in the bulk may be significant at the higher front edge.

APPENDIX 3

Derivation of Governing Equations for Time Evolution of Both Film Thickness and Surfactant Surface Concentration

At $\varepsilon \ll 1$, as noted in Chapter 3, the Navier-Stokes equations reduces to

$$\frac{dp}{dr} = \eta \frac{\partial^2 u}{\partial z^2}, \tag{A3.1}$$

$$p = p(r), \tag{A3.2}$$

$$\frac{1}{r}\frac{\partial}{\partial r}(ru) + \frac{\partial v}{\partial z} = 0, \tag{A3.3}$$

where $p(r)$, $v(r, z)$, and z are pressure, axial velocity, and axial coordinate, respectively. The following boundary condition must be satisfied:

$$u(r,0) = v(r,0) = 0, \tag{A3.4}$$

$$\eta \frac{\partial u(r,h)}{\partial z} = \frac{\partial \gamma}{\partial r} = \frac{d\gamma}{d\Gamma}\frac{\partial \Gamma}{\partial r} = -\alpha \frac{\partial \Gamma}{\partial r}, \tag{A3.5}$$

$$p = p_a - \frac{\gamma}{r}\frac{\partial}{\partial r}\left(r\frac{\partial h}{\partial r}\right), \tag{A3.6}$$

where p_a is the pressure in the ambient air. The solution of Eq. (A3.1) with boundary conditions (A3.4) and (A3.5) gives

$$u = -\frac{1}{\eta}\frac{\partial}{\partial r}\left(\frac{\gamma}{r}\frac{\partial}{\partial r}\left(r\frac{\partial h}{\partial r}\right)\right)\left(\frac{z^2}{2} - zh\right) - \frac{\alpha}{\eta}\frac{\partial \Gamma}{\partial r}z. \tag{A3.7}$$

To derive Eq. (6.94), we make use of a condition at the free liquid–air interface

$$\frac{\partial h}{\partial t} + u(r,h)\frac{\partial h}{\partial r} = v(r,h). \tag{A3.8}$$

After integration of Eq. (A3.3) over z, from 0 to h, and using the result of integration into Eq. (A3.8), we get Eq. (6.94). Substituting Eq. (A3.7) into Eqs. (6.94) and (6.95) yields Eqs. (6.96) and (6.97).

APPENDIX 4

Influence of Capillary Forces during the Initial Stage of Spreading

Let us estimate the influence of capillary forces during a short initial stage of spreading when it is significant. Later, it can be neglected everywhere except for the thin boundary layers that we consider negligible. A solution of governing Eqs. (6.96) and (6.97) is assumed in the following form:

$$h = f\left(\frac{r}{t^{\omega}}\right), \qquad \Gamma = \varphi\left(\frac{r}{t^{\omega}}\right), \tag{A4.1}$$

where f and φ are two new unknown functions; the constant ω is to be determined. Substitution of expressions (A4.1) into Eqs. (6.96) and (6.97) results in

$$\frac{\omega}{t}\xi f'(\xi) = \frac{1}{\xi}\left\{\xi\left[\frac{\beta f^3(\xi)}{3t^{4\omega}}\left(\frac{\gamma}{\xi}(\xi f'(\xi))'\right)' - \frac{f^2(\xi)}{2t^{2\omega}}\varphi'(\xi)\right]\right\}', \tag{A4.2}$$

$$\frac{\omega}{t}\xi \varphi'(\xi) = \frac{1}{\xi}\left\{\xi\varphi(\xi)\left[\frac{\beta f^2(\xi)}{2t^{4\omega}}\left(\frac{\gamma}{\xi}(\xi f'(\xi))'\right)' - \frac{f(\xi)}{t^{2\omega}}\varphi'(\xi)\right]\right\}', \tag{A4.3}$$

where $\xi = r/t^{\omega}$. There are two ways to choose ω:

1. If we require $t = t^{4\omega}$ or $\omega = 1/4$, then Eqs. (A4.1) and (A4.2) become

$$\frac{1}{4}\xi f'(\xi) = \frac{1}{\xi}\left\{\xi\left[\frac{\beta f^3(\xi)}{3}\left(\frac{\gamma}{\xi}(\xi f'(\xi))'\right)' - t^{1/2}\frac{f^2(\xi)}{2}\varphi'(\xi)\right]\right\}', \tag{A4.4}$$

$$\frac{1}{4}\xi \varphi'(\xi) = \frac{1}{\xi}\left\{\xi\varphi(\xi)\left[\frac{\beta f^2(\xi)}{2}\left(\frac{\gamma}{\xi}(\xi f'(\xi))'\right)' - t^{1/2}f(\xi)\varphi'(\xi)\right]\right\}'. \tag{A4.5}$$

Equations (A4.4) and (A4.5) show that the influence of the surfactant grows with time.

2. If we require $t = t^{2\omega}$ or $\omega = 1/2$, then Eqs. (A4.2) and (A4.3) become

$$\frac{1}{2}\xi f'(\xi) = \frac{1}{\xi}\left\{\xi\left[\frac{\beta f^3(\xi)}{3t}\left(\frac{\gamma}{\xi}(\xi f'(\xi))'\right)' - \frac{f^2(\xi)}{2}\varphi'(\xi)\right]\right\}', \tag{A4.6}$$

$$\frac{1}{2}\xi \varphi'(\xi) = \frac{1}{\xi}\left\{\xi\varphi(\xi)\left[\frac{\beta f^2(\xi)}{2t}\left(\frac{\gamma}{\xi}(\xi f'(\xi))'\right)' - f(\xi)\varphi'(\xi)\right]\right\}'. \tag{A4.7}$$

Equations (A4.6) and (A4.7) show that the influence of capillary forces decays with time. According to Eq. (A4.1), the spreading law is $r_f(t) \approx t^{1/4}$ during the period when the capillary force influence is dominant, and $r_f(t) \approx t^{1/2}$ during the period when the influence of the surfactants is dominant. Here $r_f(t)$ marks the location of the spreading front. It follows from Eqs. (A4.4) and (A4.5) that capillary force influence is dominant if $\beta \gg t^{1/2}$ or:

$$t \ll \beta^2. \tag{A4.8}$$

In the same way from Eqs. (A4.6) and (A4.7) we find that the capillary force influence is negligible and the influence of the surfactant is dominant if

$$t \gg \beta, \tag{A4.9}$$

Thus, the capillary force influence is significant only during a very short time interval $t \ll \beta$. As $\beta \ll 1$, and we can safely consider just the asymptotic behavior when the surfactant influence is dominant and condition (A4.9) is satisfied. Note that by omitting the capillary force action, we neglect the highest derivatives in Eqs. (6.96) and (6.97), hence, thin layers arise where the capillary force action is of the same order magnitude as the surfactant action.

APPENDIX 5
Derivation of Boundary Condition at the Moving Shock Front

Multiplication of Eqs. (6.106) and (6.107) by r and integration over r from r_1 to r_2, where $r_1 < r_f(t) < r_2$, where r_1, r_2 are some constant values, yields

$$\frac{d}{dt} \int_{r_1}^{r_2} rh(t,r)dr = \frac{r_2 h_2^2}{2} \frac{\partial \Gamma_2}{\partial r} - \frac{r_1 h_1^2}{2} \frac{\partial \Gamma_1}{\partial r}, \tag{A5.1}$$

$$\frac{d}{dt} \int_{r_1}^{r_2} r\Gamma(t,r)dr = r_2 \Gamma_2 h_2 \frac{\partial \Gamma_2}{\partial r} - r_1 \Gamma_1 h_1 \frac{\partial \Gamma_1}{\partial r}, \tag{A5.2}$$

Where we use the following abbreviation: $f_i = f(r_i)$. The left-hand side of Eqs. (A5.1) and (A5.2) can be transformed in the following way:

$$\frac{d}{dt} \int_{r_1}^{r_2} rf(t,r)dr = \frac{d}{dt} \left(\int_{r_1}^{r_f(t)} rf(t,r)dr + \int_{r_f(t)}^{r_2} rf(t,r)dr \right)$$

$$= \dot{r}_f r_f f_- + \int_{r_1}^{r_f(t)} r\frac{\partial f(t,r)}{\partial t}dr - \dot{r}_f r_f f_+ + \int_{r_f(t)}^{r_2} r\frac{\partial f(t,r)}{\partial t}dr,$$

where $f_\pm = f(t, r_f \pm)$. If now we consider limits r_1 tends to r_f from below ($\uparrow$) and r_2 tends to r_f from above ($\downarrow$) when both integrals in the left-hand side of this equation vanish. Then, from Eqs. (A5.1) and (A5.2) using the same limits, $r_1 \uparrow r_f, r_2 \downarrow r_f$, we conclude

$$\dot{r}_f(h_- - h_+) = \frac{1}{2}(h_+^2 \frac{\partial \Gamma_+}{\partial r} - h_-^2 \frac{\partial \Gamma_-}{\partial r}), \tag{A5.3}$$

$$\dot{r}_f(\Gamma_- - \Gamma_+) = \Gamma_+ h_+ \frac{\partial \Gamma_+}{\partial r} - \Gamma_- h_- \frac{\partial \Gamma_-}{\partial r}, \tag{A5.4}$$

which are the required boundary conditions at the shock front.

APPENDIX 6

Matching of Asymptotic Solutions at the Moving Shock Front

Let us introduce a new local variable, $\varsigma = \frac{\xi - v}{\chi(t)}$, where $\chi(t) << 1$ is a new unknown length scale to be determined. Neglecting the second curvature in Eqs. (6.96) and (6.97) and introducing unknown functions in the following form $\Gamma = \chi(t)\Phi(\xi), \quad h = F(\varsigma)$, we get

$$\chi(t) = \left(\frac{\beta}{t}\right)^{1/3}. \tag{A6.1}$$

It follows from this equation that $\chi(t) \to 0$, at $t \to \infty$. Unknown functions $\Phi(\varsigma)$ and $F(\varsigma)$ obey the following equations:

$$vF' = \left(\frac{2F^3F'''}{3} - F^2\Phi'\right)', \tag{A6.2}$$

$$v\Phi' = \left(\Phi F^2 F''' - 2F\Phi\Phi'\right)'. \tag{A6.3}$$

After integration of Eqs. (A6.2) and (A6.3) with boundary conditions $F \to 1, \quad \Phi \to 0$, at $\varsigma \to \infty$, we get

$$v\left(F - 1\right) = \frac{2}{3}F^3F''' - F^2\Phi', \tag{A6.4}$$

$$\Phi\left(v - F^2F''' + 2F\Phi'\right) = 0. \tag{A6.5}$$

Thus, either

$$\Phi = 0. \tag{A6.6}$$

and

$$F''' = \frac{3v}{2}\frac{F - 1}{F^3}, \tag{A6.7}$$

or

$$F''' = \frac{3v(F - 2)}{F^3}, \tag{A6.8}$$

and

$$\Phi' = -\frac{v(3 - F)}{F^2} \tag{A6.9}$$

From Eqs. (A6.8) and (A6.9) we conclude that

$$F \to 2, \quad \Phi' \to -\frac{v}{4}, \quad \text{at} \quad \varsigma \to -\infty. \tag{A6.10}$$

Equation (A6.10) shows that the boundary conditions (6.116) at the shock front are the only possible conditions that can be matched with the inner solution. From the above derivation, in this Appendix we conclude that in the boundary layer, $\Gamma = \left(\frac{\beta}{t}\right)^{1/3}\Phi(\varsigma) \sim \beta^{1/3}$, hence, vanishes from the point of view of the outer solution. Thus, both conditions (6.110) and (6.111) must be satisfied at the shock front.

APPENDIX 7

Solution of the Governing Equations for the Second Stage of Spreading

Putting $\beta = 0$ in Eqs. (6.126) and (6.127), we get

$$f'(\xi) = \frac{2f(\xi)}{\xi}, \qquad \varphi'(\xi) = -\frac{\xi}{4f(\xi)}, \tag{A7.1}$$

with boundary conditions that follow from Appendix 5 Eqs. (A7.3) and (A7.4)

$$f(\lambda) = 2, \qquad \varphi(\lambda) = 0, \qquad \varphi'(\lambda) = -\frac{\lambda}{8}. \tag{A7.2}$$

Then the solution of Eq. (A7.1) is

$$f(\xi) = 2\left(\frac{\xi}{\lambda}\right)^2, \qquad \varphi(\xi) = -\frac{\lambda^2}{8} \ln \frac{\xi}{\lambda}. \tag{A7.3}$$

The substitution of $f(\xi)$ in Eq. (6.128) gives

$$\lambda = 2^{5/4}. \tag{A7.4}$$

Note that the solution (A7.3) satisfies Eqs. (6.126) through (6.128) at arbitrary β.

6.6 Spreading of Aqueous Droplets Induced by Overturning of Amphiphilic Molecules or Their Fragments in the Surface Layer of an Initially Hydrophobic Substrate

The highlighted text below can be omitted at the first reading.

Let us study in this final section of Chapter 6 the spontaneous spreading of a drop of a *polar* liquid over a solid when the *amphiphilic* molecules (or their *amphiphilic* fragments) of the substrate surface layer may overturn, creating hydrophilic parts on the surface. Such a situation may occur, for example, during the contact of an aqueous drop with the surface of a polymer whose macromolecules have hydrophilic lateral groups capable of rotating around the backbone or during the wetting of polymers containing surface-active additives or Langmuir-Blodgett films composed of amphiphilic molecules. It is shown below in this section that drop spreading is possible only if there is lateral side interaction between neighboring amphiphilic molecules (or groups). This interaction leads to *tangential* transfer of the "overturned state" to some distance ahead of the advancing three-phase contact line making it partially hydrophilic. This kind of "self-organization" of its surface layer, lowers the interfacial free energy due to the emergence (or adsorption) of *polar* groups at the surface. Depending on the structure of the polymer (macro)molecules, the physical state of the polymer, and other factors, such a rearrangement may involve various forms of molecular motion, from the reorientation of individual amphiphilic lateral groups (when their rotation around the backbone is allowed) to the diffusion of the macromolecules as a whole. Each of these processes may occur on different time scales. If the characteristic time scale of self-organization is comparable with the time of measurement, this process may be observed while studying the time evolution of the contact angle (and size) of an aqueous drop over a polymer.

In a number of studies have been published (e.g., [15–18] and relevant references therein) demonstrating that interaction of a polymer with a polar liquid (first of all, with water and aqueous solutions) may result in the spontaneous rearrangement ("self-organization") of its surface layer, providing the minimization of the interfacial free energy due to the emergence (or adsorption) of polar groups at the surface. Depending on the structure of the polymer macromolecules, the physical state of the polymer, and other factors, such a rearrangement may involve various forms of molecular motion, from the reorientation of individual amphiphilic side groups (when their rotation around the backbone is allowed) to the diffusion of macromolecules as a whole. Each of these processes may occur on different time scales. If the characteristic time scale of self-organization is comparable with the time of measurement, this process may be observed while studying the contact angle (and size) of an aqueous droplet on time of contact with polymer [19]. Such a "prolonged" spreading of aqueous droplets occurs also during the study of the wettability of model systems such as nonpolar polymers containing low-molecular-weight amphiphilic additives (i.e., surfactants) capable of adsorption and/or reorientation at the polymer–liquid interface [19].

Langmuir was the first to mention the possible reorientation ("overturning") of amphiphilic molecules in contact with a polar liquid [20]. Later, experimental data were obtained that demonstrated the occurrence of such process in mono layers and Langmuir-Blodgett films composed of long-chain fatty acids [21] and at the surface of mixtures of such acids with paraffin [22] during contact with an aqueous droplet, resulting in gradual droplet spreading. Preliminary the rate of spreading was analyzed in terms of formal chemical kinetics [21,22], and the mechanism of the process (the dynamic situation in the vicinity of the three-phase contact line) was not considered at all. Recently, such a situation has also been observed for the spreading of a droplet over the surfaces of polymers, when the amphiphilic groups of their macromolecules are capable of reorientation by rotating around the macromolecule backbone.

The aim of this section is to analyze the mechanism of the spontaneous spreading of a droplet of a polar liquid induced by the overturning of amphiphilic molecules (or their fragments) in the surface layer of a solid substrate and to a develop a quantitative theory describing this process.

Theory, Derivation of Basic Equations

Let us consider the spreading of an aqueous droplet over a solid substrate ignoring the evaporation of the liquid, that is, assuming that the droplet volume remains constant during spreading. It is assumed also that the substrate is smooth, horizontal, and (which is important for the further discussion) contains in the surface layer rotationally mobile amphiphilic molecules (or amphiphilic fragments of molecules; for brevity, hereafter referred to as only molecules) capable of overturning in the plane perpendicular to the surface and incapable of lateral motions in the plane of the substrate. Hence, each of the amphiphilic molecules may be in one of two states: (i) non-overturned ("normal"), that is, when the hydrophilic head group of the molecule is oriented downward into the substrate while the hydrophobic "tail" is directed upward into the second phase (air or water) in contact with the substrate; and (ii) overturned state, that is, in the opposite (as compared to the previous case) orientation of hydrophilic and hydrophobic moieties of the molecule (Figure 6.24). Let N_∞ and N be the total number of amphiphilic molecules per surface area of the substrate capable of overturning and the number of already overturned molecules, respectively; then $p = N / N_\infty$ is the probability of finding an overturned molecule. Let us consider the process of transition of a system to the equilibrium state during the contact of the substrate surface with the air, provided that initially the system was somehow disturbed from an equilibrium state.

Let us consider for simplicity a linear chain of amphiphilic molecules, because a switch to the two-dimensional case is self-evident. Let $p_i(t)$ be the probability of the occurrence of molecule i at the instant t in the overturned state; then the same probability at the time $t + \Delta t$ will be equal to

$$p_i(t + \Delta t) = p_i(t) + P_V \Delta t - P_{SV} \Delta t + P_d \Delta t - P_b \Delta t, \tag{6.134}$$

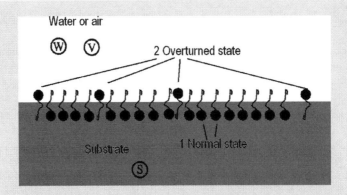

FIGURE 6.24 Polymeric substrate containing rotationally mobile amphiphilic chains in contact with air or water v, air; w, water; and s, polymeric substrate. (1) Amphiphilic chains in the "normal" state; (2) amphiphilic chains in the overturned state.

where P with subscripts are the probabilities of direct and reverse overturning per unit time: P_v is the overturning probability caused by the interaction with the air of the amphiphilic molecule that has not been overturned previously; P_{sv} is the overturning probability caused by the interaction with underlying molecules the molecule that was overturned previously; and P_d and P_b are the probabilities corresponding to the direct and the reverse overturning of a molecule due to its interaction with neighboring molecules. Only the interactions with the nearest neighbors are accounted for, that is, with the molecules having numbers $i + 1$ and $i - 1$.

All the probabilities in Eq. (6.134) may be written in the following form:

$$P_v = \alpha_v \left(1 - p_i\right), \tag{6.135}$$

$$P_{sv} = \alpha_{sv} p_i, \tag{6.136}$$

$$P_d = \beta \left[p_{i-1}\left(1 - p_i\right) + p_{i+1}\left(1 - p_i\right) \right], \tag{6.137}$$

$$P_b = \beta \left[\left(1 - p_{i-1}\right) p_i + \left(1 - p_{i+1}\right) p_i \right], \tag{6.138}$$

where constants α_v, α_{sv} and β entering the definitions of probabilities in Eqs. (6.135) and (6.138) have the dimensionality of reciprocal time and are determined using the energy of molecular interaction with surrounding phases in overturned and non-overturned states:

$$\alpha_v = \alpha \exp\left[-\Phi_v\right], \tag{6.139}$$

$$\alpha_{sv} = \alpha \exp\left[\Phi_v\right], \tag{6.140}$$

$$\beta = \alpha \left[\exp\left(\chi\right) - 1\right], \tag{6.141}$$

where $\chi = z \frac{2U_{th} - U_{tt}}{RT}$, $\Phi_v = \frac{U_{ts} + U_{hv} - U_{tv} - U_{hs}}{RT}$; subscripts t, h, v, and s correspond to the hydrophobic tail of a molecule, to its hydrophilic head group, to the air, and to the underlying molecules of the substrate, respectively; α denotes the corresponding values in the absence of any interaction, that is, it is determined only by thermal fluctuations; and z is the number of neighboring molecules, that is, $z = 2$ for a linear chain and $z = 4$ for the two dimensional case. It follows from definitions (6.139) and (6.140) that $\alpha_v \ll \alpha_{sv}$.

Let us consider Eq. (6.141) in more detail. If expressions (6.139) and (6.140) involve both overturning due to the thermal fluctuations and those caused by the interaction with the surrounding media, then, in contrast to these expressions, Eq. (6.141) involves only the overturning related to the interactions between the neighbors. Hence, random overturning caused by thermal fluctuations should not be taken into account, because they do not result in the transfer of the overturned state. This is why the unity is subtracted in Eq. (6.141).

Substituting expressions (6.135) through (6.138) into Eq. (6.134) yields:

$$\frac{p_i(t+\Delta t)-p_i(t)}{\Delta t} = \alpha_v\left[1-p_i(t)\right]-\alpha_{sv}p_i(t)$$

$$+\beta\left\{\left[p_{i-1}(t)+p_{i+1}(t)\right]\left[1-p_i(t)\right]-\left[2-p_{i-1}(t)-p_{i+1(t)}\right]p_i(t)\right\}$$

Taking the limit in the last expression at $\Delta t \to 0$ results in

$$\frac{dp_i(t)}{dt} = \alpha_v\left[1-p_i(t)\right]-\alpha_{sv}p_i(t)+\beta\left[p_{i+1}(t)+p_{i-1}(t)-2p_i(t)\right]. \tag{6.142}$$

Let a be the mean distance between molecules capable of overturning. Then Eq. (6.142) may be rewritten in the following form:

$$\frac{dp_i(t)}{dt} = \alpha_v\left[1-p_i(t)\right]-\alpha_{sv}p_i(t)+\beta a^2\frac{\left[p_{i+1}(t)+p_{i-1}(t)-2p_i(t)\right]}{a^2}. \tag{6.143}$$

This part of Eq. (6.143) is a discrete analogy of the second-order spatial derivative. Hence, using the continuous coordinate $x \approx ia$, we obtain the following second-order partial differential equation instead of Eq. (6.143):

$$\frac{\partial p}{\partial t} = \alpha_v(1-p)-\alpha_{sv}p+D\frac{\partial^2 p(t,x)}{\partial x^2}, \tag{6.144}$$

where $p(t,x)$ is the probability of the occurrence of an overturned molecule at time t at point x, and $D = \beta a^2$ is the effective diffusion coefficient of the molecule-overturned state along the surface. The extension of Eq. (6.144) to the plane (two-dimensional) case of our further interest is, as stated before, a straightforward procedure: It is reduced to the simple substitution of the partial second-order derivative with respect to one coordinate x for the sum of the partial second-order derivatives with respect to x and y because in the two dimensional case $p = p(t,x,y)$.

According to our previous consideration, the transfer of the overturned state is determined only by the interactions between adjacent molecules and should vanish in two cases: (*i*) In the absence of interactions between adjacent amphiphilic molecules, that is, at $\chi = 0$ (in this case, $\beta = 0$ and hence $D = 0$); and (*ii*) upon unlimited increase in the distance a between adjacent molecules, that is, when the surface concentration of molecules capable of overturning tends to zero. In this case, interactions between adjacent molecules also vanish. Indeed, let us assume that U_{th}, U_{tt}, and U_{hh} are determined by dispersion interactions only. In this case, $\chi(a) \sim B/a^6$, where B is a constant expressed as usual via the polarizabilities of hydrophilic head groups and hydrophobic tails: $B = (2A_{th}-A_{tt}-A_{hh})z/3$, where A_{th}, A_{tt} and A_{hh} are the corresponding Hamaker constants.

Consequently, $D(a) \sim \beta \left[\exp B / a^6 - 1 \right] a^2 \sim \beta B / a^4 \to 0$ with an increase in distance a between adjacent amphiphilic molecules.

In the equilibrium state, the probability p does not depend either on time or coordinate; this equilibrium state is further denoted by p_v, which is readily determined from Eq. (6.144):

$$p_v = \frac{\alpha_v}{\alpha_v + \alpha_{sv}} = \frac{1}{1 + \alpha_{sv} / \alpha_v}. \qquad (6.145)$$

From definitions (6.139) and (6.140), we obtain that $\frac{\alpha_{sv}}{\alpha_v} = \exp(2\Phi_v) \gg 1$, that is, p_v is a small value.

Let us discuss now the events occurring underneath the aqueous droplet at the solid–water interface. In this case, instead of equation (6.134) we arrive to

$$p_i(t + \Delta t) = p_i(t) + P_w \Delta t - P_{sw} \Delta t + P_d \Delta t - P_b \Delta t, \qquad (6.146)$$

where subscript w refers to water. Expressions for the probabilities per unit time P_w and P_{sw} are similar to the previous probabilities with corresponding constants α_w and α_{sw}, which can be obtained from relationships (6.135), (6.136), and (6.139), (6.140), respectively, by substituting subscript w for v.

A similar consideration to those used for derivation of Eq. (6.144) results in this case in the following equation:

$$\frac{\partial p}{\partial t} = \alpha_w (1 - p) - \alpha_{sw} p + D \frac{\partial^2 p(t, x)}{\partial x^2}. \qquad (6.147)$$

The equilibrium value of the probability of the occurrence of an amphiphilic molecule in the overturned state

$$p_w = \frac{\alpha_w}{\alpha_w + \alpha_{sw}}, \qquad (6.148)$$

is determined, as in the case of contact with the air, from Eq. (6.147). Unlike the case of contact with the air, the p_w probability is not a small value: On the contrary, it is close to one. It is this difference in probabilities that provides for the possibility of the aqueous droplet spreading over the initially hydrophobic surface.

Note that in the absence of lateral interactions between adjacent amphiphilic molecules, the aqueous droplet may not spread over the surface under consideration despite the effect of overturning of molecules with the hydrophilic portions upward, underneath the water. Indeed, let, in the absence of interactions between adjacent molecules, the necessary quantity of molecules be overturned with their hydrophilic portions upward (Figure 6.25a), and the substrate surface underneath the aqueous droplet becomes sufficiently hydrophilic so that the droplet edge can move into a new position as presented in Figure 6.25b. However, in the absence of lateral transfer of the overturned state of the amphiphilic molecules in the substrate, the surface both in front of the droplet edge in Figure 6.25b and behind the edge are still in the initial hydrophobic state, thus forcing the droplet edge to return immediately to the initial position (Figure 6.25a). Thus, in the absence of the lateral transfer of the overturned state described by the diffusion term in Eqs. (6.144) and (6.147), the spreading of the aqueous droplet over the surface becomes impossible. However, if the adjacent amphiphilic molecules interact with each other and the lateral transfer of the overturned state due to these interactions is possible, the droplet edge moves over the surface and this motion is determined exactly by the rate of the lateral transfer of the overturned state of the substrate molecules.

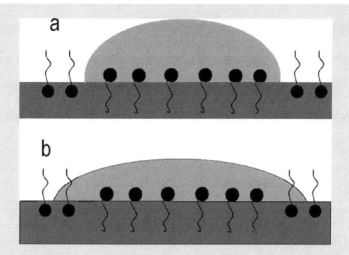

FIGURE 6.25 (a). Impossibility of a spreading of a water droplet on a hydrophobic substrate without lateral interaction between neighboring chains (explanation in the text). (b). The lateral interaction results in a possibility of spreading.

Boundary Conditions

According to the theory described in this section, the propagation of the overturned state of amphiphilic molecules along the surface under the droplet is described by Eq. (6.147) and the propagation beyond the droplet by Eq. (6.144). It is required now to formulate the boundary conditions for the probability, $p(t, r)$, at the boundary of the spreading axisymmetric droplet, that is, at $r = r_0(t)$ (Figure 6.26).

In view of the assumption of the absence of evaporation, the droplet volume, V, remains constant during the spreading and it is also assumed that the droplet is small enough, that is, the gravity action may be neglected. The time scale of the spreading process is so big enough that the

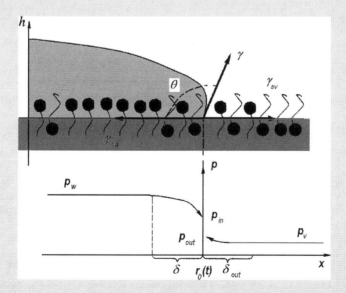

FIGURE 6.26 Spreading of a spherical droplet. $r_0(t)$, radius of the basis of a drop; $\theta(t)$, dynamic contact angle; and h, height at center of a drop.

deformations of the drop profile caused by the spreading/flow can be neglected: This means that the capillary number is enormously low, $Ca \ll 1$. According to the introduction to Chapter 3, this means, the droplet retains the shape of a spherical segment during the spreading

$$V = \frac{\pi}{6} h\left(3r_o^2 + h^2\right) = \text{const},\qquad(6.149)$$

where h and r_0 are the maximum height and radius of the base of the spreading droplet (Figure 6.26). The droplet height, h, is determined by the relationship $h = r_0 \tan(\theta/2)$, where $\theta(t)$ is the current value of the contact angle of the spreading droplet. Substituting this relationship into equation (6.149), we express the radius of the base of the spreading droplet via the current value of the contact angle as

$$r_0(t) = \left(\frac{6V}{\pi}\right)^{1/3} \frac{1}{\left(\tan\frac{\theta}{2}\left(3 + \tan^2\frac{\theta}{2}\right)\right)^{1/3}}.\qquad(6.150)$$

Let us denote surface (interfacial) tensions for the water–air (constant value), substrate–air, and substrate–water interfaces directly at the boundary of the spreading droplet by $\gamma, \gamma_{sv}(t)$, and $\gamma_{sw}(t)$, respectively. Note that the values of $\gamma_{sv}(t)$ and $\gamma_{sw}(t)$ near the droplet edge differ from the constant values of surface tensions at the polymer–air and polymer–water interfaces far from the edge of the spreading droplet and in the depth of the droplet, respectively. It is assumed that the Newman-Young condition is satisfied at any moment at the boundary of the spreading droplet

$$\cos\theta = \frac{\gamma_{sv}(t) - \gamma_{sw}(t)}{\gamma} = \cos\theta^0 + \Delta(t),\qquad(6.151)$$

where

$$\cos\theta^0 = \frac{\gamma^0_{sv} - \gamma^0_{sw}}{\gamma}, \quad \Delta(t) = \frac{\gamma_{sv}(t) - \gamma^0_{sv}}{\gamma} - \frac{\gamma_{sw}(t) - \gamma^0_{sw}}{\gamma}.$$

In this equation, superscript 0 marks corresponding initial values of interfacial tensions. Note, the $\Delta(t)$ value is positive and increases with time because the $\gamma_{sv}(t)$ value rises in time due to the possible appearance of hydrophilic head groups of the amphiphilic molecules at the substrate surface. In contrast, the $\gamma_{sw}(t)$ value decreases with time due to overturning of molecules under the droplet.

Let us emphasize once more

1. In view of the lateral interaction between adjacent amphiphilic molecules of the substrate, the overturned state may be extended beyond the boundary of the spreading droplet, resulting in an increase in surface tension of the substrate $\gamma_{sv}(t)$ in front of the moving droplet (Figure 6.26).

2. Interfacial tensions $\gamma_{sv}(t)$ and $\gamma_{sw}(t)$ do not remain constant near the droplet boundary but vary depending on the coordinate in a close vicinity of the boundary of the moving droplet. Hence, interfacial tensions in the close vicinity of the edge of the moving droplet or, in a more formal manner, the limiting values of these tensions at $r \to r_0(t)$ from the inner and outer droplet sides enter the Newman-Young equation (6.151).

The corresponding limits of the degree of overturning inside and outside the droplet are denoted by p_{in} and p_{out}, respectively (Figure 6.26). Evidently, the $\Delta(t)$ value is not an explicit function of time but depends on time in an implicit manner via the values of p_{in}, p_{out}, that is, $\Delta = \Delta(p_{in}, p_{out})$. As shown below, in this section in view of the equality of chemical potentials of amphiphilic molecules between the inner and outer boundaries of a droplet, the p_{out} value is expressed as a function of p_{in}. Hence, in view of this dependence, actually $\Delta = \Delta(p_{in})$.

Let us use an Antonov's rule to determine the unknown dependence, $\Delta (p_{in}, p_{out})$, which means the additivity of the formation of the interfacial tensions.

Let γ^∞_{sw} and γ^0_{sw} be the interfacial tensions under the droplet in the case when all molecules are overturned (all hydrophilic head groups are oriented upward) and when neither of these molecules is overturned (all hydrophilic head groups are oriented downward), respectively. Similar surface tensions outside the droplet are denoted by γ^∞_{sv} and γ^0_{sv}, respectively. According to the assumption of the additivity, the interfacial tensions in the closest vicinity of a droplet acquire the following form:

$$\gamma_{sv} = \gamma^0_{sv}(1 - p_{out}) + \gamma^\infty_{sv} p_{out},$$

$$\gamma_{sw} = \gamma^0_{sw}(1 - p_{in}) + \gamma^\infty_{sw} p_{in}.$$

(6.152)

Substitution of relationships (6.152) into the Newman-Young equation (6.151) yields the following expression for the dependence $\Delta(p_{in}, p_{out})$ under consideration:

$$\Delta(p_{in}, p_{out}) = \frac{\gamma^\infty_{sv} - \gamma^0_{sv}}{\gamma} p_{out} + \frac{\gamma^0_{sw} - \gamma^\infty_{sw}}{\gamma} p_{in}.$$

(6.153)

The obtained dependence $\Delta(p_{in}, p_{out})$ is a linear function with respect to both variables.

The spreading process under consideration is very slow (the time scale is hours). This allows employing the principle of local equilibrium accepted in non-equilibrium thermodynamics. In accordance with this principle, chemical potentials of overturned and non-overturned amphiphilic molecules remain equal from both sides of the droplet boundary, that is,

$$\mu(p_{in}, \chi) = \mu(p_{out}, \chi) + \Im, \quad \Im = \frac{U_{hv} - U_{hw}}{RT}.$$

(6.154)

The $\mu(p, \chi)$ dependence may be given, for example, in accordance with the Flory-Huggins theory [23] in the following form $\mu(p, \chi) = ln\, p + \chi (1 - p)^2$. Below in this section we consider only two limiting cases of weak and strong lateral interactions between amphiphilic molecules. At this stage, it is enough to take into account that, according to the equality of chemical potentials (6.154), p_{out} and p_{in} are interrelated by the known dependence $p_{out} = \varphi(p_{in})$. The value of Δ is dependent on only one variable, that is, $\Delta = \Delta(p_{in})$.

In view of the equality, $\tan\dfrac{\theta}{2} = \left(\dfrac{1-\cos\theta}{1+\cos\theta}\right)^{1/2}$, from Eq. (6.150), we obtain

$$r_0(p_{in}) = \left(\frac{6V}{\pi}\right)^{1/3} G(p_{in}),$$

(6.155)

where

$$G(p_{in}) = \frac{1}{\left(\dfrac{1-A}{1+A}\right)^{1/6}\left(3 + \dfrac{1-A}{1+A}\right)^{1/3}},$$

(6.156)

$$A(p_{in}) = \cos\theta^0 + \Delta(p_{in}).$$

Equation (6.155) allows determining the final droplet radius, r_∞, at the end of the spreading process using the known dependence $G(p_{in})$. To this end, it is necessary to substitute the expression for the equilibrium fraction of overturned molecules under droplet, p_w, from relationship (6.148) into Eq. (6.155), which yields

$$r_\infty = \left(\frac{6V}{\pi} \right)^{1/3} G(p_w).$$ (6.157)

It is possible to verify that $G(p)$ is an increasing function of p; that is, $r_0(t)$ increases, approaching its final value determined by Eq. (6.157).

Let us consider now the formulation of boundary conditions on a moving droplet edge. Two boundary conditions are required, because the edge itself moves and the law of this movement should be determined.

The first boundary condition expresses the balance of the number of overturned molecules at the boundary of the moving droplet, and it has the following form:

$$-D \left. \frac{\partial p}{\partial r} \right|_{r=r_0(t)-} + D \left. \frac{\partial p}{\partial r} \right|_{r=r_0(t)+} = (p_{in} - p_{out}) \frac{dr_0(t)}{dt}.$$ (6.158)

As mentioned earlier, it is necessary to set the second boundary condition relating the p_{in} and p_{out} values, because the other boundary conditions are straightforward:

the symmetry in the droplet center, that is,

$$\left. \frac{\partial p}{\partial r} \right|_{r=0} = 0,$$ (6.159)

and the tendency of the fraction of overturned molecules far from the droplet to the equilibrium value at the substrate-air interface that is determined from Eq. (6.145) as

$$p \to p_v, \quad r \to \infty.$$ (6.160)

We use the condition of equality of chemical potentials of overturned molecules to the right- and left-hand sides of the moving boundary of a droplet, $r_0(t)$, which determine the dependence

$$\Delta \mu_{in}(p_{in}) = \Delta \mu_{out}(p_{out}).$$ (6.161)

To analyze the dependences of chemical potentials, we employ the expressions resulting from the Flory-Huggins theory [23] modified to take into account the interactions of amphiphilic molecules with the environment and the underlying molecules of a substrate, namely, interactions of their hydrophilic head groups with water and their hydrophobic tails with the substrate molecules (under the droplet), and interactions of hydrophilic head groups with the air and hydrophobic tails with underlying substrate molecules (outside the droplet). This leads to the following expression:

$$\ln p_{in} + \chi (1 - p_{in})^2 = \ln p_{out} + \chi (1 - p_{out})^2 + \Im,$$ (6.162)

where χ is the known parameter of interaction of amphiphilic molecules with each other according to expression (6.141); in the case under consideration, the value of z is equal to 4.

In the case of weak interactions between adjacent molecules, that is, at $\chi \ll \Im$, from expression (6.162) we conclude $\ln p_{in} = \ln p_{out} + \Im$. As a result, we arrive to the Boltzmann distribution:

$$p_{out} = p_{in} \exp(-\Im).$$ (6.163)

In the case of strong interaction between the neighboring molecules, that is, $\chi \gg \Im$, we obtain

$$\ln p_{in} + \chi(1 - p_{in})^2 = \ln p_{out} + \chi(1 - p_{out}). \tag{6.164}$$

Hence, it results in equality of the overturned fractions: $p_{out} = p_{in}$.

The value of p_{out} is always smaller than or equal to p_{in} thus enabling us to solve Eq. (6.162) for the arbitrary case and to express p_{out} as a function of p_{in}, which was stated before by Eq. (6.161).

Thus, the problem of droplet spreading acquires the following form: the dependence $p(t, r)$ under the droplet at $0 < r < r_0(t)$ is described by equation

$$\frac{\partial p}{\partial t} = \alpha_w(1 - p) - \alpha_{sw} p + D \frac{1}{r} \frac{\partial}{\partial r} r \frac{\partial p}{\partial r}, \tag{6.165}$$

and the dependence $p(t, r)$ outside the droplet, $r > r_0(t)$, is described by the equation

$$\frac{\partial p}{\partial t} = \alpha_v(1 - p) - \alpha_{sv} p + D \frac{1}{r} \frac{\partial}{\partial r} r \frac{\partial p}{\partial r}, \tag{6.166}$$

with the following boundary conditions: condition of symmetry in the droplet center (6.159), condition (6.160) far from the droplet, condition (6.155) expressing the radius of the spreading droplet via p_{in}, condition (6.158) expressing the equality of fluxes at the droplet $r = r_0(t)$, and relationship (6.161) expressing the equality of chemical potentials of overturned molecules near the moving edge of the droplet.

Solution of the Problem

We perform the solution of the problem under consideration introducing a number of simplifying assumptions whose validity is checked now.

It is obvious that the value of p under the main part of the spreading droplet is independent of the coordinate but changes only with time due to the interaction of the amphiphilic molecules with the aqueous phase. Let us denote this coordinate-independent value by $p_d(t)$, which according to Eq. (6.165) satisfies the following equation:

$$\frac{d p_d}{d t} = \alpha_w(1 - p_d) - \alpha_{sw} p_d,$$

with the initial condition $p_d(0) = p_v$ The solution of this problem is

$$p_d(t) = p_w + (p_v - p_w)\exp(-(\alpha_w + \alpha_{sw})t). \tag{6.167}$$

It follows from Eq. (6.167) that the characteristic time scale of molecule overturning t_{tr*} is equal to $t_{tr*} = 1/(\alpha_w + \alpha_{sw})$.

In a narrow region with width δ near the droplet edge (from the inner side), the diffusion term in Eq. (6.165) becomes of the same order of magnitude as the term describing the overturning of molecules due to their interaction with water. Let us introduce dimensionless values $y = (r_0(t) - r)/\delta \lambda = \alpha_s/\alpha_w$; $\tau = t/t_*$; $\xi(t) = r(t)/r_*$, where $r_* = (6V/\pi)^{1/3}$ and a new unknown function, $g(t, y) = p(t, r) - p_d(t)$. The time scale t_* is selected below.

Rewriting Eq. (6.165) in dimensionless form using introduced variables, we obtain

$$\frac{\partial g}{\partial t} = -(\alpha_w + \alpha_{sw})g + \frac{D}{\delta^2} \frac{\partial^2 g}{\partial y^2}. \tag{6.165'}$$

Accounting for the smallness of δ/r_*, we conclude

$$\frac{\partial g}{\partial t} = \frac{1}{t_*}\frac{\partial g}{\partial \tau} + \frac{\partial g}{\partial y}\frac{r_*}{\delta t_*}\dot{\xi} \approx \frac{\partial g}{\partial y}\frac{r_*}{\delta t_*}\dot{\xi}.$$

In this case, Eq. (6.165′) acquires the following form:

$$\frac{r_*}{\delta t_*}\frac{\partial g(\tau,y)}{\partial y}\dot{\xi} = -(\alpha_w + \alpha_{sw})g(\tau,y) + \frac{D}{\delta^2}\frac{\partial^2 g(\tau,y)}{\partial y^2}, \tag{6.168}$$

All terms in Eq. (6.168) should be of the same order of magnitude; hence, the characteristic values are interrelated by the following relationships: $\delta = \sqrt{D/(\alpha_w + \alpha_{sw})}$, $t_* = r_*\delta/D$. Let us compare t_{tr*} and t_*: $t_*/t_{tr*} = t_*(\alpha_w + \alpha_{sw}) = r_*(\alpha_w + \alpha_{sw})/D = r_*/\delta \gg 1$; that is, t_* is much larger than characteristic time of overturning of amphiphilic molecules under the main portion of the droplet. This implies that under the droplet $p = p_w$ and changes occur only in the narrow region with the width δ and are described by equation

$$\frac{\partial g(\tau,y)}{\partial y}\dot{\xi} = -g(\tau,y) + \frac{\partial^2 g(\tau,y)}{\partial y^2}, \tag{6.169}$$

with boundary conditions

$$g \to 0, \quad y \to \infty, \tag{6.170}$$

and

$$g(0) = p_{in} - p_w. \tag{6.171}$$

Because the desired function $g(\tau, y)$ depends on τ as a parameter, the solution of Eq. (6.169) satisfying conditions (6.170) and (6.171) may be readily obtained and the expression for p acquires the form

$$p = p_w + (p_{in} - p_w)\exp\left(\frac{\dot{\xi} - \sqrt{\dot{\xi}^2 + 4}}{2}y\right). \tag{6.172}$$

Let us perform similar transformations in the narrow region from the outer side of the droplet front, $\delta_{out} = \sqrt{D/(\alpha_v + \alpha_{sv})}$, $y_{out} = (r - r_0(t))/\delta_{out}$, $q = p(t,r) - p_v$.

In this case, the dimensionless equation describing the probability p in the δ_{out} region outside the droplet has the following form:

$$-\sqrt{\frac{\alpha_w + \alpha_{sw}}{\alpha_v + \alpha_{sv}}}\frac{\partial q(\tau,y)}{\partial y_{out}}\dot{\xi} - q(\tau,y) + \frac{\partial^2 q(\tau,y)}{\partial y_{out}^2} \tag{6.173}$$

with boundary conditions,

$$q \to 0, \quad y_{out} \to +\infty, \tag{6.174}$$

and

$$q(0) = p_{out} - p_v \tag{6.175}$$

Assuming that $\alpha_w \sim \alpha_{sv} \gg \alpha_v \sim \alpha_{sw}$, we can conclude that $\delta = \delta_{out}$, and from Eqs. (6.174) through (6.176), we obtain

$$p = p_v + (p_{out} - p_v) \exp\left(\frac{\dot{\xi} + \sqrt{\dot{\xi}^2 + 4}}{2} y_{out}\right) \qquad (6.176)$$

Let us rewrite boundary condition (6.158) to dimensionless form

$$\left.\frac{D}{\delta}\frac{\partial p}{\partial y}\right|_{y=0} + \left.\frac{D}{\delta}\frac{\partial p}{\partial y_{out}}\right|_{y_{out}=0} = \frac{r_*}{t_*}(p_{in} - p_{out})\,\dot{\xi}. \qquad (6.158')$$

Using expressions (6.172) and (6.176), we arrive to an equation describing the motion of the droplet boundary

$$(p_{in} - p_w)\frac{\dot{\xi} - \sqrt{\dot{\xi}^2 + 4}}{2} - (p_{out} - p_v)\frac{\dot{\xi} + \sqrt{\dot{\xi}^2 + 4}}{2} = (p_{in} - p_{out})\,\dot{\xi}. \qquad (6.177)$$

Equation (6.177) has the following solution:

$$\dot{\xi} = \frac{p_w - p_{in} - p_{out} + p_v}{\sqrt{(p_w - p_{out})(p_{in} - p_v)}}. \qquad (6.178)$$

Using condition (6.155), we obtain

$$\dot{\xi} = \frac{d\xi}{d\tau} = \frac{1}{r_*}\frac{dr_o}{d\tau} = \frac{d}{d\tau}G(p_{in}) = \frac{dG}{dp_{in}}\frac{dp_{in}}{d\tau}. \qquad (6.179)$$

Let us rewrite Eq. (6.178), taking into account representation (6.179),

$$\frac{dp_{in}}{d\tau} = \frac{p_w - p_{in} - p_{out} + p_v}{G'(p_{in})\sqrt{(p_w - p_{out})(p_{in} - p_v)}}. \qquad (6.180)$$

Assuming that $p_{out} \sim p_v \ll p_{in}$, and $p_w \sim 1$, we have the final equation for p_{in}:

$$\frac{dp_{in}}{dt} = \frac{(1 - p_{in})/t_*}{G'(p_{in})\sqrt{p_{in}}}. \qquad (6.180')$$

Considering the smallness of p_{out} as compared to one, we obtain from Eq. (6.152) that the value of γ_{sv} changes slightly and remains close to the initial value γ^0_{sv}. In this case, expression (6.153) may be rewritten in the following form:

$$\Delta = \frac{\gamma^0_{sw} - \gamma^\infty_{sw}}{\gamma}p_{in} = \left[\frac{\gamma^0_{sv} - \gamma^\infty_{sw}}{\gamma} - \frac{\gamma^0_{sv} - \gamma^0_{sw}}{\gamma}\right]p_{in} = (\cos\theta^\infty - \cos\theta^0)\,p_{in}, \qquad (6.181)$$

and the function $A(p_{in})$ acquires the form

$$A(p_{in}) = \cos\theta^0 + (\cos\theta^\infty - \cos\theta^0)\,p_{in}. \qquad (6.182)$$

Thus, according to the proposed theory, the droplet spreads in a completely different manner than was suggested in [21]; that is, the equilibrium concentration of overturned amphiphilic molecules (or their fragments) is established rapidly (as compared to the characteristic time of spreading) under the main portion of the drop and retains its value over the course of the entire spreading process. All changes occur only within the narrow region in the vicinity of the perimeter of the spreading droplet.

Comparison between Theory and Experimental Data

Equation (6.180′) contains one unknown parameter, t_*, which is the characteristic time of the propagation of the overturned state, that is, the characteristic time scale of droplet spreading. Parameter t_* was used as an adjustment parameter for the comparison of the theoretical predictions according to the developed theory and experimental data reported elsewhere [21,22].

In Figure 6.27, the time dependences of the contact angle of an aqueous droplet at the surface of paraffin containing steric acid at various concentrations, C, are presented. Lines show the solutions of equation (6.180′), and the symbols denote experimental data [22] for three different concentrations, C. The values of t_* were found as follows:

$$t_* = 50.5 \text{ h at C} = 0.6 \text{ wt\%},$$

$$t_* = 27.5 \text{ h at C} = 2.0 \text{ wt\%},$$

$$t_* = 24.0 \text{ h at C} = 9.0 \text{ wt\%}.$$

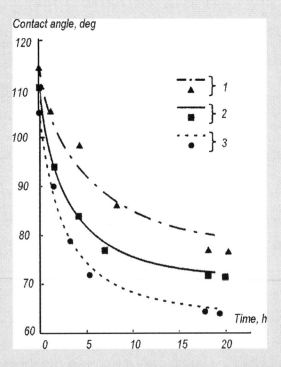

FIGURE 6.27 Time dependences of contact angle, θ, of water droplets at the surface of paraffin containing stearic acid of different concentrations in wt%. (1) $C = 0.6$; (2) $C = 2.0$; (3) $C = 9.0$. Solid lines according to Eq. (6.180′). (Experimental data from Yiannos, P.N., *J. Colloid Sci.*, 17, 334, 1962.)

These values show that as the concentration of stearic acid increases, the characteristic time of propagation of the overturned state decreases. This is due to an increase in "diffusion coefficient" D because of the decreasing average distance between acid molecules capable of overturning.

Figure 6.28 presents the time dependences of the contact angle of aqueous drops at a Langmuir-Blodgett films composed of stearic acid at various temperatures. Here, symbols represent experimental data from [21] and lines drawn correspond to the solutions of Eq. (6.180′). Figure 6.29 shows the deduced dependence of parameter t_* on temperature. Characteristic time t_* of the propagation of the overturned state decreases with temperature, which may be explained by an increase in the rotational mobility of molecules capable of overturning and simultaneously the "diffusion coefficient" D of the overturned state of stearic acid molecules increases.

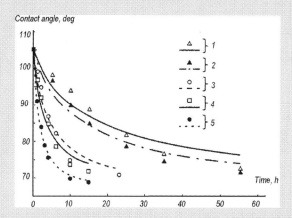

FIGURE 6.28 Time dependences of the contact angle, θ, of water droplets at the surface of a Langmuir-Blodget films formed by a stearic acid at various temperatures in Celsius (°C). (1) $t_* = 13.5\ h$; (2) $t_* = 15.5\ h$; (3) $t_* = 23.0\ h$; (4) $t_* = 25.0\ h$; (5) $t_* = 28.5\ h$. Solid lines are according to Eq. (6.180′). (Experimental data from Rideal, E. and Tadayon, J., *Proc. Roy. Soc.*, 225, 346, 1954.)

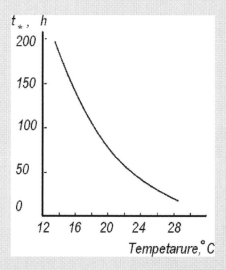

FIGURE 6.29 Dependence of characteristic time of a spreading of drops, t^*, on temperature (°C).

REFERENCES

1. Exerowa, D., and Krugliakov, P. *Foam and Foam Films: Theory, Experiment, Application*, Vol. 5, Elsevier, New York, 1988.
2. Zhdanov, S., Starov, V., Sobolev, V., and Velarde, M. Spreading of aqueous SDS solutions over nitrocellulose membranes. *J. Colloid Interface Sci.*, 264, 481–489 (2003).
3. Zolotarev, P. P., Starov, V. M., and Churaev, N. V., Kinetics of imbition of surfactant solution into narrow capillaries. *Colloid J.* (USSR Academy of Sciences, English Translation), 38, 895 (1976).
4. Churaev, N. V., Martynov, G. A., Starov, V. M., and Zorin, Z. M. Some features of capillary imbibition of surfactant solutions. *Colloid Polym. Sci.*, 259, 747 (1981).
5. Starov, V. Spontaneous rise of surfactant solutions into vertical hydrophobic capillaries. *J. Colloid Interface Sci.*, 270, 180 (2003).
6. Berezkin, V. V., Zorin, Z. M., Frolova, N. V., and Churaev, N. V., Imbibition of surfactant solutions into hydrophobic capillaries. *Colloid J.* (USSR Academy of Sciences, English Translation), 37,1040 (1975).
7. Shinoda, K., Nakagawa, T., Tamamushi, B., and Isemura, T. *Colloidal Surfactants: Some Physico-Chemical Properties*, Academic Press, New York, 1963.
8. Kumar, N., Varanasi, K., Tilton, R. D., and Garoff, S. Surfactant self-assembly ahead of the contact line on a hydrophobic surface and its implications for wetting. *Langmuir*, 19, 5366 (2003).
9. Kumar, N., Couzis, A., and Maldarelli, C. Measurement of the kinetic rate constants for the adsorption of superspreading trisiloxanes to an air/aqueous interface and the relevance of these measurements to the mechanism of superspreading. *J Colloid Interfaces Sci.*, 267, 272 (2003).
10. Starov, V. M., Zhdanov, S. A., and Velarde, M. G. Capillary imbibition of surfactant solutions in porous media and thin capillaries: partial wetting case. *J. Colloid Interface Sci.*, 273 (2), 589–595 (2004).
11. Starov, V. M., Kosvintsev, S. R., and Velarde, M. G. Spreading of surfactant solutions over hydrophobic substrates, *J. Colloid Interface Sci.*, 227, 185–190 (2000)
12. Stoebe, T., Lin, Z., Hill, R. M., Ward, M. D., and Davis, H. T., Surfactant-enhanced spreading. *Langmuir*, 12, 337 (1996), Enhanced spreading of aqueous films containing ethoxylated alcohol surfactants on solid substrates. *Langmuir*, 13, 7270 (1997), Enhanced spreading of aqueous films containing ionic surfactants on solid substrates.*Langmuir*, 13, 7276 (1997).
13. Starov, V. M., de Ryck, A., and Velarde, M. G. On the spreading of an insoluble surfactant over a thin viscous liquid layer. *J. Colloid Interface Sci.*, 190, 104 (1997).
14. Levich, V. G. *Physicochemical Hydrodynamics*, Prentice-Hall Inc., Englewood Cliffs, NJ, 1962.
15. Andrade, J. D., and Chen, W. Y. Probing polymer surface and interface dynamics. *Surf. Interface Anal.*, 8 (6), 253 (1986).
16. Andrade, J. D. (Ed.), *Polymer Surface Dynamics*, Plenum, New York, 1988.
17. Lewis, B. K., and Ratner, B. D., Observation of surface rearrangement of polymers using ESCA. *J. Colloid Interface Sci.*, 159 (1), 77 (1993).
18. Miyama, M., Yang, Y., Yasuda, T. et al. Static and dynamic contact angles of water on polymeric surfaces. *Langmuir*, 13 (20), 5494 (1997).
19. Rudoy, V. M., Stuchebryukov, S. D., and Ogarev, V. A. *Colloid J.* (English Translation), 50 (1), 199 (1988).
20. Langmuir, I. Overturning and anchoring of monolayers. *Science*, 87, 493 (1938).
21. Yiannos, P. N., Molecular reorientation of some fatty acids when in contact with water. *J. Colloid Sci.*, 17 (4), 334 (1962).
22. Rideal, E., and Tadayon, J., On overturning and anchoring of monolayers-I. Overturning and transfer. *Proc. Roy. Soc.* (*London*) *A*, 225 (1162), 346 (1954).
23. Flory, P. J. *Principles of Polymer Chemistry*, Cornell University Press, Ithaca, NY, 1990.

7

Kinetics of Simultaneous Spreading and Evaporation

Introduction

The consideration of simultaneous spreading and evaporation has been carried out for a relatively long time. Consideration of these simultaneous processes is complicated by uneven temperature distribution in the evaporating droplet, which is caused by the presence of a latent heat of evaporation. As a result, the droplet is becoming cooler as compared with the surrounding air, but the simultaneous heat exchange with the solid support makes the temperature distribution nonuniform both in the droplet volume and on the liquid–air interface. This results in the creation of a surface tension gradient, which in turn causes a flow: the Marangoni convection. This means that the interconnected processes of spreading, evaporation and heat transfer have to be considered simultaneously.

Two singularities have to be coped with from the theoretical point of view: The first problem is associated with the problem of a singularity at the moving three-phase contact line (a singularity of the viscous stress caused by an incompatibility of the non-slip condition on the solid substrate and the free surface at the liquid–air interface at the three-phase contact line). The second problem is associated with the specific behavior of the evaporation flux at the perimeter of the droplet. This singularity is caused by an incompatibility of boundary conditions at the three-phase contact line: the nonzero evaporation flux at the liquid–air interface with zero flux at the solid–air interface. Both singularities are artificial and caused by the incomplete physical model used. The first singularity can be overcome by introducing the Derjaguin's (disjoining/conjoining) pressure—which dominates in the vicinity of the apparent three-phase contact line (see Chapter 4 and [1–3])—pressure into the model. The second singularity can be overcome by consideration of the Derjaguin's pressure combined with consideration of liquid molecules adsorption on the solid–vapor interface, which takes place according to the same Derjaguin's isotherm. That is, consideration of the Derjaguin's pressure in the vicinity of the moving apparent three-phase contact line removes both singularities. Such consideration is a work for the future.

A schematic diagram of the evaporating droplet is shown in Figure 7.1. This figure shows that the vapor flow is unavoidable—both from the bulk droplet into the surrounding undersaturated vapor and along the droplet from the bulk part to the droplet edge.

Excess hydrodynamic pressure in the bulk region (region 3, Figure 7.1) is higher than in the region of thin liquid films (region 1, Figure 7.1), which are at equilibrium with the undersaturated vapor in the surrounding air surrounding the evaporating droplet. This unavoidably results in a flow of liquid from the bulk droplet (region 3, Figure 7.1) to thin liquid films at the edge (region 1, Figure 7.1). This flow is caused by the gradient of the Derjaguin's pressure inside the transition zone (region 2, Figure 7.1). Except for mentioned publications [1–3], this flow has been ignored in the literature. Unfortunately, the inclusion of Derjaguin's pressure into consideration on the current stage requires either considerable simplifications (as in [1–3]) or direct computer simulations, which is not the purpose of this book. In other words, unfortunately, the action of the Derjaguin's pressure is ignored in this chapter.

The presentation in this chapter is based mostly, but not entirely, on publications [4–13].

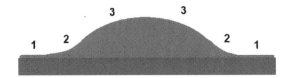

FIGURE 7.1   Schematics of an evaporating droplet on a solid substrate. (1) Adsorption layer, which is an equilibrium with *undersaturated vapor* in the surrounding air; (2) transition region, where surface forces (disjoining/conjoining pressure) act; and (3) bulk of the liquid, which can at equilibrium with *oversaturated vapor* only.

7.1 Basic Properties of Simultaneous Spreading/Evaporation

First, we must distinguish between cases of complete and partial wetting. In the case of the spreading/ evaporating droplet, the definition of complete and partial wetting requires extra effort because the usually adopted definition based on the shape of the Derjaguin's pressure isotherm is impossible: This shape in mostly unknown for the spreading/evaporating droplet. Therefore, the definition of complete and partial wetting cases is given in this chapter based on the time dependency of both the contact angle and the radius of the spreading/evaporating droplet.

The Partial Wetting Case

The spreading and evaporation behavior of a droplet in the case of partial wetting can be subdivided into three subsequent stages as shown schematically in Figure 7.2. Stage 1: During this stage, the droplet spreads relatively rapidly over the solid substrate until the radius of the droplet base reaches the maximum value, L_{ad}, and the contact angle decreases to the value of the static advancing contact angle, θ_{ad}. Stage 2: During the second stage, the three-phase contact line remains fixed at the maximum value while the contact angle decreases from the static advancing contact angle, θ_{ad}, to the static receding contact angle, θ_r, due to the loss of droplet volume caused by evaporation. Stage 3: During the third stage of

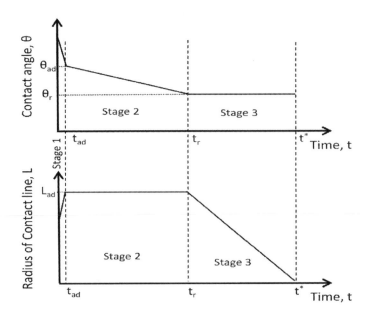

FIGURE 7.2   Three stages of spreading/evaporation of droplet in the case of partial wetting. L_{ad} is the maximum radius of droplet base; θ_{ad} is the advancing contact angle; t_{ad} is the time when θ_{ad} is reached; θ_r is the receding contact angle; t_r is the time when θ_r is reached; and t^* is the time when evaporation is finished completely.

spreading/evaporation, the drop base shrinks at an approximately constant static receding, θ_r, contact angle until the time the droplet evaporates completely.

The characteristic feature of partial wetting is the presence of contact angle hysteresis: This results in the existence of Stage 2, in which the edge of the droplet is pinned. The presence of Stage 2 allows us to conclude that we are dealing with the partial wetting case.

The Complete Wetting Case

There is no contact angle hysteresis in the case of complete wetting; therefore, Stage (2) of partial wetting is absent in the complete wetting case, and there are only two stages of spreading/evaporation (Figure 7.3): Stage 1: During this stage, the droplet spreads over the substrate, and the radius of the droplet base reaches its maximum value, L_m. Stage 3: During this stage (there is no Stage 2 in the complete wetting case), evaporation prevails over spreading, and the radius of the droplet base shrinks until complete disappearance. Note, over most of Stage 3 in the case of complete wetting, the contact angle retains the constant value, which has nothing to do with contact angle hysteresis (there is no contact angle hysteresis in the case of complete wetting, see Section 5.2); instead, the contact angle is determined by pure hydrodynamics. An interesting observation is as follows: The duration of Stage 1 in the case of partial wetting is much shorter than the duration of the same stage in the case of complete wetting. This was explained in Chapters 3 and 4: In the case of complete wetting, the spreading proceeds with a "precursor" film in front, where a lot of energy is burned. However, in the case of partial wetting, the spreading proceeds after deposition until the static advancing contact angle is reached through "caterpillar motion," which requires much less energy and, hence, results in faster spreading.

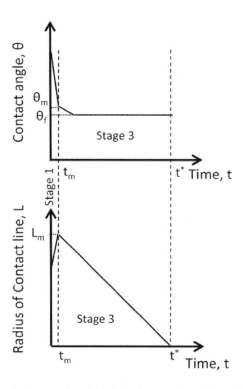

FIGURE 7.3 Two stages of spreading/evaporation of droplet in the case of complete wetting. L_m is the maximum radius of droplet base; t_m is the time when L_m is reached; θ_m is the contact angle at t_m; t^* is the time when evaporation is finished; and θ_f is the final contact angle at t^*. Note, in the case of complete wetting, stage 2 is absent (Figure 7.2).

Dependence of Evaporation Flux on Droplet Size

It is shown below in this section that the evaporation rate of a sessile droplet is proportional to the radius of the wetted area, L, and the proportionality coefficient was found in [14–20]:

$$\frac{dV}{dt} = -2\pi \frac{DM}{\rho}\left(c\left(T_{\text{surf}}\right) - Hc\left(T_\infty\right)\right)F\left(\theta\right)L, \tag{7.1a}$$

or simply:

$$\frac{dV(t)}{dt} = -\alpha L(t), \quad \alpha = 2\pi \frac{DM}{\rho}\left(c\left(T_{\text{surf}}\right) - Hc\left(T_\infty\right)\right)F\left(\theta\right), \tag{7.1b}$$

where V is the droplet volume; t is time; D, ρ, and M are vapor diffusivity in the air, density of the liquid, and the molar mass, respectively; H is the humidity of the ambient air; T_{surf} is the temperature of the droplet–air interface; T_∞ is the temperature of the ambient air; $c(T_{\text{surf}})$ and $c(T_\infty)$ are the molar concentrations of saturated vapor at the corresponding temperature; and $F\left(\theta\right)$ is a function of the contact angle, θ, which equals 1 at $\theta = \pi/2$. Equation (7.1a) was deduced for the model of evaporation and considers diffusion of only the vapor in the surrounding air, ignoring the temperature distribution along the droplet–air interface. In the case of θ independent of L (the first stage of evaporation), Eq. (7.1a) gives the evaporation rate directly proportional to the radius of the droplet base, L.

Here we show that the proportionality of the total evaporation flux, J, to the droplet perimeter has nothing to do with the distribution of the local evaporation flux, j, over the droplet surface [4]. Let us consider a stationary diffusion equation for vapor in air:

$$\frac{1}{r}\frac{\partial}{\partial r}\left(r\frac{\partial c}{\partial r}\right) + \frac{\partial^2 c}{\partial z^2} = 0,$$

where r and z are radial and vertical systems of coordinates, respectively; and c is the molar vapor concentration. The local normal flux, j, from the surface of the droplet is

$$j = -D\frac{\partial c}{\partial \vec{n}}\bigg|_{z=h(r)} = -D\left(\frac{\partial c}{\partial r}\bigg|_{z=h(r)} n_r + \frac{\partial c}{\partial z}\bigg|_{z=h(r)} n_z\right),$$

where D is the diffusion coefficient of vapor in the air; $\vec{n}$, n_r and n_z are the unit vector normal to the liquid–air interface (pointing into the air) and its radial and vertical components, respectively; and $h(r)$ is the height of the droplet surface. Let us introduce dimensionless variables using the same symbols as the original dimensional ones but with an overbar: $\bar{z} = z/L$, $\bar{r} = r/L$, $\bar{c} = c/c_\infty$, $\bar{h} = h/L$, where L is the droplet base radius; and c_∞ is the molar concentration of the vapor in the ambient air. This equation can be rewritten as

$$j = -D\frac{\partial c}{\partial \vec{n}}\bigg|_{z=h(r)} = -\frac{Dc_\infty}{L}\left(\frac{\partial \bar{c}}{\partial \bar{r}}\bigg|_{\bar{z}=\bar{h}(r)} n_r + \frac{\partial \bar{c}}{\partial \bar{z}}\bigg|_{\bar{z}=\bar{h}(r)} n_z\right) = \frac{Dc_\infty}{L}A(\bar{r},\bar{z}),$$

where $A(\bar{r},\bar{z}) = \left(\frac{\partial \bar{c}}{\partial \bar{r}}\bigg|_{\bar{z}=\bar{h}(r)} n_r + \frac{\partial \bar{c}}{\partial \bar{z}}\bigg|_{\bar{z}=\bar{h}(r)} n_z\right)$. Hence, the total flux is

$$J = 2\pi \int_0^L rj\sqrt{1+\left(\frac{\partial h}{\partial r}\right)^2}\,dr = 2\pi LDc_\infty \int_0^1 \bar{r}A(\bar{r},\bar{z})\sqrt{1+\left(\frac{\partial \bar{h}}{\partial \bar{r}}\right)^2}\,d\bar{r}.$$

These equations show that the total flux, $J \sim L$, and the local flux $j \sim 1/L$. Note, those properties do not depend on the distribution of the local evaporation flux, j, over the droplet surface. Note that those properties are valid only in the case of diffusion-controlled evaporation.

Thermal Phenomena at Evaporation

Thermal phenomena in the course of spreading evaporation have been under investigation for a long time [21–24].

Experiments by David et al. [22] showed that temperature in the bulk of a sessile evaporating droplet depends substantially on the thermal properties of the substrate and the rate of evaporation. Their measurements (Figure 7.4) showed that the temperature of an evaporating droplet is different from the ambient temperature and is almost constant during the spreading/evaporation process. This observation is used below in this section.

In [4] the dependence of the total vapor flux, J, on the radius of the droplet base, L, and the contact angle, θ, was investigated using numerical simulations. All calculations were performed with the effects of both local heat of vaporization (LHV) and Marangoni convection (MC) included. The results were obtained for substrates made of materials of various thermal conductivity and compared with those calculated for the isothermal cases reported by Hu and Larson [24] and Schonfeld et al. [23]. In the case of a highly heat-conductive solid support (e.g., copper), the difference between the present simulations and the results from [23,24] for the isothermal case do not exceed 3% [4]. This is because of the small temperature change at the droplet surface, which is close to isothermal conditions. However, if other materials are used with lower heat conductivity (down to the heat conductivity of air), then the evaporation flux is substantially reduced as compared with the isothermal case [4]. Such flux reduction is connected to the noticeable temperature decrease of the droplet surface.

In [4] the mean temperature of the droplet surface, $T_{surf} = \frac{1}{S}\int_S T_s ds$ was introduced, where S is the surface area of the evaporating droplet. The dimensionless total flux $J/J_{\pi/2} (L, T_{surf})$ was plotted in [4], where $J_{\pi/2}$ is the total flux in the case, then the contact angle is equal to $\pi/2$. All calculated total fluxes [4] for all substrates turned out to be universally dependent on total vapor flux, J, versus the contact angle, θ: This universal dependency coincides with the dependency for the isothermal case if T_{surf} is used as a temperature of the droplet–air interface. This shows that the variation of the surface temperature is the major phenomenon influencing the evaporation rate (Figures 7.5 and 7.6). Note that $J_{\pi/2}$ in Figure 7.5 was calculated by taking the temperature on the droplet surface to be equal to the temperature of the substrate, $T_{substr} = T_\infty + 5K$, whereas in Figure 7.6, the average temperature, T_{surf}, was used.

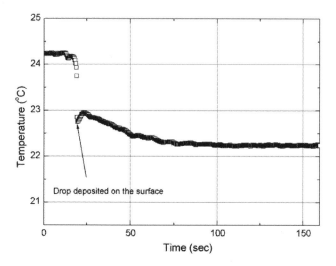

FIGURE 7.4 Evolution of temperature inside a water droplet after the droplet is deposited on a PTFE substrate. (Redrawn from David, S. et al., *Colloid Surf. A*, 298, 108, 2007.)

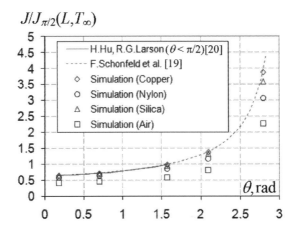

FIGURE 7.5 Rescaled dependence of the total vapor flux from the droplet surface, J, on contact angle θ, $L = 1$ mm. Both latent heat of vaporization and Marangoni convection were taken into account. (Redrawn from Semenov, S. et al., *Colloids Surf. A*, 372, 127–134, 2010.)

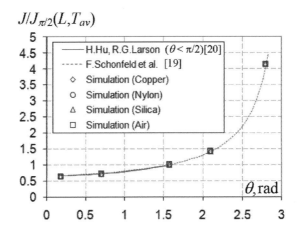

FIGURE 7.6 Universal behavior: rescaled dependence of the total vapor flux from the droplet surface, J, on contact angle θ, $L = 1$ mm. Both LHV and MC are taken into account. All points from Figure 7.5 are on the universal isothermal curve (the same as in Figure 7.5) and indistinguishable, when T_{surf} used as the temperature of the droplet–air interface. (Redrawn from Semenov, S. et al., *Colloids Surf. A*, 372, 127–134, 2010.)

7.2 Spreading and Evaporation of Sessile Droplets: Universal Behavior in the Case of Complete Wetting

Introduction

In the situation of complete wetting on a nonporous substrate, a volatile liquid droplet undergoes two competing mechanisms that occur until the droplet has completely evaporated: (a) spreading, which results in an extension of the droplet base, and (b) evaporation, which results in a shrinkage of the droplet base. This process alters the dynamics of droplet spreading as compared with the non-evaporating case through the corresponding changes to the radius of the base and the contact angle.

A number of experimental investigations of the evaporation and spreading droplets have been undertaken in [25–29] using alkanes and other liquids on nonporous, complete-wetting surfaces.

In Section 7.1, we showed that the following equation for the evaporation rate of a sessile droplet is valid:

$$\frac{dV}{dt} = -2\pi \frac{DM}{\rho}\left(c(T_{surf}) - Hc(T_\infty)\right)F(\theta)L, \tag{7.1a}$$

or simply:

$$\frac{dV(t)}{dt} = -\alpha\, L(t), \quad \alpha = -2\pi\frac{DM}{\rho}\left(c(T_{surf}) - Hc(T_\infty)\right)F(\theta), \tag{7.1b}$$

where V is the droplet volume; t is time; D, ρ, and M are the vapor diffusivity in the air, the density of the liquid, and the molar mass, respectively; H is the humidity of the ambient air; T_{surf} is the average temperature of the droplet–air interface, $T_{surf} = \frac{1}{s}\int_S T_s(s,t)ds$ (Section 7.1), where S is the liquid–air interface, $T_s(s, t)$, is the temperature of the surface at the position s on the interface; and T_∞ is the temperature of the ambient air; $c(T_{surf})$ and $c(T_\infty)$ are the molar concentrations of saturated vapor at the corresponding temperature; and $F(\theta)$ is a function of contact angle, θ, which equals 1 at $\theta = \pi/2$ [16]. According to [16], there is a simple polynomial fitting of this function in two ranges of contact angle as follows: $F(\theta) = \left(0.6366 \cdot \theta + 0.09591 \cdot \theta^2 - 0.06144 \cdot \theta^3\right)/\sin\theta$ if $\theta < \pi/18$ and $F(\theta) = \left(0.00008957 + 0.6333 \cdot \theta + 0.116 \cdot \theta^2 - 0.08878 \cdot \theta^3 + 0.01033 \cdot \theta^4\right)/\sin\theta$ if $\theta > \pi/18$. Equation (7.1a) was deduced for the model of evaporation, which considers diffusion only of the vapor in the surrounding air. In the case of a small contact angle, θ, less than 10°, the aforementioned polynomial fitting gives $F(\theta) \cong 0.6366$ and, hence, is independent of the contact angle. In this case, Eq. (7.1a) gives the evaporation rate directly proportional to the radius of the droplet base, L. Hence,

$$\frac{dV(t)}{dt} = -\alpha L(t), \tag{7.1c}$$

where α is a constant independent of the contact angle according to (7.1a) and (7.1b).

The main focus of the developed theoretical model described here is as follows: If the reduced radius, reduced volume and the reduced contact angle are plotted against the reduced time, the whole array of experimental data follows a corresponding universal curve.

Theory

Let us consider a small liquid droplet on a complete wetting solid substrate. The smallness of the droplets means that their size is small compared with the capillary length and, hence, the action of gravity on the process can be neglected. However, the droplet size is still bigger than 1 μm (see Section 7.5). The dynamics of the droplet is defined by two competing mechanisms: spreading, which results in an extension of the droplet base, and evaporation, which results in a shrinkage of the droplet base. Both processes occur simultaneously throughout the experiment. The spreading stage of the droplet over the solid substrate is initially the major factor in the change of the droplet's shape, but its effect lessens over time. Once the spreading effect lessens, evaporation becomes the key factor in determining the droplet's dimensions.

First, it is shown that in the experiments presented below, in this Chapter 2 as well as in experiments in [25–29], the droplet should retain a spherical shape in the course of spreading/evaporation. Recall that we ignore the presence of the transition zone in the vicinity of the apparent three-phase contact line because the action of surface forces (Derjaguin's pressure) is not taken into account.

Consider, for simplicity, an example of spreading of two-dimensional (cylindrical) droplets. In the case of water, $\gamma = 72.5$ dyn/cm, $\rho = 1$ g/cm³; hence, the capillary length $a \sim 0.27$ cm. All the droplets under consideration are expected to be smaller than this length.

The two important relevant parameters are the Reynolds number, Re, and the capillary number, Ca. We show below in this section that if both numbers are small, then the droplet should be of a spherical shape.

The Reynolds number characterizes the importance of inertial forces as compared with viscous forces. In the case of a two-dimensional droplet, the Navier–Stokes equations with the incompressibility condition take the following form:

$$\rho\left(\frac{\partial u}{\partial t}+u\frac{\partial u}{\partial x}+v\frac{\partial u}{\partial y}\right)=-\frac{\partial p}{\partial x}+\eta\left(\frac{\partial^2 u}{\partial x^2}+\frac{\partial^2 u}{\partial y^2}\right),\tag{7.2}$$

$$\rho\left(\frac{\partial v}{\partial t}+u\frac{\partial v}{\partial x}+v\frac{\partial v}{\partial y}\right)=-\frac{\partial p}{\partial y}+\eta\left(\frac{\partial^2 v}{\partial x^2}+\frac{\partial^2 v}{\partial y^2}\right),\tag{7.3}$$

$$\frac{\partial u}{\partial x}+\frac{\partial v}{\partial y}=0,\tag{7.4}$$

where $\vec{v}=(u,v)$ is the velocity vector; the gravity action is neglected; (x,y) are the coordinates; and ρ, η, and p are the liquid density, dynamic viscosity, and pressure, respectively. Let U^* and v^* be scales of the velocity components in the tangential and the vertical directions, respectively; r^* and h^* are the corresponding scales in the horizontal and vertical directions. Using the incompressibility condition, it is concluded that $\frac{U^*}{r^*}=\frac{v^*}{h^*}$, or $v^*=\varepsilon U^*$, $\varepsilon=\frac{h^*}{r^*}$. The derivative $\partial u/\partial t$ can be estimated as U^*/t^*, where $t^*\sim 1$ s is the time of spreading over a distance of about the droplet radius and $U^*\sim r^*/t^*$. Therefore, one obtains the following estimation:

$$\frac{\rho\dfrac{\partial u}{\partial t}}{\rho u\dfrac{\partial u}{\partial x}}\sim\frac{\dfrac{U^*}{t^*}}{\dfrac{\left(U^*\right)^2}{r^*}}=\frac{r^*}{t^*U^*}\sim 1.$$

The terms $\rho\frac{\partial u}{\partial t}$ and $\rho\frac{\partial v}{\partial t}$ do not influence the obtained results because both considered terms are small as compared with the viscous terms (see below in this section).

If the droplet has a low slope, then $\varepsilon\ll 1$ and, hence, the velocity scale in the vertical direction is much smaller than the velocity scale in the tangential direction. Using the first Navier–Stokes equation it is estimated that

$$\rho u\frac{\partial u}{\partial x}\sim\rho v\frac{\partial u}{\partial y}\sim\frac{\rho U^{*2}}{r^*},\tag{7.5}$$

$$\eta\frac{\partial^2 u}{\partial x^2}\sim\varepsilon^2\eta\frac{\partial^2 u}{\partial y^2}\ll\eta\frac{\partial^2 u}{\partial y^2},\quad\eta\frac{\partial^2 u}{\partial y^2}\sim\frac{\eta U^*}{h^{*2}}.\tag{7.6}$$

These estimations show that all derivatives in the low-slope approximation in the tangential direction, x, can be neglected as compared with derivatives in the axial direction y. The Reynolds number can be estimated as

$$Re\sim\frac{\rho u\dfrac{\partial u}{\partial x}}{\eta\dfrac{\partial^2 u}{\partial y^2}}\sim\frac{\dfrac{\rho U^{*2}}{r^*}}{\dfrac{\eta U^*}{h^{*2}}}=\frac{\rho U^* h^{*2}}{\eta r^*}=\varepsilon^2\frac{\rho U^* r^*}{\eta},\tag{7.7}$$

or

$$Re=\varepsilon^2\frac{\rho U^* r^*}{\eta}.\tag{7.8}$$

This expression shows that the Reynolds number under the low-slope approximation is proportional to ε^2. Hence, during the initial stage of spreading, when $\varepsilon \sim 1$, the Reynolds number is not small; however, as soon as the low-slope approximation is valid, Re becomes small even if $\rho U^* r^*/\eta$ is not small enough. This means that during the short initial stage of spreading, neither the low-slope approximation nor the low Reynolds number approximation is valid. However, only the main part of the spreading/evaporation process, after the short initial stage is over, is of our concern in this section. It is shown [8] that Re should be calculated only in the close vicinity of the moving contact line, where the low-slope approximation is valid, because in the main part of the spreading droplet, the liquid moves much slower than the liquid located close to the edges moves. Hence, the inertial terms in the Navier–Stokes equations can be safely omitted after the short initial stage and only the Stokes equations should be used:

$$0 = -\frac{\partial p}{\partial x} + \eta\left(\frac{\partial u^2}{\partial x^2} + \frac{\partial^2 u}{\partial y^2}\right), \tag{7.9}$$

$$0 = -\frac{\partial p}{\partial y} + \eta\left(\frac{\partial^2 v}{\partial x^2} + \frac{\partial^2 v}{\partial y^2}\right), \tag{7.10}$$

$$\frac{\partial u}{\partial x} + \frac{\partial v}{\partial y} = 0. \tag{7.11}$$

The capillary number, $Ca = U\eta/\gamma$, characterizes the relative influence of the viscous forces as compared with the capillary forces. To estimate possible values of Ca, let us adopt $r_* \sim 0.1$ cm, $\gamma \sim 30$ dyn/cm, and $\eta \sim 10^{-2}$ P (oils), which are close to our experiments below in this chapter. Let the droplet edge move outward a distance equal to its radius over 1 s, which can be considered as a very high velocity of spreading. This gives the following estimation: $Ca \sim 3 \times 10^{-5} \ll 1$. Therefore, it should be expected for Ca to be even less than 10^{-5} over the duration of the spreading/evaporation process. According to the former, it is assumed that both the capillary and Reynolds numbers are very small except for the very short initial stage of spreading. The duration of the initial stage of spreading, t_0, was estimated in Section 5.1 as beginning immediately after the droplet is deposited on the solid substrate. In the case of aqueous droplets, this time is around $t_0 \sim 10^{-2}$ s.

Let us consider the consequence of the smallness of the capillary number, $Ca \ll 1$, using the same example of the spreading/evaporation of a two-dimensional (cylindrical) droplet. Let the length scales in both the x and y directions in the main part of the spreading droplet be r^*, then the pressure has the order of magnitude of the capillary pressure inside the main part of the droplet, that is, $p \sim \gamma/r^*$. Using the incompressibility condition, it is perceived that the velocity in both directions, u and v, has the same order of magnitude, U^*.

Let us introduce the following dimensionless variables, which are marked by an overbar:

$$\bar{p} = \frac{p}{\gamma/r^*}, \quad \bar{x} = \frac{x}{r^*}, \quad \bar{y} = \frac{y}{r^*}, \quad \bar{u} = \frac{u}{U^*}, \quad \bar{v} = \frac{v}{U^*}.$$

Using these variables, the Stokes equations can be rewritten as

$$\frac{\partial \bar{p}}{\partial \bar{x}} = Ca\left(\frac{\partial^2 \bar{u}}{\partial \bar{x}^2} + \frac{\partial^2 \bar{u}}{\partial \bar{y}^2}\right), \tag{7.12}$$

$$\frac{\partial \bar{p}}{\partial \bar{y}} = Ca\left(\frac{\partial^2 \bar{v}}{\partial \bar{x}^2} + \frac{\partial^2 \bar{v}}{\partial \bar{y}^2}\right). \tag{7.13}$$

It has already been shown that $Ca \ll 1$, which means that the right-hand side of both the above equations (7.12) and (7.13) is very small. Hence, these equations can be rewritten as

$$\frac{\partial \overline{p}}{\partial \overline{x}} = 0 \text{ and } \frac{\partial \overline{p}}{\partial \overline{y}} = 0,$$

which means that the pressure remains constant inside the main part of the spreading droplet.

The normal stress balance on the main part of the spreading droplet is

$$\overline{p} = \frac{\overline{h}''}{\left(1+\overline{h}'^{2}\right)^{3/2}} + Ca\left[-\frac{2}{\left(1+\overline{h}'^{2}\right)}\left\{-\overline{h}'\left(\frac{\partial \overline{u}}{\partial \overline{y}} + \frac{\partial \overline{v}}{\partial \overline{x}}\right) - \frac{\partial \overline{v}}{\partial \overline{y}} - \overline{h}'^{2}\frac{\partial \overline{u}}{\partial \overline{x}}\right\}\right]. \tag{7.14}$$

Using the condition $Ca \ll 1$, the equation simplifies to

$$\overline{p} = \frac{\overline{h}''}{\left(1+\overline{h}'^{2}\right)^{3/2}} = \text{const} \tag{7.15}$$

even in the case where the droplet profile does not satisfy the low-slope approximation, that is, even if $\overline{h}'^{2} \sim 1$ is not small. This shows that the spreading droplet keeps its spherical shape over the main part of the droplet. Note that the radius of the droplet base, $R(t)$, changes over time, and this change results in quasi-steady-state changes of the droplet profile. In the low-slope approximation, the capillary pressure inside the main part of the droplet is $p \sim \gamma/R^{*}$, where $R^{*} \sim r^{*2}/h^{*}$. Thus, the capillary pressure is much smaller than $p \sim \gamma/r^{*}$. Using the dimensionless variables $\overline{p} = \frac{p}{\gamma/R^{*}}$, $\overline{x} = \frac{x}{r^{*}}$, $\overline{y} = \frac{y}{h^{*}}$, $\overline{u} = \frac{u}{U^{*}}$, and $\overline{v} = \frac{1}{\varepsilon}\frac{v}{U^{*}}$ it can be shown that the right-hand side of Eq. (7.12) increases by a factor $(r^{*}/h^{*})^{3}$, and the right-hand side of Eq. (7.13) increases by a factor (r^{*}/h^{*}). Nevertheless, because of the very small value of the capillary number ($<10^{-5}$), the conclusion on the constancy of the pressure inside the main part of the spreading droplet remains valid. During spreading, the ratio $h{:}r$ decreases, and the capillary pressure decreases; however, the velocity of spreading also decreases and, therefore, this above conclusion remains valid for all stages of spreading.

The smallness of Ca means that the surface tension is much more powerful over the main part of the droplet and, hence, the droplet has a spherical shape everywhere except for the vicinity of the apparent three-phase contact line. The size of this region, l^{*}, was estimated in Section 2.3 (Eq. 7.49). It is shown that the following inequality is satisfied: $h^{*} \ll l^{*} \ll r^{*}$. Hence, $\delta = h^{*}/l^{*} \ll 1$ is a small parameter inside the vicinity of the moving contact line. This means that the curvature of the liquid interface inside the vicinity of the moving contact line can be estimated as

$$\frac{\gamma h''}{\left(1+h'^{2}\right)^{3/2}} \sim \frac{\gamma \frac{h^{*}}{l^{*2}}\overline{h}''}{\left(1+\frac{h^{*2}}{l^{*2}}\overline{h}'^{2}\right)^{3/2}} = \frac{\gamma \frac{h^{*}}{l^{*2}}\overline{h}''}{\left(1+\delta^{2}\overline{h}'^{2}\right)^{3/2}} \approx \gamma \frac{h^{*}}{l^{*2}}\overline{h}''. \tag{7.16}$$

Hence, the low-slope approximation is valid inside the vicinity of the moving contact line even if the droplet profile is not very low, that is, even if $\overline{h}'^{2} \sim 1$ is not small. Therefore, the low-slope approximation can always be used inside the vicinity of the moving contact line except for the case when the slope is close to $\pi/2$.

Following arguments developed in Sections 5.1 and 5.2, the whole droplet profile can be subdivided into an "outer" spherical region and an "inner" region in the vicinity of the moving three-phase contact line. The "outer" solution (under a low-slope approximation) is

$$h(t,r) = \frac{2V}{\pi L^{4}}(L^{2}-r^{2}), \quad r < L(t). \tag{7.17}$$

This expression shows that the droplet surface profile remains spherical during the spreading process except for a short initial stage. Equation (7.17) gives the following value of the dynamic contact angle, θ, $(\tan \theta \approx \theta)$:

$$\theta = \frac{4V}{\pi L^3}, \tag{7.18}$$

or

$$L = \left(\frac{4V}{\pi \theta} \right)^{1/3}. \tag{7.19}$$

Note, the contact angle θ is an apparent macroscopic contact angle because it is related to the "outer" spherical region and does not take into account the shape of the "inner" region.

The droplet motion is a superposition of two motions: (a) the spreading of the droplet over the solid substrate, which causes expansion of the droplet base, and (b) shrinkage of the base caused by the evaporation and so the following equation can be written:

$$\frac{dL}{dt} = v_+ - v_-, \tag{7.20}$$

where v_+, v_- are unknown velocities of the expansion and the shrinkage of the droplet base, respectively. The derivative of both sides of Eq. (7.19) gives:

$$\frac{dL}{dt} = -\frac{1}{3} \left(\frac{4V}{\pi \theta^4} \right)^{1/3} \frac{d\theta}{dt} + \frac{1}{3} \left(\frac{4}{\pi V^2 \theta} \right)^{1/3} \frac{dV}{dt}. \tag{7.21}$$

Over the whole duration of the spreading/evaporation, both the contact angle and the droplet volume can only decrease with time. Accordingly, the first term on the right-hand side of Eq. (7.21) is positive, and the second is negative. The comparison of these two equations yields

$$v_+ = -\frac{1}{3} \left(\frac{4V}{\pi \theta^4} \right)^{1/3} \frac{d\theta}{dt} > 0, \tag{7.22}$$

$$v_- = -\frac{1}{3} \left(\frac{4}{\pi V^2 \theta} \right)^{1/3} \frac{dV}{dt} > 0. \tag{7.23}$$

There are two substantially different characteristic time scales in the problem under consideration: $t_\eta^* \ll t^*$, where t_η^* and t^* are time scales of the viscous spreading and the evaporation, respectively; $\lambda = t_\eta^*/t^* \ll 1$ is a smallness parameter (around 0.1 under the chosen experimental conditions). Both time scales are calculated below in this section. Hence, $L = L(T_\eta, T_e)$, where T_η is a fast time of the viscous spreading and T_e is the slower time of the evaporation. The time derivative of $L(T_\eta, T_e)$ is [30]

$$\frac{dL}{dt} = \frac{\partial L}{\partial T_\eta} + \lambda \frac{\partial L}{\partial T_e}. \tag{7.24}$$

Comparison of Eqs. (7.21) through (7.24) shows that

$$v_+ = \frac{\partial L}{\partial T_\eta} = -\frac{1}{3} \left(\frac{4V}{\pi \theta^4} \right)^{1/3} \frac{d\theta}{dt} \tag{7.25}$$

$$v_- = -\lambda \frac{\partial L}{\partial T_p} = -\frac{1}{3}\left(\frac{4}{\pi V^2 \theta}\right)^{1/3}\frac{dV}{dt} \tag{7.26}$$

The decrease of the droplet volume, V, with time is determined solely by the evaporation, hence, the droplet volume, V, depends only on the slow time scale.

According to the previous consideration, the spreading process can be subdivided into two stages:

1. A first fast but short stage, when evaporation can be neglected, and the droplet spreads with an approximately constant volume. This stage goes in the same way as the spreading over dry solid substrates, and the arguments developed earlier for that case can be used here.
2. A second slower stage, when the spreading process is almost over, and the evolution is determined by evaporation.

During the first stage, the dependency of the droplet base radius can be rewritten in the following form Sections 4.1 and 4.2:

$$L(t) = \left[\frac{10\gamma\omega}{\eta}\left(\frac{4V}{\pi}\right)^3\right]^{0.1}(t+t_0)^{0.1}, \tag{7.27}$$

where t_0 is the duration of the initial stage of spreading, when the capillary regime of spreading is not applicable; and ω is an effective lubrication parameter, which was discussed and estimated in Section 4.2. Note, the parameter ω is independent of the droplet volume. According to Eq. (7.27), the characteristic time scale of the first stage of spreading is

$$t_\eta^* = \frac{\eta L_0}{10\gamma\omega}\left(\frac{\pi L_0^3}{4V_0}\right)^3, \tag{7.28}$$

where $L_0 = L(0)$ is the radius of the droplet base in the end of the very fast initial stage of spreading.

Combining Eqs. (7.27) and (7.18) gives

$$\theta = \left(\frac{4V}{\pi}\right)^{0.1}\left(\frac{\eta}{10\gamma\omega}\right)^{0.3}(t+t_0)^{-0.3}. \tag{7.29}$$

Substituting this expression into Eq. (7.22) gives the following expression for the velocity of the droplet base expansion, v_+:

$$v_+ = 0.1\left(\frac{4V}{\pi}\right)^{0.3}\left(\frac{10\gamma\omega}{\eta}\right)^{0.1}\frac{1}{(t+t_0)^{0.9}}. \tag{7.30}$$

According to Eq. (7.1), the rate of evaporation is proportional to the radius of the droplet. Substitution of Eqs. (7.1c), (7.25), (7.26) and (7.30) into Eq. (7.20) results in

$$\frac{dL}{dt} = 0.1\left(\frac{4V}{\pi}\right)^{0.3}\left(\frac{10\gamma\omega}{\eta}\right)^{0.1}\frac{1}{(t+t_0)^{0.9}} - \frac{\alpha L^2}{3V}. \tag{7.31}$$

Note, according to Eq. (7.31), both spreading and evaporation proceed simultaneously. This gives a system of two differential equations (Eqs. 7.1c and 7.31) with the following boundary conditions:

$$V(0) = V_0, \tag{7.32}$$

$$L(0) = L_0 = \left[\frac{10\gamma\omega}{\eta} \left(\frac{4V_0}{\pi} \right)^3 \right]^{0.1} t_0^{0.1},$$ (7.33)

where V_0 is the initial droplet volume; and L_0 is the radius of the droplet after the very fast initial stage is over. Let the system of differential equations (Eqs. 7.1c and 7.31) be made dimensionless using new scales:

$$\bar{L} = L / L_{max}, \quad \bar{t} = t / t_{max}, \quad \bar{V} = V / V_0, \quad \bar{t}_m = t_m / t_{max},$$

where L_{max} is the maximum value of the droplet base, which is reached at the time instant t_m, which is to be determined; and t_{max} is the total duration of the process, that is, the moment when the droplet completely evaporates. Using Eq. (7.1c) in dimensionless form, we conclude $d\bar{V}/d\bar{t} = -t_{max}\alpha L_{max}\bar{L}/V_0$. This equation shows that the characteristic time scale of the evaporation process is $t^* = V_0/\alpha L_{max}$. Hence, the total duration of the whole process can differ from this characteristic time scale only by a constant: $t_{max} = \beta t^*$, where β is a dimensionless number, which is estimated below in this section. Note that t^* and t_{max} (and, hence, β) are calculated using experimental data.

The value of L_{max} is determined using Eq. (7.31) at the moment $t = t_m$, when $dL/dt = 0$. This gives

$$L_{max}^2 = \frac{0.3}{\alpha} \left(\frac{4}{\pi} \right)^{0.3} V_0^{1.3} \left(\frac{10\gamma\omega}{\eta} \right)^{0.1} \frac{1}{\left(t_m + t_0 \right)^{0.9}}.$$ (7.34)

Using this definition, Eqs. (7.1c) and (7.31) can be rewritten as

$$\frac{d\bar{V}}{d\bar{t}} = -\beta\bar{L},$$ (7.35)

$$\frac{d\bar{L}}{d\bar{t}} = \beta \frac{\left(\bar{t}_m + \bar{\tau} \right)^{0.9}}{3(\bar{t} + \bar{\tau})^{0.9}} \bar{V}^{0.3} - \beta \frac{\bar{L}^2}{3\bar{V}}$$ (7.36)

with boundary conditions

$$\bar{V}(0) = 1,$$ (7.37)

and

$$\bar{L}(0) = \bar{L}_0 = \frac{10}{3} \beta \left(\bar{t}_m + \bar{\tau} \right)^{0.9} \bar{\tau}^{0.1},$$ (7.38)

where $\bar{\tau} = t_0 / t_{max} \ll 1$, $\bar{L}_0 = L_0 / L_{max} < 1$.

The system of two ordinary differential equations (Eqs. 7.35 and 7.36) includes the three following dimensionless parameters: $\bar{\tau} = t_0 / t_{max} \lll 1$, $\beta < 1.5$, $\bar{t}_m = t_m / t_{max} < 1$ (see an estimation of the parameter β). The same symbols with an overbar are used for the dimensionless variables as for corresponding dimensional variables. This four parameters should be selected using the following four conditions:

$$\bar{L}(1) = 0,$$ (7.39)

$$\bar{V}(1) = 0,$$ (7.40)

$$\bar{L}(\bar{t}_m) = 1,$$ (7.41)

$$\frac{d\bar{L}(\bar{t}_m)}{dt} = 0. \tag{7.42}$$

Note that these two conditions are used to determine only one parameter, $\bar{t}_m$: The position of the maximum on the dependency $\bar{L}(\bar{t})$ should coincide with that given by Eq. (7.41). That is, there are only three independent conditions in (7.39) through (7.42).

Differential Eqs. (7.35) and (7.36), initial conditions (7.37) and (7.38) and conditions (7.39) through (7.42) do not include any dimensional parameters or their combination. Hence, $\bar{\tau}, \beta, \bar{t}_m$ should be dimensionless universal numbers. This shows, that the solution of Eqs. (7.35) and (7.36), that is, the radius of the droplet base, $\bar{L}(\bar{t})$; the volume, $\bar{V}(\bar{t})$; and the contact angle should be universal functions of dimensionless time.

According to the system of Eqs. (7.35) and (7.36), the dimensionless velocities of spreading and shrinkage caused by the evaporation are as follows:

$$\bar{v}_+ = \beta \frac{(\bar{t}_m + \bar{\tau})^{0.9}}{3(\bar{t} + \bar{\tau})^{0.9}} \bar{V}^{0.3}, \ \bar{v}_- = \beta \frac{\bar{L}^2}{3\bar{V}}. \tag{7.43}$$

Figure 7.7 shows the dimensionless velocity $\bar{v}_+$ and $\bar{v}_-$ calculated according to Eq. (7.43) and the system of Eqs. (7.35) and (7.36). Figure 7.7 shows that

- The duration of the first stage is short. The capillary spreading prevails on this stage over the droplet base shrinkage caused by the liquid evaporation.
- The spreading of the droplet almost stops after the first stage of spreading and the shrinkage of the droplet base is determined by the evaporation of the liquid from the droplet.

Let us consider the asymptotic behavior of system (7.35) and (7.36) during the second stage of the process, when evaporation prevails. Velocity of the expansion of the droplet, v_+, decreases over the second stage of the spreading according to Figure 7.7. To understand the asymptotic behavior, the term corresponding to v_+ in the right-hand side of Eq. (7.36) is neglected. This gives

$$\frac{d\bar{L}}{dt} = -\beta \frac{\bar{L}^2}{3\bar{V}}, \tag{7.44}$$

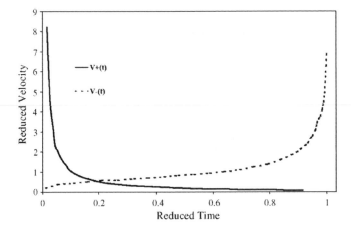

FIGURE 7.7 Dimensionless spreading (v_+) and evaporation (v_-) velocities according to Eq. (7.43).

whereas Eq. (7.35) is left unchanged. The system of differential Eqs. (7.35) and (7.44) can be solved analytically. For this purpose, Eq. (7.44) is divided by Eq. (7.35), giving $d\bar{L}/d\bar{V} = \bar{L}/3\bar{V}$. This equation can be easily integrated, and the solution is,

$$\bar{V} = C\,\bar{L}^3,\tag{7.45}$$

where C is an integration constant.

Rewriting Eq. (7.18) using the same dimensionless variables results in

$$\bar{V} = \frac{\pi L_{\max}^3}{4V_0}\,\theta\,\bar{L}^3.\tag{7.46}$$

Comparing Eqs. (7.45) and (7.46) shows that the dynamic contact angle asymptotically remains constant over the duration of the second stage of evaporation. This constant value is marked as θ_f. Introducing $\theta_m = 4V_0/\pi L_{\max}^3$, which is the value of the dynamic contact angle at the time instant when the maximum value of the droplet base is reached. Then Eq. (7.46) can be rewritten as

$$\frac{\theta}{\theta_m} = \frac{\bar{V}}{\bar{L}^3}\tag{7.47}$$

and this relationship should be a universal function of the dimensionless time, $\bar{t}$. This equation shows that the integration constant in Eq. (7.45) is $C = \theta_f/\theta_m$.

Let us estimate the constant β. For this purpose, we solve equations that describe the time evolution during the second stage and completely neglect the presence of the first stage. That is, from Eqs. (7.35) and (7.36),

$$\frac{d\bar{V}}{d\bar{t}} = -\beta\bar{L},\tag{7.48}$$

$$\frac{d\bar{L}}{d\bar{t}} = -\beta\frac{\bar{L}^2}{3\bar{V}},\tag{7.49}$$

with the following conditions:

$$\bar{L}(0) = 1,\tag{7.50}$$

$$\bar{L}(1) = 0.\tag{7.51}$$

Note, the first condition, Eq. (7.50), is not valid during the first stage of spreading. The solution of the problem (7.48) and (7.49) with boundary condition (7.51) is given by

$$\bar{L}(\bar{t}) = \left(\frac{2\beta\theta_m}{3\theta_f}\right)^{0.5}(1-\bar{t})^{0.5}.$$

This dependency coincides with the predicted and experimentally confirmed earlier in Sections 7.2. Using now the boundary condition (7.50), we conclude

$$\beta = \frac{3\theta_f}{2\theta_m} < 1.5$$

because obviously $\theta_f < \theta_m$. This conclusion is in good agreement with experimental data.

Experimental

Materials

The alkanes used in our experiments were *n*-heptane and *i*-octane purchased from Sigma-Aldrich, UK. The solid substrates used were microscope glass slides, which were cleaned prior each experiment according to the following protocol: (i) soaking with isopropyl alcohol to remove organic contaminants for 30 min and rinsed with deionized water; (ii) soaking in chromic acid for removal of inorganic contaminants for 50 min; (iii) extensive rinsing with distilled and deionized water; (iii) drying in a regulated oven. The cleaning procedure was repeated after each experiment.

Methodology

The time evolution of the radius of a drop base, $L(t)$, and the dynamic contact angle, $\theta(t)$, were monitored simultaneously (Figure 7.8).

The substrate was placed in a hermetically closed and insulated chamber, allowing strict control of the chamber's environment. The chamber was also equipped with a fan (1,000 rpm). An experimental chamber used was described in Chapter 4. A high precision 10-μL Hamilton syringe (Hamilton GB Ltd., UK) was used to inject 3-μL droplets of alkanes onto the solid substrate. A mechanical manipulator was designed to enable the gentle placement of the droplet on the substrate while minimizing the kinetic impact. Experiments were carried out at these conditions for both *n*-heptane and *i*-octane: (i) at 25°C without the fan switched on, (ii) at 40°C without the fan switched on, and (iii) at 25°C with the fan switched on.

The spreading/evaporation processes were captured at 60 frames per second for the whole experiment duration using two video cameras, which simultaneously captured a side view and a view from above. The side-view video images were analyzed using Scion Image software to measure the contact angle and height of the droplet, whereas the top view images were analyzed using Olympus i-Speed software to measure the radius of the droplet base. The experiment was repeated to produce at least five sets of data for each individual condition. Hence, each experimental point plotted is an average of five experimental points.

Results and Discussions

Our experimental observations, as well as those presented in [25–29], show that the whole spreading/evaporation process in the case of complete wetting can be divided into two stages: (a) the fast first stage, when the drop spreads out until a maximum radius of the drop base, L_m, is reached, which is followed by the second slower stage (b) when the radius of the drop base shrinks because of evaporation until it disappears completely at the moment t_{max}. According to the theoretical requirements, from the previous

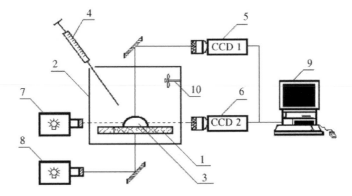

FIGURE 7.8 Schematic diagram of experimental setup. (1) Glass slide; (2) enclosed chamber; (3) tested drop; (4) syringe; (5,6) CCD cameras; (7,8) illuminators; (9) CPU unit; (10) power-assisted fan.

TABLE 7.1

Comparison of Initial Spreading, t_m, and Total Time, t_{max}

Materials	T (°C)	Fan Speed (rpm)	t_m (s)	t_{max} (s)	$\bar{t}_m$
Heptane	25	0	4	85	0.047
	40	0	3	35	0.086
	25	1,000	4	80	0.05
Octane	25	0	3	73	0.041
	40	0	2	29	0.069
	25	1,000	4	66	0.06

section. the droplets must be of a spherical shape throughout the spreading/evaporation process, and the duration of the first stage of spreading, t_m, much smaller compared to t_{max}. As observed in our experiments and in [25–29], both of these requirements were met. Table 7.1 shows that values of t_m were always less than 10% of those of t_{max} regardless of the experimental conditions used.

Extracting Theoretical Parameters from Experimental Data

Experimental values of the total duration of the spreading/evaporation process, t_{max}, of the duration of the first stage of the process, t_m, the maximum value of the spreading radius, L_m, and the contact angle, θ_m, were extracted directly from experimental data. The volume was calculated by rearranging Eq. (7.18) to $V(t) = \frac{\pi}{4} L^3(t)\theta(t)$.

The parameter α in Eq. (7.1) was calculated in the following way. Eq. (7.1c) was integrated, resulting in:

$$V = V_0 - \alpha \int_0^t L(t)dt. \tag{7.52}$$

The experimental dependencies of the volume of the droplet on time, $V(t)$, were plotted against $\int_0^t L(t)dt$. In all cases, $V(t)$ showed a linear trend (Figure 7.9). Using these linear dependencies, the proportionality coefficient, α, was fitted. The fitted values of α are presented in Table 7.2. After that, all necessary parameters were known.

Note that the parameter α for experiments conducted at 25°C is significantly lower as compared with that at 40°C (Table 7.2). This is expected because the evaporation rate increases at higher temperatures. However, the introduction of a fan in our experiments does not appear to have affected the evaporation rate at all.

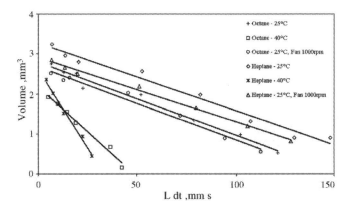

FIGURE 7.9 Comparison of the trends in volume changes between the different sets of variables used.

TABLE 7.2

Fitted Values of Parameter α in Eq. (7.52)

Operating Conditions	Fitted α	
	Heptane	Octane
25°C Fan Speed 0	0.017	0.018
40°C Fan Speed 0	0.078	0.045
25°C Fan Speed 1,000rpm	0.016	0.018

Note: In all cases, $V(t)$ dependency showed a linear trend for V vs. $\int_0^t L(t)dt$.

Comparison of Experimental Data and Predicted Universal Behavior

The solid lines in Figures 7.10 through 7.12 represent theoretical universal dependencies, which were calculated according to Eqs. (7.35) and (7.36), with boundary conditions (7.37) and (7.38). The four unknown parameters were selected according to conditions (7.39) through (7.42). The selected parameters are $\bar{\tau} = 2 \times 10^{-3}$, $\beta = 1.281$ (which is in good agreement with the previous estimation), $\beta = \frac{3\theta_f}{2\theta_m} < 1.5$, and the selected value of $\bar{t}_m \approx 0.261$ in Eq. (7.36) coincides with the maximum position of the theoretical dependency, $\bar{L}(\bar{t})$ (Figure 7.10). Experimental data on the spreading/evaporation of different liquids extracted from the literature has been plotted in the same dimensionless form. We managed to extract only dimensionless dependences, $\bar{L}(\bar{t})$, from the literature [25–29].

Figure 7.10 shows our theoretical model plotted against data of different liquids ranging from water, alkane, and silicon oil from the literature [25–29] in the form of the dimensionless radius of the droplet base, $\bar{L}$, versus dimensionless time, $\bar{t}$.

Figure 7.11 displays the data obtained from our experiments for the contact angle measurement and the data obtained in [25–29] compared with the predicted universal curve. As before, the data are plotted as dimensionless contact angle, $\bar{\theta} = \theta/\theta_m$, against $\bar{t}$ (Figure 7.11). The contact angle dropped very quickly once it was placed on the substrate. The contact angle then slowed its decline and reached a steady-state contact angle, θ_f. Like the radius, the contact angle follows a universal trend (Figure 7.11), which agrees well with the theoretical prediction as well as with the literature data.

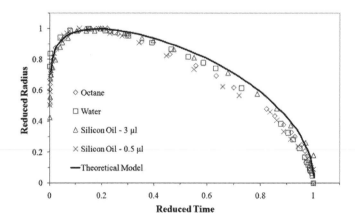

FIGURE 7.10 Dimensionless radius against dimensionless time curve for the behavior of the droplet radius comparing different liquids spreading/evaporating on solid substrates extracted from literature [25–29] and theoretical prediction. The solid line is calculated according to Eqs. (7.35) and (7.36).

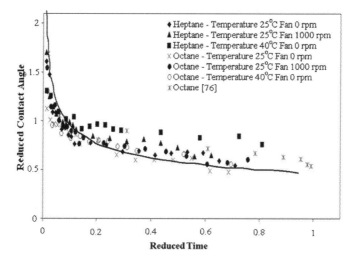

FIGURE 7.11 Dimensionless contact angle, $\bar{\theta} = \theta / \theta_m$, compared against dimensionless time curve for the behavior of the droplet radius comparing theoretical, our experimental, and literature data. The solid line is calculated according to Eqs. (7.35) and (7.36).

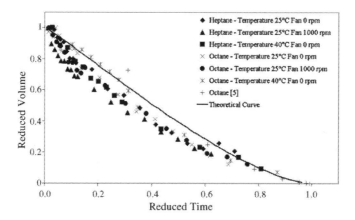

FIGURE 7.12 Dimensionless volume compared against dimensionless time curve for the behavior of the droplet radius comparing theoretical, our experimental, and literature data. The solid line is calculated according to Eqs. (7.35) and (7.36).

The presented theory also predicted a dimensionless universal behavior of volume on time dependency. Experimental results are presented in Figure 7.12. Like the previous two universal curves, both our experimental data and the literature data agree well with the theory predictions.

Conclusions

Experiments were designed to allow monitoring the radius and the contact angle of spreading/evaporating droplets in the case of complete wetting. During the spreading/evaporating process, the droplet retained a spherical form. The whole process is clearly divided into two stages: the fast first stage, when the process is mostly determined by the kinetics of spreading and a second slower stage, when the process is mostly determined by evaporation. The duration of the first stage compared to the overall duration of the process time taken by the droplet to evaporate was found to be considerably smaller. We also confirmed the change in the drop volume with respect to the radius of the drop base to be a linear relation (Figure 7.9).

A theoretical model was developed to account for the spreading/evaporation process, which produced universal theoretical curves for the radius, contact angle, volume, and spreading/evaporation velocities on time.

We compared experimental results, both extracted from the literature and ours, to that of the theoretical predictions. We found that the experimental data corresponds well with the predicted universal behavior regardless of the different conditions under which the experiments were conducted.

List of Main Symbols

Latin

C	Integration constant
a	Capillary length
g	Gravity acceleration
h	Height of droplet
L	Radius of the droplet base
p	Pressure
t	Time
u, v	Tangential and vertical velocity component
V	Volume
x, y	Coordinates

Greek

α	Proportionality constant
γ	Interfacial tension
δ	h^*/l^*, parameter inside the vicinity of the moving contact line
ε	h^*/r^*, slope of the droplet
η	Dynamic viscosity of a liquid
θ	Dynamic contact angle of the droplet
λ	$t_\eta^* / t_{\max}$
ρ	Droplet density
τ	$t_0/t_{\max}$
ω	Effective lubrication parameter

Subscripts

0	Initial
e	Evaporation stage
η	Viscous
m	Correspond to the moment when the droplet reaches its maximum radius
	max Maximum
+	Spreading
−	Shrinkage

Superscripts

*	Characteristic value
−	Dimensionless

7.3 Evaporation of Sessile Droplets: Universal Behavior in the Presence of Contact Angle Hysteresis

In this section, we present a theory describing the diffusion-limited evaporation of sessile water droplets in the presence of contact angle hysteresis. Theory describes two stages of evaporation process: (stage 1) evaporation with a constant radius of the droplet base; and (stage 2) evaporation with a constant contact angle. During stage 1, the contact angles decrease from the static advancing contact angle to the static receding contact angle. During stage 2, the contact angle remains equal to the static receding contact angle. Universal dependences are deduced for both evaporation stages. Obtained universal curves are validated against available in the literature experimental data.

Introduction

We mentioned in the introduction to this chapter that, from the scientific point of view, the problem of evaporating of a sessile droplet is of substantial interest because of singularity problems arising at the three-phase contact line: (i) a singularity of evaporation flux due to an incompatibility of boundary conditions at the liquid–gas interface and at the solid–liquid and solid–gas interface; and (ii) the viscous stress singularity also appears at the apparent three-phase contact line due to no-slip boundary condition at the solid surface that is usually used in continuum hydrodynamics. The solution of these problems on a microscale level and obtaining corresponding macroscopic boundary conditions is one of the research goals in the field that can be resolved based on consideration of surface forces action (Derjaguin's pressure) in the vicinity of the apparent three-phase contact line.

It was established in Section 7.1 that, in the presence of contact angle hysteresis, the evaporation of a sessile droplet in nonsaturated vapor atmosphere goes through four consecutive stages: (1) a spreading until the value of the static advancing contact angle, θ_{ad}, is reached; (2) evaporation proceeds with a constant contact area and decreasing contact angle, θ, until the contact angle reaches the static receding value, θ_r; (3) evaporation with a constant contact angle, θ_r, and decreasing radius of the contact line, L; (4) evaporation with decrease of both L and θ until droplet disappears. The duration of stage 1 is very short because advancing goes via "caterpillar motion" (see Chapter 3) and, hence, the evaporation can be neglected during stage 1 and can be described using a conventional hydrodynamic approach (see Chapter 3). Only stages 2 and 3 are considered below in this section.

In the introduction and Sections 7.1 and 7.2, it was shown that the evaporation of sessile droplets obeys the following equation:

$$\frac{dV}{dt} = -2\pi \frac{DM}{\rho}\left(c_{\text{sat}}\left(T_{\text{surf}}\right) - H c_{\text{sat}}\left(T_{\infty}\right)\right)F\left(\theta\right)L, \tag{7.1a}$$

where V is the droplet volume; t is time, D, ρ, and M are vapor diffusivity in the air, density of the liquid, and the molar mass, respectively; H is the humidity of the ambient air; T_{surf} is the temperature of the droplet–air interface; T_{∞} is the temperature of the ambient air; $c_{\text{sat}}(T_{\text{surf}})$ and $c_{\text{sat}}(T_{\infty})$ are the molar concentrations of saturated vapor at the corresponding temperature; and $F\left(\theta\right)$ is a function of the contact angle, θ, with value 1 at $\theta = \pi/2$. T_{surf} was introduced in Section 7.1 as the average temperature over the surface of evaporating droplet.

In Section 7.1, we investigated using computer simulations the evaporation of a droplet of a pure liquid, taking into account the heat transfer in the whole system, vapor diffusion into the ambient air, and the Marangoni convection inside the droplet. Computer simulations allowed us to investigate the influence of the latent heat of vaporization and thermal Marangoni convection on the evaporation rate, J. It was shown that one of the important parameters influencing the process of evaporation is the average temperature, T_{surf}, of the surface of evaporating droplet.

Now, we deduce a universal law of evaporation for sessile droplets of pure liquid in the presence of contact angle hysteresis. This law of evaporation is based on the results of computer simulations that were briefly discussed in Section 7.1.

Theory

Problem Statement

The droplets in question are supposed to have size $L \ll a$, where $a = \sqrt{\gamma/\rho g}$ is the capillary length, γ is the surface tension of liquid–air interface, ρ is the liquid density, and g is the gravitational acceleration. For water droplets $a \approx 2.7$ mm. Thus, gravity force can be neglected, and the liquid–air interface has the shape of a spherical cap (Figure 7.13) as a direct consequence of the action of surface tension and low Ca number (see the introduction to Chapter 7). The evaporation process is governed by diffusion in the ambient air, which imposes the lower limit for the droplet size, which is on the order of 10^{-7} m (see Section 7.5).

According to the Kelvin equation, the concentration of saturated vapor above the droplet's surface depends on the curvature of this surface. Estimations show that for droplet size more than 10^{-7} m, variation in the concentration due to surface curvature are less than 1%. Therefore, variation of the concentration of the saturated vapor is not taken into account here. The problem is solved using a steady-state approximation.

Two Stages of Evaporation of a Sessile Droplet

In the presence of contact angle hysteresis, the whole duration of the evaporation process can be subdivided into three stages (Figure 7.2). We consider only stages 2 and 3 (Figure 7.2).

1. The second stage: The radius of the contact line, L, remains constant and equal to its initial value L_0, which is the maximum value of the radius of the droplet base. At the same time, the contact angle, θ, decreases from its initial value, which is equal to the static advancing contact angle, θ_{ad}, to the final value, which is equal to the static receding contact angle, θ_r. The moment when the second stage started is adopted as a zero moment, $t = 0$. At this moment, the droplet base reaches its maximum value and the contact angle is equal to the static advancing contact angle, θ_{ad}.
2. The third stage: The contact angle remains constant and equal to θ_r while the radius of the contact line decreases from L_0 to almost zero value.

In the same way as in Section 7.2, we can show that during both stages of evaporation the droplet retains the spherical shape because of a very small capillary number, Ca. That is, the volume of the droplet, V, can be presented as follows:

$$V = L^3 f(\theta), \qquad f(\theta) = \frac{\pi}{3} \frac{(1-\cos\theta)^2 (2+\cos\theta)}{\sin^3 \theta}. \tag{7.53}$$

Taking into account the results of computer simulations (Section 7.1), we can rewrite Eq. (7.1a) as

$$\frac{dV}{dt} = -\beta F(\theta) L, \tag{7.54}$$

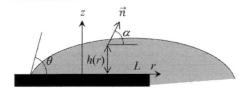

FIGURE 7.13   Illustration of droplet parameters: L, $h(r)$, θ and $\vec{n}$ in the case $\theta < \pi/2$.

where

$$\beta = 2\pi \frac{DM}{\rho} \left[c_{\text{sat}} \left(T_{\text{surf}} \right) - c_\infty \right], \text{ where } T_{\text{surf}} = \int_S T \, dS. \tag{7.55}$$

According to our previous consideration, during both stages of evaporation the mass conservation law has the form given by Eq. (7.54). Taking into account the consideration in Section 7.2, the parameter β is adopted as a constant.

The Second Stage of Evaporation

During the second stage of evaporation (Figure 7.2), the radius of the contact line remains constant, and Eqs. (7.53) and (7.54) can be rewritten as $L_0^3 f'(\theta) \frac{d\theta}{dt} = -\beta F(\theta) L_0$, or

$$L_0^2 f'(\theta) \frac{d\theta}{dt} = -\beta F(\theta), \tag{7.56}$$

with the initial condition

$$\theta\big|_{t=t_0} = \theta_{ad}. \tag{7.57}$$

Let us introduce the following dimensionless time $\tau = \frac{t}{t_{ch}}$, where $t_{ch} = \frac{L_0^2}{\beta}$ is the characteristic time of the process. Eq. (7.56) now takes the following form:

$$f'(\theta) \frac{d\theta}{d\tau} = -F(\theta). \tag{7.58}$$

Direct integration of this equation with the boundary condition (7.57) results in

$$A(\theta, \theta_{ad}) = \tau, \tag{7.59}$$

where $A(\theta, \theta_{ad}) = \int_\theta^{\theta_{ad}} \frac{f'(\theta)}{F(\theta)} d\theta$. This equation shows that the deduced dependency should be universal and does not depend on the nature of the liquid and the droplet volume. The only parameter left is the initial contact angle (static advancing contact angle), which is supposed to be independently determined (see Chapter 3).

The first stage proceeds until the contact angle reaches its final value equal to the static receding contact angle. Using Eq. (7.59), we conclude that the end of the first stage, $\tau_r = \frac{t_r}{t_{ch}}$, is determined as

$$A(\theta_r, \theta_{ad}) = \tau_r. \tag{7.60}$$

The Third Stage of Evaporation

During the third stage of evaporation (Figure 7.2), the contact angle remains constant, but the radius of the contact line varies. Hence, Eqs. (7.53) and (7.54) can now be rewritten as $3L^2 f(\theta_r) \frac{dL}{dt} = -\beta F(\theta_r) L$. Let us introduce the same dimensionless time as before and dimensionless radius of the contact line: $\ell = L/L_0$. Hence this equation can be rewritten as

$$\frac{d\ell^2}{d\tau} = -\frac{2}{3} \frac{F(\theta_r)}{f(\theta_r)}, \quad \tau > \tau_r \tag{7.61}$$

with the following initial condition: $\ell(\tau_r) = 1$. Direct integration of this equation results in $\ell^2(\tau) = 1 - \frac{2F(\theta_r)}{3f(\theta_r)}(\tau - \tau_r)$, or

$$\ell(\tau) = \sqrt{1 - \frac{2F(\theta_r)}{3f(\theta_r)}(\tau - \tau_r)}. \tag{7.62}$$

This dependence gives a universal dependence during the third stage of evaporation.

The resulting curves are schematically represented in Figure 7.14.

Validation against Experimental Data

The universal laws of evaporation of a sessile droplet, represented by Eqs. (7.59) and (7.62), are validated against available experimental data extracted from literature sources [31–33]. The characteristic time of the process can be calculated using Eq. (7.62): $t_{ch} = \frac{t_r}{A(\theta_r, \theta_{ad})}$. This can be calculated using experimental values of θ_{ad} and θ_r. The obtained value of t_{ch}, as well as θ_{ad} and θ_r, can be used to plot the dependencies $\theta(\tau)$ and $\ell(\tau)$ corresponding to the second (Figure 7.15) and third (Figure 7.16) stages of evaporation, respectively. Figures 7.15 and 7.16 show good agreement of the proposed theory and the available experimental data.

Using the proposed theory, we can estimate the average temperature, T_{surf}, of the droplet's surface. Substituting the experimentally calculated value of $t_{ch} = \frac{t_r}{A(\theta_r, \theta_{ad})}$ into the theoretical equations $t_{ch} = \frac{L_0^2}{\beta}$, and using the initial value of contact line radius, L_0, we can obtain the value of β. After that, using Eq. (7.55) for β and the dependence of concentration of saturated vapor, $c_{sat}(T)$, on temperature (the Clausius-Clapeyron equation) [11], we can calculate the average temperature, T_{surf}, of the droplet's surface. Thus, for the experimental data represented in Figures 7.15 and 7.16, we calculated the following values of T_{surf}: for water on corning glass $T_{surf} = T_\infty - 2.21\text{K}$, for water on PMMA $T_{surf} = T_\infty - 2.76\text{K}$, and for water on PET $T_{surf} = T_\infty - 0.52\text{K}$. It was impossible to calculate T_{surf} for the experiments presented in paper [32] (water on polished epoxy surface) because the humidity of ambient air was not specified in their experiments. Average temperatures of the droplet surface, T_{surf}, calculated using the suggested algorithm, are smaller than the ambient temperature, T_∞, which indicates the surface cooling. We can see that calculated temperature drops, $T_{surf} - T_\infty$, are of the same order of magnitude as the experimentally measured temperature drop inside the droplet bulk [22] (Figure 7.4). This confirms the validity of the proposed assumptions. Note, a real substrate used in any experiment is of a finite size, and the calculated temperature drop, $T_{surf} - T_\infty$, can be used for the estimation of thermal conductivity of an equivalent semi-infinite substrate, giving the same cooling as found for the real substrate of finite thickness.

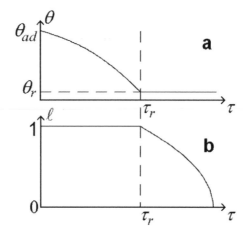

FIGURE 7.14 Dependencies on the dimensionless time: (a) contact angle; and (b) radius of the contact line.

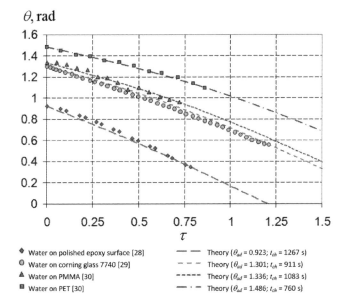

FIGURE 7.15   Second stage of evaporation, Eq. (7.59). Dependence of contact angle, θ, on dimensionless time, τ.

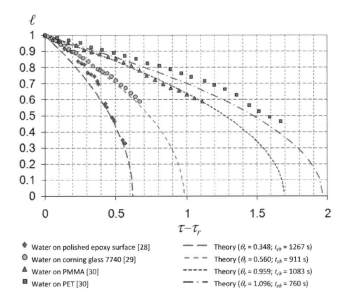

FIGURE 7.16   Third stage of evaporation, Eq. (7.62). Dependence of dimensionless radius of contact line, ℓ, on dimensionless time, τ. τ_r is the duration of the second stage.

Let us introduce new dimensionless times. Eq. (7.59) can be rewritten as

$$\int_{\theta}^{\pi/2} \frac{f'(\theta)}{F(\theta)} d\theta = \tau + \int_{\theta_{ad}}^{\pi/2} \frac{f'(\theta)}{F(\theta)} d\theta, \tag{7.63}$$

or

$$B(\theta) = \tilde{\tau}, \tag{7.64}$$

where $B(\theta) = \int_\theta^{\pi/2} \frac{f'(\theta)}{F(\theta)} d\theta = A(\theta, \pi/2)$, and $\tilde{\tau} = \tau + B(\theta_{ad})$ is a new dimensionless time. Function $A(\theta_1, \theta_2)$ now can be expressed as follows: $A(\theta_1, \theta_2) = B(\theta_1) - B(\theta_2)$. According to its derivation, Eq. (7.64) is supposed to be the universal dependence describing the first stage of evaporation. This is completely confirmed by comparison with available experimental data in Figure 7.17.

The introduction of the new dimensionless time $\overline{\tau} = \frac{2F(\theta_r)}{3f(\theta_r)}(\tau - \tau_r)$ to the Eq. (7.62) gives

$$\ell(\overline{\tau}) = \sqrt{1 - \overline{\tau}}. \tag{7.65}$$

Again, according to its derivation, Eq. (7.65) is supposed to represent the universal curve describing the second stage of evaporation. Experimental data in Figure 7.18 confirm that conclusion.

In the presented consideration, we used a pure diffusion model of evaporation.

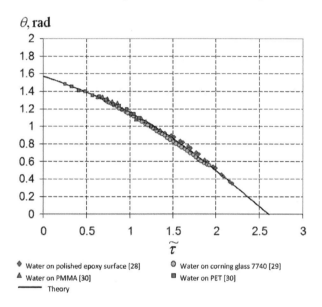

FIGURE 7.17 First stage of evaporation, Eq. (7.62). Dependence of contact angle, θ, on dimensionless time, $\tilde{\tau}$.

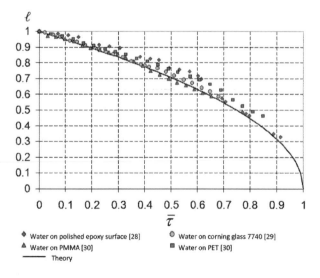

FIGURE 7.18 Second stage of evaporation, Eq. (7.65). Dependence of dimensionless radius of contact line, ℓ, on dimensionless time, $\overline{\tau}$.

Conclusions

In this section, we presented a theory describing the evaporation of sessile water droplets. In the presence of contact angle hysteresis, the evaporation process can be subdivided into three stages. Introducing a dimensionless contact line radius and dimensionless times for the longest stages, stages 2 and 3, allowed us to deduce universal laws describing these two stages of the evaporation process in the case of contact angle hysteresis. The theory describes two stages (L = const at the second stage, and $\theta = \theta_r$ = const at the third stage) and shows good agreement with experimental data available in the literature.

7.4 Spreading and Evaporation of Surfactant Solutions

Introduction

Simultaneous spreading and evaporation of droplets of trisiloxane solutions (super spreaders) over hydrophobic substrates were studied both experimentally, using a video-microscopy technique, and theoretically. The experiments were carried out over a wide range of surfactant concentration at three different temperatures and at three values of relative humidity. Four different stages were detected: the first stage corresponds to spreading until the contact angle reaches the value of the static advancing contact angle, θ_{ad}. The duration of this stage is rather short because spreading occurs through "caterpillar motion" in the case of partial wetting (see Chapter 3). Hence, the evaporation during the short first stage can be neglected. The evaporation is essential during the next two stages. The second stage takes place at a constant perimeter and ends when the contact angle reaches the static receding contact angle, θ_r. During the third stage, the perimeter decreases at a constant contact angle, θ_r. During the final short fourth stage, both the perimeter and the contact angle decrease until the drop disappears. The physical phenomena behind the appearance of the fourth stage are yet to be understood. Note, the fourth stage usually does not appear during the spreading/evaporation of droplets of pure liquids in the case of partial wetting (Figure 7.2).

The developed theory predicts universal curves for the longest stages: the contact angle during the second stage and the droplet perimeter during the third stage. A good agreement between the theory predictions and experimental data has been found for the second stage of evaporation, and for the third stage for concentrations above the critical aggregation concentration (CAC), where the surface tension is not concentration dependent. However, some deviations were found for the low concentration range where there is a sharp change of the surface tension as the surfactant concentration increases in the course of evaporation. The agreement becomes better as the concentration approaches the CAC.

In Section 7.3, we established some basic results for the evaporation of a drop under partial wetting conditions:

1. If droplet evaporation is limited by vapor diffusion into surrounding gas, then the evaporation rate is proportional to the radius of the drop onto the substrate, L.

2. The spreading and evaporation process is composed of four stages: (1) L increases while the contact angle, θ, decreases down to the static advancing contact angle value, θ_{ad}. (2) The contact angle decreases from θ_{ad} down to static receding contact angle value, θ_r, at constant L. (3) The contact angle remains constant and equal its receding value, θ_r, while the radius of the base droplet, L, decreases (Figure 7.2).

However, in the case of spreading/evaporation of surfactant solutions, there is one extra stage: (4) Both the contact angle and L decrease until the drop completely disappears. This stage probably cannot be understood without consideration of surface forces (Derjaguin's pressure) action in the vicinity of the moving apparent three-phase contact line.

In Section 7.1, we presented the results of computer simulations of the evaporation of a drop of pure fluid. The results showed that the evaporation process of small (no gravity effects, but still macroscopic, that is, the droplets are bigger than 1 μm [see Section 7.5]) sessile droplets in the presence of contact angle hysteresis can be subdivided into the several stages mentioned. Introducing a dimensionless contact line radius and dimensionless times for the second and third stages of evaporation allowed us to deduce universal laws describing the experimental data for water droplets onto various substrates in the presence of contact angle hysteresis (Section 7.3). Based on these results, a model was proposed capable of explaining quantitatively the second and third stages of the evaporation of pure fluids.

The model presented in Section 7.3 allowed us to build universal curves of the time dependencies of contact angle and droplet base radius, L in the case of spreading/evaporation of surfactant solution droplets.

The aim of this section is to perform a detailed experimental study of the time dependencies of the contact angle, the volume and the radius of aqueous surfactant solution drops onto a hydrophobic TEFLON-AF substrate. We used drops of an aqueous solution of a super-spreader surfactant (Silwet L77) over a wide concentration range both below and above the CAC. Only the experimental data obtained for the second and third stages are compared with the theoretical and computer simulation results here.

Experimental

SILWET L77 was purchased from Sigma-Aldrich (Germany) and used as received. Poly(4,5-difluoro-2,2-bis(trifluorimethyl)-1,3-dioxole-co-tetrafluoroethylene), hereinafter TEFLON-AF, was purchased from Sigma-Aldrich (Germany) as powder; the Flourinet F-77 solvent was bought from 3M (USA); and the silicon wafers were obtained from Siltronix (France). Ultrapure deionized water (Younglin Ultra 370 Series, Korea) with a resistivity higher than 18 MΩ and TOC (Total Organic Compound) lower than 4 ppm was used for preparing all the surfactant solutions.

All the surfactant solutions were prepared by weight using a balance precise to ± 0.01 mg. A pH 7.0 buffer was used as the solvent to prevent hydrolysis of the SILWET L77. The solutions were used immediately after preparation. The silicon wafers were cleaned using piranha solution for 20 min (caution, piranha solution is highly oxidizing!). The solid substrates were prepared as follows: the TEFLON-AF powder was suspended in the Flourinet F77 and spin-coated onto the silicon wafers. The average roughness of the 20×20 µm surface was ≈ 1.0 nm as measured by AFM (Atomic Force Microscope) (tapping mode). The macroscopic contact angle of pure water was ($104° \pm 2°$) on those substrates. Drops of 4 mm^3 were deposited onto the substrate for measurements. Five independent measurements were performed for each experimental point reported in this work, and the average was used.

The experimental technique used was similar to the one used earlier in Sections 5.2 and 5.5 (see also [34,35]) with some modifications that allowed us to continuously monitor the temperature and the relative humidity inside the experimental set up. Figure 7.19 shows a diagram of the experimental device. The cameras were calibrated using a micro ruler with a precision of ± 0.5 µm.

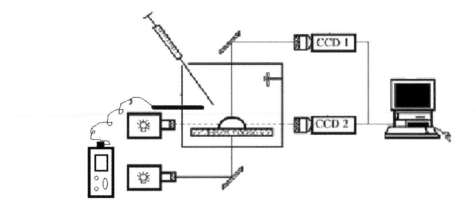

FIGURE 7.19 Diagram of the experimental device used. CCD1 and CCD2 are the cameras to capture the drop profiles from the top and side view. Inside the chamber both the temperature and the relative humidity were controlled and continuously monitored.

Sessile drops were deposited onto the substrate inside a chamber attached to a thermostat; their shape and size were captured by the charged coupled device (CCD) camera (side view) at 30 frames per second. The initial drop volumes used were about 4 mm³ to ensure that the gravity effects can be neglected, and the drops always had a spherical cap shape. The images captured were analyzed using drop tracking and evaluation analysis software (Micropore Technologies, UK) that allowed us to monitor the time evolution of the drop base diameter, height, radius of the curvature, and contact angle of the drops. The precision of the contact angle measurements was ±2° under dynamic conditions, that is, spreading and evaporation, those of height and diameter were ±1 μm and that of the temperature was ±0.5°C. The relative humidity was maintained constant by placing a saturated salt solution inside the measuring chamber, and it was measured with a precision of ±2%.

Results and Discussion

Figure 7.20 shows the typical behavior of contact angle, θ, the radius of the base of the droplet, L, and the volume to the power 2/3, $V^{2/3}$, for a SILWETT L77 solution of concentration $c = 0.25 \cdot$CAC (CAC being the critical aggregation concentration of the surfactant, CAC = 0.1 g/L) at ambient temperature 18°C and 90% relative humidity. The end of each of the three first stages of the spreading/evaporation process is marked by a vertical bar in Figure 7.20. Note the presence of the fourth stage of spreading/evaporation when both the radius of the droplet base and the contact angle change simultaneously.

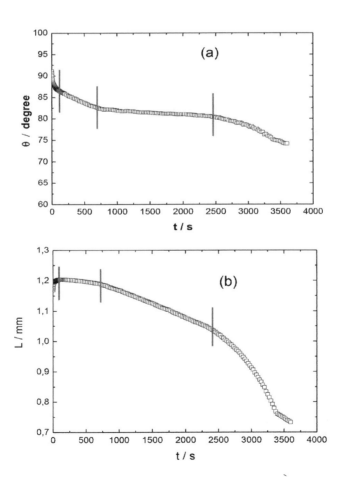

FIGURE 7.20 Process of spreading/evaporation of a sessile droplet with the dependence of (a) contact angle, (b) contact radius. *(Continued)*

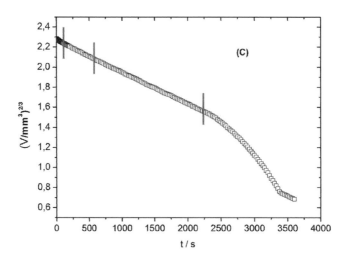

FIGURE 7.20 (Continued) Process of spreading/evaporation of a sessile droplet with the dependence of (c) droplet volume to the power 2/3 for a surfactant critical aggregation concentration (CAC) of 0.25 at ambient temperature 18°C and 90% relative humidity. The vertical red bars mark ends of stages 1 to 3 of the spreading/evaporation process, respectively.

The characteristic times of the spreading behavior were found to be in the range of 20–50 s, values similar to those of [35] for trisiloxanes. The experimental results for the droplet volume show that $V^{2/3}$ decreases linearly with time during first three stages of the process, thus it can be inferred that evaporation is diffusion controlled, as in the case of the sodium dodecyl sulfate (SDS) solutions [36].

Figure 7.21 shows the dependence of advancing, θ_{ad}, and receding, θ_r, contact angles on surfactant concentration, C, for aqueous solutions of surfactant SILWET L-77. One can see from Figure 7.21 that the receding contact angle, θ_r, levels off approximately at the CAC; and the advancing contact angle, θ_{ad}, levels off approximately at the critical wetting concentration (CWC). This conclusion is an agreement with earlier observations. Note CAC = 0.1 g/L, CWC = 0.40 mmol/L = 0.25 g/L.

Theory Presented in Section 7.3 and Used in This Section for the Spreading/Evaporation of Surfactant Solutions

Coefficient $\beta = 2\pi \frac{DM}{\rho} \left[c_{sat} \left(T_{surf} \right) - c_\infty \right]$, where $T_{surf} = \int_S T \, dS$ is the average temperature of the liquid–air droplet surface, S. In this equation, D is the diffusion coefficient of the liquid vapor in air; ρ is the liquid density and M its molecular weight; c_{sat} is a concentration of the saturated vapor; $c_\infty = H \cdot c_{sat} \left(T_\infty \right)$; and H is the relative humidity of the ambient air. It was shown in Section 7.1 that for pure fluids the average temperature of the droplet–air interface, T_{surf}, can be taken as a constant during the evaporation process. Thus, at constant values of the ambient temperature, T_∞, and relative humidity, H, coefficient β remains constant during the evaporation process. It is possible to show that β decreases with increasing relative humidity, H, at constant T_∞. Numerical modeling presented in [10] was used to predict the dependence of β on the ambient temperature, T_∞. Note that this model does not consider the dependence of diffusion coefficient, D, on temperature.

Note, in the case of the spreading/evaporation of the droplets of the surfactant solutions investigated there are four stages of spreading evaporation as compared with three stages (Figure 7.2) in the case of pure liquids: the fourth final stage is present when both the radius of the droplet base and the contact angle change simultaneously. The nature of this fourth stage is yet to be understood.

Comparison of the Experimental Data for Evaporation of Surfactant Solutions with the Theoretical Predictions for Pure Liquids

It was shown in [12] that the β parameter does not depend on the concentration of surfactants in the range investigated and within the rage of experimental error.

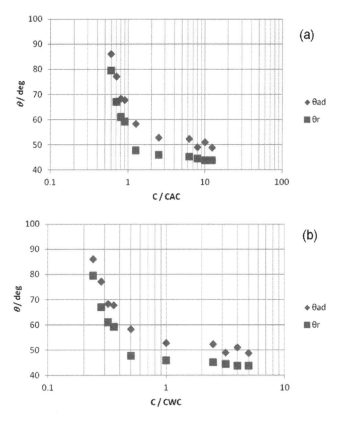

FIGURE 7.21 Dependence of advancing, θ_{ad}, and receding, θ_r, contact angles on surfactant concentration, C, for aqueous solutions of SILWET L-77: (a) reduced concentration is normalized by the critical aggregation concentration (CAC), 0.1 g/L; (b) reduced concentration is normalized by critical wetting concentration (CWC), 0.25 g/L.

Figure 7.22a and b shows the comparison of the universal behavior predicted for the second and third stages of evaporation with the values calculated from the experimental data for all the temperatures, relative humidities and concentrations measured (72 sets of θ, V and L vs. t data [12]). It can be observed that the experimental data follow the predicted universal curve in the second stage over a broad range of the reduced time, which is the same behavior shown in the previous section for pure water. However, the situation is more complex for the third stage of the spreading/evaporation process, although the agreement with the theory predictions is still rather good. The degree of agreement becomes more pronounced as the surfactant concentration increases, and the agreement between the theory and experiment is very good for concentrations close to and above the CAC. This may be understood considering that for the $0 < C < $ CAC range, the air–liquid and solid–liquid interfacial tensions would change as the evaporation progressed due to the increase of surfactant concentration. The surface tension strongly decreases as concentration increases in the range $C < $ CAC and, therefore, also θ and the evaporation rate. The decrease of the droplet volume is higher during the second evaporation stage, and, hence, the increase of surfactant concentration. This phenomenon was not included in either the computer simulations or the theory. This may also explain why the agreement between theory and experiment for pure water is similar to that of the more concentrated solutions of surfactant because in this case the air–solution interfacial tension remains constant.

Doganci et al. [36] carried out evaporation experiments using droplets of aqueous solution of SDS at various concentrations both above and below the critical micelle concentration (CMC). They found that evaporation is diffusion controlled as indicated by the time dependence of the drop volume, V, observed throughout the experiment: $V^{2/3} = a + bt$, where a and b are constants. These data will be compared with our data on the evaporation of trisiloxane solutions.

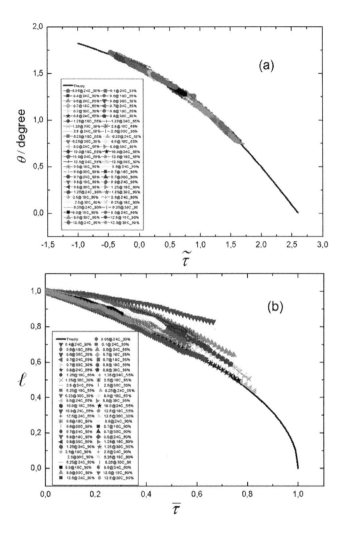

FIGURE 7.22 Comparison of the experimental results with the universal curves predicted by the theory for SILWET L77 solutions over the wide concentration range concentration range, at three different temperatures and relative humidity for the (a) first stage; and (b) second stage of evaporation.

In Figure 7.23, the experimental data published by Doganci et al. [36] for their experiments using SDS surfactant (55% relative humidity, 21°C) together with our results for SILWET L-77 (90% relative humidity, 18°C) are presented. Figure 7.23 shows that the agreement with theory is similar for both surfactants although the scattering around the universal curve for the second evaporation stage seems to be higher for the SDS data. Note, solid substrates used for the SILWET L77 and for the SDS solutions were different.

Note, there is no theoretical description for the last fourth evaporation stage.

Conclusions

The kinetics of a simultaneous spreading/evaporation process of droplets of aqueous solutions of the super-spreader SILWET L77 (commercial super-spreader surfactant) have been studied experimentally and compared with the theoretical predictions from Section 7.3 for pure water. In this case, the static advancing/receding contact angles are concentration dependent. Increasing the surfactant concentration reduces the static advancing contact angle from 104° ± 2° down to 83° ± 2°, as expected. Four different stages of process were found: (a) The first one corresponds to the spreading stage, which ends when the

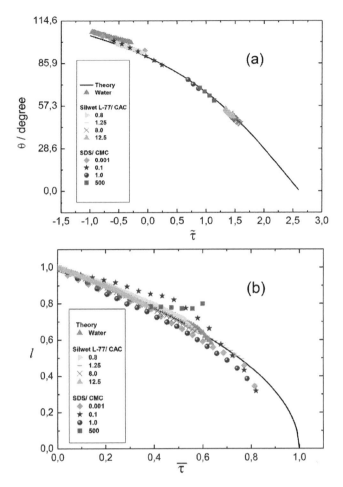

FIGURE 7.23 Comparison of the universal behavior predicted by the theory and experimental results of sodium dodecyl sulfate (SDS) [36] and of SILWET L77 solutions for (a) first stage of evaporation; and (b) second stage of evaporation.

contact angle reaches the value of static advancing contact angle θ_{ad}. The characteristic times found for all the concentrations, temperatures and relative humidity are in agreement with those reported in [35] for solutions of pure trisiloxanes. (b) The second stage takes place at constant droplet perimeter and decreasing contact angle. This stage ends when the contact angle reaches the value of static receding contact angle θ_r. In this stage a very good agreement was observed between our experimental data (time dependence of θ for the pinned droplet) with the universal behavior predicted by the theory (see Section 7.3). In addition, the data of Doganci et al. [36] for SDS solutions also show a very satisfactory agreement with the theoretical predictions. (c) The third stage takes place at constant $\theta = \theta_r$ while the droplet perimeter decreases. In this case, the theory also predicts a universal curve for the time dependence of the droplet perimeter. The agreement of the experiments with the theory is less satisfactory than for the second stage, especially for concentrations in which the surface tension still shows strong surfactant concentration dependence, that is, below the CAC in the case of the SILWET L77 solutions. During the second and third stages of spreading/evaporation, the volume-to-power 2/3, $V^{2/3}$, shows a linear dependence on time. This confirms the dominance of the vapor diffusion from the droplet surface over the kinetic aspects related to the mass transfer through the vapor–solution interface. (d) The fourth stage is characterized by a simultaneous decrease of the perimeter and the contact angle until the complete disappearance of the droplet. So far, there is no theoretical description for this stage, which for a droplet radius small enough should include the contributions of the surface forces (Derjaguin's pressure) because the droplet height becomes comparable with the range of surface forces action.

7.5 Evaporation of Microdroplets

The aim of the numerical study presented in this section is to investigate the influence of individual effects (kinetic effects, latent heat of vaporization, Marangoni convection, Stefan flow, droplet's surface curvature) on the rate of evaporation of a water droplet placed on a heat conductive substrate for different sizes of the droplet (down to submicrometer sizes). Simulations were performed for one particular set of parameters: Ambient relative air humidity is set to 70%, ambient temperature is 20°C, contact angle is 90°, and the substrate material is copper. The suggested model combines both diffusive and kinetic models of evaporation. The obtained results allow estimating the characteristic droplet sizes where each of the aforementioned phenomena become important or can be neglected. Presentation in this section is based on [37,13].

Introduction

The aim of the computer simulations presented below in this section is to show how the evaporation of pinned sessile submicrometer droplets of water on a solid surface differs from the evaporation of millimeter-sized droplets. The obtained results prove the importance of kinetic effects, whose influence becomes more pronounced for submicrometer droplets.

The model used here includes both diffusive and kinetic models of evaporation simultaneously. The Hertz-Knudsen-Langmuir equation [38,39] is used as a boundary condition at the liquid–air interface instead of a condition of saturated vapor. The overall evaporation rate is limited either by the rate of vapor diffusion into the ambient air or by the rate of molecule transfer across the liquid–air interface. As a result, the vapor concentration at the liquid–air interface falls between its saturated value and its value in the ambient air. This intermediate value of vapor concentration at the liquid–air interface drives both the transition of molecules from liquid to air (kinetic flux) and the vapor diffusion into the ambient air (diffusive flux). Thus, the rate of evaporation is limited by the slower of these two processes.

Computer simulations are performed using the software COMSOL Multiphysics. The dependence of total molar evaporation flux, J_c, on the radius of the contact line, L, of pinned droplets is studied in the next section using a quasi-steady-state approximation as in Section 7.1.

The evaporation of small droplets is under consideration below in the next section, and surface forces (Derjaguin's pressure) could play an important role: the smaller the droplet, the more important the influence of these forces. Unfortunately, on this stage we are unable to include the action of the Derjaguin's pressure into consideration and it is a problem for future investigation.

Problem Statement

Only droplets of a size less than 1 mm are under consideration here. That is, the influence of gravity is neglected. It is assumed that axisymmetric sessile droplet forms a sharp three-phase contact line and maintains a spherical-cap shape of the liquid–air interface due to the action of liquid–air interfacial tension (very small Ca number).

Due to the small size of the evaporating droplets under consideration, the diffusion of vapor in the air phase dominates its convective transport. This is confirmed by small values of both thermal and diffusive Peclet numbers: $Pe_\kappa = Lu/\kappa < 0.05$; $Pe_D = Lu/D < 0.04$, where L is the radius of the contact line, u (<1 mm/s) is the characteristic velocity of vapor convection due to evaporation (the Stefan flow) and the Marangoni convection, κ is the thermal diffusivity of the surrounding air, and D is the diffusion coefficient of vapor in air at standard conditions.

The problem is solved under a quasi-steady-state approximation. That is, all time derivatives in all equations are neglected. The quasi-steady solution of the problem gives a simultaneous distribution of the heat and mass fluxes in the system.

The volume of the droplet diminishes because of evaporation. Therefore, the liquid–air interface moves with some velocity and can be calculated based on knowledge of the evaporation rate and two particular assumptions: that the cap of the droplet preserves its spherical shape and that the contact line is pinned (L = const). The regime of the moving contact line is not considered here.

The parameters of the following materials are used in the computer simulations here: copper as the substrate, water as the liquid in a droplet, and a humid air as a surrounding medium. The pressure in the surrounding air is equal to the atmospheric pressure, the ambient temperature is 20°C, the ambient air humidity is 70%, and the contact angle is assumed 90°.

Governing Equations in the Bulk Phases

The following governing equations describe the heat and mass transfer in the bulk phases:
heat transfer in a solid phase:

$$\Delta T = 0,$$

where Δ is the Laplace operator; and T is the temperature; the heat transfer inside the fluids (liquid or air) is

$$\vec{u} \cdot \nabla T = \kappa \, \Delta T,$$

where $\vec{u}$ is the fluid velocity, ∇ is the gradient operator, and κ is the thermal diffusivity of the fluid; and incompressible Navier-Strokes equations are used to model hydrodynamic flows in both fluids (liquid or air),

$$\rho \, \vec{u} \cdot \nabla \vec{u} = \nabla \cdot \mathbf{T},$$

where ρ is the fluid density, $\nabla \vec{u}$ is the gradient of the velocity vector, $\mathbf{T}$ is the full stress tensor, and $\nabla \cdot \mathbf{T}$ is the dot-product of nabla operator and the full stress tensor. The full stress tensor is expressed via hydrodynamic pressure, p, and viscous stress tensor, π, as

$$\mathbf{T} = -p\mathbf{I} + \pi,$$

where $\mathbf{I}$ is the identity tensor. The continuity equation for fluids is also used:

$$\nabla \cdot \vec{u} = 0.$$

The diffusion equation for vapor in air phase:

$$\Delta c = 0, \tag{7.66}$$

where c is the molar concentration of the liquid vapor.

Boundary Conditions

No-slip and no-penetration boundary conditions are used for Navier–Stokes equations at the liquid–solid and air–solid interfaces, resulting in zero velocity at these interfaces:

$$\vec{u} = 0$$

Let Γ be the liquid–air interface. Let also j_c be a density of a molar vapor flux across the liquid–air interface, then a density of mass vapor flux, j_m, across this interface is $j_m = j_c \cdot M$, where M is the molar mass of an evaporating substance (water). Let u_Γ be the normal velocity of the liquid–air interface itself (Figure 7.24). Then the boundary condition for the normal velocity of liquid at the liquid–air interface reads:

$$\rho_l \left(\vec{u}_l \cdot \vec{n} - u_\Gamma \right) = j_m, \tag{7.67}$$

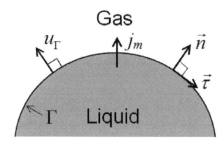

FIGURE 7.24 Notations: Γ is the liquid–vapor interface; u_Γ is the normal velocity of the interface Γ in the direction of the normal unit vector $\vec{n}$ (from the liquid phase to the gaseous one); $\vec{\tau}$ is the unit vector tangential to the interface Γ; and j_m is the mass flux across the interface Γ.

where ρ_l is the liquid density; and subscript l stands for liquid. Expressions for evaporation flux, j_m, and normal interfacial velocity, u_Γ, are to be specified. A condition of the stress balance at the liquid–air interface is used to obtain boundary conditions for the pressure and tangential velocity:

$$\left(\mathbf{T}\cdot\vec{n}\right)_l - \left(\mathbf{T}\cdot\vec{n}\right)_g = -\gamma\left(\nabla\cdot\vec{n}\right)_\Gamma \vec{n} + \gamma_T'\left(\nabla_\Gamma T\right), \tag{7.68}$$

where subscripts l and g stand for liquid and air, respectively; $\mathbf{T}$ is the full stress tensor; γ is the interfacial tension of the liquid–air interface; γ_T' is the derivative of the interfacial tension with the temperature; $\nabla_\Gamma T$ is the surface gradient of temperature; and $\left(\nabla\cdot\vec{n}\right)_\Gamma$ is the divergence of the normal vector at the liquid–air interface, which is equal to the curvature of the interface. Boundary condition (7.68) is a vector one. The boundary conditions in normal and tangential direction are deduced as follows: (i) the boundary condition for thermal Marangoni convection, which determines the tangent component of the velocity vector is obtained by multiplying Eq. (7.68) by the tangential vector $\vec{\tau}$ (Figure 7.24) and neglecting the viscous stress in the air phase (due to smallness of the air viscosity compared to the liquid viscosity); (ii) the similar procedure with normal vector $\vec{n}$ (Figure 7.24) results in a boundary condition for pressure in the liquid at the liquid–air interface.

The normal flux of vapor at the air–solid interface is zero because there is no penetration into the solid surface:

$$\nabla c\cdot\vec{n} = 0,$$

where c is the molar concentration of the vapor in the air; and $\vec{n}$ is the unit vector, perpendicular to the solid–air interface.

We consider the air phase as a mixture of vapor and dry air. Note that due to the mass conservation law, the mass flux of vapor, j_m, perpendicular to the liquid–air interface in the vapor phase should be equal to the mass flux of liquid perpendicular to the interface in the liquid phase:

$$j_m = \rho_l\left(\vec{u}_l\cdot\vec{n} - u_\Gamma\right) = \rho_v\left(\vec{u}_g\cdot\vec{n} - u_\Gamma\right) - D_{\text{vapour in air}}\nabla\rho_v\cdot\vec{n}. \tag{7.69}$$

The density of the mass flux of dry air, $j_{m,\text{air}}$, across the liquid–air interface is assumed to be zero:

$$j_{m,\text{air}} = \rho_{\text{air}}\left(\vec{u}_g\cdot\vec{n} - u_\Gamma\right) - D_{\text{air in vapour}}\nabla\rho_{\text{air}}\cdot\vec{n} = 0, \tag{7.70}$$

where ρ_v and ρ_{air} are densities of vapor and dry air, respectively; D is the diffusion coefficient; $\vec{u}_l$ and $\vec{u}_g$ are velocity vectors of liquid and air, respectively; and u_Γ and $\vec{n}$ are shown in Figure 7.24. As the air (water vapor + dry air) under consideration includes more than one species of molecules, then the mass flux for each species in the air phase consists of two components: convective part, $\rho\vec{u}_g$, and diffusive one, $D\nabla\rho$. Flux in the pure liquid includes only a convective term, $\rho_l\vec{u}_l$. Normal fluxes at the liquid–air interface are considered relative to the liquid–air interface. This results in an additional term: $-\rho u_\Gamma$. Let us adopt the following assumption: $\rho_g = \rho_{air} + \rho_v = \text{const}$. This assumption means incompressibility of the air phase. This assumption results in

$$\nabla \rho_{air} = -\nabla \rho_v. \tag{7.71}$$

A direct consequence of the above assumption is (7.71)

$$D_{air \ in \ vapour} = D_{vapour \ in \ air} = D. \tag{7.72}$$

After substitution of Eqs. (7.71) and (7.72) into Eqs. (7.69) and (7.70) and simple algebraic manipulations, we arrive to an expression for the density of mass flux across the liquid–air interface, j_m, as a function of molar vapor concentration, c, in the air phase:

$$j_m = \frac{-D\nabla\rho_v \cdot \vec{n}}{1 - \rho_v/\rho_g} = \frac{-D\nabla c \cdot \vec{n}}{1/M - c/\rho_g}, \tag{7.73}$$

where the following relation has been used: $\rho_v = cM$, M is the molar mass of an evaporating substance (water). Equation (7.73) connects the evaporation flux, j_m, with both the gradient of vapor concentration in the normal direction and the concentration itself. On the other hand, the rate of mass transfer across the liquid–air interface is given by the Hertz-Knudsen-Langmuir equation [38,39]:

$$j_m = \alpha_m \sqrt{\frac{MRT}{2\pi}} \left(c_{sat}(T) - c\right), \tag{7.74}$$

where α_m is the mass accommodation coefficient (probability that uptake of vapor molecules occurs upon collision of those molecules with the liquid surface); R is the universal air constant; T and c are the local temperature in K and molar vapor concentration at the liquid–air interface, respectively; and c_{sat} is the molar concentration of saturated vapor. The molar concentration of saturated vapor is taken as a function of a local temperature and a local curvature of the liquid–air interface according to Clausius-Clapeyron and Kelvin equations.

Combining Eqs. (7.73) and (7.74), we obtain a boundary condition for diffusion equation (7.66) at the liquid–air interface:

$$\frac{-D\nabla c \cdot \vec{n}}{1/M - c/\rho_g} = \alpha_m \sqrt{\frac{MRT}{2\pi}} \left(c_{sat}(T) - c\right).$$

Summing Eqs. (7.69) and (7.70) and taking into account the assumption $\left(\rho_g = \rho_{air} + \rho_v = \text{const}\right)$, we arrive to the boundary condition for the normal velocity of air at the liquid–air interface:

$$\rho_g \left(\vec{u}_g \cdot \vec{n} - u_\Gamma\right) = j_m. \tag{7.75}$$

The tangent velocity of air at the liquid–air interface is determined by the no-slip condition:

$$\vec{u}_g \cdot \vec{\tau} = \vec{u}_l \cdot \vec{\tau}.$$

Boundary conditions of temperature continuity are applied at all interfaces (liquid–air, liquid–solid and air–solid):

$$T_l = T_g, \quad T_l = T_s, \quad T_g = T_s,$$

where subscripts l, g and s stand for liquid, air and solid, respectively. Continuity of the heat flux is applied on solid–liquid and solid–air interfaces:

$$-k_l(\nabla T)_l \cdot \vec{n} + k_s(\nabla T)_s \cdot \vec{n} = 0,$$

$$-k_g(\nabla T)_g \cdot \vec{n} + k_s(\nabla T)_s \cdot \vec{n} = 0,$$

where k is the thermal conductivity of the corresponding phase; and $\vec{n}$ is the unit vector, perpendicular to a corresponding interface. Note, at all these interfaces the convective heat flux is zero due to the no-penetration conditions. At the liquid–air interface, heat flux experiences discontinuity caused by the latent heat of vaporization and there is also a convective heat flux through this interface:

$$\left[\rho_l c_{pl} T \left(\vec{u}_l \cdot \vec{n} - u_\Gamma \right) - k_l(\nabla T)_l \cdot \vec{n} \right] - \left[\rho_g c_{pg} T \left(\vec{u}_g \cdot \vec{n} - u_\Gamma \right) - k_g(\nabla T)_g \cdot \vec{n} \right] = j_c \Lambda, \qquad (7.76)$$

where c_{pl} and c_{pg} are specific heat capacities at constant pressure for the liquid and the air, respectively; Λ is the latent heat of vaporization (or enthalpy of vaporization, units: J/mol); $\vec{n}$ is the unit vector, normal to the liquid–air interface, and pointing into the air phase; and j_c is the surface density of the molar flux of evaporation (mol·s^{-1}·m^{-2}) at the liquid–air interface. Using Eqs. (7.67) and (7.75) and the relation between molar and mass fluxes, $j_m = j_c \cdot M$, we derive the following expression:

$$\rho_l c_{pl} T \left(\vec{u}_l \cdot \vec{n} - u_\Gamma \right) - \rho_g c_{pg} T \left(\vec{u}_g \cdot \vec{n} - u_\Gamma \right) = c_{pl} T \, j_m - c_{pg} T \, j_m = j_c M T \left(c_{pl} - c_{pg} \right).$$

Using this expression transforms Eq. (7.76) into the required boundary condition for the heat flux discontinuity at the liquid–air interface:

$$-k_l(\nabla T)_l \cdot \vec{n} + k_g(\nabla T)_g \cdot \vec{n} = j_c \left(\Lambda - MT \left(c_{pl} - c_{pg} \right) \right).$$

At the axis of symmetry ($r = 0$) the following boundary conditions are satisfied:

$$\left. \frac{\partial c}{\partial r} \right|_{r=0} = 0, \left. \frac{\partial T}{\partial r} \right|_{r=0} = 0, \left. \frac{\partial u_z}{\partial r} \right|_{r=0} = 0, \left. u_r \right|_{r=0} = 0, \left. \frac{\partial p}{\partial r} \right|_{r=0} = 0,$$

where u_r and u_z are radial and vertical components of the velocity vector; and p is the hydrodynamic pressure.

At the outer boundary of the computational domain values of temperature, T_∞, and vapor concentration, c_∞, are imposed. In the case when air convection (the Stefan flow) is taken into account, the condition of open boundary (zero normal stress) is imposed at the outer boundary of the air domain:

$$\mathbf{T} \cdot \vec{n} = 0,$$

where $\mathbf{T}$ is the full stress tensor, and $\vec{n}$ is the unit normal vector. This boundary condition allows air to both enter and leave the domain.

In our computer simulations, we assume that the droplet under consideration retains a spherical-cap shape in the course of evaporation and that the contact line is pinned ($L = $ const). Then, knowing the total mass evaporation flux, $J_m = \int_\Gamma j_m \, dA$ (dA is the element of area of the interface Γ), we can calculate the normal velocity of the liquid–air interface, u_Γ, at any point of the interface:

$$u_\Gamma = \frac{-J_m}{\pi \, \rho_l \, L^2} \cdot \frac{z}{(z + \delta z) \cdot n_z + r \cdot n_r} \cdot \frac{1 + \cos\theta}{1 - \cos\theta},$$

where $\delta z = L \cdot \cos\theta / \sin\theta$; θ is the contact angle; n_r and n_z are radial and vertical components of the vector $\vec{n}$, respectively, shown in Figure 7.24; and the origin of the cylindrical coordinates (r, z) is supposed to be at the point of intersection of the axis of droplet symmetry with the liquid–solid interface (Figure 7.24).

Computer Simulations

Computer simulations were performed using commercial software COMSOL Multiphysics. The numerical method used in COMSOL is a finite-element method with quadratic Lagrangian elements. The software transforms all equations into their weak form before discretization. The method of Lagrange multipliers is used to apply boundary conditions as constraints.

The computational domain is selected as a circle in (r, z) coordinates of the cylindrical system of coordinates. The center of this circle is located at the origin of the system of coordinates. The radius of the computational domain is 100 times bigger than the radius of the contact line, L, which prevents the influence of the proximity of outer boundaries on the solution inside of the droplet.

The choice of contact angle $90°$ allows avoiding the problem of singularity of diffusive evaporation flux at the three-phase contact line; but the problem of viscous stress singularity remains. The current model does not include surface forces action (Derjaguin's pressure) to overcome this problem (see introduction to Chapter 7). Instead, the mesh is refined around the contact line to reduce the area of singularity influence. Thus, the size of the computational mesh elements around the three-phase contact line is chosen to be 100 times smaller than the droplet size.

The growth rate for mesh elements is less than 1.1 in the whole computational domain.

Results and Discussion

Isothermal Evaporation

The model described above in this section is an extension of the previous sections of Chapter 7, which was developed for diffusion-limited evaporation of aqueous or aqueous solution droplets. In distinction to the previous model, the present one takes into account additional phenomena: Stefan flow in air, effect of curvature of the droplet's surface on saturated vapor pressure (Kelvin's equation), and kinetic effects (also known as Knudsen effects). A numerical model allows switching individual phenomena on/off in order to understand their contribution to the overall process of heat and mass transfer in the course of evaporation.

The value of mass accommodation coefficient α_m is taken as 0.5, which is the average experimentally measured value of α_m for water according to [38].

When Kelvin's and kinetic effects are switched off, then the model represents the case of diffusion-limited evaporation. This allows us to validate the present model against the previous one in Sections 7.1 through 7.4, which in turn were validated against available experimental data for the diffusion-limited case. Computer simulations showed agreement with the previous results (Figure 7.25) for sufficiently big droplets.

In Figure 7.25, the total molar fluxes of droplet's evaporation, $J_{c,i}$, are presented for various isothermal cases (index i stands for the "isothermal"). The flux was computed according to the developed in this section model, in which the Stefan flow, heat transfer and the thermal Marangoni convection were omitted. Figure 7.25 shows that kinetic effects change the slope of curves (see triangles and circles) for submicrometer droplets ($L < 10^{-6}$ m) only. The influence of the curvature of the droplet's surface (Kelvin's equation) becomes significant only for droplet sizes less than 10^{-7} m. However, for such low sizes the surface forces action (Derjaguin's pressure) has to be taken into account (not included in the present model).

Note once more that the presented model is valid only for droplet sizes bigger than the radius of surface forces action, which is around 10^{-7} m = 0.1 μm. That is, the data in Figure 7.25 for the droplet size smaller than 10^{-7} m are presented only to show the trend. Figure 7.25 shows that if the radius of the droplet base is bigger than 10^{-7} m then (i) deviation of the saturated vapor pressure caused by the droplet curvature (Kelvin's equation) can be neglected, (ii) a deviation from the pure diffusion model of evaporation can be neglected for the droplet size bigger than 10^{-6} m, and (iii) this deviation becomes noticeable only if the droplet size is less than 10^{-6} m. This deviation is caused by an increasing influence of the kinetic effects at the liquid–air interface (Hertz-Knudsen-Langmuir equation) and this theory should be applied together with the diffusion equation of vapor in the air if the droplet size is less than 10^{-6} m.

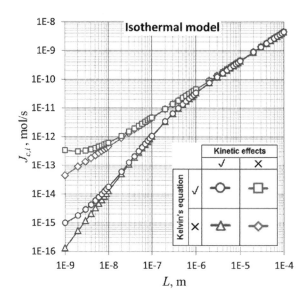

FIGURE 7.25 Dependence of the droplet's molar evaporation flux, $J_{c,i}$, on the droplet size, L, for isothermal model of evaporation. Parameters used: $\alpha_m = 0.5$; $\theta = 90°$; relative air humidity is 70%. Note, results for $L < 10^{-7}$ m do not have physical meaning, as additional surface forces must be included into the model. These points are shown to demonstrate the trends of curves.

These conclusions show that a consideration of evaporation of microdroplets, completely covered by the surface forces action (that is less than 10^{-7} m), should take into account both the deviation of the saturated vapor pressure caused by the droplet curvature and the kinetic effects.

According to the model of diffusion-limited evaporation, the evaporation flux, $J_{c,i}$, must be linearly proportional to the droplet size, L, that is $J_{c,i} \sim L$. This is in agreement with the data presented in Figure 7.25 for droplets bigger than 10^{-6} m. However, for a pure kinetic model of evaporation (no vapor diffusion, uniform vapor pressure in the air) flux $J_{c,i}$ is supposed to be proportional to the area of the droplet's surface, that is, in the case of pinned droplets (constant contact area), $J_{c,i} \sim L^2$ should be satisfied. To check the validity of this model at various droplet sizes, let us assume that the dependency of the evaporation flux on the droplet radius has the following form $J_{c,i} = A(\theta) \cdot L^n$, where n is the exponent to be extracted from our model and A is a function of the contact angle, θ. In general it is necessary to calculate the partial derivative

$$n = \partial\left(\ln J_{c,i}\right) / \partial\left(\ln L\right), \tag{7.77}$$

in order to compute the exponent n using the present model. However, in the case of a pinned droplet, the only varying parameter for each individual curve in Figure 7.25 is the contact line radius, L. Hence, the partial derivatives in Eq. (7.77) can be replaced by ordinary derivatives, and in this way n was calculated using data presented in Figure 7.25. The calculated values of n are presented in Figure 7.26.

One can see from Figure 7.26 that the exponent n, as expected, is equal to 1 for a pure diffusive isothermal model of evaporation within the whole studied range of L values (diamonds in Figure 7.26). In the case when kinetics effects and Kelvin's equation are both taken into account in addition to the pure diffusion, Figure 7.26 shows that the diffusion model of evaporation dominates for droplets with the size bigger than 10^{-5} m, that is, for droplets bigger than 10 μm.

Taking into account kinetic effects only additional to the diffusion without Kelvin's equation (triangles in Figure 7.26), results in a smooth transition from the linear dependence $J_{c,i} \sim L$, that is, $n = 1$ (diffusive model) to the quadratic one $J_{c,i} \sim L^2$, that is, $n = 2$ (kinetic model) as the size of the droplet decreases down to $L = 10-9$ m (Figure 7.26). This shows that $J_{c,i}$ is tending to be proportional to L^2 as the size of the droplet decreases, which means that evaporation flux becomes proportional to the area of the liquid–air

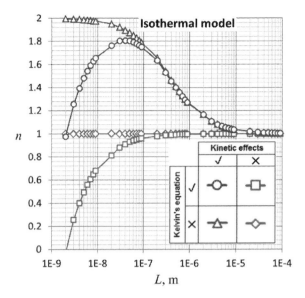

FIGURE 7.26 Exponent n for the dependence $J_{c,i} = A(\theta) \cdot L^n$ for the isothermal model of evaporation. Parameters used: $\alpha_m = 0.5$; $\theta = 90°$; and relative air humidity is 70%. Note, results for $L < 10^{-7}$ m do not have physical meaning, as surface forces action must be included into the model here. These points are shown to demonstrate the trends of curves.

interface. Influence of the curvature (Kelvin's equation) on the saturated vapor pressure results in a substantially lower exponent n as compared with the kinetic theory (Figure 7.26). However, this happens only for a droplet completely in the range of surface forces action, that is, less than 10^{-7} m. Below this limit, the droplet no longer has a spherical cap shape, even on the droplet's top (microdroplets according to Chapter 2). The evaporation process in this case should be substantially different from the one considered above in this section. Thus, the range of sizes less than 10^{-7} m is not covered by the theory presented here.

Influence of Thermal Effects

Computer simulations were also performed including both Kelvin's equation and kinetic effects in the case when the thermal effects were taken into account. This was made to show the influence of the latent heat of vaporization, Marangoni convection, and Stefan flow on the evaporation process. The evaporation rates, J_c, of the droplets were normalized using those, $J_{c,i}$ (circles in Figure 7.25), from the isothermal model. Results are presented in Figure 7.27.

Figure 7.27 shows that the latent heat of vaporization reduces the evaporation flux as compared to the isothermal case (that is $J_c{:}J_{c,i} < 1$) in all cases considered. The reason is a temperature decrease at the droplet's surface due to heat consumption during the evaporation process. This reduces the value of the saturated vapor pressure at the droplet's surface and, subsequently, the rate of vapor diffusion into the ambient air. The relative reduction of the evaporation rate (caused by the latent heat of vaporization) reaches the maximum for droplets with $L \sim 10^{-5}$ m. This size, according to Figures 7.25 and 7.26, is in the range of diffusion-limited evaporation. For smaller droplets, when the kinetic effects come into play, the influence of the latent heat on evaporation rate, J_c, decreases. The reason for that is that the vapor above the droplet surface according to the kinetic model of evaporation is not saturated and therefore its pressure is less influenced by the local temperature but is instead more influenced by the relative humidity of the ambient air. If we exclude kinetic effects and Kelvin's equation from both nonisothermal (J_c) and isothermal ($J_{c,i}$) models, and include only latent heat of vaporization, then the ratio $J_c{:}J_{c,i}$ becomes independent on L and equal its value as for the diffusion-limited case (millimeter-sized droplets).

Taking into account kinetic effects, Kelvin's equation, latent heat of vaporization, and thermal Marangoni convection (for water droplets) affects droplets of size $L > 10^{-5}$ m (triangles in Figure 7.27).

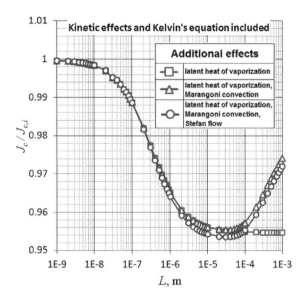

FIGURE 7.27 Influence of latent heat of vaporization, Marangoni convection, and Stefan flow on the evaporation rate in the case when kinetic effects and Kelvin's equation are included into the model. J_c and $J_{c,i}$ are total molar flux of evaporation and the one in the isothermal case, respectively.

In this case, the evaporation rate, J_c, is lower than that for isothermal model and higher than that if the latent heat were included but without Marangoni convection (squares in Figure 7.27).

For water droplets of the size $L < 10^{-5}$ m, the influence of Marangoni convection is negligible and evaporation rate, J_c, coincides with the one for a model that includes only latent heat of vaporization (squares in Figure 7.27).

The effect of Stefan flow in the surrounding air slightly changes the evaporation rate in the present model (circles in Figure 7.27) and makes it lower due to an appearance of an outward convective heat flux in the air above the droplet. This reduces the heat flux from the ambient environment to the droplet's surface through the air phase. Thus, the temperature of the droplet's surface becomes lower, reducing the evaporation rate. In our particular case, this effect appeared to be much weaker than the effects of latent heat of vaporization. The Stefan flow effect is also weaker than the effect of thermal Marangoni convection for $L > 10^{-4}$ m, but a bit stronger for $L < 10^{-4}$ m. However, in any case the influence of the Stefan flow is small and can be neglected.

Note that the influence of the thermal effects on the kinetics of evaporation is less than 5% (according to Figure 7.27).

Conclusions

Computer simulations of evaporation of small sessile droplets of water were performed. The presented model combines diffusive and kinetic models of evaporation. The effect of the latent heat of vaporization, thermal Marangoni convection and Stefan flow in the surrounding air were investigated for a particular system: a water droplet on a heat-conductive substrate (copper) in air at standard conditions. The results of modeling allow estimation of the characteristic droplet sizes when each of the aforementioned phenomena become important or can be neglected.

The presented model is valid only for droplet size bigger than the radius of surface forces action, which is around 10^{-7} m = 0.1 μm. That is, the data in Figures 7.25 through 7.27 for the droplet size smaller than 10^{-7} m are presented only to show the trend. When the radius of the droplet base, L, is bigger than 10^{-7} m, then (i) deviation of the saturated vapor pressure caused by the droplet curvature (Kelvin's equation) can be neglected, (ii) a deviation from the pure diffusion model of evaporation can be neglected for the droplet size bigger than 10^{-6} m, and (iii) this deviation becomes noticeable only if the droplet size is less than 10^{-6} m. This deviation is caused by an increasing influence of the kinetic effects at the liquid–air interface (Hertz-Knudsen-Langmuir equation) and this theory should be applied together with the diffusion equation of vapor in the air if the droplet size is less than 10^{-6} m.

These conclusions show that a consideration of evaporation of microdroplets completely covered by the surface forces action (that is less than 10^{-7} m) should include both the deviation of the saturated vapor pressure caused by the droplet curvature and the kinetic effects.

The latent heat of vaporization results in a temperature decrease at the surface of the droplet. Due to that, the evaporation rate is reduced. This effect is more pronounced in the case of diffusion-limited evaporation ($L > 10^{-5}$ m), when vapor pressure at the droplet's surface is saturated and determined by local temperature. The effect of Marangoni convection in water droplets is negligible for droplets of size $L < 10^{-5}$ m. For the system considered, the Stefan flow effect appeared to be weaker than the effect of thermal Marangoni convection for $L > 10^{-4}$ m, but stronger for $L < 10^{-4}$ m. However, in all cases, its influence is small and can be neglected. According to Figure 7.27, the influence of the latent heat of vaporization on the kinetics of evaporation is less than 5%.

The presented model can be applied for evaporation of any other pure simple liquids, not only water.

REFERENCES

1. Eggers, J., and Pismen, L. M. Nonlocal description of evaporating drops. *Phys. Fluids*, 22, 112101 (2010).
2. Moosman, S., and Homsy, G. M. Evaporating menisci of wetting fluids. *J. Colloid Interface Sci.*, 73, 212–223 (1980).
3. Ajaev, V. S., Gambaryan-Roisman, T., and Stephan, P. Static and dynamic contact angles of evaporating liquids on heated surfaces. *J. Colloid Interface Sci.*, 342, 550–558 (2010).
4. Semenov, S., Starov, V. M., Rubio, R. G., and Velarde, M. G. Instantaneous distribution of fluxes in the course of evaporation of sessile liquid droplets: Computer simulations. *Colloids Surf. A*, 372, 127–134 (2010).
5. Semenov, S., Starov, V. M., and Rubio, R. G. Evaporation of pinned sessile microdroplets of water on a highly heat-conductive substrate: Computer simulations (MICRO DROPLETS). *Eur. Phys. J. Special Topics*, 219, 143–154 (2013).
6. Lee, K. S., Cheah, C. Y., Copleston, R. J., Starov, V. M., and Sefiane, K. Spreading and evaporation of sessile droplets: Universal behaviour in the case of complete wetting. *Colloids Surf. A: Physicochem. Eng. Aspects*, 323, 63–72 (2008).
7. Semenov, S., Starov, V. M., Velarde, M. G., and Rubio, R. G. Droplets evaporation: Problems and solutions. *Eur. Phys. J. Spec. Top.*, 197, 265–278 (2011).
8. Semenov, S., Trybala, A., Rubio, R. G., Kovalchuk, N., Starov, V., and Velarde, M. G. Simulteneous spreading and evaporation: Recent developments. *Adv. Colloid Interface Sci.*, 206, 382–398 (2014).
9. Semenov, S., Starov, V. M., Rubio, R. G., and Velarde, M. G. Computer simulations of evaporation of pinned sessile droplets: Influence of kinetic effects. *Langmuir*, 28, 15203–15211 (2012); *Langmuir*, 28, 16724–16724 (2012).
10. Semenov, S., Starov, V. M., Rubio, R. G., Agogo, H., and Velarde, M. G. Evaporation of sessile droplets: Universal behaviour in the case of contact angle hysteresis. *Colloids Surf. A*, 391, 135–144 (2011).
11. Semenov, S., Starov, V. M., Rubio, R. G., and Velarde, M. G. Instantaneous distribution of fluxes in the course of evaporation of sessile liquid droplets: Computer simulations. *Colloids Surf. A*, 372, 127–134 (2010).
12. Agogo, H., Semenov, S., Ortega, F., Rubio, R. G., Starov, V. M., and Velarde, M. G. Spreading and evaporation of surfactant solution droplets. *Progr. Colloid Polym. Sci.*, 139, 1–6 (2012).
13. Semenov, S., Starov, V. M., Rubiob, R. G., and Velarde, M. Computer simulations of evaporation of pinned sessile droplets: Influence of kinetic effects. *Langmuir*, 28, 15203 (2012).
14. Deegan, R. D. Pattern formation in drying drops. *Phys. Rev. E*, 61, 475 (2000).
15. Girard, F., Antoni, M., and Sefiane, K. On the effect of Marangoni flow on evaporation rates of heated water drops. *Langmuir*, 24, 9207–9210 (2008).

16. Picknett, R. G., and Bexon, R. The evaporation of sessile or pendant drops in still air. *J. Colloid Interface Sci.*, 61, 336–350 (1977).

17. Girard, F., Antoni, M., Faure, S., and Steinchen, A. Evaporation and Marangoni driven convection in small heated water droplets. *Langmuir*, 22, 11085–11091 (2006).

18. Girard, F., Antoni, M., Faure, S., and Steinchen, A. Numerical study of the evaporating dynamics of a sessile water droplet. *Microgravity Sci. Technol.*, XVIII (3/4), 42–46 (2006).

19. Girard, F., Antoni, M., Faure, S., and Steinchen, A. Influence of heating temperature and relative humidity in the evaporation of pinned droplets. *Colloids Surf. A*, 323, 36–49 (2008).

20. Girard, F., and Antoni, M. Influence of substrate heating on the evaporation dynamics of pinned water droplets. *Langmuir*, 24, 11342–11345 (2008).

21. Dunn, G. J., Wilson, S. K., Duffy, B. R., David, S., and Sefiane, K. The strong influence of substrate conductivity on droplet evaporation. *J. Fluid Mech.*, 623, 329 (2009).

22. David, S., Sefiane, K., and Tadrist, L. Experimental investigation of the effect of thermal properties of the substrate in the wetting and evaporation of sessile drops. *Colloid Surf. A*, 298, 108 (2007).

23. Schonfeld, F., Graf, K. H., Hardt, S., and Butt, H. J. Evaporation dynamics of sessile liquid drops in still air with constant contact radius. *Int. J. Heat Mass Transfer*, 51, 3696 (2008).

24. Hu, H., and Larson, R. G. Evaporation of a sessile droplet on a substrate. *J. Phys. Chem. B*, 106, 1334 (2002).

25. Guéna, G., Allancon, P., and Cazabat, A. M. Receding contact angle in the situation of complete wetting: Experimental check of a model used for evaporating droplets. *Colloids Surf. A*, 300, 307 (2007).

26. Guéna, G., Poulard, C., Voué, M., De Coninck, J., and Cazabat, A. M. Evaporation of sessile liquid droplets. *Colloids Surf. A*, 291, 191 (2006).

27. Guéna, G., Poulard, C., and Cazabat, A. M. The dynamics of evaporating sessile droplets. *Colloid J.* (Russian Academy of Sciences, English Translation), 69 (1), 1 (2007).

28. Guéna, G., Poulard, C., and Cazabat, A. M. Evaporating drops of alkane mixtures. *Colloids Surf. A*, 298, 2 (2007).

29. Guéna, G., Poulard, C., and Cazabat, A. M. The leading edge of evaporating droplets. *J. Colloid Interface Sci.*, 312, 164 (2007).

30. Nayfeh, A. H. *Perturbation Methods*, Wiley-Interscience, New York, 1973.

31. Bourges-Monnier, C., and Shanahan, M. E. R. Influence of evaporation on contact angle. *Langmuir*, 11, 2820–2829 (1995).

32. Chin-Tai, C., Fan-Gang, T., and Ching-Chang, C. Evaporation evolution of volatile liquid droplets in nanoliter wells. *Sens. Actuators A*, 130–131, 12–19 (2006).

33. Yildirim Erbil, H., McHale, G., Rowan, S. M., and Newton, M. I. Determination of the receding contact angle of sessile drops on polymer surfaces by evaporation. *Langmuir*, 15, 7378–7385 (1999).

34. Ritacco, H., Ortega, F., Rubio, R. G., Ivanova, N., and Starov, V. M. Equilibrium and dynamic surface properties of trisiloxane aqueous solutions. Part 1. Experimental results. *Colloids Surf. A*, 365, 199–203 (2010).

35. Ivanova, N., Starov, V., Johnson, D., Hilal, N., and Rubio, R. G. Spreading of aqueous solutions of trisiloxanes and conventional surfactants over PTFE AF coated silicon wafers. *Langmuir*, 25, 3564–3570 (2009).

36. Doganci, M. D., Sesli, B. U., and Erbil, H. Y. Diffusion-controlled evaporation of sodium dodecyl sulfate solutions drops placed on a hydrophobic substrate. *J. Colloid Interf. Sci.*, 362, 524–531 (2011).

37. Sergey, S., Starov, V. M., and Rubio, R. G. Evaporation of pinned sessile microdroplets of water on a highly heat-conductive substrate: computer simulations. *Eur. Phys. J. Spec. Top.*, 197, 265 (2011).

38. Kryukov, A. P., Levashov, V. Y., and Sazhin, S. S. Evaporation of diesel fuel droplets: kinetic versus hydrodynamic models. *Int. J. Heat Mass Transfer*, 47, 2541 (2004).

39. Sazhin, S. S., Shishkova, I. N., Kryukov, A. P., Levashov, V. Y., and Heikal, M. R. Evaporation of droplets into a background gas: Kinetic modeling. *Int. J. Heat Mass Transfer*, 50, 2675 (2007).

8

Main Problems in Kinetics of Wetting and Spreading to Be Solved

1. Development of the theory of structural components of the Derjaguin's pressure.
2. Measurements of the Derjaguin's pressure isotherm in all regions of thicknesses, including $P_e < 0$ and regions of unstable flat films.
3. Development of the theory of the Derjaguin's pressure in the case of curved and non-flat liquid films.
4. Examination of equilibrium profiles (of droplets/menisci), taking into account both normal and tangential stress balances at the liquid–air interface as well as excess free energy of the system, including not only the normal stress balance as used currently but also the tangential stress balance.
5. Examination of equilibrium droplets profile in the case of aqueous surfactant solutions: inside the transition zone the surface concentration of surfactants cannot remain constant, which results in formation of tangential stress. The only way to counterbalance this tangential stress is inclusion of tangential component of Derjaguin's pressure.
6. Analysis of surface forces and equilibrium droplet/meniscus shapes in the case of nonwetting, where the equilibrium contact angle is above 90°.
7. Analysis of hysteresis of contact angles in the case of nonwetting through the Derjaguin's pressure.
8. Investigation of the influence of surface roughness and/or inhomogeneity on equilibrium and hysteresis contact angles in the case of partial, complete wetting and non-wetting through the Derjaguin's pressure isotherms.
9. The presence of thick β-films behind the receding menisci has been experimentally confirmed (see Section 3.2). However, experimental proof of the presence of thick β-films behind receding droplets is to be made.
10. Consideration of the role of surface forces in the aggregation processes apparently underlying (i.e., at the physicochemical core of) Alzheimer's disease and amyotrophic lateral sclerosis (Lou Gehrig's disease).
11. Consideration of spreading of surfactant solutions if the Derjaguin's pressure is taken into account in the vicinity of the apparent three-phase contact line.
12. Investigation of the kinetics of simultaneous spreading/evaporation when the action of the Derjaguin's pressure is taken into account in the vicinity of the apparent three-phase contact line.

Index

Note: Page numbers in italic and bold refer to figures and tables, respectively.

Made in United States
Orlando, FL
23 March 2022